GEOMETRIC ASYMPTOTICS

To our teacher

I. M. GEL′FAND

who has been a guide and inspiration to us we dedicate this book as an expression of our love and admiration.

MATHEMATICAL SURVEYS · Number 14

GEOMETRIC ASYMPTOTICS

BY

VICTOR GUILLEMIN AND SHLOMO STERNBERG

1977
AMERICAN MATHEMATICAL SOCIETY
PROVIDENCE, RHODE ISLAND

Library of Congress Cataloging in Publication Data

CIP

Guillemin, V 1937-
Geometric asymptotics.

(Mathematical surveys ; no. 14)
Includes bibliographies and index.
1. Geometry, Differential. 2. Asymptotic expansions. 3. Optics, Geometrical. I. Sternberg, Shlomo, joint author. II. Title. III. Series: American Mathematical Society. Mathematical surveys ; no. 14.
QA649.G86 516'.36 77-8210
ISBN 0-8218-1514-8

AMS (MOS) subject classifications (1970). Primary 20–XX, 22–XX, 35–XX, 44–XX, 49–XX, 53–XX, 58–XX, 70–XX, 78–XX, 81–XX.

Preface

There has been a great deal of active research recently in the fields of symplectic geometry and the theory of Fourier integral operators. In symplectic geometry there has been much progress in understanding the geometry of dynamical systems, and the process of "quantization" as applied not only to the theory of dynamical systems, but also as a key tool in the analysis of group representations. Fourier integral operators have made possible a much more systematic analysis of the singularities of solutions of linear partial differential equations than existed heretofore, together with a good deal of geometric information associated with the spectra of such operators. These two subjects are modern manifestations of themes that have occupied a central position in mathematical thought for the past three hundred years—the relations between the wave and the corpuscular theories of light. The purpose of this book is to develop these themes, and present some of the recent advances, using the language of differential geometry as a unifying influence.

We cannot pretend to do justice to the history of our subject, either in the short space of this preface, or in the body of the book. Yet some very brief mention of the early history is in order. The contributions of Newton and Huygens to the theory of light are known to all students. (Huygens' "envelope construction"—wherein the superposition of singularities distributed along a family of surfaces accumulates as singularity along the envelope—makes its appearance in the modern setting as a formula for the behaviour of wave front sets under functorial operations; cf. Chapter VI, equation (3.6).) In Fresnel's prize memoir of 1818, he combined Huygens' envelope construction with Young's "principle of interference" to explain not only the rectilinear propagation of light but also diffraction effects. Thus Fresnel applied Huygens' method to the superposition of oscillations instead of disturbances and was able to calculate the diffraction caused by straight edges, small aperatures and screens. In Chapter VI we study the geometry of the superposition of "disturbances" while in Chapter VII we study the geometry of the superposition of (high frequency) oscillations. In 1828

Hamilton published his fundamental paper on geometrical optics, introducing his "characteristics", as a key tool in the study of optical instruments. It wasn't until substantially later that Hamilton realized that his method applied equally well to the study of mechanics. Hamilton's method was developed by Jacobi and has been a cornerstone of theoretical mechanics ever since. In Chapter IV we discuss the modern version of the Hamilton-Jacobi theory—the geometry of symplectic manifolds. It is interesting to note that although Hamilton was aware of the work of Fresnel, he chose to ignore it completely in his fundamental papers on geometrical optics. Nevertheless, he made a theoretical prediction in the wave theory—the possibility of "conical refraction"—which was experimentally verified soon thereafter. Unfortunately, we shall have nothing to say in this book on the issue of conical refraction (the problem of multiple characteristics in hyperbolic differential equations) but hope that some of the methods that we develop will prove useful in this connection. In 1833, Airy published a paper dealing with the behavior of light near a caustic. (A caustic is a set where geometrical optics predicts an infinite intensity of light. An object placed at a caustic will become quite hot, and hence the name.) In this paper, he introduced the functions now known as Airy functions. In Chapter VII, we show how the theory of singularities of mappings can be applied to understand and extend Airy's results. In 1887, Kelvin introduced the method of stationary phase—a technique for the asymptotic evaluation of certain types of definite integrals—in order to explain the V shaped wake that trails behind a ship as it moves across the water. The method of stationary phase, in a form expounded by Hörmander, will be our principal analytical tool.

We now turn to a chapter by chapter description of the contents of the book.

Chapter I. Introduction. The Method of Stationary Phase

The solution of the reduced wave equation $\Delta\mu + k^2\mu = 0$ with initial data prescribed on a hypersurface $S \subset \mathbf{R}^3$ is given by an integral of the form

$$\int \frac{e^{ik|x-y|}}{|x-y|} a(y)\, dS. \qquad (\star)$$

In order to see how $(\star)$ behaves in the range of large frequencies we first consider more general integrals of the form

$$\int e^{ik\phi(z)} a(z)\, dz.$$

The method of stationary phase, for evaluating such integrals for k large, is discussed in detail following Hörmander's idea of making use of Morse's lemma. It is then applied to $(\star)$ to explain the phenomenon of shift in phase, when light rays pass through caustics. In the Appendix to this Chapter, we present a proof of Morse's lemma and its extensions to the existence of polynomial normal forms for functions satisfying the Milnor condition. The technique of proof, due to Moser and Palais, will recur several times in various contexts where normal forms of different kinds of geometric objects are required. This chapter is rather short and is intended mainly to motivate the type of asymptotic developments we consider in Chapters II and VII.

Chapter II. Differential Operators and Asymptotic Solutions

We discuss the Luneburg-Lax-Ludwig technique for constructing asymptotic solutions for differential equations $P(x, D, \tau)$ involving a large parameter τ. This technique involves solving the characteristic or eikonal equation and then inductively solving a series of transport equations. We show how these can be solved using methods of symplectic geometry and in doing so also show how the characteristic and transport equation, viewed symplecticly, make sense in the vicinity of caustics even though the asymptotic expansions blow up at caustics. This enables us, in §6, to analytically "continue" our asymptotic expansions through caustics making the appropriate phase adjustments. In §7, we apply these results to the time independent Schrödinger equation and derive the Bohr-Sommerfeld quantization conditions of Keller and Maslov. We also derive an asymptotic formula for the fundamental solution of the time dependent Schrödinger equation and show that it can be interpreted, as Feynmann does, as the evaluation of a Feynmann integral by means of stationary phase[1].

Chapter III. Geometrical Optics

We indicate how the techniques of the preceding chapter can be applied to optics. The point of view we emphasize here is that geometrical optics is just the study of the bicharacteristics of Maxwell's equation. In particular we develop the elementary laws of refraction and reflection from this point of view, and discuss focusing and magnification. We spend more time than the reader may think is warranted discussing Gaussian optics. However, this section is partly intended to motivate our study of the linear symplectic group a little later on. We also want to emphasize that the abstract mathematical considerations of Chapters II and IV are quite closely connected to concrete physical applications. However the treatment is not meant to be complete. There are two excellent and highly readable texts on the subject: Born and Wolf, "Principles of Physical Optics" and Luneburg, "Mathematical Theory of Optics", and the reader is referred to them for a detailed treatment. We finally discuss Maxwell's equations themselves.

The first three chapters can be viewed as principally motivational. The main formal development starts with Chapter IV.

---

[1] The techniques we describe in this chapter have been further developed by the mathematicians of the NYU school to obtain deep results in physical optics, acoustics and scattering theory. We regret very much that the limitations of this book prevent us from touching on their remarkable work.

Chapter IV. Symplectic Geometry

We have already made sporadic use of symplectic methods. Our intention here is to provide a much more systematic treatment than in the earlier chapters. We begin by giving Weinstein's proof of Darboux's theorem (and the Kostant-Weinstein generalization which says that near a Lagrangian submanifold Λ a given symplectic manifold looks locally like $T^* \Lambda$.) In §2 we discuss linear symplectic geometry, the Langrangian Grassmannian and its universal covering space, and define the Maslov index following the ideas of Leray and Souriau. In §3 we discuss Hörmander's cross ratio construction of the Maslov class. This material is based on §3.3 of Hörmander's Acta paper on Fourier integral operators, with some simplifications suggested to us by Kostant. In §§4 and 5 we develop the main facts we will need in Chapters VI and VII about Lagrangian manifolds. In preparation for Chapters VI and VII we have put considerable stress on their "functorial" properties. In §6 we discuss periodic orbits of Hamiltonian systems and make an application to the Kepler problem (following Moser). In §7 we discuss symplectic manifolds on which a group G acts transitively. Finally, in §8 we discuss relations between symplectic geometry and the calculus of variations.

Chapter V. Geometric Quantization

In this chapter we discuss the current state of knowledge concerning the geometry of quantization as introduced (independently) by Kostant and Souriau. The main function of this chapter, as far as the rest of the book is concerned, is the introduction of metalinear structures, half-forms, and metaplectic structures. These notions will be of crucial importance for us in Chapters VI and VII where we develop the symbol calculus in the metalinear category. The first two sections deal with the concept, due to Kostant and Souriau, of prequantization. This has the effect of selecting, among symplectic manifolds, those satisfying certain integrality conditions. (In the case of relativistic particles this has the effect of constraining the "spin" to take on half integral values. In the case of the hydrogen atom this constrains the negative energy levels to their appropriate discrete values.) In §3 we introduce the notion of polarization, a concept introduced independently by Kostant and Souriau in the real case, while Auslander and Kostant introduced the notion of a complex polarization. The rest of the chapter represents joint research by Blattner, Kostant, and Sternberg, dealing with half-forms and a pairing between sections associated to different polarizations, together with some physical examples due to Simms. The concept of a metaplectic structure is introduced together with the pairing, mentioned above which can be viewed as a generalization of the Fourier transform. The metaplectic representation of Segal-Shale-Weil is constructed by geometric methods using the pairing, and is applied to the construction of "symplectic spinors". In a sense, this chapter is the most incomplete in the book. It is clear that the quantization scheme proposed here is not broad enough to include all interesting cases. A preliminary computation, outlined at the end of the chapter, indicates that the requirement that two polarizations be unitarily related—an essential ingredient in the scheme—is only slightly less restrictive than they be "Heisenberg related", i.e. that the pairing be geometrically equivalent to the classical Fourier transform. Thus more sophisticated procedures must be developed, presumably along the lines of closer and closer approximations to the Feynmann path integral method. Even within the current framework many basic questions remain unanswered: What are the appropriate conditions guaranteeing the convergence of the integrals involved in the pairing? What is the correct formulation of the pairing in the case of complex polarizations, and its relation with such objects as the Bergmann kernel? Is there a discrete analogue of the notion of polarization, with an appropriate pairing so as to tie in with the theory of theta functions? Must one look at generalized sections, associated polarizations, or higher cohomologies in order to construct the quantized group representations? To what extent are the recent examples, of Vergne and Rothschild-Wolf, where the group representation is not independent of the polarization, a consequence of geometrical pathology of the polarization in question? What is the relation between polarizations and pairings as introduced here geometrically to formally similar objects arising in the p-adic theory, especially in Weil's fundamental papers? In short, it is quite clear that the subject matter of this chapter is only in its preliminary stages of development, and it is to be hoped that the present chapter will be completely out of date within a few years.

Chapter VI. Geometric Aspects of Distributions

This is perhaps the central chapter of the book from the mathematical point of view. In it we develop the theory of generalized functions (or sections of a vector bundle) in terms of their behavior under smooth maps. A *density* on a manifold X is an object which locally looks like a function, but transforms, under change of coordinates, in such a way that integration makes sense. If μ is a density and v is a smooth function of compact support on a manifold X, we define $\mu(v)$ by $\int_X v\mu$, so that densities define linear functionals on $C_0^\infty(X)$. By a generalized density we then mean any continuous linear functional on $C_0^\infty(X)$.

Let X and Y be manifolds and $f\colon X \to Y$ a proper mapping. If μ is a generalized density on X the "push-forward" $f_*\mu$ is defined by the formula

$$f_*\mu(v) = \mu(f^*v), \qquad v \in C_0^\infty(Y).$$

It turns out that in certain instances one can define the "pull-back" $f^*\mu$ of a generalized function on Y. (For example this is always the case if f is a fiber mapping.) The purpose of this chapter is to develop systematically the theory of distributions in such a way as to permit maximum interplay between these two functors.

In §1 we indicate the basic properties of these two functors, and use them to define a special class of distributions called δ-distributions and derive a fixed point formula. In §2 we give a few applications of the ideas discussed in §1; in particular we prove a few theorems about characters of induced representations (in the theory of group representations) and sketch a proof of the Atiyah-Bott fixed point formula. In §3 we define the wave front set of a distribution. It turns out that from our functorial point of view it is convenient to do this using the Radon transform rather than the Fourier transform as Hörmander does. (This is because the Radon transform can be defined as a composition of a "push-forward" and a "pull-back".) Therefore, we have included in this section a discussion of the basic properties of the Radon transform. The idea of using the Radon transform was pointed out to us by Dave Schaeffer in a lecture in which he expounded some fundamental papers by Ludwig who had developed the theory of singularities from this point of view. Ludwig's papers have served as a guide to us in constructing much of the theory of this chapter. In addition we discuss a few other Radon-like transforms which are defined by push-pull operations and show that the only "elliptic" examples have properties very similar to the classical Radon transforms on rank one homogeneous spaces.

In §4 we discuss a larger class of distributions than those of §1. This class is obtained by starting with the δ-function on the real line and applying the operations of pull-back, push-forward, differentiation and multiplication by smooth functions. What one gets is approximately the class of distributions discussed by Hörmander in his basic paper on Fourier integral operators (though his way of defining these distributions is quite different from ours.) In §5 we develop a symbol calculus for these distributions, again making heavy use of the two functorialities. Our symbols differ a little from those of Hörmander in that instead of using the Maslov line bundle, as Hörmander does, to handle phase adjustments we use the metaplectic structure discussed in Chapter V. In particular, our symbols are "half-forms" instead of half-densities. We get a slightly less general theory than that of Hörmander. (The manifolds we consider have to satisfy $w_1(X)^2 = 0$, $w_1(X)$ being the first Stiefel-Whitney class.) However, we feel this disadvantage is outweighed by the advantage that the symbols are less complicated. We indicate how our theory needs to be modified to cover the more general situation. In §6 we discuss the calculus of Fourier integral operators and derive some applications to partial differential equations.

In §7 we describe how our distributions behave under composition with differential operators and show that on the symbol level $\sigma(P\mu)$ is given by applying the transport equation to $\sigma(\mu)$ just as we did in Chapter II in a somewhat more elementary setting. As an application we describe the progressing and the regressing fundamental solutions of the wave equation

$$\frac{d^2\mu}{dt^2} - \Delta\mu = 0$$

for the Laplace-Beltrami operator Δ on a compact manifold. In §8 we use the results of §7 to obtain some theorems in spectral theory. In particular we obtain Hörmander's asymptotic formula for the spectral function of a positive self-adjoint elliptic operator on a compact manifold, and a result of Chazarain-Duistermaat-Guillemin on the singularities of the Fourier transform of the spectral function. This result asserts, roughly speaking, that the knowledge of the eigenvalues of the Laplacian gives the lengths of the closed geodesics. To use the phraseology of Mark Kac, one can "hear" the lengths of the closed geodesics.[2] (This result is connected with a physical observation of Hermholtz in his study of stringed instruments. Helmholtz was puzzled by the following problem: Why is it that a bowed string, which is a "forced vibration", should yield (approximately) the same note as a plucked string, which is a "free vibration". The frequency of a forced vibration should be the same as the frequency of the forcing term (the bow). There must be some mechanism whereby the string triggers the bow to execute a forcing action at exactly the free frequency of the string.) Helmholtz examined the motion of strings using his 'vibrating microscope' (today we would use a stroboscope) and discovered the following result: The plucked string vibrates between the extreme positions as indicated:

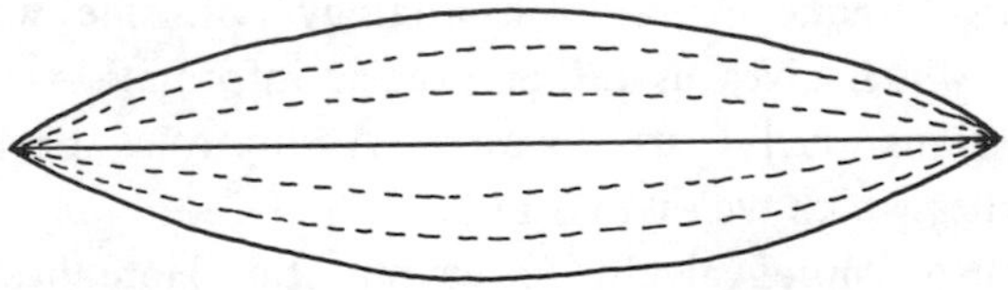

where the central line represents the rest position of the string. On the other hand, the instantaneous positions of the bowed string are of the form

[2] The earliest reference that we were able to find in the scientific literature to this inverse problem is in a report to the British Association by Sir Arthur Schuster in 1882 on spectroscopy. Until that time, the primary function of the analysis of the spectra of atoms and molecules was to identify the chemicals in question. (Indeed the most striking results were those of Kirchoff in determining the chemistry of the sun's atmosphere by analysing the absorbtion lines of the solar spectrum.) In fact, the study of spectra was known as "spectrum analysis". In this report, Schuster suggested that the primary function of the study of spectra in the future would be to analyze the structure of atoms and molecules, and coined the name "spectroscopy" for this new science. He writes:

> "but we must not too soon expect the discovery of any grand and very general law, for the constitution of what we call a molecule is no doubt a very complicated one, and the difficulty of the problem is so great that were it not for the primary importance of the result which we may finally hope to obtain, all but the most sanguine might well be discouraged to engage in an inquiry which, even after many years of work, may turn out to have been fruitless. We know a great deal more about the forces which produce the vibrations of sound than about those which produce the vibrations of light. To find out the different tunes sent out by a vibrating system is a problem which may or may not be solvable in certain special cases, but it would baffle the most skillful mathematician to solve the inverse problem and to find out the shape of a bell by means of the sounds which it is capable of sending out. And this is the problem which ultimately spectroscopy hopes to solve in the case of light. In the meantime we must welcome with delight even the smallest step in the desired direction."

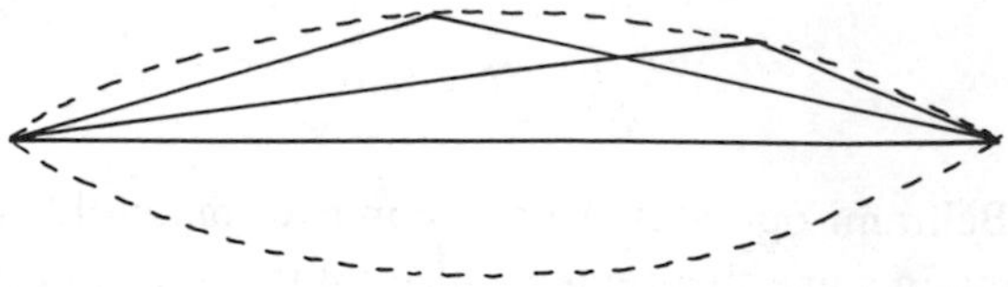

where the kink in the string moves around the circuit with the velocity of sound along the string, and it is this motion of the kink which triggers the bow to adhere and detach from the string. Thus, in a bowed string, we hear the period of the kink, i.e. the length of the "closed geodesic". In an appendix to this Chapter we describe Gelfand's celebrated results on the Plancherel formula for the complex semi-simple Lie groups in terms of the method of stationary phase.

Chapter VII. Compound Asymptotics

We go back to the type of problems discussed in Chapter II. We have more machinery at our disposal now, so we can formulate the results of Chapter II more systematically. We define objects on manifolds which we call *asymptotics*. They are functions (densities, half-densities, etc.) which depend on a large parameter τ. Any two such objects are identified if they have the same asymptotic growth as $\tau \to \infty$. Following Leray we define the Fourier transform of an asymptotic on $\mathbf{R}^n$, and use it to analyze the singularities of the asymptotic. More generally given an asymptotic $[\gamma]$ on a manifold X we define a subset $F[\gamma]$ of $T^* X$ which we call its "frequency set" (in analogy with the wave front set of Hörmander) and which gives us rather precise information about where high frequency oscillations of $[\gamma]$ are located. Asymptotics have some obvious functorial properties which we discuss in §3.

In §4 we develop a symbol calculus for asymptotics. Here these are two parallel theories—one associated to "exact" Lagrangian manifolds (for which $(1/2\pi)\alpha$ is exact) and one associated to "integral" Lagrangian manifolds for which $(1/2\pi)\alpha \in H^1(\Lambda, Z)$, i. e. $1/(2\pi)\alpha$ has integral cycles. The first of these is associated to asymptotics depending on a continuous parameter and the second to asymptotics associated to a discrete parameter. (These subtleties did not enter in Chapter VI because there we dealt with homogeneous Lagrangian manifold and α vanishes when restricted to a homogeneous Lagrangian manifold.) We then discuss the subprincipal symbol and the transport equation for asymptotics and indicate how integral asymptotics should be used, in conjuction with half forms, to obtain quantization conditions. In particular, our point of view is quite different from that espoused in Chapter II.

Even in dimension zero the asymptotic property of an arbitrary asymptotic can be quite complicated. The only general result known is Bernstein's theorem which describes asymptotic properties of integrals of the form

$$\int a(z)e^{i\tau\alpha(z)}\,dz$$

for functions α with isolated singularities. This is discussed in §5.

In §§6-9 we discuss asymptotic properties of certain kinds of generic asymptotics, such as those occurring in optics in the vicinity of a simple caustic. The simple caustic case is discussed in §6, and used to obtain results of Ludwig on uniform asymptotic expansions. To obtain analogous results for more complicated types of caustics we need canonical form theorems for the associated phase functions. These canonical forms are discussed in §§7 and 8. Finally in §9 we indicate some generalizations of the results of §6. The results of §§6-9 are joint research of Guillemin and Schaeffer, the results being obtained in 1972. Since then, two articles have appeared covering similar grounds, one by Duistermaat and one by Arnol'd. We feel that the point of view developed here, and the results obtained are sufficiently different from the above-mentioned articles to warrent publication in their original form.

For the reader who prefers to deal directly with the mathematical theory, unencumbered by historical allusions or physical applications we recommend that he begin with Chapter VI, and refer back to those sections of Chapters IV and V as they become necessary.

As we indicated in the above chapter by chapter analysis, much of the present book represents joint work with others. Many of the results of Chapter IV were obtained jointly with Kostant; Chapter V represents joint work of Blattner, Kostant, and Sternberg. The second half of Chapter VII is joint work of Guillemin and Schaeffer and many comments by Schaeffer were extremely helpful in the development of Chapter VII. As many of the results are presented here for the first time, we are grateful to Bob Blattner, Bert Kostant and Dave Schaeffer for allowing the material to be published as part of this book.

The book itself is based on an intensive joint Harvard-MIT course in the fall of 1973, together with seminars over the past few years. We wish to thank John Guckenheimer and Marty Golubitsky for their contribution to the course and seminar, and to Molly Scheffé for taking careful notes on which a lot of the current manuscript is based. Above all, we wish to thank Mary McQuillin for her assistance in seeing this book through its many editorial stages.

Both authors were supported in part by the National Science Foundation and the second author by the John Simon Guggenheim Memorial Foundation to whom we wish to express our thanks.

Notation

In general, we follow the notation used in the text *Advanced Calculus* by Loomis and Sternberg, Addison-Wesley Publishing Co., Reading, Mass. 1968, with some slight changes. The letters X, Y, Z, W, M will be used for differentiable manifolds; usually M will denote a general differentiable manifold, and X a manifold carrying some additional structures. Points on these manifolds will be denoted by lower case letters such as x, y, z, etc. Local coordinates will be written as $(x^1, \ldots, x^n)$. If M is a manifold and T^*M is its cotangent bundle, then the local coordinates on T^*M associated with the local coordinates $(x^1, \ldots, x^n)$ on M will be $(x^1, \ldots, x^n, \xi_1, \ldots, \xi_n)$ and, sometimes $(q^1, \ldots, q^n, p_1, \ldots, p_n)$. The fundamental one form on the cotangent bundle will be written as α, so that, in terms of these local coordinates, $\alpha = \xi_1\, dx^1 + \cdots + \xi_n\, dx^n$ or $\alpha = p_1\, dq^1 + \cdots + p_n\, dq^n$. We will use bold face greek letters, usually $\boldsymbol{\xi}$, $\boldsymbol{\eta}$, or $\boldsymbol{\zeta}$ to denote either tangent vectors or vector fields. The tangent space to a manifold M at a point x will be denoted by TM_x, and so a typical element of this tangent space will be $\boldsymbol{\xi} \in TM_x$. If $f\colon M \to N$ is a smooth map between differentiable manifolds, its differential at the point x will be denoted by df_x, so that $df_x\colon TM_x \to TN_{f(x)}$. The "pull back" of a differential form θ on N by f will be denoted by $f^*\theta$. Thus, for example, if θ is a *linear* differential form on N, and $\theta_y \in T^*N_y$ is its value at $y \in N$, then for $\boldsymbol{\xi} \in TM_x$ we have

$$\langle \boldsymbol{\xi}, (f^*\theta)_x \rangle = \langle df_x \boldsymbol{\xi}, \theta_{f(x)} \rangle,$$

where $\langle\ ,\ \rangle$ gives the pairing between vectors and covectors. We will use the symbol $D_{\boldsymbol{\xi}}$ to denote the Lie derivative with respect to the vector field $\boldsymbol{\xi}$. Thus, if Ω is a k-form, $D_{\boldsymbol{\xi}}\Omega$ is its Lie derivative with respect to $\boldsymbol{\xi}$; if η is another vector field, $D_{\boldsymbol{\xi}}\eta = [\boldsymbol{\xi}, \eta]$ is the Lie bracket; if u is a function, we frequently write $\boldsymbol{\xi}u$ for $D_{\boldsymbol{\xi}}u$.

In an oriented Riemannian or psuedoriemmanian manifold of dimension n, we have the "star operator", denoted by $\star$, mapping k forms into $n - k$ forms. Thus, for example, if S is an oriented surface in Euclidean three space, $\mathbf{R}^3$ and u is a smooth function on $\mathbf{R}^3$, we write $\iint_S \star\, du$ for the expression $\iint_S (\partial u/\partial n)\, dS$ that occurs in many of the older analysis texts.

TABLE OF CONTENTS

Chapter I. Introduction. The Method of Stationary Phase

One of the early conclusive experiments verifying the wave nature of light was the double mirror experiment of Fresnel. In this experiment, two plane mirrors are placed so as to form an angle of slightly less than 180 degrees between them. If light is incident from a source S, then interference fringes are observed in the region common to the two beams reflected by the mirrors. The two beams can be assumed as coming from S_1 and S_2, the images of S in the two mirrors. These sources can be thought of as synchronous and homogeneous since they are derived from the same source S. If a screen is placed in the region common to the two beams, then the points equidistant from S_1 and S_2 are highly illuminated, and dark and light regions then alternate giving the interference pattern.

It was discovered by Gouy (*Comp. Rend.* 110 (1890), p. 1251) that if one of the plane mirrors is replaced by a concave mirror, and the screen placed beyond the focus, then the center is dark rather than light and, in fact, the entire interference pattern is reversed. Thus it appears that the light goes through a phase shift of π when passing through a focus. A focus can be thought of as the coincidence of two focal lines. Thus, as Gouy points out, we can formulate the above result as saying that light goes through a phase shift of $\pi/2$ when passing through a focal line. This phenomenon was explained by Poincaré in his lectures of 1891-1892 on the theory of light, cf. [2]. His method was to apply an asymptotic evaluation of certain integrals arising in the solution of the wave equation. This asymptotic evaluation was invented earlier by Kelvin and is known as the method of stationary phase. We now describe this discussion. Instead of dealing with mirrors, we shall first assume that we have a surface emitting radiation of high frequency. Also, to simplify the discussion, we will treat the scalar wave equation rather than the vector equations of Maxwell. The discussion extends easily to the vector case.

Let us consider spherically symmetric solutions of the wave equation

$$\left(\frac{\partial^2}{\partial t^2} - \Delta\right)u = 0 \qquad \text{where } \Delta = \frac{\partial^2}{\partial x^2} + \frac{\partial^2}{\partial y^2} + \frac{\partial^2}{\partial z^2}$$

is the usual Laplacian in Euclidean three dimensional space. In terms of polar coordinates we have

$$\Delta u = \frac{1}{r^2}\frac{\partial}{\partial r}r^2\frac{\partial u}{\partial r} + \frac{1}{r^2 \sin\theta}\frac{\partial}{\partial\theta}\sin\theta\frac{\partial u}{\partial\theta} + \frac{1}{r^2\sin^2\theta}\frac{\partial^2 u}{\partial\phi^2}$$

so that if u is spherically symmetric, $u = u(r, t)$, the wave equation becomes

$$\frac{\partial^2 u}{\partial t^2} = \frac{1}{r^2}\frac{\partial}{\partial r}r^2\frac{\partial u}{\partial r} = \frac{1}{r}\left[2\frac{\partial u}{\partial r} + r\frac{\partial^2 u}{\partial r^2}\right] = \frac{1}{r}\frac{\partial^2}{\partial r^2}(ru).$$

Thus $v = ru$ satisfies the one dimensional wave equation

$$\left(\frac{\partial^2}{\partial t^2} - \frac{\partial^2}{\partial r^2}\right)v = 0.$$

The general solution of this equation is given by

$$v(r, t) = f(r + t) + g(r - t)$$

and so the general solution of the symmetric wave equation is given by

$$u(r, t) = \frac{f(r + t)}{r} + \frac{g(r - t)}{r}.$$

Here the first term represents an incoming wave and the second term represents an outgoing wave. In particular, if we take $f = 0$ and $g(s) = e^{iks}$ then

$$w_k(t, r) = \frac{e^{ik(r-t)}}{r}$$

represents an outgoing (sinusoidal) wave of frequency k. Indeed, up to normalizing constants, it is easy to check that

$$E_k(r) = \frac{e^{ikr}}{r}$$

is the fundamental solution to the reduced wave operator $\Delta + k^2$, i.e.,

$$(\Delta + k^2)E_k = C\delta$$

for a suitable constant C, in fact $C = -4\pi$.

Thus, let y be a point in $\mathbf{R}^3$. The function

$$c(y)w_k(t, |x - y|), \qquad x \neq y,$$

then describes a steady emission of radiation from y of frequency k. Here the complex number $c(y)$ gives the amplitude and phase of the emitted radiation.

Now suppose that radiation is steadily being emitted from all points y on a surface S, with density $c(y)dy$ where we assume that the $c(y)$ all have the same phase; we thus may as well assume that $c(y)$ is real. Then for any $x \notin S$, the radiation at x will be of the form

$$e^{-ikt} I_k(x)$$

where the $I_k(x)$ is an integral over S;

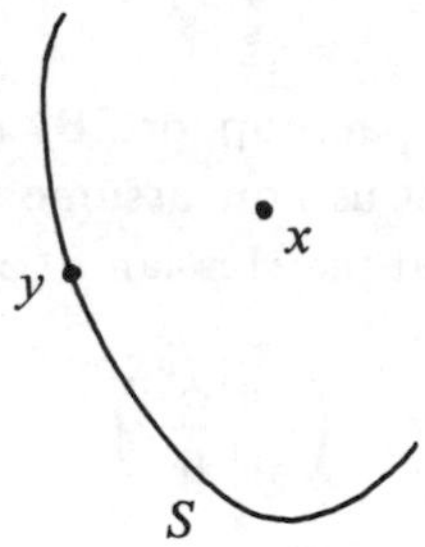

explicitly,

$$I_k(x) = \int_S \frac{e^{ik|x-y|}}{|x-y|} c(y)\, dA.$$

We wish to evaluate this integral asymptotically for large values of k. We thus are interested in an integral of the form

$$\int a(y) e^{ik\phi(y)}\, dy$$

for large values of k. The evaluation of integrals of this type is known as the method of stationary phase. Observe that the major contribution will come from a neighborhood of the points where $d\phi = 0$. In fact we can break up a into a sum of pieces of small support by using a partition of unity. Suppose that $d\phi \neq 0$ on supp a. Then we claim that integration by parts shows that the above integral is $O(k^{-N})$ for any N, if a is a C^∞ function of y. To see this let ξ be the vector field

$$\xi = \sum \frac{\partial \phi}{\partial y^i} \frac{\partial}{\partial y^i}$$

so that

$$\xi e^{ik\phi} = ik\left[\sum \left(\frac{\partial \phi}{\partial y^i}\right)^2\right] e^{ik\phi} = ik|\xi|^2 e^{ik\phi}$$

where $|\xi|^2 \neq 0$. Let

$$\eta = \frac{-i}{k|\xi|^2}\xi.$$

Then

$$\eta e^{ik\phi} = e^{ik\phi}$$

so that

$$\int ae^{ik\phi}\,dy = \int a[\eta e^{ik\phi}]\,dy = \int (\eta^t a)e^{ik\phi}\,dy$$
$$= \frac{-i}{k}\int \xi^t\left(\frac{a}{|\xi|^2}\right)e^{ik\phi}\,dy = \frac{1}{k}\int be^{ik\phi}\,dy,$$

for some C^∞ function b of compact support. By repeating this operation we see that the integral is $O(k^{-N})$. Let us now assume that each critical point of ϕ is non-degenerate. This means that the Hessian $d^2\phi$, i.e., the matrix

$$\left(\frac{\partial^2\phi}{\partial y^i\,\partial y^j}\right)$$

is non-singular at each critical point. Let p be a critical point of ϕ. By a lemma of Morse, we can introduce coordinates $z_1, \ldots, z_n$ about p so that

$$\phi = \phi(p) + [-z_1^2 \cdots - z_\ell^2 + z_{\ell+1}^2 + \cdots + z_n^2]/2$$
$$= \phi(p) + Q(z)/2$$

where ℓ is the index of the quadratic form $d^2\phi_p$. (See Appendix I to this chapter for a proof of Morse's lemma.)

If

$$\left(\frac{\partial y}{\partial z}\right)$$

is the Jacobian matrix of the change of coordinates then, if supp a lies in the desired coordinate neighborhood,

$$\int ae^{ik\phi}\,dy = e^{ik\phi(p)}\int ae^{ikQ(z)/2}\left|\det\frac{\partial y}{\partial z}\right|\,dz.$$

Notice that if Q is the matrix of $Q(z)$ then at p we have

$$Q = \left(\frac{\partial y}{\partial z}\right)^t\left(\frac{\partial^2\phi}{\partial y^i\,\partial y^j}\right)\left(\frac{\partial y}{\partial z}\right)$$

so that

$$\left|\det\frac{\partial y}{\partial z}\right|(p) = \left|\det\frac{\partial^2\phi}{\partial y^i\,\partial y^j}(p)\right|^{-\frac{1}{2}}$$

We are thus reduced to considering an integral of the form

$$\int f(z)e^{ikQ(z)1/2}\,dz$$

for large k. Here $z = (z_1, \ldots, z_n)$. We claim that this integral has the asymptotic expansion

$$\left(\frac{2\pi}{k}\right)^{n/2} e^{i\pi(n-2\ell)/4} f(p) + O(k^{-n/2-1}).$$

We prove this as follows: We can write $f(z) = f(0) + \sum z_i f_i(z)$ where the f_i are smooth, but, of course, not necessarily of compact support. If we could establish that $\int e^{ikQ(z)/2}g(z)dz$ were well defined for suitable functions not having compact support, we could write

$$\int e^{ikQ(z)/2} z_j f_j(z)\,dz = \pm\frac{1}{ik}\int \frac{\partial}{\partial z_j}(e^{ikQ(z)/2}) f_j\,dz$$

$$= \mp\frac{1}{ik}\int e^{ikQ(z)/2}\frac{\partial f_j}{\partial z_j}\,dz$$

and this last integral is again of the type we are considering. Thus the highest order term in any asymptotic expansion will come from the constant term, and we are left with the task of evaluating the integral $\int e^{ikQ(z)/2}dz$. This is a product of one dimensional integrals, and we must evaluate the integral $\int e^{iku^2/2}du$. Before evaluating this integral let us go to the problem of making sense of such integrals.

We are thus interested in making sense of integrals of the form $\int e^{ikQ(z)/2}h(z)\,dz$. We can reduce the question of convergence of the multiple integral to that of the iterated integral, and hence to a problem in one real variable. Let $h(t)$ be a C^2 function of the real variable t which is bounded and with bounded first two derivatives. Consider the integral

$$\int_{-\infty}^{\infty} e^{-\lambda t^2/2} h(t)\,dt.$$

We claim that this integral is uniformly convergent for $\operatorname{Re}\lambda \geq 0$, $|\lambda| \geq 1$. Indeed, for $0 < R < S$ we have

$$\int_R^S e^{-\lambda t^2/2}h(t)\,dt = -\lambda^{-1}\int_R^S \frac{1}{t}(e^{-\lambda t^2/2})'h(t)\,dt$$

$$= -\lambda^{-1}e^{-\lambda t^2/2}(h(t)/t)\Big|_R^S + \frac{1}{\lambda}\int_R^S e^{-\lambda t^2/2}(h(t)/t)'\,dt$$

$$= -\lambda^{-2}e^{-\lambda t^2/2}[\lambda(h(t)/t) - (1/t)(h(t)/t)']\Big|_R^S$$

$$+\lambda^{-2}\int_R^S e^{-\lambda t^2/2}[(1/t)(h(t)/t)']'\,dt.$$

The integral on the right is absolutely convergent and the end terms tend to zero as $R \to \infty$. Furthermore if M is a bound for h and its first two derivatives then the above expressions can be estimated purely in terms of M. Thus if h depends on some auxiliary parameters and is uniformly bounded together with its first two derivatives in terms of these parameters then the integral

$$\int_{-\infty}^{\infty} e^{-\lambda t^2/2} h\, dt$$

converges uniformly with respect to these parameters. (In particular, for Re $\lambda = 0$ the integral $\int e^{-\lambda Q(z)/2} f(z)\, dz$ is well defined.) Furthermore, we see that $\int e^{-\lambda t^2/2} h(t)\, dt$ is a holomorphic function of λ for Re $\lambda > 0$ and continuous for Re $\lambda \geq 0$, $\lambda \neq 0$. In particular we can take $h \equiv 1$. Now $\int_{-\infty}^{\infty} e^{-u^2/2}\, du = \sqrt{2\pi}$ and a change of variables for λ real together with analytic continuation for λ complex shows that

$$\int_{-\infty}^{\infty} e^{-\lambda u^2/2}\, du = \left(\frac{2\pi}{\lambda}\right)^{1/2},$$

for Re $\lambda > 0$, where the square root is computed by continuation from the positive real axis. Letting $\lambda \to \mp ik$ gives

$$\int e^{\pm iku^2/2}\, dy = \left(\frac{2\pi}{k}\right)^{1/2} e^{\pm \pi i/4}.$$

This completes the argument and establishes the formula. Assuming that ϕ has only finitely many critical points in supp a, we can therefore assert that

$$\int a(y) e^{ik\phi(y)}\, dy = \left(\frac{2\pi}{k}\right)^{n/2} \sum_{y \mid d\phi(y)=0} e^{\pi i \operatorname{sgn} H(y)/4} \frac{e^{ik\phi(y)} a(y)}{\sqrt{|\det H(y)|}} + O(k^{-n/2-1}), \tag{S.P.}$$

where $H(y)$ is the Hessian of ϕ at y.

Here sgn H denotes the signature of the quadratic form H. (For Q the signature is the number of $+$'s less the number of $-$'s, i.e., $n - 2\ell$.)

This formula is the method of stationary phase. We shall give a more invariant formulation of this result later on. We wish to apply this formula to our present circumstance. Observe that if $\phi \neq 0$ then $d\phi = 0$ if and only if $d\phi^2 = 0$ and

$$d^2(\tfrac{1}{2}\phi^2) = \phi d^2 \phi$$

at each critical point. Since it is easier to deal with $\frac{1}{2}|y - x|^2$ than with $|y - x|$ we can use the above remark to simplify the calculations. It is pretty clear that for any fixed x the function $\psi^x(y) = \frac{1}{2}|y - x|^2$ has a critical point if and only if the line joining x to y is orthogonal to the y surface. The Hessian of ψ^x at any critical point y is related to the first and second fundamental forms of the surface. Let us do the computation in somewhat greater generality, as this involves only slightly more effort (cf. Milnor[3]).

Let M be an ℓ-dimensional submanifold of $\mathbf{R}^m$. Thus each point of M has a neighborhood with coordinates $u = (u^1, \ldots, u^\ell)$ and a map $y = y(u) = (y^1(u), \ldots, y^m(u))$ into $\mathbf{R}^m$, where the Jacobian matrix $(\partial y/\partial u)$ has rank ℓ. For simplicity we will regard M as a subset of $\mathbf{R}^m$ (although all that we have to say works for immersed submanifolds as well as embedded ones).

We can thus consider the normal bundle, $N(M)$, as consisting of those pairs (y, w) where $y \in M$ and $w \in \mathbf{R}^m$ is orthogonal to the tangent space to M at y. Thus $w \perp TM_y$. Notice that from this point of view we are regarding $N(M)$ as lying in $\mathbf{R}^m + \mathbf{R}^m$. It is easy to check that $N(M)$ is an m-dimensional immersed submanifold. Indeed let U be a coordinate patch on M with coordinates $u_1, \ldots, u_\ell$. The vectors

$$\frac{\partial y}{\partial u_1}, \ldots, \frac{\partial y}{\partial u_\ell}$$

are all tangent to M at each point of U. Extend to a basis of $\mathbf{R}^m$ at some $p \in U$ by adding vectors $v_{\ell+1}, \ldots, v_m$. Thus

$$\frac{\partial y}{\partial u_1}(p), \ldots, \frac{\partial y}{\partial u_\ell}(p), v_{\ell+1}, \ldots, v_m$$

is a basis of $\mathbf{R}^m$ at p and hence

$$\frac{\partial y}{\partial u_1}, \ldots, \frac{\partial y}{\partial u_\ell}, v_{\ell+1}, \ldots, v_m$$

is a basis of $\mathbf{R}^m$ at all points of U near enough to p. By shrinking we may also call this neighborhood U. Orthonormalize this basis to get

$$t_1(u), \ldots, t_\ell(u), n_{\ell+1}(u), \ldots, n_m(u).$$

Then $n_{\ell+1}(u), \ldots, n_m(u)$ span the normals to M at $y(u)$. Thus $U \times \mathbf{R}^{m-\ell}$ gives a parametrization of the normal bundle over U where

$$(u_1, \ldots, u_\ell, s_{\ell+1}, \ldots, s_m) \to (y(u), s_{\ell+1} n_{\ell+1}(u) + \cdots + s_m n_m(u)).$$

Define the map E: $N(M) \to \mathbf{R}^m$ by

$$E(y, w) = y + w.$$

Notice that

$$T(N(M))_{(y,w)} = TM_y + NM_y \subset \mathbf{R}^m + \mathbf{R}^m.$$

Therefore its differential,

$$dE_{(y,0)}\colon T(N(M))_{(y,0)} \to \mathbf{R}^m,$$

given by

$$dE_{(y,0)}(v, w) = v + w,$$

is surjective. Thus, by the inverse function theorem, E is a diffeomorphism of a neighborhood of $(y, 0)$ into $\mathbf{R}^m$. Of course, E will not, in general, be a diffeomorphism of $N(M) \to \mathbf{R}^m$. Indeed *a critical value of E is called a focal point of M.*

Thus $x \in \mathbf{R}^m$ is *not* a focal point if, for all (y, w) such that $E(y, w) = x$, the rank of dE is m. Now the differential of E with respect to w always has rank $m - \ell$. Thus x is a critical value if and only if there is some (y, w) with $E(y, w) = x$ and such that the differential of E with respect to y has rank less than ℓ.

Now consider the function ψ^x given as above by

$$\psi^x(y) = \tfrac{1}{2}|y - x|^2 = \tfrac{1}{2}(y, y - 2x) + \tfrac{1}{2}|x|^2.$$

Then

$$d\psi^x = (dy, y - x)$$

so that $d\psi^x = 0$ if and only if $y - x$ is normal to TM_y, and then $x = E(y, y - x)$. The figure below suggests the following:

(i) $d^2\psi^x$ is singular if and only if x is a focal point.

(ii) Let L be the line through y and x. If there is no focal point between y and x then ψ^x has a minimum at y, and, indeed, $d^2\psi^x$ is positive definite.

(iii) If there is a focal point of the form $E(y, v)$, $v \in L$, where $(y, v) \in N(M)$ is a singular point, and v lies between 0 and $x - y$ then ψ^x is not a minimum, and the index of $d^2\psi^x$ is related to the number of such focal points. (In the figure, ψ^{x^1} and ψ^{x^3} have a minimum at y while ψ^{x^2} has a maximum.)

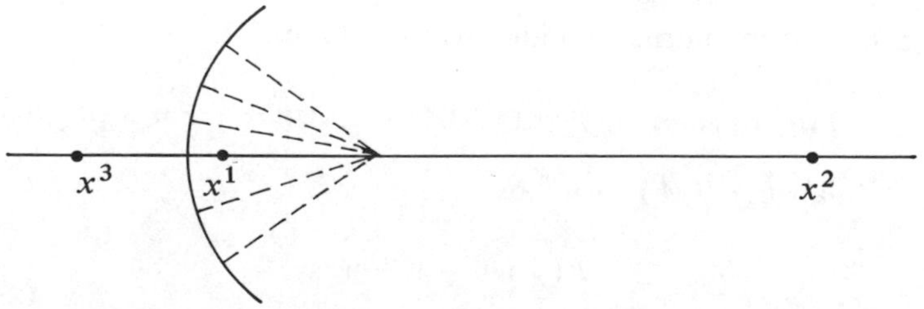

Indeed, we shall prove *the Hessian $d^2\psi^x(y)$ is nondegenerate if and only if $x = E(y, v)$ and* rank $dE_{(y,v)} = m$, i.e., *x is not a focal point of a neighborhood of*

y. Furthermore, the index of $d^2\psi^x$, i.e., the number of negative eigenvalues of this quadratic form, is given by

$$\operatorname{ind}(d^2\psi^x) = \sum_{0<t<1} \operatorname{corank} dE_{(y,tv)}.$$

This result is a special case of the Morse index theorem.

Loosely speaking, the formula says that the index of $d\psi^x$ is the number of focal points between y and x, counted with multiplicity.

Let us introduce local coordinates as above, and let $F\colon U \times \mathbf{R}^{m-\ell} \to \mathbf{R}^m$ be the map E expressed in terms of these coordinates; thus

$$F(u,s) = y(u) + s \cdot n(u) = y(u) + s_{\ell+1} n_{\ell+1}(u) + \cdots + s_m n_m(u).$$

Thus

$$\frac{\partial F}{\partial u_i} = \frac{\partial y}{\partial u_i} + \sum_{r=\ell+1}^{m} s_r \frac{\partial n_r}{\partial u_i}, \qquad i = 1, \ldots, m,$$

$$\frac{\partial F}{\partial s_r} = n_r, \qquad r = \ell + 1, \ldots, m.$$

Now the point $(y,v) \in N(M)$ whose coordinates are (u,s) will be a regular point for E if and only if the m vectors

$$\frac{\partial F}{\partial u_i}, \frac{\partial F}{\partial s_r}$$

are linearly independent. The m vectors $(\partial y/\partial u_i)$, n_r are always linearly independent by construction. Taking the scalar product with the $(\partial F/\partial u_i)$, $(\partial F/\partial s_r)$ we get the matrix

$$\left[\begin{array}{c|c} \left(\dfrac{\partial y}{\partial u_i}, \dfrac{\partial y}{\partial u_j}\right) + \sum s_r \left(\dfrac{\partial n_r}{\partial u_i}, \dfrac{\partial y}{\partial u_j}\right) & \sum s_r \left(\dfrac{\partial n_r}{\partial u_i} \cdot n_t\right) \\ \hline 0 & 1 \cdots 1 \end{array}\right]$$

which is non-singular if and only if the upper left hand block is non-singular. With no loss of generality we may assume that we have chosen $n_{\ell+1} = n$ to point in the direction of v so that $v = sn$ with $s > 0$. Thus the corank of $dE_{(x,v)}$ is the same as the corank of the matrix

$$\left(\left(\frac{\partial y}{\partial u_i}, \frac{\partial y}{\partial u_j}\right) + s\left(\frac{\partial n}{\partial u_i}, \frac{\partial y}{\partial u_j}\right)\right).$$

Now

$$0 = \frac{\partial}{\partial u_i}\left(n, \frac{\partial y}{\partial u_j}\right) = \left(\frac{\partial n}{\partial u_i}, \frac{\partial y}{\partial u_j}\right) + \left(n, \frac{\partial^2 y}{\partial u_i \partial u_j}\right).$$

Thus the (i, j)th entry of the matrix can be written as

$$\left(\frac{\partial y}{\partial u_i}, \frac{\partial y}{\partial u_j}\right) - \left(\frac{\partial^2 y}{\partial u_i \partial u_j}, v\right) - \frac{\partial}{\partial u_j}\left(\frac{\partial y}{\partial u_i}, v\right) \quad \text{since} \quad v = x - y$$

$$= \frac{\partial^2 \psi^x}{\partial u_i \partial u_j}.$$

This shows that corank $dE_{(x,v)}$ = nullity $d^2\psi^x$. In particular, if (y, v) is a regular point of E then $d^2\psi^x(y)$ is non-degenerate, establishing the first part of the proposition. Let

$$g_{ij} = \left(\frac{\partial y}{\partial u_i}, \frac{\partial y}{\partial u_j}\right).$$

The quadratic form on $\mathbf{R}^\ell$ whose matrix is (g_{ij}) is called the *first fundamental form.* It is just the Euclidean scalar product on TM_y considered as a bilinear form on $\mathbf{R}^\ell$ when we identify $\mathbf{R}^\ell$ with TM_y. It is positive definite.

The bilinear form whose matrix is

$$(r_{ij}(n)) = \left(\frac{\partial^2 y}{\partial u_i \partial u_j}, n\right), \quad n \in N(M)_y,$$

is called the *second fundamental form in the direction n.*

We let I_y and II_y denote the first and second fundamental forms at y.

We have shown that

$$\frac{\partial^2(\psi^x)}{\partial u^i \partial u^j} = g_{ij} - s r_{ij}(n) \quad \text{where } v = sn.$$

Let us compute the index of the right hand side. By a linear change of coordinates in the u's, we may assume that $(g_{ij}) = (\delta_{ij})$; this amounts to introducing an orthogonal basis for the first fundamental form. Then, $(r_{ij}) = (r_{ij}(n))$ is a symmetric matrix with eigenvalues $\mu_1, \ldots, \mu_m$ and pairwise orthogonal eigenvectors. By a further orthogonal change of the u variables we can diagonalize the matrix $(r_{ij}(n))$. Thus we get

$$\frac{\partial^2 \psi^x}{\partial u^i \partial u^j} = \delta_{ij} - s\mu_i \delta_{ij} = (1 - s\mu_i)\delta_{ij}.$$

Thus the index of $d^2\psi^x$ is the number of μ_i such that $s\mu_i > 1$. This is the same as

$$\sum_{0<t<1} (\text{number of } \mu_i \text{ with } st\mu_i = 1) = \sum_{0<t<s} \text{corank } (I - t(r_{ij}))$$

$$= \sum_{0<t<s} \text{corank } dE_{y,tv},$$

which proves the second part of the proposition.

The eigenvectors of $(r_{ij}(n))$ are called the *principal directions of curvature* of M in the direction n. The numbers $K_i = \mu_i^{-1}$ are called the *principal radii of curvature.*

If we set $\varphi^x(y) = |x - y|$ so that $\psi^x = \frac{1}{2}[\varphi^x]^2$ then the Hessians satisfy

$$H\varphi = \frac{1}{\varphi} H\psi.$$

In our case M is two dimensional so that

$$\frac{1}{|\det H\varphi|^{1/2}} = \frac{|y - x|}{|\det H\psi|^{1/2}} = \frac{|y - x|}{|(1 - |y - x|\mu_1)(1 - |y - x|\mu_2)|^{1/2}}.$$

For each x and y let d denote the distance from x to y. If $x - y$ is normal to M let # denote the number focal points between y and x.

Substituting into our original integral gives

$$I_k = \frac{2\pi}{k} \sum_{y|E(y,v)=x} e^{(1-\#)\pi i/2} e^{ikd} \frac{a(y)}{|(1 - \mu_1 d)(1 - \mu_2 d)|^{1/2}} + O(k^{-2}).$$

So far we have been considering the rather artificial situation where each point on a surface radiates uniformly in all directions. We will now show that similar considerations apply to physically interesting solutions of the wave equation, with the real surface replaced by an imaginary surface and use made of Stokes' theorem. We begin by recalling Green's formula: Let u and v be two functions on $\mathbf{R}^3$; then

$$d(u * dv - v * du) = ud * dv - vd * du$$

so, by Stokes' theorem

$$\iiint_D u\Delta v - v\Delta u dV = \iint_{\partial D} (u * dv - v * du),$$

if D is any bounded region with smooth boundary. If u satisfies $\Delta u + k^2 u = 0$ and v satisfies $\Delta v + k^2 v = \delta_P$ then the left hand side becomes $u(P)$, if $P \in D$ and 0 if $P \notin \overline{D}$. Thus we have the formula of Helmholtz

$$\frac{1}{4\pi} \iint_{\partial D} \left[\frac{e^{ikr}}{r} * du - u * d\left(\frac{e^{ikr}}{r} \right) \right] = \begin{cases} u(P) & \text{if } P \in D, \\ 0 & \text{if } P \notin \overline{D}, \end{cases}$$

if u is a solution of $(\Delta^2 + k^2)u = 0$ and r denotes the distance from P.

In many applications we are interested in the situation where D, instead of being bounded, represents the exterior to some surface S. Let us first apply the formula to the bounded region D_R, consisting of the intersection of D with a ball of radius R centered at P.

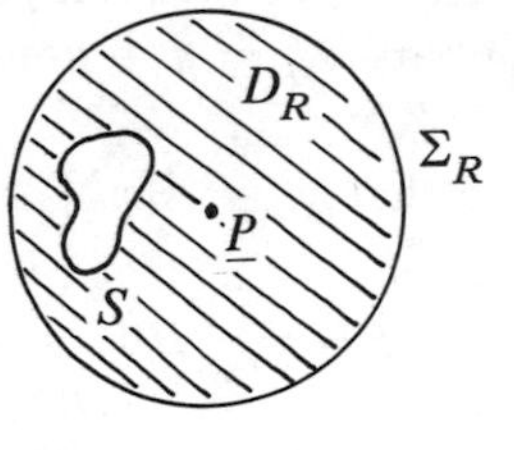

If R is taken large enough we get

$$\frac{1}{4\pi}\iint_S + \frac{1}{4\pi}\iint_{\Sigma_R} = \begin{cases} u(P) & \text{if } P \in D, \\ 0 & 0 \text{ if } P \notin D. \end{cases}$$

where Σ_R is the sphere of radius R. Now

$$d\left(\frac{e^{ikr}}{r}\right) = \frac{e^{ikr}}{r}\left[ik - \frac{1}{r}\right]dr$$

and $* \, dr = R^2 d\omega$ on Σ_R, where $d\omega$ is the element of solid angle on Σ_R. Thus the second integral becomes

$$\iint e^{ikR}\left[r\left(\frac{\partial u}{\partial r} - iku\right) + u\right]_{r=R} d\omega.$$

Thus the integral over the sphere will go to zero as $R \to \infty$ if

$$\iint |u| \, d\omega = o(R^{-2}) \quad \text{and} \quad \iint \left|\frac{\partial u}{\partial r} - iku\right| d\omega = o(R^{-3}).$$

where the integrals are evaluated for $r = R$. These conditions are known as the Sommerfeld radiation conditions. Their significance is that they represent the condition that u consists of expanding waves radiating outward and no incoming waves[1]. Let us assume that this condition is satisfied. Then the value of u outside some surface S is given by

$$u(P) = \frac{1}{4\pi}\iint_S \left[\frac{e^{ikr}}{r} * du - u * d\left(\frac{e^{ikr}}{r}\right)\right]. \qquad \text{(H)}$$

[1] For a precise mathematical explanation of the Sommerfeld radiation conditions see the book by Lax and Phillips [4, pp. 120-128.] The gist of what they prove is the following: Let $f = \{f_1, f_2\}$ be Cauchy data for the wave equation; we thus seek a solution of the wave equation

$$\frac{\partial^2 w}{\partial t^2} - \Delta w = 0$$

with $w(x, 0) = f_1$ and $(\partial w/\partial t)(x, 0) = f_2$. We say f is *eventually outgoing* if there is some constant c such that $w = 0$ for $|x| < t - c$. If we seek a solution of fixed frequency, then the appropriate Cauchy data are $\{w, ikw\}$. Suppose that w is a solution of the reduced wave equation outside some bounded domain. Then $\{w, ikw\}$ is eventually outgoing if and only if the Sommerfeld radiation conditions are satisfied.

In this way, the solution exterior to S is described in terms of "radiation emitted from S".[2] It was Huygens who originally had the idea that propagated disturbances in the wave theory could be represented as the superposition of "secondary disturbances" along an intermediate surface such as S; but he did not have an adequate explanation of why there was no "backward wave", i. e. why the propagation was only in the outward direction. The idea that the backward waves would cancel one another out because of phase differences was due to Fresnel. Fresnel believed that if all the sources were inside S, the "secondary radiation" (i.e. the integrand in Helmholtz's formula) from each separate surface element would produce a null effect at each interior point due to interference. The above argument, due essentially to Helmholtz, was the first rigorous mathematical treatment of the problem, and shows that the internal cancellation is due to the total effect of the boundary. Nevertheless, as we shall see below, an application of stationary phase shows that (under suitable hypotheses) Fresnel was right, up to terms of order $1/k$. We now apply stationary phase to Helmholtz's formula. The function u, which occurs on the right hand side of the formula will itself be oscillatory, and we must make some assumptions about its form before we can proceed. We shall assume that, near S, $u = ae^{ik\varphi}$ where a and φ are smooth, and that $\|\text{grad } \varphi\| = 1$. This would be the case, for example, if u represented radiation from a single point, Q, lying inside S, where $\varphi(y) = y - Q$. Also, we shall see in the next chapter how to construct "approximate solutions" to the reduced wave equation which are of this form with a and φ arbitrarily prescibed along S (with φ subject to the constraint that grad φ not be tangent to S). These "approximate solutions" satisfy the wave equation up to an error of order k^{-N} for any large N; and we can apply our calculations to these approximate solutions. Indeed, we shall carry out the stationary phase calculations here only to order $1/k$.

We shall assume that we are sufficiently far from S so that $1/r^2$ is negligible in comparison with k, and that a and da are also negligible in comparison with k. Substituting into (H), we find that the top order term (relative to powers of k) is

$$\frac{ik}{4\pi}\iint_S (a/r)e^{ik(\varphi+r)}(* \, d\varphi - * \, dr).$$

Now the points of stationary phase are those points, y, on S where

[2] So far, we have dealt with "monochromatic radiation", u, corresponding to the time dependent function v where $v(x,y,z,t) = u(x,y,z)e^{-ikt}$. For a fixed point, P, let v_P denote the function $v_P(x,y,z,t) = v(x,y,z,t-r)$, where r is the distance from P to (x,y,z). Then substitution into Helmholtz's formula shows that

$$v(P,t) = \frac{1}{4\pi}\iint_S (v_P * d(1/r) - (1/r)(\partial v/\partial t_P) * dr - (1/r) * dv_P).$$

This is Kirchhoff's formula. Since it is linear in v, and does not explicitly involve the frequency, it is true for any superposition of monochromatic waves of varying frequencies, and hence for an arbitrary solution of the wave equation. In this form, the relation with Huygens' principle is very apparent, cf. [5].

grad $\varphi(y)$ + grad $r(y)$ is normal to S. There are two possible situations:

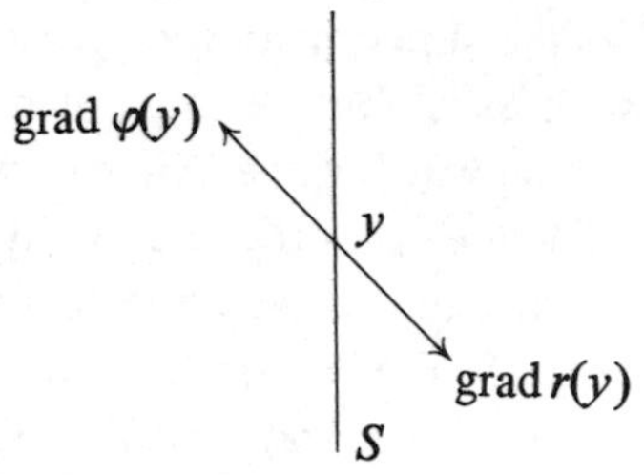

$$\text{grad } \varphi(y) = -\text{ grad } r(y)$$
$$* \, d\varphi(y) = - * dr(y)$$
$$P = y + r \text{ grad } \varphi(y)$$

(a)

case (a)

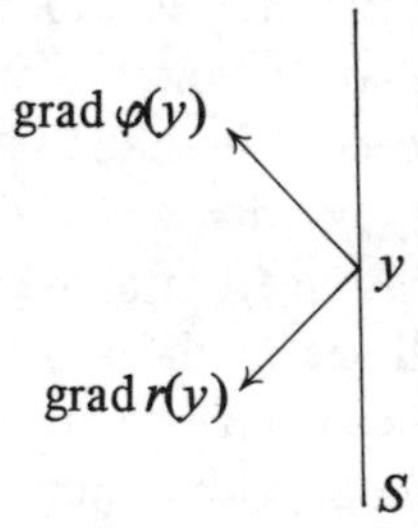

$$\text{grad } \varphi(y) = 2(\text{grad } \varphi(y), n)n - \text{grad } r(y)$$
$$* \, d\varphi(y) = * \, dr(y)$$
$$P = y - r(2(\text{grad } \varphi(y), n)n - \text{grad } \varphi(y))$$

(b)

case (b)

Let us suppose for the moment that y is a non-degenerate critical point of type (b). The top order term in the stationary phase formula will vanish, and the total contribution coming from y in Helmholtz's formula will be of order $1/k$. (Notice that if S were convex and grad φ pointed outward, then for any P inside S, all the critical points would be of type (b). This, in a sense, justifies Fresnel's view that there is "local cancellation" of the backward wave.) For nondegenerate critical points of type (a) since $* \, d\varphi(y) = - * dr(y)$, we may, in computing the highest order contribution to the stationary phase formula, replace the above integral by

$$\frac{ik}{2\pi} \iint (a/r) e^{ik(\varphi + r)} d * r.$$

This shows, that (up to order $1/k$) the induced "secondary radiation" along S behaves as if it

(i) has an amplitude equal to $1/\lambda$ times the amplitude of the primary wave where $\lambda = 2\pi/k$ is the wave length, and

(ii) its phase is one quarter of a period ahead of the primary wave. (This is a way of interpreting the factor i.)

Fresnel made these two assumptions directly in his formulation of Huygens' principle, and this led many to regard his theory as being *ad hoc*. As we have seen, they are a consequence of the method of stationary phase and Helmholtz's formula.

We still must discuss the question of when the critical points are non-degenerate. We shall treat points of type (a); the points of type (b) can be treated in an identical manner. Actually, the discussion is almost the same as in our treatment of emitted radiation: Let us define the "exponential map" $E\colon S \times \mathbf{R}^+ \to \mathbf{R}^3$ by

$$E(y,r) = y + r \operatorname{grad} \varphi(y).$$

Then the critical points on S associated with a point P consist precisely of those y such that $E(y,r) = P$, where $r = \|y - P\|$. If grad $\varphi(y)$ is not tangent to S, then E is a diffeomorphism near $(y, 0)$. Under this assumption we proceed as on page 9. (We can consider the situation on page 9 as the special case where grad φ is everywhere normal to S.) One shows that y is a degenerate critical point for $P = E(y,r)$ if and only if (y,r) is a point at which the map E is singular. In this case we call P a focal point of the map E at y. As before, if P is not a focal point, then the index of the Hessian of $\varphi + r$ at y is the number of focal points on the ray segment from y to P (counted with multiplicity). We leave the details to the reader.

Finally, we should observe that if grad φ is close to the normal to a surface S, then the focal points of the map E associated to φ will be close to the corresponding focal points of the surface. In this way we get an explanation of the Gouy double mirror experiment mentioned at the beginning of the Chapter.

References, Chapter I

1. L. G. Gouy, C. R. Acad. Sci. Paris **110** (1890), 1251.
2. H. Poincaré, *Theorie mathématique de la lumière*. II, George Carré, Paris, 1892, pp. 168–174.
3. J. W. Milnor, *Morse theory*, Ann. of Math. Studies, no. 51, Princeton Univ. Press, Princeton, N. J., 1963. MR **29** #634.
4. P. D. Lax and R. S. Phillips, *Scattering theory*, Pure and Appl. Math., vol. 26, Academic Press, New York and London, 1967. MR **36** #530.
5. B. B. Baker and E. T. Copson, *The mathematical theory of Huygens' principle*, Clarendon Press, Oxford, 1953.
6. J. Larmor, Proc. London Math. Soc. (2) **19** (1919), 169–180.
7. M. Born and E. Wolf, *Principles of optics*, 4th ed., Pergamon Press, Oxford and New York, 1970.
8. R. S. Palais, Bull. Amer. Math. Soc. **75** (1969), 968–671. MR **40** #6593.
9. M. Golubitsky and V. Guillemin, Advances in Math. **15** (1975), 2.

Appendix I. Morse's Lemma and Some Generalizations

Let f be a smooth function defined near the origin in a vector space V. Here we shall take V to be some finite dimensional vector space, but our assertions work just as well in an Banach space. Suppose that $f(0) = 0$, that $df(0) = 0$, and that $\frac{1}{2}d^2f(0)$ is a nonsingular quadratic form Q. Morse's lemma asserts that we can make a change of coordinates near the origin so that f is quadratic in the new variables. More precisely, it asserts that there is some neighborhood, U, of 0 and some diffeomorphism $\varphi: U \to V$ such that

$$(f \circ \varphi) = Q(x, x).$$

The proof that we present is due to Palais [4] based on an earlier idea of Moser. We shall encounter other uses of the same techniques of proof later on.

Let us set

$$f^t(x) = Q(x, x) + t(f(x) - Q(x, x))$$

so that

$$f^1 = f,\ f^0 = Q \text{ and } \dot{f}^t = \frac{df^t}{dt} = f - Q.$$

We shall seek a one parameter family of diffeomorphisms φ^t, such that

$$f^t \circ \varphi^t = f^0 \tag{I.1}$$

Then clearly φ^1 does the job. Let ξ^t be the vector field tangent to φ^t so that

$$\xi^t(\varphi^t(x)) = \frac{d\varphi^t}{dt}(x).$$

Differentiating (I.1) gives

$$(\dot{f}^t + \xi^t f^t) \circ \varphi^t = 0.$$

If we could find a time dependent vector field ξ^t such that $\dot{f}^t + \xi^t f^t = 0$ then we could integrate it to get a one parameter family of diffeomorphisms satisfying (I.1). We are thus looking for a vector field satisfying

$$f^1 - f^0 + \xi^t f^t = 0$$

or

$$df^t(\xi^t) = f^0 - f^1. \tag{I.2}$$

Now for any x near 0 and for any $v \in V$ we have

$$df_x^t(v) = \int_0^1 \frac{d}{ds} df_{sx}^t(v)\,ds$$

since $df_0^t = 0$. Thus

$$df_x^t(v) = \int_0^1 \frac{d}{ds} df_{sx}^t(v)\,ds = \int_0^1 d^2 f_{sx}^t(x,v)\,ds = B_x^t(x,v)$$

where B_x^t is the quadratic form defined by

$$B_x^t(u,v) = \int_0^1 d^2 f_{sx}^t(u,v)\,ds$$

so that

$$B_x^t = B_x^0 + t(B_x^1 - B_x^0)$$

where $B_x^0 = 2Q$ does not depend on x. Now at $x = 0$ we have $B_0^t = 2Q$ is nonsingular for all $0 \leq t \leq 1$. Therefore B_x^t is nonsingular for all x in some neighborhood of 0 and all $0 \leq t \leq 1$. We can now rewrite (I.2) as

$$B_x^t(x, \xi^t) = f^0 - f^1. \tag{I.3}$$

Now $(f^0 - f^1)(0) = 0$ and $(df^0 - df^1)(0) = 0$. Thus, setting $g = f^0 - f^1$, we have

$$\begin{aligned} f^0 - f^1 &= \int_0^1 \frac{d}{ds} g(sx)\,ds = \int_0^1 dg_{sx}(x)\,ds \\ &= \int_0^1 \int_0^1 d^2 g_{rsx}(sx, x)\,dr\,ds = C_x(x,x) \end{aligned}$$

where C_x is the quadratic form

$$C_x(u,v) = \int_0^1 \int_0^1 d^2 g_{rsx}(su, v)\,dr\,ds.$$

We now choose ξ^t to be the unique solution to

$$B_x^t(u, \xi^t) = C_x(u,x)$$

for all $u \in V$. The ξ^t so obtained is clearly smooth and $\xi^t(0) = 0$ for all t. By restricting, if necessary, to a smaller neighborhood of the origin, we can integrate the ξ^t to obtain the desired φ^t. Notice that our procedure is such that if f depends smoothly on some parameters the φ will depend smoothly on these same parameters.

We can use the same proof to obtain the following very useful extension of Morse's lemma; cf. Golubitsky and Guillemin [5].

Lemma. *Let f and g be smooth functions vanishing together with their first derivatives at* $0 \in \mathbf{R}^n$. *Suppose that*

$$g = f + \sum h_{ij} \frac{\partial f}{\partial x_i} \frac{\partial f}{\partial x_j}$$

where the $h_{ij} = h_{ji}$ *are smooth functions defined near* 0 *and where*

$$\sum_i h_{ij} \frac{\partial^2 f}{\partial x_i \partial x_k}(0) = 0, \qquad i, k = 1, \ldots, n.$$

Then there is a locally defined diffeomorphism φ *of* $\mathbf{R}^n$ *such that* $\varphi(0) = 0$ *and*

$$\varphi^* f = g.$$

If f, g, and h depend smoothly on parameters so does φ.

Proof. As before, let us set

$$f^t = f + t \sum h_{ij} \frac{\partial f}{\partial x_i} \frac{\partial f}{\partial x_j}$$

and look for a vector field $\xi^t = (w_t^1, \ldots, w_t^n)$ such that $\xi^t(0) = 0$ and $\dot{f}^t + \xi^t f^t = 0$. Now

$$\frac{\partial f^t}{\partial x_k} = \frac{\partial f}{\partial x_k} + t \sum_j \left[\sum_i \frac{\partial h_{ij}}{\partial x_k} \frac{\partial f}{\partial x_i} + 2 \sum_i h_{ij} \frac{\partial^2 f}{\partial x_i \partial x_k} \right] \frac{\partial f}{\partial x_j}$$

$$= \frac{\partial f}{\partial x_k} + t \sum b_{kj} \frac{\partial f}{\partial x_j}$$

where b_{kj} is the expression in brackets. By hypothesis, $b_{kj}(0) = 0$. Thus letting B denote the matrix (b_{kj}) we see that $I + tB$ is invertible for small values of x. Let $C = (I + tB)^{-1}$ where $C = (c_{ij})$. Then

$$\frac{\partial f}{\partial x_i} = \sum c_{ij} \frac{\partial f^t}{\partial x_j}$$

so that

$$\dot{f}^t = \sum h_{\ell i} \frac{\partial f}{\partial x_\ell} \frac{\partial f}{\partial x_i} = \sum_{h\ell} h_{\ell i} \frac{\partial f}{\partial x_k} c_{ij} \frac{\partial f^t}{\partial x_j}.$$

Thus if we set

$$w_t^j = - \sum_{\ell, i} h_{\ell i} \frac{\partial f}{\partial x_\ell} c_{ij}$$

and $\xi^t = (w_t^1, \ldots, w_t^n)$ we see that $\xi^t(0) \equiv 0$ and $\dot{f}^t + \xi^t f^t = 0$. As ξ^t depends explicitly on h and f we obtain the smooth dependence on parameters proving the lemma.

It is easy to check that if we take f to be a non-singular quadratic form we recover the standard Morse lemma. Let us describe some other applications. Let m denote the maximal ideal of the ring of germs of C^∞ functions vanishing at the origin. It is generated by the coordinate functions $x_i, \ldots, x_n$. A (germ of a) function f has a non-degenerate critical point at 0 if and only if the ideal

$$\left(\frac{\partial f}{\partial x_i}, \ldots, \frac{\partial f}{\partial x_n}\right) = m.$$

Here

$$\left(\frac{\partial f}{\partial x_1}, \ldots, \frac{\partial f}{\partial x_n}\right)$$

denotes the ideal generated by the first partial derivatives of f. More generally, we say that a function f having a critical point at 0 satisfies the Milnor condition of order s if

$$m^s \subset \left(\frac{\partial f}{\partial x_1}, \ldots, \frac{\partial f}{\partial x_n}\right).$$

We then have the following result:

(*MATHER, TOUGERON*) *Let f satisfy the Milnor condition to order s and suppose that $g = f + u$ where $u \in m^{2s+1}$. Then there is a local diffeomorphism φ with $\varphi^* g = f$.*

PROOF. Since

$$u \in m^{2s+1} \subset m \cdot \left(\frac{\partial f}{\partial x_1}, \ldots, \frac{\partial f}{\partial x_n}\right)^2$$

we can write

$$u = \sum h_{ij} \frac{\partial f}{\partial x_i} \frac{\partial f}{\partial x_j}$$

with $h_{ij}(0) = 0$, and the lemma applies.

Notice that this shows, in particular, that any f satisfying the Milnor condition is equivalent to a polynomial. A holomorphic version of the lemma will yield a celebrated result of Arnold to the effect that a holomorphic function with an isolated critical point is holomorphically equivalent to a polynomial.

We have further use of the lemma in Chapter IV.

Chapter II. Differential Operators and Asymptotic Solutions

§1. Differential operators.

In the preceding chapter we analyzed the solutions of the reduced wave equation in the high frequency limit. Our procedure was to find an exact solution to the equation, and then to use the method of stationary phase to give an approximation to the solution. For a general linear differential equation we do not expect to be able to solve the equation in closed form. Nevertheless, we may expect to be able to find a "high frequency approximation" to the solution directly, without actually solving the equation. The development of this idea will occupy us for some time.

We begin the discussion by recalling the notion of a linear differential operator. Let E and F be vector bundles of the differentiable manifold M. A smooth vector bundle map from E to F is, by definition, a smooth section of $\mathrm{Hom}(E, F)$. If T is such a vector bundle map and s is a smooth section of E, then Ts is a smooth section of F. Thus T induces a map, L, from $C^\infty(E) \to C^\infty(F)$ (where $C^\infty(E)$ is the space of smooth sections of E). Notice that L satisfies

$$L(fu) = fLu$$

for any function f. Conversely, any map L: $C^\infty(E) \to C^\infty(F)$ which commutes with multiplication by functions in the above sense defines a section of $\mathrm{Hom}(E, F)$. Indeed, if $u(x) = 0$ then we can write $u = \sum f_i v_i$ where $f_i(x) = 0$, and therefore $(Lu)(x) = \sum f_i(x)(Lv_i)(x) = 0$. Thus the value of Lw at x depends only on $w(x)$. The map $w(x) \to (Lw)(x)$ is an element of $\mathrm{Hom}(E, F)_x$, and thus L determines a section of $\mathrm{Hom}(E, F)$. It is easy to check that this section is smooth.

We shall use f to denote the map of $C^\infty(E) \to C^\infty(E)$ (and $C^\infty(F) \to C^\infty(F)$) given by multiplication by f. Then the above condition on L can be written as

$$[L,f] = L \circ f - f \circ L = 0.$$

We shall call such operators, L, differential operators of degree zero, for reasons which will soon be apparent. We shall denote the space of such operators by $\mathfrak{D}_0(E,F)$.

A differential operator of degree (at most) one is a linear operator

$$L\colon C^\infty(E) \to C^\infty(F)$$

such that

$$[L,f] \in \mathfrak{D}_0(E,F).$$

Notice that $[L,f]$ determines a section of $\operatorname{Hom}(E,F)$. Let us denote this section by $\sigma(f)$. Since L is linear, we know that $[L,c] = 0$ for any constant c. Thus $\sigma L(f+c) = \sigma L(f)$. Notice also that

$$\begin{aligned}[L,fg] &= L \circ fg - fg \circ L \\ &= (L \circ f - f \circ L) \circ g + f \circ (L \circ g - g \circ L) \\ &= [L,f]g + f[L,g],\end{aligned}$$

and since $[L,f] \in \mathfrak{D}_0(E,F)$ this last expression equals $g[L,f] + f[L,g]$. Thus, if f and g both vanish at x then $\sigma L(fg)(x) = 0$. This shows that if df vanishes at x then $\sigma L(f)(x) = 0$. Thus $\sigma L(f)(x)$ depends only on $df(x)$. Therefore, we get a well defined map, $\sigma(L)\colon T^*M_x \to \operatorname{Hom}(E,F)_x$. This map is known as the symbol of the differential operator. Let $u_1, \ldots, u_m$ and $v_1, \ldots v_n$ be local sections providing bases of E and F near x and let $x_1, \ldots x_k$ be coordinates near x. Then

$$L(\sum s_j u_j) = \sum A^r_{ij} \frac{\partial s_j}{\partial x_r} v_i + \sum B_{ij} s_j v_i.$$

If we write A^r for the matrix (A^r_{ij}) and B for the matrix (B_{ij}) then we can shorten the above to read

$$Ls = \sum A^r \frac{\partial s}{\partial x_r} + Bs$$

and it is clear that the matrix representation of $\sigma(L)(df)$ is given by

$$\sigma(L)(df) = \sum A^r \frac{\partial f}{\partial x_r}.$$

In terms of local coordinates and trivializations it is clear that

$$Lu = \sum A_i \frac{\partial u}{\partial x_i} + Bu$$

where the A_i and B are sections of $\mathrm{Hom}(E, F)$ and

$$\sigma(L)(df) = \sum A_i\left(\frac{\partial f}{\partial x_i}\right).$$

We can proceed inductively to define differential operators of degree k: A linear map $L\colon C^\infty(E) \to C^\infty(F)$ is called a differential operator of degree (at most) $k + 1$ if

$$[L,f] \in \mathfrak{D}_k(E, F)$$

where $\mathfrak{D}_k(E, F)$ denotes the space of differential operators of degree k. We clearly have $\mathfrak{D}_k(E, F) \subset \mathfrak{D}_{k+l}(E, F)$ and we set

$$\mathfrak{D}(E, F) = \bigcup_k \mathfrak{D}_k(E, F).$$

We can compose $M \in \mathfrak{D}_\ell(F, G)$ and $L \in \mathfrak{D}_k(E, F)$ to obtain

$$M \circ L \in \mathfrak{D}_{k+\ell}(E, G),$$

if E, F, and G are three vector bundles over X.

Let $L \in \mathfrak{D}(E, F)$ and let f and g be C^∞ functions. Then

$$[[L,f],g] = [[L,g],f] + [L,[f,g]] = [[L,g],f]$$

since $[f,g] = 0$. Thus, if $L \in \mathfrak{D}_k(E, F)$ and if $f_1, \ldots f_k$ are C^∞ functions then

$$[[L,f_1], \ldots, f_k]$$

is an operator of degree zero depending symmetrically on the f's. If we consider it as a function of f_k we can apply the result for a first order differential operator to conclude that it depends only on df_k. By symmetry, it depends only on $df_1, \ldots, df_k$. We have thus defined a symmetric linear function on

$$T^* \otimes \cdots \otimes T^*$$

with values in $\mathrm{Hom}(E, F)$. The corresponding homogeneous polynomial on T^* is called the symbol of L and denoted by $\sigma(L)$; thus

$$\sigma(L)(df) = \frac{1}{k!}\Big[[L,f], \ldots, f\Big] \qquad (k \text{ brackets})$$

or, if we denote $[L,f]$ by $(\mathrm{ad} f)L$ we can write

$$\sigma(L)(df) = \frac{1}{k!}(\mathrm{ad}\, f)^k L.$$

We now apply a standard argument from the elementary theory of Lie algebras: Let s be a (real or complex) variable and consider the expression $e^{-sf}Le^{sf}$ as a function of s. Differentiating with respect to s gives

$$\frac{d}{ds}\left(e^{-sf}Le^{sf}\right) = e^{-sf}(Lf - fL)e^{sf} = e^{-sf}[L,f]e^{sf}.$$

Now $[L,f]$ is again a differential operator and we can repeat the process. Thus

$$\left(\frac{d}{ds}\right)^{\ell}(e^{-sf}Le^{sf}) = e^{-sf}((\operatorname{ad} f)^{\ell}L)e^{sf}.$$

If L is of degree k and $\ell > k$ then the right hand side vanishes. Thus $e^{-sf}Le^{sf}$ is a polynomial in s and we can apply Taylor's formula to obtain

$$e^{-sf}Le^{sf} = \sum_{0}^{k} \frac{(\operatorname{ad} f)^{j}L}{j!}s^{j}. \tag{1.1}$$

From this formula we can read off the explicit expression for $\sigma(L)(df)$ in terms of local coordinates: Suppose we have chosen local trivializations of E and F and have chosen local coordinates so that L has the expression

$$L = \sum_{|\alpha|\leq k} A_{\alpha}\frac{\partial^{\alpha}}{\partial x^{\alpha}} \quad \text{where} \quad \alpha = (\alpha_1, \ldots, \alpha_n), \quad \frac{\partial^{\alpha}}{\partial x^{\alpha}} = \frac{\partial^{\alpha_1 + \cdots + \alpha_n}}{(\partial x_1)^{\alpha_1}\cdots(\partial x_n)^{\alpha_n}}.$$

Here we have $E_{|U} \sim V \times U$ and $F_{|U} \sim W \times U$ where U is some neighborhood and the A_{α} are functions from U to $\operatorname{Hom}(V, W)$. The only contribution from $e^{-sf}Le^{sf}$ that can involve s^k comes from the terms α with $|\alpha| = k$, using all the differentiations to hit the exponential. Thus

$$\sigma(L)(df) = \sum_{|\alpha|=k} A_{\alpha}(df)^{\alpha} \quad \text{where} \quad (df)^{\alpha} = \left(\frac{\partial f}{\partial x_1}\right)^{\alpha_1}\cdots\left(\frac{\partial f}{\partial x_n}\right)^{\alpha_n},$$

or

$$\sigma(L)(\xi) = \sum_{|\alpha|=k} A_{\alpha}\xi^{\alpha}. \tag{1.2}$$

Let us compute $(\operatorname{ad} f)^{k-1}L/(k-1)!$. There will be one contribution coming from the term of degree $k-1$ in L and one from the term of degree k. We claim that the entire expression is

$$\frac{(\operatorname{ad} f)^{k-1}L}{(k-1)!} = \Bigg(\sum \frac{\partial\sigma(L)}{\partial\xi^{i}}(df)\frac{\partial}{\partial x_i} + \tfrac{1}{2}\sum_{i,j}\frac{\partial^2\sigma(L)}{\partial\xi^{i}\partial\xi^{j}}(df)\frac{\partial^2 f}{\partial x_i\partial x_j} + \sum_{|\beta|=k-1} A_{\beta}(df)^{\beta}\Bigg). \tag{1.3}$$

The third term is obviously the contribution from the component of order $k-1$ in the local description of L. We shall give two proofs of (1.3). The first is by induction on k, the formula being clearly true for $k = 1$ (in which case the middle term on the right vanishes). Assume the formula is correct for operators of degree k, and we wish to prove it for operators of degree $k + 1$. By linearity we may assume that our operator is of the form

$$A \frac{\partial}{\partial x_r} \frac{\partial^\alpha}{\partial x^\alpha} \qquad \text{where } |\alpha| = k,$$

and since the A is irrelevant we may as well deal with $(\partial/\partial x_r)(\partial^\alpha/\partial x^\alpha)$. We are thus looking for the coefficient of s^k in

$$e^{-sf}\left(\frac{\partial}{\partial x_r}\frac{\partial^\alpha}{\partial x^\alpha}\right)e^{sf} = e^{-sf}\left(\frac{\partial}{\partial x_r}\right)e^{sf}e^{-sf}\left(\frac{\partial^\alpha}{\partial x^\alpha}\right)e^{sf}$$
$$= \left[\frac{\partial}{\partial x_r} + s\frac{\partial f}{\partial x_r}\right]e^{-sf}\frac{\partial^\alpha}{\partial x^\alpha}e^{sf}.$$

Now, by induction,

$$e^{-sf}\frac{\partial^\alpha}{\partial x^\alpha}e^{sf} = (df)^\alpha s^k + \left(\sum \alpha_i (df)^{\alpha-\delta_i}\frac{\partial}{\partial x_i}\right.$$
$$\left. + \frac{1}{2}\sum_{i,j} \alpha_i(\alpha_j - \delta_{ij})(df)^{\alpha-\delta_i-\delta_j}\frac{\partial^2 f}{\partial x_i \partial x_j}\right)s^{k-1} + \cdots$$

where the remaining terms are of lower order in s. Here $\delta_i = (0, \ldots, 1, 0, \ldots)$ where the 1 is in the ith position and zeros elsewhere. Substituting into the preceding equation and extracting the coefficient of s^k gives

$$(df)^\alpha \frac{\partial}{\partial x_r} + \frac{\partial}{\partial x_r}(df)^\alpha + \sum \alpha_i (df)^{\alpha-\delta_i+\delta_r}\frac{\partial}{\partial x_i}$$
$$+ \frac{1}{2}\sum \alpha_i(\alpha_j - \delta_{ij})(df)^{\alpha-\delta_i-\delta_j+\delta_r}\frac{\partial^2 f}{\partial x_i \partial x_j}$$
$$= \sum (\alpha + \delta_r)_i (df)^{\alpha+\delta_r-\delta_i}\frac{\partial}{\partial x_i}$$
$$+ \tfrac{1}{2}\sum (\alpha + \delta_r)_i((\alpha + \delta_r)_j - \delta_{ij})df^{\alpha+\delta_r-\delta_i-\delta_j}\frac{\partial^2 f}{\partial x_i \partial x_j}$$

proving the result.

Our second proof will use the Fourier transform and yield a formula for all the terms in (1.1). When using the Fourier transform it is convenient to use a slightly different local expression for a differential operator. Let us set

$$D^\alpha = \left(\frac{1}{i}\frac{\partial}{\partial x_1}\right)^{\alpha_1} \cdots \left(\frac{1}{i}\frac{\partial}{\partial x_n}\right)^{\alpha_n}$$

so that

$$D^\alpha e^{i\xi\cdot x} = \xi^\alpha e^{i\xi\cdot x}.$$

If f is any C^∞ function of compact support or any function in the Schwartz space $\mathcal{S}$, its Fourier transform, $\hat{f}$, is defined by

$$\hat{f}(\xi) = \int e^{-i\xi\cdot x} f(x)\,dx$$

so that

$$f(x) = (2\pi)^{-n} \int e^{i\xi\cdot x} \hat{f}(\xi)\,d\xi \quad \text{and} \quad D^\alpha f = (2\pi)^{-n} \int e^{i\xi\cdot x} \xi^\alpha \hat{f}(\xi)\,d\xi.$$

We can now write the most general linear differential operator locally as

$$P(x, D) = \sum_{|\alpha| \le k} a_\alpha(x) D^\alpha = \sum_{j=1}^{k} P_j(x, D)$$

where the P_j are homogeneous polynomials in D,

$$P_j(x, D) = \sum_{|\alpha| = j} a_\alpha D^\alpha,$$

so that

$$P_j(x, D) f = (2\pi)^{-n} \int e^{i\xi\cdot x} P_j(x, \xi) \hat{f}(\xi)\,d\xi.$$

and

$$P(x, D) f = (2\pi)^{-n} \int e^{i\xi\cdot x} P(x, \xi) \hat{f}(\xi)\,d\xi.$$

Let us write

$$P^\gamma = \frac{\partial^\gamma}{\partial \xi^\gamma} P$$

so, for example, $P^\gamma = 0$ for $|\gamma| > k$. For any smooth function φ set

$$h_\varphi(x, z) = \varphi(z) - \varphi(x) - \sum (z_j - x_j) \frac{\partial \varphi}{\partial x_j}(x).$$

We claim that

$$[e^{-i\tau\varphi} P(x, D) e^{i\tau\varphi}] u = \sum_\gamma \frac{1}{\gamma!} P^\gamma(x, \tau d\varphi) [D_z^\gamma (e^{i\tau h_\varphi(x,z)} u(z))]|_{z=x} \tag{1.4}$$

where D_z^γ means that the operator D^γ is applied in the z variables. Before proving (1.4) observe that the coefficient of $(i\tau)^j$ on the left of (1.4) is exactly the jth coefficient in (1.1). Notice also that h_φ vanishes quadratically at $z = x$. Thus the terms involving γ with $|\gamma| = 1$ which involve differentiating the exponential will vanish when we set $z = x$. Thus the coefficient of τ^k in (1.4) is $P_k(x, d\varphi)$ while the coefficient of τ^{k-1} is

$$P_{k-1}(x, d\varphi)u + \sum_{|\gamma|=1} P_k^\gamma(x, d\varphi) D^\gamma u + \sum_{|\gamma|=2} \frac{1}{\gamma!} P_k^\gamma(x, d\varphi) D^\gamma \varphi(x) u.$$

(The last term obtains because $h_\varphi(x, z)$ and $\varphi(z)$ differ by constant and linear terms and hence have the same second derivatives.) If we take into account the factor of i we notice that this gives (1.3).

The proof of (1.4) is basically Taylor's formula: Write

$$v(x) = e^{i\tau\varphi(x)} u(x)$$

so that

$$P(x, D)(e^{i\tau\varphi} u) = P(x, D)v = (2\pi)^{-n} \int e^{i\xi\cdot x} P(x, \xi) \hat{v}(\xi)\, d\xi$$

where $\hat{v}(\xi) = \int e^{-i\xi\cdot y} e^{i\tau\varphi(y)} u(y)\, dy$. Thus

$$\begin{aligned} e^{-i\tau\varphi(x)} P(x, D)(e^{i\tau\varphi} u) &= (2\pi)^{-n} \iint e^{i\xi\cdot(x-y)} e^{i\tau[\varphi(y)-\varphi(x)]} P(x, \xi) u(y)\, dy\, d\xi \\ &= (2\pi)^{-n} \iint e^{i(\xi - \tau d\varphi(x))\cdot(x-y)} P(x, \xi) e^{i\tau h(x, y)} u(y)\, dy . d\xi. \end{aligned}$$

Let us make the change of variables $\eta = \xi - \tau d\varphi(x)$. Then this last expression becomes

$$(2\pi)^{-n} \int e^{i\eta\cdot(x-y)} P(x, \eta + \tau d\varphi) e^{(i\tau h(x, y))} u(y)\, dy\, d\eta$$

and applying Taylor's formula to P about the point $\tau d\varphi(x)$ yields (1.4).

§2. Asymptotic sections.

In order to discuss the notion of an asymptotic solution to a partial differential equation we must first recall some basic facts from the theory of asymptotic series and reformulate some of the standard definitions in a geometrical setting. Let f and g be two real valued functions on $\mathbf{R}^+$. We say that f and g are asymptotically equal (at infinity) if, for every $N > 0$, we have

$$\lim_{t\to\infty} t^N (f(t) - g(t)) = 0. \tag{2.1}$$

This clearly defines an equivalence relation and we shall call an equivalence class such as $[f]$ an *asymptotic number*. We say that an asymptotic number $[f]$ has an asymptotic expansion, $f \sim \sum a_n \tau^{-n}$, if there exist real numbers a_n such that for any integer $N > 0$ we have

$$\tau^N\Big(f(\tau) - \sum_{n \le N} a_n \tau^{-n}\Big) \to 0. \tag{2.2}$$

Here the n are all required to be integers. It is clear that the coefficients a_n are completely determined by $[f]$. Of course, not every $[f]$ has an asymptotic expansion. On the other hand, given any sequence $\{a_n\}$ we can always find a smooth function f such that $f \sim \sum a_n \tau^{-n}$. This is an old theorem of E. Borel whose proof is as follows: Choose some C^∞ function ρ such that $\rho(\tau) = 0$ for $\tau \le 1, 0 \le \rho \le 1$ and $\rho(\tau) = 1$ for $\tau \ge 2$. Then let $f(\tau) = \sum a_n \rho(2^{-n-|a_n|}\tau)\tau^{-n}$, for $\tau \ge 1$. Notice that

$$|a_n \rho(2^{-n-|a_n|}\tau)\tau^{-n}| \le \frac{|a_n|}{2^{n+|a_n|}}$$

since the left hand side vanishes for $\tau < 2^{n+|a_n|}$. This shows that the series converges for all τ and also that (2.2) holds.

Notice that (2.1) makes sense if f and g, instead of being real valued functions, take these values in any topological vector space V. We then talk of an asymptotic vector instead of an asymptotic number. Similarly we say that $[f]$ has an asymptotic expansion if (2.2) holds where the a's are now vectors in V. For example we might take $V = C_0^\infty(E)$ where E is a vector bundle over some manifold M, and where $C_0^\infty(E)$ is given its standard topology (in which a sequence of sections converge if they have a fixed common support and converge uniformly together with various derivatives of any finite order relative to some (and hence any) choice of local trivializations and coordinates.)

Let V and W be topological vector spaces and $L: V \to W$ a continuous linear map. Then L clearly induces a linear map from asymptotic vectors of V to asymptotic vectors of W. For example, if V is the space of smooth densities of compact support on a manifold M then we have a linear map $\int: V \to \mathbf{R}$ given by integration. Thus, if $[\rho]$ is an asymptotic density then $\int [\rho] = [\int \rho]$ is an asymptotic number. If $L: C_0^\infty(E) \to C_0^\infty(F)$ is a differential operator and $[s]$ is an asymptotic section of E then $L[s] = [Ls]$ is an asymptotic section of F and we say that s is an asymptotic solution of the corresponding homogeneous equation if $L[s] = 0$. If $L: V \to W$ is a continuous map and $[f]$ has an asymptotic expansion $f \sim \sum a_n \tau^{-n}$ then $L[f]$ has the asymptotic expansion $Lf \sim \sum (La_n)\tau^{-n}$.

If U, V, and W are three topological vector spaces and $B: U \times V \to W$ is a continuous bilinear map then we can obviously apply the previous considerations so that $B([u],[v]) = [B(u,v)]$ is a well defined asymptotic element of W if $[u]$ and $[v]$ are asymptotic elements of U and V. If $u \sim \sum a_n\tau^{-n}$ and $v \sim \sum b_n\tau^{-n}$ then $B(u,v) \sim \sum c_n\tau^{-n}$ where

$$c_n = \sum_{i+j=n} B(a_i, b_j).$$

In particular, let $L(V, W)$ denote the space of continuous linear maps from V to W endowed with the strong topology so that the evaluation map $L(V, W) \times V \to W$ is continuous. Then an asymptotic element $[A]$ of $L(V, W)$ can be called an asymptotic operator from V to W, and $[A][v] = [w]$ is well defined if $[v]$ is an asymptotic element of V.

Let E and F be smooth vector bundles over M. By an asymptotic differential operator from E to F we shall mean an asymptotic operator from $C_0^\infty(E)$ to $C_0^\infty(F)$ which has the expansion

$$L \sim \sum \frac{L_k}{(i\tau)^k}$$

where the L_k are differential operators from E to F. An equation of the form $[L][u] = [v]$, where $[u]$ and $[v]$ are asymptotic sections of E and F is then called an asymptotic differential equation. For example, the reduced wave equation,

$$(\Delta^2 + \tau^2)u = 0,$$

can be considered as an asymptotic differential equation if we write it as

$$\left(1 + \frac{1}{\tau^2}\Delta^2\right)u = 0.$$

Notice that in our mathematical discussion we will consider τ as a parameter used in describing an equivalence relation. In actual practice (for example in the wave equation) one is interested in a large but fixed value of τ.

In our definition of an asymptotic differential operator we shall want to restrict the L_k somewhat further. We shall require that

$$\deg L_k \leq k \tag{2.3}$$

and

$$\textit{for any } l \textit{ we have } \deg L_k < k - l \textit{ except for finitely many } k. \tag{2.4}$$

Condition (2.3) is just for bookkeeping convenience. Condition (2.4) is a little more essential in that it allows for a recursive procedure for attempting to find an asymptotic solution to the equation $[L][u] = 0$, which we describe in the next section.

Let

$$D_{\tau j} = \frac{1}{i\tau}\frac{\partial}{\partial x_j} \quad \text{and} \quad D_\tau^\alpha = \left(\frac{1}{i\tau}\right)^{|\alpha|}\frac{\partial^{|\alpha|}}{\partial x^\alpha}.$$

Then the local expression for an asymptotic differential operator can be written as

$$\sum_\alpha a_\alpha(x,\tau) D_\tau^\alpha \tag{2.5}$$

where

$$a_\alpha(x,\tau) \sim \sum A_{\alpha j}(x)\tau^{-j}.$$

Condition (2.4) says that in (2.5) only finitely many $A_{\alpha j}$ are unequal to zero for any fixed j.

§3. The Lüneburg-Lax-Ludwig technique.

It turns out that instead of trying to look for the most general asymptotic section to solve an asymptotic partial differential equation, there is a subclass of asymptotic sections with rather nice geometrical properties. It will take a bit of machinery to describe the class properly; however these asymptotic sections will, in a sense, be generalizations of the simple asymptotics that we now introduce. We say that $[u]$ is a *simple asymptotic section* of the vector bundle E if there exist a function φ and sections u_n such that

$$u \sim e^{i\tau\varphi} \sum \frac{u_n}{(i\tau)^n}. \tag{3.1}$$

The function φ is then called the phase function of $[u]$. Suppose $[L]$ is an asymptotic differential operator from E to F and suppose that we wish to find a simple asymptotic solution. That is we look for a $[u]$ given by (3.1) which satisfies $[L][u] = 0$. Now $[L][u]$ is obviously a simple asymptotic section of F with the same phase φ,

$$[L][u] = e^{i\tau\varphi} \sum \frac{v_n}{(i\tau)^n}$$

and we may attempt to find the u_n's recursively so that $v_n = 0$. For this purpose let us define $\sigma([L])$ by

$$\sigma([L]) = \sum_k \sigma(L_k) \textit{ where } L_k \textit{ is considered as an operator of order } k. \tag{3.2}$$

Thus, in (3.2), if $\deg L_k < k$ then $\sigma(L_k) = 0$. In view of (2.4) the sum on the right of (3.2) is finite. It is then clear that

$$v_0 = \sigma([L])(d\varphi)u_0$$

and thus u_0 must satisfy the algebraic equation

$$\sigma([L])(d\varphi)u_0 = 0. \tag{3.3}$$

Now $\sigma([L])(d\varphi)$ is a vector bundle map from E to F and for a general φ there is no reason to expect it to have a nontrivial kernel. If, for any $m \in M$, we want to be able to find solutions of (3.3) which do not vanish near m then we must demand that

$$\ker \sigma([L])(d\varphi) \neq 0 \qquad \text{for any } x. \tag{3.4}$$

Notice that this is a first order (nonlinear) partial differential equation for φ. This equation is called the characteristic equation of $[L]$. For example, in the case of the reduced wave equation (where $[L]$ is a scalar operator) the equation is

$$\sum \left(\frac{\partial \varphi}{\partial x^i}\right)^2 - 1 = 0.$$

Thus the first step in the process is to choose a φ which is a solution of the characteristic equation. We shall describe the procedure for finding solutions to such equations in the next section. It will turn out that we can only expect to find local solutions φ on account of geometrical considerations. Suppose we have found a solution φ to (3.4). Our next step is to choose u_0 to be a section of $\ker \sigma([L])$.

To proceed further observe that, just as in §1, we have

$$e^{-i\tau\varphi}[L]e^{i\tau\varphi} \sim \sum_k \sum_j \frac{(\operatorname{ad} \varphi)^j L_k (i\tau)^j}{j!\,(i\tau)^k}$$

where the coefficient of $(i\tau)^{-\ell}$ involves only a finite sum because of (2.4). Collecting the coefficients of $(i\tau)^{-\ell}$ we can write this as

$$e^{-i\tau\varphi}[L]e^{i\tau\varphi} \sim \sum \frac{A_\ell(L, \varphi)}{(i\tau)^\ell} \tag{3.5}$$

where $A_\ell([L], \varphi)$ is the differential operator of order l given by

$$A_\ell([L], \varphi) = \sum_k \frac{(\operatorname{ad} \varphi)^{k-\ell} L_k}{(k-\ell)!}, \tag{3.6}$$

the sum being finite as we have observed. Here

$$A_0([L], \varphi) = \sigma([L])(d\varphi)$$

and, by (1.3), the local expression for A_1 is

$$A_1([L], d\varphi) = \sum \frac{\partial\sigma([L])(d\varphi)}{\partial\xi^i} \frac{\partial}{\partial x^i} + \text{zeroth order terms} . \qquad (3.7)$$

Now

$$[L](e^{i\tau\varphi} \sum u_j(i\tau)^{-j}) = e^{i\tau\varphi} \sum_{\ell,j} A_\ell([L], \varphi)u_j(i\tau)^{-(j+\ell)}$$

so that the equation $[L][u] = 0$ becomes

$$A_0([L], \varphi)u_0 = 0,$$

$$A_0([L], \varphi)u_1 + A_1([L], \varphi)u_0 = 0,$$

$$A_0([L], \varphi)u_2 + A_1([L], \varphi)u_1 + A_2([L], \varphi)u_0 = 0,$$

and so on.

We have exhausted the first of these equations. In order to analyze the second of these equations we make an additional assumption about $\sigma([L])(d\varphi)$; namely that ker $\sigma([L])(d\varphi)$ is a vector subbundle of E. (In particular, we want the rank of $\sigma([L])(d\varphi)$ not to vary.) Let K_φ denote this kernel and let C_φ denote its cokernel, which is now also a vector bundle. Thus we have the exact sequence

$$0 \to K_\varphi \xrightarrow{\iota_\varphi} E \xrightarrow{A_0([L],\varphi)} F \xrightarrow{\rho_\varphi} C_\varphi \to 0,$$

where ι_φ and ρ_φ are the obvious injection and projection. Let us apply the projection ρ_φ to the second of the above equations. We get

$$\rho_\varphi A_1([L], \varphi)u_0 = 0.$$

This is a first order linear partial differential equation for u_0, called the transport equation. More precisely, let us define the first order linear differential operator $R_\varphi \colon K_\varphi \to C_\varphi$ by

$$R_\varphi = \rho_\varphi \circ A_1([L], \varphi) \circ \iota_\varphi . \qquad (3.8)$$

Then u_0 must satisfy the algebraic constraint

$$u_0 \text{ is a section of } K_\varphi$$

and the homogeneous first order partial differential equation

$$R_\varphi u_0 = 0.$$

If u_0 meets these requirements, then $A_1([L], \varphi)u_0$ lies in the image of $A_0([L], \varphi)$ and thus we can find a section $\bar{u}_1$ of E such that

$$A_0([L], \varphi)\bar{u}_1 + A_1([L], \varphi)u_0 = 0.$$

Notice that $\bar{u}_1$ is determined up to adding an arbitrary section, $\bar{\bar{u}}_1$, of K_φ. Let us set $u_1 = \bar{u}_1 + \bar{\bar{u}}_1$ and examine the next equation:

$$A_0([L], \varphi)u_2 + A_1([L], \varphi)u_1 + A_2([L], \varphi)u_0 = 0.$$

Applying ρ_φ we see that

$$R_\varphi \bar{\bar{u}}_1 + \rho_\varphi(A_1([L], \varphi)\bar{u}_1 + A_2([L], \varphi)u_0) = 0.$$

This is an inhomogeneous linear first order partial differential equation for $\bar{\bar{u}}_1$ called the second order transport equation. It is now clear how to proceed inductively. At each stage we solve a first order partial differential equation of the form

$$R_\varphi \bar{\bar{u}}_k + (\cdots) = 0$$

where the $(\cdots)$ involves the u's of lower order and $\bar{u}_k$. This allows us to solve algebraically for $\bar{u}_{k+1}$ which is determined up to a section $\bar{\bar{u}}_{k+1}$ of K_φ, and $\bar{\bar{u}}_{k+1}$ is then determined by the next order transport equation. This then gives a procedure for finding an asymptotic solution. Furthermore, as we shall see in the next few sections, under reasonable hypotheses concerning $[L]$ the solution of the characteristic equation and the transport equations can be reduced to the solution of *ordinary* differential equations.

Before passing to a discussion of the characteristic equation we would like to point out that the above procedure can be applied to the problem of the propagation of discontinuities of solutions of partial differential equations. We shall only give a sketchy treatment now, since we will discuss the relation between asymptotics and discontinuous solutions later on. We will also assume that the reader is familiar with the elementary notions about distributions, and in particular, the concept of a generalized solution to a partial differential equation. As these notions will be presented in detail further on, the reader may prefer to skip to the end of this section for the time being.

Let $L\colon E \to F$ be a first order linear differential operator and let u be a generalized section of E having a jump discontinuity. More precisely, we assume that

$$u = H(\varphi)u_+ + H(-\varphi)u_-$$

where u_+ and u_- are smooth sections of E, where φ is a smooth function such that $d\varphi \neq 0$ when $\varphi = 0$ and where H is the Heaviside jump function: $H(x) = 0$ for $x < 0$ and $H(x) = 1$ for $x > 0$. Thus u jumps from u_- to u_+ across the surface $\varphi = 0$. Then

$$\begin{aligned} Lu &= \sigma(L)(dH(\varphi))u_+ + \sigma(L)(dH(-\varphi))u_- + H(\varphi)Lu_+ + H(-\varphi)Lu_- \\ &= \delta(\varphi)\sigma(L)(d\varphi)(u_+ - u_-) + H(\varphi)Lu_+ + H(-\varphi)Lu_-. \end{aligned}$$

If $Lu = 0$ then we must have $Lu_+ = 0$ for $\varphi > 0$ and $Lu_- = 0$ for $\varphi < 0$ and, in addition

$$\sigma(L)(d\varphi) \cdot (u_+ - u_-) = 0 \qquad \text{at} \qquad \varphi = 0.$$

From this equation we can derive two sorts of conclusions. First of all the equation limits the kinds of jumps that are possible: We must have

$$u_+ - u_- \in \ker \sigma(L)(d\varphi) \qquad \text{at} \qquad \varphi = 0.$$

Secondly, we see that the surface of discontinuity, $\varphi = 0$, is limited by the condition

$$\ker \sigma(L)(d\varphi) \neq \{0\} \qquad \text{at} \qquad \varphi = 0.$$

Notice that this is not quite a partial differential equation for φ since $\ker \sigma(L)(d\varphi) \neq \{0\}$ is required to hold only for $\varphi = 0$.

There is an important case where this does reduce to a characteristic equation. Suppose that $X = M \times \mathbf{R}$; that E is trivial in the $\mathbf{R}$ direction, and that

$$L = A\frac{\partial}{\partial t} + K$$

where $K: E \to F$ is a differential operator whose symbol does not depend on t, and A does not depend on t. Furthermore, suppose that $\varphi = \psi - t$ where ψ does not depend on t. Then

$$\sigma(L)(d\varphi) = \sigma(K)(d\psi) - A$$

and the equation

$$\ker(\sigma(K)(d\psi) - A) \neq \{0\} \tag{3.9}$$

does not involve t. Thus if it holds whenever $t = \psi$ it must hold identically. Thus if the jumps propagate in time along the surfaces $t = \psi$ then ψ must satisfy (3.9) which is called the *eikonal* equation. For example, if $F = E$ and $A = \mathrm{id}$ then (3.9) can be written as

$$\det(\sigma(K)(d\psi) - \mathrm{id}) = 0.$$

§4. The method of characteristics.

In this section we review the theory of the integration of first order partial differential equations such as (3.4). Let $\mathrm{Ch}([L])$ denote the set of covectors γ for which $\sigma([L])(\gamma)$ has a non-trivial kernel. Thus $\mathrm{Ch}([L])$ is a subset of the cotangent bundle

$$\mathrm{Ch}([L]) \subset T^*M$$

and is called the characteristic variety of $[L]$. We can thus rewrite (3.4) as

$$(d\varphi)(m) \in \mathrm{Ch}(L) \qquad \text{for all } m.$$

If we think of $d\varphi$ as a map from M to T^*M then this equation says that the image of $d\varphi$ lies entirely in $\mathrm{Ch}[L]$. The most general first order partial differential equation can be thought of in this way. We are given some subset, $\mathcal{V}$, of T^*M and we seek a function φ such that $d\varphi(M) \subset \mathcal{V}$. Now the map $d\varphi$: $M \to T^*M$ is a section of T^*M, but it is not the most general type of section. We can break the analysis of our partial differential equation up into two parts:

(A) We seek a section s: $M \to T^*M$ such that $s(M) \subset \mathcal{V}$.

(B) The section s should be of the form $s(m) = d\varphi(m)$, for all m where φ is some function.

Now condition (B) has a number of equivalent reformulations. Recall (Loomis-Sternberg [4, Chapter 13]) that the cotangent bundle has a natural linear differential form, α, defined as follows: Let π: $T^*M \to M$ be the natural projection and let $\eta_\gamma \in T_\gamma(T^*M)$ where $\pi\gamma = m$. Then

$$\langle \eta_\gamma, \alpha_\gamma \rangle = \langle d\pi_\gamma \eta_\gamma, \gamma \rangle.$$

In particular, if s is a section of T^*M, so that $\pi \circ s = \mathrm{id}$, then, for $\zeta_m \in T_m(M)$,

$$\langle \zeta_m, (s^*\alpha)_m \rangle = \langle d\pi \circ ds(\zeta_m), s(m) \rangle = \langle \zeta_m, s(m) \rangle$$

or, for short,

$$s^*\alpha = s.$$

In particular $s = d\varphi$ if and only if

$$s^*\alpha = d\varphi.$$

We might relax this condition in two ways. First of all we could use the weaker differential condition $ds^*\alpha = 0$ or $s^*d\alpha = 0$. If we let $\omega = d\alpha$ then this equation reads $s^*\omega = 0$. Locally, of course, this is the same as $s^*\alpha = d\varphi$. Secondly we could drop the requirement that s be a map from M to T^*M. Let Y be a manifold and s: $Y \to T^*M$. We say that s is *isotropic* if $s^*\omega = 0$. If $\dim Y = \dim M$ and s is an immersion (i.e., ds is injective at all points) then we say that Y (together with s) is an *(immersed) Lagrangian submanifold* of T^*M.

Observe that a Lagrangian submanifold is *maximally isotropic*. In other words if $z = s(y)$ and $\xi_z \in T(T^*M)_z$ is such that $\langle \xi_z \wedge \eta_z, \omega_z \rangle = 0$ for all $\eta_z \in ds_y TY_y$ then $\xi_z \in ds_y TY_y$. Indeed, since ω_z is non-singular it cannot vanish on any $n + 1$ dimensional subspaces (where $n = \dim M$), and by assumption, it vanishes on $ds_y TY_y$.

Notice that if $\iota\colon \Lambda \to T^*M$ is an immersed Lagrangian submanifold and if $\pi \circ \iota$ is a diffeomorphism then $\iota \circ (\pi \circ \iota)^{-1} = s$ is a section of T^*M satisfying $s^*\omega = 0$. We can thus analyze the problem of solving first order partial differential equations as follows: We are given a subset $\mathcal{V} \subset T^*M$. (Usually $\mathcal{V}$ will be a submanifold near points of interest to us.) We seek a Lagrangian submanifold $\iota\colon \Lambda \to T^*M$ such that:

(i) $\iota(\Lambda) \subset \mathcal{V}$;
(ii) $\pi \circ \iota$ is a diffeomorphism;
(iii) $\iota^*\alpha = d\varphi$.

The Hamilton-Jacobi method of characteristics, which we shall soon describe, will give a procedure for solving (i). It will also be possible to solve (ii) *locally*. The failure to be able to solve (ii) globally is due to a phenomenon generalizing the notion of focal points and is thus intrinsic to the geometry of the problem. As to (iii), again one always has local solutions, since $d\iota^*\alpha = 0$ implies $\iota^*\alpha = d\varphi$, locally. However the global problem again will not, in general, have a solution. Indeed, as we shall see in §7, a modified version of this global problem is intimately related to the celebrated Bohr-Sommerfeld quantization conditions.

Before proceeding, let us give some examples of Lagrangian submanifolds of $X = T^*M$.

(a) As we have already remarked, if φ is any C^∞ function then $\Lambda = d\varphi(M)$ is Lagrangian since $(d\varphi)^*\alpha = d\varphi$.

(b) Let $W \subset M$ be a submanifold. Then $N(W) \subset T^*M$ is a Lagrangian submanifold. In fact not only is $\iota^*d\alpha = 0$ but actually $\iota^*\alpha = 0$. For if $z \in N(W)$ then $\langle \zeta_x, z\rangle = 0$ for $\zeta_x \in TW_x$ where $x = \pi z$. Since $d\pi\boldsymbol{\eta}_z \in TW_x$ for $\boldsymbol{\eta}_z \in TN(W)_z$ we see that $\langle \boldsymbol{\eta}_z, \alpha_z\rangle = \langle d\pi\boldsymbol{\eta}_z, z\rangle = 0$.

More generally, let Λ be a *homogeneous* Lagrangian submanifold, i.e., stable under multiplication by positive real numbers. Let μ_t denote the one parameter group of maps of T^*M into itself given by multiplication by t, and let $\boldsymbol{\eta}$ be the corresponding vector field. Since $\pi \circ \mu_t = \pi$ we see that

$$\mu^*_t\alpha = t\alpha$$

and therefore

$$D_{\boldsymbol{\eta}}\alpha = \frac{d}{dt}\mu^*_t\alpha_{|t=0} = \alpha.$$

But

$$D_{\boldsymbol{\eta}}\alpha = \boldsymbol{\eta}\rfloor d\alpha + d(\boldsymbol{\eta}\rfloor\alpha).$$

Since $d\pi\boldsymbol{\eta} = 0$ the last term vanishes and we have

$$\boldsymbol{\eta}\rfloor\omega = \alpha.$$

Now to say that Λ is stable under μ_t means that $\boldsymbol{\eta}$ is tangent to the image of Λ, so that $\iota^*\eta$ is a well defined vector field on Λ and $\iota^*\alpha = \iota^*\eta \lrcorner \iota^*\omega = 0$. We have thus proved that

if Λ is a homogeneous Lagrangian manifold then $\iota^*\alpha = 0$.

For example, let W be a submanifold of $\mathbf{R}^m$. In Chapter I we have identified $N(W)$ with the set of all $(y, v) \in \mathbf{R}^m + \mathbf{R}^{m*}$ with $y \in W$ and $v \cdot dy = 0$. On $\mathbf{R}^m + \mathbf{R}^{m*}$ the form α is given by $\alpha_{(x,v)} = v \cdot dx$ and so $\alpha|_{N(W)} = 0$.

Still in $\mathbf{R}^m$, consider the image of $N(W)$ under the exponential map E. We can view the exponential map as follows. Let $N_1(W)$ denote the unit normal bundle consisting of (y, v) with v orthogonal to W at y and $\|v\| = 1$. Let G: $N_1(W) \times \mathbf{R} \to \mathbf{R}^m \oplus \mathbf{R}^m$ be given by $G(y, v, t) = (y + tv, v)$. Then G is an immersion. Furthermore

$$G^*\alpha = v \cdot d(y + tv) = v \cdot dy + dt + tv \cdot dv.$$

But $v \cdot dy = 0$ by definition and $v \cdot dv = 0$ since $v \cdot v \equiv 1$. Thus

$$G^*\alpha = dt$$

showing that G: $N_1(W) \times \mathbf{R} \to \mathbf{R}^m \oplus \mathbf{R}^m$ is Lagrangian. Notice that the exponential map can be thought of as the composition $\pi \circ G$ where π projects $T^*\mathbf{R}^m$ onto $\mathbf{R}^m$. The focal points are exactly the critical points of $\pi \circ G$.

For the rest of this section it will be somewhat convenient to operate in the slightly broader context of *symplectic manifolds*. A manifold X is called a symplectic manifold if X comes equipped with a nonsingular closed two form, ω. Thus we are given a two form ω on X such that $d\omega = 0$ and the map from vectors to covectors given by $\xi_x \to \xi_x \lrcorner \omega$ is an isomorphism at each $x \in X$. For example, the form $\omega = d\alpha$ on T^*M makes T^*M into a symplectic manifold. (This is most easily checked by using the expression for α in terms of local coordinates: If $q^1, \ldots, q^n$ are coordinates on M and $p^1, \ldots, p^n$ the corresponding coordinates on T^*M then $\alpha = p^1 dq^1 + \cdots + p^n dq^n$ so that

$$\omega = dp^1 \wedge dq^1 + \cdots + dp^n \wedge dq^n$$

is clearly closed and nonsingular). Symplectic manifolds will be studied in detail in Chapter IV. The notions of an isotropic submanifold and of a Lagrangian submanifold clearly continue to make sense in the context of symplectic manifolds.

We now generalize the construction of the preceding example. Let X be a symplectic manifold. For any smooth function f define the vector field ξ_f by $\xi_f \lrcorner \omega = -df$. Now let H be a smooth function such that $dH \neq 0$ when $H = 0$. (In the preceding example $X = T^*\mathbf{R}^m$ and $H(x, v) = v \cdot v - 1$.) Let φ_t be the flow generated by ξ_H, and let $\mathcal{V}$ denote the submanifold defined by $H = 0$.

PROPOSITION 4.1. *Let $Y \subset \mathcal{V}$ be a submanifold which is isotropic and which is transversal to ξ_H (i.e., ξ_H is nowhere tangent to Y). Then*

$$\varphi\colon Y \times \mathbf{R} \to X$$

given by $\varphi(y,t) = \varphi_t(y)$ is an isotropic immersion of $Y \times \mathbf{R}$ into $\mathcal{V}$.

PROOF. Since $\xi_H H = 0$ we conclude that ξ_H is tangent to $\mathcal{V}$ and so $\varphi_t \mathcal{V} \subset \mathcal{V}$ and thus $\varphi(Y \times \mathbf{R}) \subset \mathcal{V}$. Now

$$d\varphi_{(y,t)}\left(\eta_y + c\frac{\partial}{\partial t}\right) = d\varphi_t \eta_y + c\xi_H(\varphi(y,t)) = d\varphi_t(\eta_y + c\xi_H(y)).$$

Since φ_t is a diffeomorphism, this last expression cannot vanish unless $\eta_y + c\xi_H(y) = 0$. By assumption this cannot happen unless $\eta_y = 0$ and $c = 0$. Thus φ is an immersion. Finally we must show that

$$\langle \zeta \wedge \eta, \varphi^*\omega \rangle = 0$$

for all tangent vectors, ζ and η, to $Y \times \mathbf{R}$. It clearly suffices to consider the two cases:

(i) ζ and η are both tangent to Y and

(ii) $\zeta = \partial/\partial t$ and η is tangent to Y.

In case (i) we have, for $\zeta, \eta \in T(Y \times \mathbf{R})_{y,t}$,

$$\begin{aligned}\langle \zeta \wedge \eta, \varphi^*\omega \rangle &= \langle d\varphi_{(y,t)}\xi \wedge d\varphi_{(y,t)}\eta, \omega \rangle = \langle d\varphi_t \zeta \wedge d\varphi_t \eta, \omega \rangle \\ &= \langle \zeta \wedge \eta, \varphi_t^*\omega \rangle = \langle \zeta \wedge \eta, \iota^*\omega \rangle = 0\end{aligned}$$

since Y is isotropic. In case (ii) the computation becomes

$$\begin{aligned}\langle \xi_H(\varphi) \wedge d\varphi_t \eta, \omega \rangle &= \langle \xi_H \wedge \eta, \varphi_t^*\omega \rangle = \langle \xi_H \wedge \eta, \iota^*\omega \rangle \\ &= \langle \eta, \iota^*(\xi_H \rfloor \omega) \rangle = \langle \eta, dH \rangle = 0\end{aligned}$$

because $H = 0$ on $\mathcal{V}$ and $Y \subset \mathcal{V}$ proving the proposition.

An induction on the preceding argument will give the basic integration theorem yielding Lagrangian submanifolds satisfying prescribed conditions. First let us introduce some notation.

On any symplectic manifold we can introduce an anticommutative operation on smooth functions known as the Poisson bracket: For smooth functions f and g set

$$\{f, g\} = \xi_f g.$$

Since $\xi_f g = \langle \xi_f, dg \rangle = \langle \xi_f, \xi_g \rfloor \omega \rangle = \langle \xi_g \wedge \xi_f, \omega \rangle$ we see that $\{f, g\} = -\{g, f\}$. It is also clear that

$$\{f, gh\} = \{f, g\}h + \{f, h\}g.$$

Furthermore, if we take the Lie derivative with respect to ξ_f of the equation

$$\xi_g \rfloor \omega = dg$$

we get

$$[\xi_f, \xi_g] \rfloor \omega + \xi_g \rfloor D_{\xi_f} \omega = -d(\xi_f g) = -d\{f, g\}.$$

But $D_{\xi_f} \omega = \xi_f \rfloor d\omega + d(\xi_f \rfloor \omega) = 0$ so

$$[\xi_f, \xi_g] = \xi_{\{f,g\}}.$$

Let $\mathcal{V}$ be a submanifold of a symplectic manifold X and let z be a point of $\mathcal{V}$. We say that $\mathcal{V}$ is *integrable at z* if, for any pair of functions f and g which vanish on $\mathcal{V}$ we have $\{f, g\}(z) = 0$. In other words, if $\mathcal{I}(\mathcal{V})$ denotes the ideal of all smooth functions vanishing on $\mathcal{V}$ then $f \in \mathcal{I}(\mathcal{V})$ and $g \in \mathcal{I}(\mathcal{V})$ implies $\{f, g\}(z) = 0$. We say that $\mathcal{V}$ is *integrable* if it is integrable at all its points. Thus to say that $\mathcal{V}$ is integrable means that $\{\mathcal{I}(\mathcal{V}), \mathcal{I}(\mathcal{V})\} \subset \mathcal{I}(\mathcal{V})$. To understand the significance of integrability we notice the following:

Suppose that there exists a Lagrangian manifold Λ *such that* $z \in \Lambda \subset \mathcal{V}$. *Then* $\mathcal{V}$ *is integrable at* z.

Indeed, if $f \in \mathcal{I}(\mathcal{V})$ then $df|_{\mathcal{V}} = 0$ so certainly $df|_{\Lambda} = 0$. Thus $\xi_f \rfloor \omega$ vanishes on any vector tangent to Λ. Since $T\Lambda$ is maximal isotropic (at all points) this implies that ξ_f is tangent to Λ. But then if $g \in \mathcal{I}(\mathcal{V})$ we get $\{f, g\} = \langle \xi_f, dg \rangle = 0$, at all points of Λ, in particular at z.

Suppose that $\mathcal{V}$ is an integrable submanifold of X. If $f \in \mathcal{I}(\mathcal{V})$ then $\langle \xi_f, dg \rangle_z = 0$ for all $g \in \mathcal{I}(\mathcal{V})$, and for all $z \in \mathcal{V}$. Since such dg span the annihilator space of $T\mathcal{V}_z$ we conclude that

if $\mathcal{V}$ *is integrable then* ξ_f *is tangent to* V *for all* $f \in \mathcal{I}(\mathcal{V})$.

Since $[\xi_f, \xi_g] = \xi_{\{f,g\}}$ we conclude that the set of vector fields of the form ξ_f where $f \in \mathcal{I}(\mathcal{V})$ is a Lie algebra. If $\mathcal{V}$ has codimension k then the set of such ξ_f span a k dimensional subspace of $T\mathcal{V}_z$ at each z. We can thus assert:

THEOREM 4.1. *Let* $\mathcal{V}$ *be an integrable submanifold of codimension* k *in a symplectic manifold* X *of dimension* $2n$. *Let* $\Lambda_0 \subset \mathcal{V}$ *be an* $(n-k)$*-dimensional isotropic submanifold which is transversal to all the* ξ_f, $f \in \mathcal{I}(\mathcal{V})$. *Then there exists a Lagrangian manifold,* Λ, *with* $\Lambda_0 \subset \Lambda \subset \mathcal{V}$. *Furthermore* Λ *is essentially unique.*

PROOF. As we have already remarked, any Lagrangian submanifold of $\mathcal{V}$ must be tangent to all the ξ_f. In terms of the language of foliations, the ξ_f determine a foliation on $\mathcal{V}$ and we must take Λ to be the union of the leaves of the foliation passing through Λ_0. The fact that Λ is Lagrangian then follows from a repeated application of Proposition 4.1. In more detail, suppose that we can find $f_1, \ldots, f_k$ such that $\xi_{f_1}, \ldots, \xi_{f_k}$ are linearly independent on $\mathcal{V}$. (We can always do this locally. The passage from the local argument to the global will be left to the reader.) Then let Λ_1 be the submanifold swept out by Λ_0 under the flow generated by ξ_{f_1}. By Proposition 4.1 this is isotropic. Let Λ_2 be the submanifold swept out by Λ_1 under the flow generated by ξ_{f_2} etc. The $\Lambda = \Lambda_k$ that is finally obtained is thus Lagrangian.

Theorem 4.1 gives us a procedure for finding Lagrangian submanifolds of $\mathcal{V}$, starting with isotropic submanifolds. Let us now briefly discuss the problem of projecting down to M. Suppose that $\mathcal{V}$ is locally described by equations $f_1 = 0, \ldots, f_k = 0$. Furthermore, suppose that the vector fields

$$\xi_1 = \xi_{f_1}, \ldots, \xi_k = \xi_{f_k}$$

have the property that $d\pi_z \xi_{1z}, \ldots, d\pi_z \xi_{kz}$ are linearly independent vectors of TM_x where $x = \pi z$. If we could then choose the initial isotropic manifold, Λ_0, so that

$$\{d\pi_z(T\Lambda_0)_z, d\pi_z \xi_{1z}, \ldots, d\pi_z \xi_{kz}\} = TM_x$$

then, by our construction of Λ we see that $d\pi_z T\Lambda_z = T_z M$, i.e., $\pi \circ \iota$ is a diffeomorphism near z.

Let us describe this procedure in terms of local coordinates. Let $x_1, \ldots, x_n$ be local coordinates on M, and let $q_i = x_i \circ \pi$ and p_i be the corresponding local coordinates on T^*M. If f is any function then

$$\xi_f = \sum \frac{\partial f}{\partial p_i}\frac{\partial}{\partial q_i} - \frac{\partial f}{\partial q_i}\frac{\partial}{\partial p_i}.$$

Suppose that $\mathcal{V}$ is given by the equations $f_1 = 0, \ldots, f_k = 0$ and suppose that these equations have been solved for $p_1, \ldots, p_k$. In other words, suppose that

$$f_1 = p_1 - g_1(q_1, \ldots, q_n, p_{k+1}, \ldots, p_n), \ldots,$$

$$f_k = p_k - g_k(q_1, \ldots, q_n, p_{k+1}, \ldots, p_n).$$

Then

$$\xi_{f_i} = \frac{\partial}{\partial q_i} - \sum_{k+1}^{n} \frac{\partial g_i}{\partial p_j}\frac{\partial}{\partial q_j} + \sum_{1}^{n} \frac{\partial g_i}{\partial q_j}\frac{\partial}{\partial p_j}.$$

It is clear that $d\pi_z \xi_1, \ldots, d\pi_z \xi_k$ are linearly independent. Now let $z = (q_1^0, \ldots, q_n^0, p_1^0, \ldots, p_n^0)$ be a point of $\mathcal{V}$. Let Λ_0 be the submanifold described by

$$p_i = g_i(q_1^0, \ldots, q_k^0, q_{k+1}, \ldots, q_n, p_{k+1}^0, \ldots, p_n^0),$$

$i = 1, \ldots, k$, where $q_{k+1}, \ldots, q_n$ are close to $q_{k+1}^0, \ldots, q_n^0$. Thus Λ_0 is an $(n - k)$ dimensional submanifold of $\mathcal{V}$ which is clearly transversal to $\xi_{f_1}, \ldots, \xi_{f_k}$

and which satisfies the desired conditions with respect to projection onto M. To see that Λ_0 is isotropic observe that $\omega = \sum dp_i \wedge dq_i$ and, upon restricting to Λ_0, we have $dq_1 = \cdots = dq_k = 0$ and $dp_{k+1} = \cdots = dp_n = 0$. Thus $\iota^*\omega = 0$.

§5. Bicharacteristics.

We now wish to apply the techniques of the preceding section to the characteristic equation and to the transport equations. Our main result is that, at least locally, these equations reduce to the solution of ordinary differential equations involving differentiation along a line element field on the characteristic variety , providing that $[L]$ has (real) "simple characteristics", a concept to be discussed. (Roughly speaking, it is the condition that the characteristic variety be a nice submanifold of codimension one in a rather strong sense.)

Let us recall the setting of §3. To each point $z \in T^*M$ we are given a linear map $\sigma(z)$: $E_{\pi z} \to F_{\pi z}$, where $\sigma(z) = \sigma([L], z)$ is the symbol of $[L]$ evaluated at z . We can consider the vector bundles E and F as being defined on $X = T^*M$ via π. (Strictly speaking we should use a different symbol such as $\pi^{\#}E$ for the pull-back bundle.) We are thus in the following situation: $X (= T^*M)$ is a symplectic manifold and σ: $E \to F$ is a vector bundle morphism. At each $z \in X$ we have the exact sequence

$$0 \to K_z \to E_z \overset{\sigma(z)}{\to} F_z \to C_z \to 0$$

where $K_z = \ker \sigma(z)$ and $C_z = \operatorname{coker} \sigma(z)$. The characteristic variety, $\mathcal{V}$, consists of those z for which $K_z \neq \{0\}$. A solution of the characteristic equation is then a Lagrangian submanifold of $\mathcal{V}$ (which, in addition, projects diffeomorphically onto M). We claim that at each z there is a well defined linear map

$$\mathcal{R}_z: K_z \otimes T_z^* \to C_z$$

called the bicharacteristic symbol associated with σ. It is defined using the notion of *intrinsic derivative*, a concept that makes sense on any manifold X, and which we now define. After a somewhat lengthy detour to discuss the notion of intrinsic derivative we shall return to the definition of $\mathcal{R}_z$. Let σ: $E \to F$ be a vector bundle map over X, and let $z \in X$. Choose trivializations $E \sim E_z \times U$ and $F \sim F_z \times U$ valid near z. In terms of this trivialization, σ determines a map A: $E_z \to F_z$ at each $x \in U$. For each $e \in E_z$ choose a function s: $U \to E_z$ with $s(z) = e$. Now As is a function with values in the vector space F_z. We can compute its differential at z. Let ρ_z be the projection of F_z onto C_z. If ξ is any tangent vector at z define

$$I_z(e \otimes \xi) = \rho_z(\langle \xi, d(As) \rangle).$$

We claim that if $e \in K_z$ then the right hand side does not depend on the choice of s or the choice of trivializations and hence defines a map

$$I_z\colon K_z \otimes T_z \to C_z$$

called the *intrinsic derivative*:

(i) *Independence of the choice of s*. Choose any other s'.Then $s - s' = \sum s_i e_i$ where $s_i(z) = 0$. Then $A(s - s') = \sum s_i A e_i$ so that

$$d(A(s - s'))_z = \sum (ds_i)_z (Ae_i)_z$$

so that

$$\rho_z \langle \xi, dA(s - s') \rangle = \rho_z (\sum \langle \xi, ds_i \rangle A e_i) = \rho_z A (\sum \langle \xi, ds_i \rangle e_i) = 0.$$

(ii) *Independence of choice of trivialization*. Changing the trivialization means replacing A by BAC, where C is a function defined near z with values in $\operatorname{Hom}(E_z, E_z)$ and B with values in $\operatorname{Hom}(F_z, F_z)$, with $C(z) = \mathrm{id}$ and $B(z) = \mathrm{id}$. By (i) we may assume that $s \equiv e$ is constant. Then

$$d(BACs)_z = (dB)_z A_z e + dA_z e + A_z (dC)_z e.$$

The first term vanishes since $e \in \ker \sigma_z = \ker A_z$. The third term lies in F_z and will vanish when projected onto C_z. Thus $\rho_z d(BACe)_z = \rho_z dA_z e$. This defines the notion of intrinsic derivative.

For example, if E and F are line bundles, then a choice of non-zero section of E and of F (which amounts to a trivialization) allows σ to be represented as multiplication by a function f. Changing the sections replaces f by fg with $g \neq 0$. Now at points where $f(z) = 0$, the kernel and cokernel are one dimensional, and the intrinsic derivative is given by df, that is, for each ξ, we have $I_z(e, \xi) = \langle \xi, df \rangle e$ in terms of the trivializations. (Here $E_z = K_z$ and $F_z = C_z$.) Notice that if $df_z \neq 0$, then $\mathcal{V}$ is a submanifold near z whose normal bundle is generated by df.

We can generalize this example: Suppose that $\dim E = \dim F$ and that $\dim K_z = 1$ and $I_z \neq 0$. We claim that *then* $\mathcal{V}$ *is a submanifold of codimension one near* z *and that* $\dim K_x = 1$ *for* $x \in \mathcal{V}$ *near* z. Then clearly $\dim C_x = 1$ and $I_x \in \operatorname{Hom}(K_x \otimes T_x, C_x)$ which is isomorphic to $T_x^* \otimes \operatorname{Hom}(K_x, C_x)$. Thus since $I_x \neq 0$ and $\dim \operatorname{Hom}(K_x, C_x) = 1$ we see that I_x determines a line in T_x^*, and we claim that *this line is just the normal bundle to* $\mathcal{V}$.

To prove the above assertions, let us choose a complement L_z to K_z and extend it near z to be a sub-bundle, $L \subset E$. Then σ is non-singular when restricted to L and determines a sub-bundle, $Q = \sigma L$, of F. The map σ induces a map, $\sigma'\colon E/L \to F/Q$ where E/L and F/Q are now line bundles. At z we clearly have $K_z \sim E_z/L_z$ and $C_z \sim F_z/Q_z$ and, with this identification, σ and σ' have the same intrinsic derivative at z. Thus the intrinsic derivative of σ' does not vanish at z, so we can apply the preceding discussion to σ' and conclude that the set where $\sigma'(x) = 0$ is a submanifold near z. But notice that if $\sigma'(x) = 0$, then $\sigma'(x)$ and hence $\sigma(x)$ is not surjective. But, since $\dim E = \dim F$, this means that $K_x \neq \{0\}$ i.e., that $x \in \mathcal{V}$. Thus $\mathcal{V}$ is a submanifold near z and the remaining assertions follow easily by a similar argument.

If $\dim E = \dim F$ and $\dim K_z = 1$ with $I_z \neq 0$ we say that σ is *simple* at z. We say that the asymptotic differential operator $[L]$ has real simple characteristics if $\sigma([L])$ is simple at all z. For a differential operator (where σ is homogeneous) we relax this condition to require that σ be simple at $z \neq 0$.

Notice that *if* $\mathcal{V}$ is a submanifold near z and if K' is a sub-bundle of E such $K'_x \subset K_x$ for $x \in \mathcal{V}$ (near z) then $I(k \otimes \xi) = 0$ for ξ tangent to $\mathcal{V}$ and $k \in K'_z$. Indeed, we may choose s so that $As \equiv 0$ on $\mathcal{V}$.

As we have said the notion of intrinsic derivative makes sense on an arbitrary manifold. Let us now suppose that the manifold X is symplectic. The two form ω allows us to identify TX with T^*X: If $\xi \in TX_z$ is a tangent vector then $-\xi \lrcorner \omega$ is a covector at z, and this map is an isomorphism since ω is nonsingular. Thus $I_z: K_z \otimes T_z \to C_z$ determines a map $\mathcal{R}_z: K_z \otimes T_z^* \to C_z$, which is the desired *bicharacteristic symbol*. Suppose that E and F are line bundles, and $\mathcal{R}_z \neq 0$. Then $\mathcal{R}_z$ determines a section of $\mathrm{Hom}(K \otimes T^*, C) \simeq T \otimes \mathrm{Hom}(K, C)$ along $\mathcal{V}$. Since $\mathrm{Hom}(K, C)$ is a line bundle, this determines a one dimensional sub-bundle of T along $\mathcal{V}$, i.e., a line element field. If, locally, σ is represented as multiplication by some function f, then $\mathcal{V}$ is given locally by the equation $f = 0$, the normal bundle to $\mathcal{V}$ is generated by df and the corresponding line field on $\mathcal{V}$ is generated by ξ_f. As $\mathcal{V}$ has codimension one, it is automatically integrable and ξ_f is exactly the vector field used in the constructions in §4. Of course ξ_f itself is not invariantly defined, but the line field it generates is, and this line field is called the *bicharacteristic field*. Exactly the same considerations apply in the case of simple characteristics. In the more general case, where codim $\mathcal{V} > 1$ integrability must be added as a hypothesis. The question of the relationship between the integrability of $\mathcal{V}$ and properties of the operator L is an interesting one; see Guillemin, Quillen and Sternberg [5].

Suppose we have solved the characteristic equation. Thus we have a Lagrangian manifold $\Lambda \subset \mathcal{V}$ which projects diffeomorphically onto M, so that, locally, $\Lambda = d\varphi(M)$. Notice that the (assumption and) definition in §3 asserts that

$$(K_\varphi)_m = K_{d\varphi(m)},$$

i.e., that $K|_\Lambda$ is a vector bundle and is in fact the pull-back to Λ, via π, of the bundle K_φ and similarly for C_φ. Now (3.8) defined a first order differential operator, R_φ. Let $\sigma(R_\varphi)$ be its symbol. Thus

$$\sigma(R_\varphi): K_\varphi \otimes T^*M \to C_\varphi.$$

We claim that, for any $\gamma \in T_m^* M$ and $k \in K_{d\varphi(m)}$,

$$\sigma(R_\varphi)(k \otimes \gamma) = \mathfrak{R}_{d\varphi(m)}(K \otimes d\pi^* \gamma), \tag{5.1}$$

where we have identified $K_{d\varphi(m)}$ with $(K_{|\Lambda})_{d\varphi(m)}$, and similarly for the C's.

To see this, let us trivialize the bundles E and F in a neighborhood U of m, which trivializes them over $\pi^{-1}(U)$, and let x be coordinates on U with x, ξ the induced coordinates on $\pi^{-1}U$. Let k be an element of K_z (where $z = d\varphi(m)$) considered as a constant function. If

$$\boldsymbol{\eta} = a\frac{\partial}{\partial x} + b\frac{\partial}{\partial \xi} = \sum a_i \frac{\partial}{\partial x_i} + b_i \frac{\partial}{\partial \xi_i}$$

is a tangent vector at z then

$$I_z(k \otimes \boldsymbol{\eta}) = \rho\left(a\frac{\partial \sigma}{\partial x}k + b\frac{\partial \sigma}{\partial \xi}k\right).$$

Also $-\boldsymbol{\eta} \rfloor \omega = -bdx + ad\xi$ and so

$$\mathfrak{R}_z(k \otimes (bdx - ad\xi)) = \rho\left(a\frac{\partial \sigma}{\partial x}k + b\frac{\partial \sigma}{\partial \xi}k\right).$$

If we set $\gamma = df$ then $\pi^* \gamma = \dfrac{\partial f}{\partial x}dx$ and the above formula becomes

$$\mathfrak{R}_z(k \otimes \pi^* \gamma) = \rho\left(\frac{\partial \sigma}{\partial \xi}\frac{\partial f}{\partial x}k\right).$$

Comparing this with the local expression,

$$\rho \circ \left[\frac{\partial \sigma}{\partial \xi}\frac{\partial}{\partial x} + \cdots\right],$$

for R_φ proves (5.1).

Notice that the differential operator R_φ depends, for its definition, on the function φ, and hence on the fact that Λ projects diffeomorphically onto M. On the other other hand $\mathfrak{R}$ is defined at all points of $\mathcal{V}$ and thus the symbol of R_φ is well defined at all points of Λ. The zeroth order term in R_φ *does* depend on the existence of φ. We shall see how to deal with this problem in the next section.

In the case of simple real characteristics K and C are line bundles and $\mathfrak{R}$ defines the bicharacteristic field on $\mathcal{V}$. What we have just proved is *that the transport equations all reduce to solving ordinary linear differential equations*. Thus we have locally solved the problem of finding asymptotic solutions in the case of simple real characteristics. The solution is only local because of the fact that the solution of the characteristic equation is only valid up to the point where Λ no longer projects diffeomorphically onto M.

Let us now describe these results in local coordinates for the case of differential operators. Let us assume that E and F are trivial vector bundles, and that coordinates $t, x_1, \ldots, x_m$ have been introduced on M. We will assume that the differential operator L takes the form

$$Lu = \frac{\partial u}{\partial t} - \sum_1^n A_i \frac{\partial u}{\partial x_i} + Bu,$$

where the $A_i = A_i(t,x)$. (Notice that this really amounts only to the assumption that the coefficient of $\partial/\partial t$ in L be non-singular, i.e., that $\sigma(L)(dt)$ be non-singular, in other words that dt be nowhere characteristic.) If $\tau, \xi^1, \ldots, \xi^n$ are the dual coordinates then

$$\sigma(L) = \tau I - \sum \xi^i A_i$$

and the characteristic variety is given by

$$\det(\tau I - \sum \xi^i A_i) = 0.$$

The differential operator L is said to be strictly hyperbolic if, for each $\xi = (\xi^1, \ldots, \xi^n) \neq 0$, the matrix $\sum \xi^i A_i$ has n distinct real eigenvalues, $\tau_1(t,x,\xi), \ldots, \tau_n(t,x,\xi)$. Then

$$\det \sigma(L) = (\tau - \tau_1(t,x,\xi)) \cdots (\tau - \tau_n(t,x,\xi)),$$

and the (non-zero portion of) the characteristic variety splits up into n sheets, given by $\tau = \tau_j$.

All non-zero characteristics of $\sigma(L)$ are simple. Indeed, for $\xi \neq 0$, we can write

$$E = E_1 \oplus \cdots \oplus E_n \tag{5.2}$$

where $E_j = E_j(t,x,\xi)$ is the subspace spanned by the jth eigenvalue of $\sum \xi^i A_i$. Then at $z = (t,x,\tau_j,\xi)$ we have $\ker \sigma(L) = \operatorname{coker} \sigma(L) = E_j$. If we choose a non-zero section $s = e_j$ of E_j then $\sigma(L)e_j = (\tau - \tau_j(t,x,\xi))e_j$,

$$I_z = d(\tau - \tau_j(t,x,\xi)) \neq 0.$$

(Here we have chosen a basis, $e_j(z)$, of K_z and C_z and hence I_z becomes identified with a linear differential form.) The jth bicharacteristic line element field is then spanned by the vector field

$$\boldsymbol{\eta}_j = \frac{\partial}{\partial t} - \frac{\partial \tau_j}{\partial \xi^i}\frac{\partial}{\partial x_i} + \frac{\partial \tau_j}{\partial t}\frac{\partial}{\partial \tau} + \frac{\partial \tau_j}{\partial x_i}\frac{\partial}{\partial \xi^i}.$$

Notice that $\boldsymbol{\eta}_j$ is always transversal to the hypersurface $t = \text{const}$. Thus, if $\Lambda_{j,0}$ is an m-dimensional isotropic submanifold contained in the jth sheet of the characteristic variety, then the flow generated by $\boldsymbol{\eta}_j$ will sweep out a Lagrangian manifold Λ_j lying in the jth sheet of the characteristic variety. In particular, a function $\varphi(x)$ will determine an initial isotropic submanifold $\Lambda_{j,0}$ by

$$t = 0, \quad \xi^i = \frac{\partial\varphi}{\partial x_i}, \quad \tau = \tau_j(x, 0, \xi).$$

For small values of t the Lagrangian manifold Λ_j will project diffeomorphically onto M (for any compact set of x). This then determines a solution $\varphi_j(t, x)$ to the characteristic equation with the initial conditions $\varphi_j(0, x) = \varphi(x)$.

For future applications it will be important for us to be able to assert that under suitable conditions we can choose φ_j to be linear in t, i.e., that

$$\varphi_j(t, x) = \varphi(x) - ct \tag{5.3}$$

for some suitable constant c. (We have already encountered solutions of this type when we studied the propagation of singularities.) Let us write ψ for φ_j. To say that ψ has the desired form (up to an unessential additive constant) means that

$$\frac{\partial^2\psi}{\partial t\,\partial x_i} = 0 \quad \text{for all } i \quad \text{and} \quad \frac{\partial^2\psi}{\partial t^2} = 0. \tag{5.4}$$

Now the image of the vector field $\partial/\partial t$ under the map $d\psi\colon M \to T^*M$ is

$$\frac{\partial}{\partial t} + \frac{\partial^2\psi}{\partial t^2}\frac{\partial}{\partial\tau} + \sum \frac{\partial^2\psi}{\partial t\,\partial x_i}\frac{\partial}{\partial\xi^i}.$$

This is the unique vector field tangent to Λ which projects onto $\partial/\partial t$ if Λ projects diffeomorphically onto M. If (5.4) is to hold this last vector field is just $\partial/\partial t$. Thus (5.3) is equivalent to the requirement that if we consider $\partial/\partial t$ as a vector field on T^*M then

$$\frac{\partial}{\partial t} \textit{ is tangent to } \Lambda. \tag{5.5}$$

Notice that condition (5.5) makes sense even when Λ does not project diffeomorphically onto M. Now the vector field $\boldsymbol{\eta}_j$ is tangent to $\Lambda = \Lambda_j$ by construction. Suppose that τ_j does not depend on t. Then,

$$\boldsymbol{\eta}_j = \frac{\partial}{\partial t} - \sum\left(\frac{\partial\tau_j}{\partial\xi^i}\frac{\partial}{\partial x_i} - \frac{\partial\tau_j}{\partial x_i}\frac{\partial}{\partial\xi^i}\right) \quad \text{and} \quad \left[\boldsymbol{\eta}_j, \frac{\partial}{\partial t}\right] = 0 = \left[\boldsymbol{\eta}_j, \boldsymbol{\eta}_j - \frac{\partial}{\partial t}\right].$$

Thus $\partial/\partial t$ will be tangent to Λ if $\boldsymbol{\eta}_j - \partial/\partial t$ is tangent to Λ at $t = 0$, and this will happen if

$$\zeta = -\sum\left(\frac{\partial\tau_j}{\partial\xi^i}\frac{\partial}{\partial x_i} - \frac{\partial\tau_j}{\partial x_i}\frac{\partial}{\partial\xi^i}\right)$$

is tangent to Λ_0. This, of course, is a constraint on our choice of initial manifold Λ_0. For example, if $\partial\tau_j/\partial\xi = 0$ and $\partial\tau_j/\partial x \neq 0$ then $\zeta \neq 0$ while $d\pi\zeta = 0$. This would preclude the possibility of choosing Λ_0 with ζ tangent to Λ_0 and such that $\pi|_{\Lambda_0}$ is an immersion. Notice that τ_j is a homogeneous function of ξ. Therefore by Euler's theorem

$$\tau_j = \sum \frac{\partial\tau_j}{\partial\xi^i}\xi^i, \quad \text{and} \quad \frac{\partial\tau_j}{\partial\xi} = 0 \Rightarrow \tau_j(x,\xi) = 0$$

Let us assume that this does not occur, i.e., that $\tau(x,\xi) \neq 0$ for $\xi \neq 0$. Then $d\pi\zeta \neq 0$ and we may apply the method of characteristics to find a Λ_0 lying in $t = 0$, $\tau_j =$ const. Notice, by the way, that this fixes the constant in (5.3) as

$$c = \frac{\partial\varphi_j}{\partial t} = \tau = \tau_j.$$

We have thus proved: *If $\tau_j(x,\xi) \neq 0$ for $\xi \neq 0$, and is independent of t then the jth characteristic equation has a solution of the form $\varphi_j(t,x) = \varphi(x) - \tau_j t$.*

Suppose we are given an asymptotic initial section

$$u \sim \sum \frac{b_k}{\tau^k} e^{i\tau\varphi(x)}, \qquad d\varphi \neq 0.$$

We can then find n solutions, $\varphi_1(t,x), \ldots, \varphi_n(t,x)$ of the characteristic equation corresponding to the different sheets of the characteristic variety, all with $\varphi_j(0,x) = \varphi(x)$. We can write $u(x) = u_1(x) + \cdots + u_n(x)$ corresponding to the direct sum decomposition of E and then apply the procedure described above to each component, solving ordinary differential equations with the initial conditions being provided by $b_k(x) = b_{k_1}(x) + \cdots + b_{k_n}(x)$. This gives an asymptotic solution to the initial value problem. Of course, it is limited to small values of t.

We now briefly sketch how our construction of asymptotic solutions to the initial value problem for hyperbolic equations can be used to establish the existence and uniqueness of actual solutions (for small values of t). When we have developed more machinery an existence and uniqueness theorem for all t will follow from some rather general results. Let

$$u(x) = \left(\frac{1}{2\pi}\right)^n \int \hat{u}(\xi)e^{i\xi\cdot x}\,d\xi$$

be the initial value, where $\hat{u}$ is the Fourier transform of u. We can write this as

$$u(x) = c_n \int_0^\infty \tau^{n-1}\left(\int_{S^{n-1}} \hat{u}(\tau\eta)e^{i\tau\eta\cdot x}\,dS\right)d\tau.$$

Now for any constant vector v, we can find an asymptotic solution to the initial value problem with initial condition $ve^{i\tau\eta\cdot x}$. In other words, we can find some function $w(t, x; v, \eta, \tau)$ such that

$$Lw = O(\tau^{-N}) \quad \text{for any } N \quad \text{and } w(0, x; v, \eta, \tau) = ve^{i\tau\eta\cdot x}.$$

It is clear from our construction that we may arrange that w depends smoothly on v and η. (Although our procedure defines w for large τ we can obviously arrange that w be defined for all $\tau \geqslant 0$ so that the second condition above holds.) Choose a basis $v_1, \ldots, v_n$ and write

$$\hat{u}(\tau\eta) = \sum a_i(\tau\eta)v_i.$$

Consider

$$\bar{u}(t, x) = c_n \sum_i \int_0^\infty \int_{S^{n-1}} a_i(\tau\eta)w(t, x, v_i, \eta, \tau)\tau^{n-1}\,dS\,d\tau \overset{\text{def}}{=} V(t)u.$$

It is then clear that

$$\bar{u}(0, x) = u(x) \quad \text{and} \quad L\bar{u} = Ku$$

where

$$K(t)u = c_n \sum_i \iint a_i(\tau\eta)z_i(t, x, \eta, \tau)\,dS\,d\tau$$

with $z_i = O(\tau^{-N})$ for all N. Thus K is a smoothing operator, i.e., the Fourier inversion formula shows that $K(t)$ is given by a smooth integral kernel. In particular $K(t)$ is a compact operator. We have thus solved the equation up to a smoothing operator. If we write

$$L = \frac{\partial}{\partial t} - A(t)$$

then our construction above gives us an operator, V, such that

$$LV = V' - AV = K \quad \text{or} \quad V' = AV + K, \qquad V(0) = I.$$

We are interested in finding a solution to

$$U' = AU, \qquad U(0) = I.$$

Suppose that we actually had found a $U(t)$ and were given $K(t)$, and wished to determine V. We would do this by variation of constants, i.e., we set $V = UB$ so that

$$V' = U'B + UB' = AV + UB'$$

so that

$$B' = U^{-1}K, \quad B(0) = I, \quad \text{or} \quad B = I + \int_0^t U^{-1}(s)K(s)ds$$

and

$$V(t) = U(t)\left[I + \int_0^t U^{-1}(s)K(s)ds\right].$$

Now we are given V and wish to determine U. It is clear by differentiating that (for t small enough so that the operator in brackets is invertible) if we have a $U(t)$ satisfying the above equation then $U' = AU$ and $U(0) = I$. But we can write the above equation as

$$U^{-1}(t) = \left[I + \int_0^t U^{-1}(s)K(s)ds\right]V^{-1}(t).$$

The explicit form of V shows that it is invertible for small t. This last equation can clearly be solved by iteration, providing us with a solution to our initial value problem. The uniqueness can be proved by applying an existence theorem to the adjoint operator. As we are not primarily concerned with these type of questions we leave the details to the reader, and refer the reader to the very readable original papers of Lax [2] and Ludwig [3].

Let us now make some remarks about some extensions of the method of high frequency approximation as observed by Ludwig. The fact that we are looking for solutions of the form

$$u \sim \sum u_k \frac{e^{i\tau\varphi}}{(i\tau)^k}$$

can be reformulated as follows: Consider the sequence of functions, $\{\gamma_k\}$ on $\mathbf{R}^1$,

$$\gamma_k(s) = \frac{e^{i\tau s}}{(i\tau)^k}.$$

Then

$$\gamma'_{k+1} = \gamma_k \quad \text{and} \quad u \sim \sum u_k \varphi^* \gamma_k$$

where $\varphi^* \gamma_k = \gamma_k(\varphi)$.

Now suppose that we replace the functions $e^{i\tau s}/(i\tau)^k$ by an arbitrary sequence of functions (or possibly distributions on $\mathbf{R}^1$ satisfying $\gamma'_{k+1} = \gamma_k$). We can then attempt to apply the preceding method. For this we need to

(a) attach some meaning to the symbol $\sim$ when we write $u \sim \sum u_k \varphi^* \gamma_k$,

(b) be sure that $u = 0$ implies that the individual coefficients of the $\varphi^* \gamma_k$ vanish separately.

Here are some examples of the method:

(1) The sum is finite (i.e., only finitely many u_k are different from zero). Here $\sim$ means $=$. Of course we expect to be able to find solutions of this kind for only special equations. However, interesting examples do arise; for instance spherically symmetric solutions of the wave equation in an odd number of space dimensions can be written in this form.

(2) $\gamma_k = s^k/k!$ and $\sim$ means $=$. If the equations have analytic coefficients this gives a method (originally due to Hadamard) for finding analytic solutions to hyperbolic equations. For a convergence proof see Ludwig [3].

(3) Let $x_+ = 0$ for $x < 0$ and $x_+ = x$ for $x > 0$. Then (as distributions) we have

$$\left(\frac{x_+^{k+1}}{(k+1)!}\right)' = \frac{x_+^k}{k!}$$

and hence $\gamma_k = x_+^k/k!$ is a suitable candidate. Suppose that for two generalized sections u and v we let $u \sim v$ mean that $u - v$ is smooth. Then $u \sim \sum u_k \varphi^* \gamma_k$ in this case implies that the singularity of u occurs along $\varphi = 0$. We essentially discussed the top order term relating to this situation at the end of §3. As before the simplest procedure is to look at φ of the form $t - \psi$ in the product (time independent) case. We leave the development to the reader as a very instructive exercise. In particular, it is interesting to observe the fact that a singularity propagates along bicharacteristics and cannot disappear.

§6. The transport equation.

In this section we shall examine in more detail the transport equation

$$R_\phi u = 0$$

where the transport operator R_ϕ is given by (3.8) and (3.7). We have seen in the preceding section that the *symbol* of the first order differential operator R_ϕ actually arises as an invariantly defined object on the characteristic variety. Suppose that Λ is a Lagrangian solution variety of our original asymptotic differential operator, and that $\pi|_\Lambda$ fails to be a diffeomorphism at a point z with $\pi z = x$. Suppose, in fact, that z is on the boundary of some region U of Λ which does project diffeomorphically onto an open set, $\pi U = \mathcal{O}$ of M. Then U can be represented as graph $d\phi$, so that R_ϕ is defined on $\mathcal{O}$. As we approach x the zero-th order term in R_ϕ becomes singular. It is the purpose of this section to reinterpret R_ϕ as a well defined differential operator on Λ in such a way that it makes sense even through the points where π is singular. Before discussing the general case we begin with an instructive example. Let us discuss a version of the

"Tricomi equation." Consider the differential operator D defined by

$$D = \frac{1}{2}\left[\frac{\partial^2}{\partial x^2} + (x^2 - 1)\frac{\partial^2}{\partial x_0^2}\right]. \tag{6.1}$$

This equation can be regarded as a wave equation for $|x| < 1$, where the "index of refraction" approaches zero as $|x| \to 1$. It can thus be regarded as a crude one dimensional model of "light being totally refracted by two parallel mirages." Then the symbol is given by

$$\sigma = \tfrac{1}{2}[\xi^2 + (x^2 - 1)\xi_0^2]$$

and the corresponding Hamiltonian vector field is given by

$$\xi_\sigma = (x^2 - 1)\xi_0 \frac{\partial}{\partial x_0} + \xi\frac{\partial}{\partial x} - x\xi_0^2\frac{\partial}{\partial \xi}. \tag{6.2}$$

In the present situation M is two dimensional and so we may choose as our Λ_0 the line

$$(x_0, x, \xi_0, \xi) = (\ell, 0, 1, 1).$$

The characteristic curves are thus given as the solutions of the ordinary differential equations and initial conditions

$$\frac{d\xi_0}{dt} = 0, \quad \xi_0(0) = 1, \quad \frac{d\xi}{dt} = -x\xi_0^2, \quad \xi(0) = 1,$$

$$\frac{dx}{dt} = \xi, \quad x(0) = 0, \quad \frac{dx_0}{dt} = (x^2 - 1)\xi_0, \quad x_0(0) = \ell,$$

and are, explicitly,

$$\xi_0 \equiv 1, \qquad \xi = \cos t,$$

$$x = \sin t, \qquad x_0 = -\tfrac{1}{2}(t + \sin t \cos t) + \ell.$$

The manifold Λ is thus the cylinder

$$x^2 + \xi^2 = 1, \qquad \xi_0 = 1,$$

lying over the strip $|x| \leqslant 1$ in the (x_0, x) plane. Notice that the projection, π, becomes singular over the lines $x = \pm 1$.

As we are dealing with a trivial line bundle, the expression for the transport operator is given by (1.3). There are no first order terms in the operator, and the cross terms in the second derivative of $\sigma(L)$ (i.e., in the second term on the right of (1.3)) vanish. Furthermore, $d\phi = \xi_0 dx_0 + \xi dx = dx_0 \pm (\sqrt{1 - x^2})dx$ for

$|x| < 1$. Thus (1.3) becomes, for $\xi > 0$,

$$(x^2 - 1)\frac{\partial u_0}{\partial x_0} + \xi\frac{\partial u_0}{\partial x} - \tfrac{1}{2}\left(\frac{x}{\sqrt{1 - x^2}}\right)u_0$$

or, if we think of $u_0(x_0, x)$ as a function of all four variables depending only on the first two,

$$\xi_\sigma u_0 = \tfrac{1}{2}\left(\frac{x}{\sqrt{1 - x^2}}\right)u_0$$

where ξ_σ is the Hamiltonian vector field given by (6.2). This is an ordinary differential equation for u along each trajectory which can be written as

$$\frac{d(\log u_0)}{dt} = \tfrac{1}{2}\frac{\sin t}{\cos t}.$$

A similar equation holds for $\xi < 0$. Solving the equation yields

$$\log u_0 = -\tfrac{1}{2}\log|\cos t| + f$$

where the constant, f, may depend on the trajectory, i.e., on ℓ.

$$u_0(t) = \frac{f(\ell)}{\sqrt{|\cos t|}} = \frac{f(\ell)}{\sqrt[4]{1 - x^2}},$$

where (we shall see) f is periodic of period 2π. To explain the interpretation of this last expression let us introduce the angular coordinate $\theta = \arcsin x$ so that (ℓ, θ) can be used as coordinates on the cylinder Λ. Then

$$|d\theta| = \frac{1}{\sqrt{1 - x^2}}|dx|$$

and, formally,

$$|d\theta|^{1/2} = \frac{1}{\sqrt[4]{1 - x^2}}|dx|^{1/2}.$$

Notice that the transport equation and its solution both blow up at $x = \pm 1$. However, let us regard the expression

$$u_0(x_0, x) = \frac{f(\ell)}{\sqrt[4]{1 - x^2}}$$

as the coefficient of the formal expression $u_0(x_0, x)|dx_0\, dx|^{1/2}$. If we introduce coordinates ℓ, θ on the cylinder Λ then we can regard our solution as the "half density" on the cylinder

$$f(\ell)|d\ell d\theta|^{1/2}.$$

Since the same trajectory crosses the ℓ-axis at ℓ and $\ell + 2\pi$, we see that f must be periodic of period 2π. The important lessons to be learned from this example are that (i) the transport equation in x becomes singular at caustics, and (ii) if we reinterpret the transport equation as an equation for "half densities" on Λ it is no longer singular.

Thus our problem is how to extend the method of §3 past a caustic, at least as far as the first term in the expansion is concerned. As we shall see, with suitable reinterpretation, the transport equation makes perfectly good sense on all of Λ. We will then have to deal with the problem of how to pass from information "upstairs" on Λ back down to M.

Let us now examine the transport equation in general for scalar operators. To simplify the notation let us write σ for $\sigma(L)$, and χ for the symbol of the term next order down in the local expansion of L so that

$$\sigma = \sum_{|\alpha|=k} A_\alpha \xi^\alpha \quad \text{and} \quad \chi = \sum_{|\alpha|=k-1} A_\alpha \xi^\alpha$$

where

$$L = \sum_{|\alpha|\leqslant k} A_\alpha \left(\frac{\partial}{\partial x}\right)^\alpha .$$

The transport equation $R_\phi u = 0$ becomes (for each solution ϕ of the characteristic equation)

$$\sum \frac{\partial \sigma}{\partial \xi^i}(x, d\phi(x)) \frac{\partial u}{\partial x^i} + \left[\tfrac{1}{2} \sum \frac{\partial^2 \sigma}{\partial \xi^i \partial \xi^j} \frac{\partial^2 \phi}{\partial x^i \partial x^j} + \chi(x, d\phi(x))\right] u = 0.$$

Here u is considered as a function of x alone. If we think of u as being defined on T^*M (and constant in the ξ direction) then we can write the transport equation as

$$D_{\xi_\sigma} u + \left[\tfrac{1}{2} \sum \frac{\partial^2 \sigma}{\partial \xi^i \partial \xi^j} \frac{\partial^2 \phi}{\partial x^i \partial x^j} + \chi\right] u = 0 \tag{6.3}$$

on $\Lambda = \mathit{graph}\ d\phi$. To explain the meaning of (6.3) let us consider first the local situation where Λ and M are both oriented and $\pi_{|\Lambda}$ is a diffeomorphism.

We recall some basic facts about densities. A density ρ of order s is a rule which assigns a number, $\rho(\boldsymbol{\eta}^1, \ldots, \boldsymbol{\eta}^n)$, to each n-tuplet of vectors $\boldsymbol{\eta}^1, \ldots, \boldsymbol{\eta}^n$ at each point of Λ so that

$$\rho(A\boldsymbol{\eta}^1, \ldots, A\boldsymbol{\eta}^n) = |\det A|^s \rho(\boldsymbol{\eta}^1, \ldots, \boldsymbol{\eta}^n)$$

for any linear transformation A. If $F : \Lambda \to M$ is a diffeomorphism and ρ is a density on M, then $F^*\rho$ is defined by

$$(F^*\rho)(\boldsymbol{\eta}^1, \ldots, \boldsymbol{\eta}^n) = \rho(dF\boldsymbol{\eta}^1, \ldots, dF\boldsymbol{\eta}^n).$$

Notice that

$$\begin{aligned}(F^* \rho)(A\boldsymbol{\eta}^1, \ldots, A\boldsymbol{\eta}^n) &= \rho(dF(A\boldsymbol{\eta}^1), \ldots, dF(A\boldsymbol{\eta}^n)) \\ &= \rho(dF \circ A \circ dF^{-1} dF\boldsymbol{\eta}^1, \ldots, dF \circ A \circ dF^{-1} dF\boldsymbol{\eta}_n) \\ &= |\det A|^s F^* \rho(\eta^1, \ldots, \eta^n).\end{aligned}$$

Thus $F^* \rho$ is again a density. If ξ is a smooth vector field and ρ a smooth density of order s then the Lie derivative $D_\xi \rho$ is defined by

$$D_\xi \rho = \frac{d}{dt} (\exp t\xi)^* \rho_{|t=0}.$$

The space of densities evaluated at a single point is clearly one dimensional and it makes sense to say that a density does not vanish at a given point. If ρ does not vanish at any point then

$$D_\xi \rho = f\rho$$

for some function f and we shall call f the logarithmic derivative of ρ and write

$$f = D_\xi(\log \rho).$$

If ρ_1 is a density of order s_1 and ρ_2 is a density of order s_2 then $\rho_1 \rho_2$ is a density of order $s_1 + s_2$ where

$$(\rho_1 \rho_2)(\boldsymbol{\eta}^1, \ldots, \boldsymbol{\eta}^n) = \rho_1(\boldsymbol{\eta}^1, \ldots, \boldsymbol{\eta}^n)\rho_2(\boldsymbol{\eta}^1, \ldots, \boldsymbol{\eta}^n).$$

Since $F^*(\rho_1 \rho_2) = F^*(\rho_1) F^*(\rho_2)$ we have

$$D_\xi(\rho_1 \rho_2) = (D_\xi \rho_1)\rho_2 + \rho_1 D_\xi \rho_2$$

and if ρ_1 and ρ_2 are both nowhere zero we have

$$D_\xi(\log(\rho_1 \rho_2)) = D_\xi \log \rho_1 + D_\xi \log \rho_2.$$

We now propose to perform the following local computation. In terms of the coordinates $x^1, \ldots, x^n$ we let $|dx|^s$ denote the density of order s on M which assigns the value one to the n-tuplet $(\partial/\partial x^1, \ldots, \partial/\partial x^n)$ at each point of M. Then $u|dx|^{\frac{1}{2}}$ is a density of order $\frac{1}{2}$ on X. Let $\iota : \Lambda \to T^*(M)$ denote immersion of Λ as a submanifold of $T^*(M)$ so that $\pi \circ \iota$ is just the restriction of π to Λ. Then $\pi \circ \iota$ is a diffeomorphism and

$$\rho = (\pi \circ \iota)^* (u|dx|^{1/2})$$

is a density of order $\frac{1}{2}$ on Λ. We wish to compute $D_\xi(\log \rho)$ where ξ is the restriction of ξ_σ to Λ. Now

$$D_\xi \log \rho = \tfrac{1}{2} D_\xi \log \rho^2 = \tfrac{1}{2} D_\xi \log[(\pi \circ \iota)^* u^2 |dx|]$$
$$= D_\xi(\log(\pi \circ \iota)^* u) + \tfrac{1}{2} D_\xi \log((\pi \circ \iota)^* |dx|).$$

Now ξ_σ is a vector field defined on all of $T^*(M)$ and $\pi^* \log u$ (which we simply write as $\log u$) is a function defined on all of $T^*(M)$. Therefore the first term in the above expression is just $D_{\xi_\sigma}(\log u)_{|\Lambda}$. To compute the second term we must take into account the orientations of Λ and M.

If $\pi \circ \iota$ preserves orientation then (identifying n-forms with densities)

$$(\pi \circ \iota)^* |dx| = (\pi \circ \iota)^* (dx^1 \wedge \cdots \wedge dx^n)$$
$$= \iota^* \pi^* (dx^1 \wedge \cdots \wedge dx^n).$$

Again, $\pi^*(dx^1 \wedge \cdots \wedge dx^n)$ is a well defined n-form on $T^*(M)$ and therefore we can write

$$D_{\xi_\sigma}(\iota^* \pi^* (dx^1 \wedge \cdots \wedge dx^n)) = \iota^* D_{\xi_\sigma}(dx^1 \wedge \cdots \wedge dx^n),$$

where, on the right, ξ_σ is regarded as a vector field on $T^*(M)$ and we have written $dx^1 \wedge \cdots \wedge dx^n$ for $\pi^*(dx^1 \wedge \cdots \wedge dx^n)$. Then

$$D_{\xi_\sigma}(dx^1 \wedge \cdots \wedge dx^n) = \sum_i dx^1 \wedge \cdots \wedge D_{\xi_\sigma} dx^i \wedge \cdots \wedge dx^n$$
$$= \sum_i dx^1 \wedge \cdots d\frac{\partial \sigma}{\partial \xi_i} \wedge \cdots \wedge dx^n$$
$$= \sum_i \frac{\partial^2 \sigma}{\partial \xi^i \partial x^i} dx^1 \wedge \cdots \wedge dx^n$$
$$+ \sum_{ij} dx^1 \wedge \cdots \wedge \frac{\partial^2 \sigma}{\partial \xi_i \partial x^j} d\xi_j \wedge \cdots \wedge dx^n.$$

Upon restriction to Λ we must substitute $d\xi_i = d(\partial\varphi/\partial x^i)$ and we get

$$\iota^*(D_{\xi_\sigma}(dx^1 \wedge \cdots \wedge dx^n))$$
$$= \left(\sum_{ij} \frac{\partial^2 \sigma}{\partial \xi_i \partial \xi_j} \frac{\partial^2 \varphi}{\partial x^i \partial x_j} + \sum \frac{\partial^2 \sigma}{\partial x^i \partial \xi_i} \right) dx^1 \wedge \cdots \wedge dx^n.$$

Thus

$$D_{\xi_\sigma} \log(\pi \circ \iota)^* u |dx|^{1/2}$$
$$= D_{\xi_\sigma} \log u + \frac{1}{2} \sum \frac{\partial^2 \sigma}{\partial \xi_i \partial \xi_j} \frac{\partial^2 \varphi}{\partial x^i \partial x^j} + \frac{1}{2} \sum \frac{\partial^2 \sigma}{\partial \xi^i \partial x^i},$$

under the assumption that the orientations of Λ and M are consistent. If we compare this with equation (6.3) we see that the transport equation can be written as

$$D_{\xi_\sigma} \log \rho = -\chi + \frac{1}{2} \sum \frac{\partial^2 \sigma}{\partial \xi_i \, \partial x^i} . \tag{6.4}$$

If the orientation of Λ and M are reversed, the same argument goes through. In this case

$$\begin{aligned} D_{\xi_\sigma} \log(\pi \circ \iota)^* (u|dx|^{1/2}) &= \iota^* D_{\xi_\sigma} \log u + \tfrac{1}{2} D_{\xi_\sigma} \log \pi^* \iota^* |dx| \\ &= \iota^* D_{\xi_\sigma} \log u + \tfrac{1}{2} D_{\xi_\sigma} \log(-\pi^* \iota^* |dx|) \end{aligned}$$

since the logarithmic derivative is clearly insensitive to multiplication by a constant.

This suggests the following change of viewpoint concerning the original differential equation. Let us think of u as being the coefficient of the half density $u(x)\,|dx|^{1/2}$ and consider D as an operator on half densities. Then

$$\rho = (\pi \circ \iota)^* (u|dx|^{1/2})$$

is a half density on Λ. Thus the transport equation is to be regarded as an equation for a half density on Λ. Furthermore, we shall now see that the right hand side of (6.4) is an invariantly defined function on $T^*(M)$, provided that L is regarded as an operator on half densities. (We are indebted to Keith Hannabus for help in the following discussion.)

Indeed, let $|\mathbf{A}^n|^{1/2}(M)$ denote the space of smooth half densities on M, and let $|\mathbf{A}^n|_c^{1/2}(M)$ denote the space of smooth half densities of compact support. Notice that $|\mathbf{A}^n|_c^{1/2}(M)$ has the structure of a pre-hilbert space: If ρ_1, ρ_2 are two half densities of compact support then $\rho_1 \bar{\rho}_2$ is a density of compact support and hence can be integrated. We can therefore define

$$(\rho_1, \rho_2) = \int \rho_1 \bar{\rho}_2$$

(where $\bar{\rho}(\xi_1, \ldots, \xi_n) = \overline{\rho(\xi_1, \ldots, \xi_n)}$). Then any differential operator D has a formal adjoint, D^*, which is the unique differential operator satisfying

$$(D\rho_1, \rho_2) = (\rho_1, D^* \rho_2).$$

Notice that if D is a differential operator of order m (with real coefficients) then

$$\sigma(D) = (-1)^m \sigma(D^*)$$

so that

$$B = \tfrac{1}{2}(D - (-1)^m D^*) \tag{6.5}$$

is a differential operator of order $m - 1$. Thus its symbol is a globally defined function on $T^*(M)$. We claim that in terms of local coordinates

$$\sigma(B) = \chi - \frac{1}{2}\sum \frac{\partial^2\sigma}{\partial\xi^i\,\partial x^i}. \tag{6.6}$$

To see this observe that B is linear in D and that $\sigma(B)$ can only involve the top two terms in D. We notice that if $\sigma(D) = 0$ then $D - (-1)^m D^* = 2\chi + \cdots$. To complete the proof we need only check the result for the case that D has the local expression

$$D = a(x)\frac{\partial^{|\alpha|}}{\partial x^\alpha}, \qquad |\alpha| = m.$$

We may also assume that $\rho_1 = f|dx|^{1/2}$ and $\rho_2 = g|dx|^{1/2}$. Then

$$(D\rho_1, \rho_2) = \int \left(a(x)\frac{\partial^{|\alpha|}}{\partial x^\alpha} f\right)\bar{g}\,dx = (-1)^{|\alpha|}\int f\frac{\partial^{|\alpha|}}{\partial x^\alpha}(\overline{ag})dx$$

and it is clear from direct computation that

$$\sigma(B) = -\frac{1}{2}\frac{\partial^2\sigma}{\partial\xi^i\,\partial x^i}.$$

With this interpretation, (6.6) is indeed a well defined function on T^*M called the *subprincipal symbol* of the original differential operator.

The definition of the subprincipal symbol as a function on all of T^*M *does* depend on the fact that we are viewing the differential operator as an operator on half densities. In other words, we assume that there is some differential operator on half densities which becomes identified with our original differential operator when we choose the particular trivialization given by $|dx|^{1/2}$. Let $|\wedge|^{1/2}M$ be the bundle of half densities. If $D : |\wedge|^{1/2}M \to |\wedge|^{1/2}M$ is a differential operator, then $\sigma(D)$ is a well defined function on T^*M. Indeed, since $\operatorname{Hom}(E, E)$ is canonically trivial for any line bundle, E, in particular for $E = |\wedge|^{1/2}M$ (considered as a bundle over T^*M), $\sigma(D)$, which is a section of $\operatorname{Hom}(|\wedge|^{1/2}M, |\wedge|^{1/2}M)$, is well defined as a function on T^*M. We have just seen that the subprincipal symbol is then also a well-defined function on T^*M, and have given an interpretation to the transport equation.

Let us now examine the situation where L is a differential operator from $E_1 \otimes |\wedge|^{1/2}$ to $E_2 \otimes |\wedge|^{1/2}$ where E_1 and E_2 are line bundles. Then L^* is a differential operator from $E_2^* \otimes |\wedge|^{1/2}$ to $E_1^* \otimes |\wedge|^{1/2}$ and so the expression (6.6) does not make sense. Nevertheless, we can give an invariant interpretation to the transport operator. To see this, suppose that s and s' are two non-vanishing sections of E_1 with $s' = fs$, and suppose that $t' = gt$ where t and t' are non-vanishing sections of E_2.

Let L be a differential operator from $E_1 \otimes |\wedge|^{1/2}$ to $E_2 \otimes |\wedge|^{1/2}$. If r is a section of E_1 then $r = us = u's'$ where $u = u'f$ and we can write $z = wt = w't'$ where $w = w'g$, if z is a section of E_2. The choice of s and t determines a differential operator $D : |\wedge|^{1/2} \to |\wedge^{1/2}|$ by $L(s \otimes \rho) = t \otimes D(\rho)$, and the choice of s', t' gives the operator D'. The two operators are clearly related by

$$D' = g^{-1} \circ D \circ f.$$

Thus,

$$\sigma' = g^{-1}\sigma f \quad \text{so that} \quad \xi' = g^{-1}f\xi, \qquad \text{when } \sigma = 0,$$

and

$$\chi' = g^{-1}\chi f + g^{-1}\frac{\partial\sigma}{\partial\xi^i}\frac{\partial f}{\partial x^i}\,.$$

Thus if c and c' are the subprincipal parts of D and D',

$$\begin{aligned} c' &= g^{-1}\chi f + g^{-1}\frac{\partial\sigma}{\partial\xi^i}\frac{\partial f}{\partial x^i} - \frac{1}{2}g^{-1}\frac{\partial^2\sigma}{\partial\xi^i\partial x^i}f - \frac{1}{2}\frac{\partial g^{-1}}{\partial x^i}\frac{\partial\sigma}{\partial\xi^i}f - \frac{1}{2}g\frac{\partial\sigma}{\partial\xi^i}\frac{\partial f}{\partial x^i} \\ &= g^{-1}cf + \tfrac{1}{2}g^{-1}(\xi f) - \tfrac{1}{2}f(\xi g)^{-1}. \end{aligned}$$

Now if η is a vector field and Ω is a $\frac{1}{2}$ density and h is a function, we have the formula

$$D_{h\eta}\Omega = hD_\eta\Omega + \tfrac{1}{2}(D_\eta h)\Omega. \tag{6.7}$$

Therefore, on $\frac{1}{2}$ densities along Λ,

$$D_{\xi'} = g^{-1}fD_\xi + \tfrac{1}{2}fD_\xi g^{-1} + \tfrac{1}{2}g^{-1}D_\xi f$$

and therefore

$$\begin{aligned} D_{\xi'} + c' &= g^{-1}fD_\xi + g^{-1}cf + g^{-1}D_\xi f \\ &= g^{-1} \circ [D_\xi + c] \circ f. \end{aligned}$$

This is precisely the transition law required for the definition of a (first order) partial differential operator on Λ from $E_1 \otimes |\wedge|^{1/2}\Lambda$ to $E_2 \otimes |\wedge|^{1/2}\Lambda$.

§7. The Maslov cycle and the Bohr-Sommerfeld quantization conditions.

Let us put together the results of the past few sections to discuss the first term in the asymptotic solution (3.1) of an asymptotic partial differential operator, as described in §3. For simplicity we will assume the asymptotic differential operator in question is from $|\wedge|^{1/2}M$ to $|\wedge|^{1/2}M$, as discussed in the preceding section. We assume that we have solved the characteristic equation, i.e., that we have picked out a Lagrangian submanifold, Λ, of the characteristic variety. We assume also that we have solved the transport equation on Λ, so that we have a half density, v, defined on Λ. Our question is how to pass from v back down to a half density on M.

Let $\iota: \Lambda \to T^*M$ be the immersion of Λ as a submanifold and let us assume that $\pi \circ \iota$ is proper. If x is a regular value of $\pi \circ \iota$ then

$$(\pi \circ \iota)^{-1}(x) = \{y_1, \ldots, y_k\}$$

where each y_j has a neighborhood, W_j, which maps diffeomorphically onto a neighborhood, U_j, of x. Let us assume that $k \neq 0$, i.e., that $(\pi \circ \iota)^{-1}x \neq \emptyset$. For each j, $W_j =$ graph $d\varphi_j$ where φ_j is defined on U_j and is determined up to additive constant. We can use the diffeomorphism $\pi \circ \iota_{|W_j}$ to identify $v_{|W_j}$ with a half density, u_j, on U_j. In accordance with the procedure of §3, this should give a contribution $u_j e^{i\tau\varphi_j}$. If we let $U = \bigcap U_j$ then we should expect that on U we get the sum

$$u_1 e^{i\tau\varphi_1} + \cdots + u_k e^{i\tau\varphi_k}. \tag{7.1}$$

The trouble is that this expression is ambiguous due to the arbitrariness of the choice of phases for the φ_j. If there would be only one term in the sum, then this arbitrariness would simply lead to a (relatively harmless) overall phase factor. However, since a sum is involved, the choice of the relative phases will have an enormous effect.

In order to deal with this problem it is simpler to examine what is happening up on Λ. At each regular point, λ, of $\pi \circ \iota$ on Λ we have the function $\varphi \circ (\pi \circ \iota)$ defined up to the arbitrary phase factor, where graph $d\varphi$ is a local parametrization of Λ near λ. If we fix the phase at λ then the phase is fixed in any simply connected neighborhood of λ on which $\pi \circ \iota$ remains regular. Indeed, by definition, $d[\varphi \circ \pi \circ \iota] = \iota^*\alpha$, where α is the fundamental form on T^*M. The problem is how to change the phase as we go across the singular set.

The procedure we shall use to handle this problem is to apply the method of stationary phase. We shall give a parametrization of the Lagrangian manifold Λ which is valid in the neighborhood of a point of the singular set, and we shall write down an asymptotic half density u', which is also valid near the image of the singular set (and in particular does not have singularities at the caustic). This half density coincides with an expression of the type (7.1) at regular values, up to terms of order τ^{-1}, and hence fixes the relative phases. It will turn out that this procedure will not depend on the particular choice of u'. We will sketch the procedure here. A detailed justification of some of the steps depends on some considerations of symplectic geometry which we shall present in Chapter IV.

The method is as follows: Let $x = (x^1, \ldots, x^n)$ be coordinates on M. We introduce some auxiliary variables θ and a function $\varphi = \varphi(x, \theta)$. We assume that the set $C_\varphi = \{(x, \theta) | d_\theta \varphi = 0\}$ is a submanifold and that the differentials $d(\partial\varphi/\partial\theta_j)$ are linearly independent along C. At C_φ the differential $d_x \varphi$ is well defined and we assume that *the map*

$$\mu : (x, \theta) \to d_x \varphi(x, \theta)$$

maps C_φ *into* Λ. The fact that the differentials $d(\partial\varphi/\partial\theta_j)$ are linearly independent implies that this map is an *immersion*. Indeed in local coordinates the map is the restriction of the map

$$(x, \theta) \rightsquigarrow \left(x, \frac{\partial\varphi}{\partial x}\right) = \left(x^1, \ldots, x^n, \frac{\partial\varphi}{\partial x^1}, \ldots, \frac{\partial\varphi}{\partial x^n}\right)$$

to C_φ. Thus any tangent vector with a $\partial/\partial x$ component which is non-zero can not be sent into zero. On the other hand the image of $\boldsymbol{\eta} = \sum c_j \partial/\partial\theta^j$ is

$$\sum c_j \frac{\partial^2\varphi}{\partial\theta^j \partial x^i} \frac{\partial}{\partial\xi_i}$$

and if $\boldsymbol{\eta}$ is tangent to C_φ we have

$$0 = \left\langle \boldsymbol{\eta}, d\frac{\partial\varphi}{\partial\theta^k} \right\rangle = \sum c_j \frac{\partial^2\varphi}{\partial\theta^j \partial\theta^k}.$$

Since the $d(\partial\varphi/\partial\theta^j)$ are assumed to be independent this implies that $c_j = 0$.

The existence of such φ follows from considerations of symplectic geometry which we shall discuss in Chapter IV. (There we shall see that we can always choose the θ to be some of the ξ's and arrange that $C_\varphi = \Lambda$.)

The choice of φ, together with the density $|dx d\theta|$ induces a positive density on C_φ. It is the unique density, σ, such that

$$\left|\det\left(\left\langle \boldsymbol{\eta}_i, d\left(\frac{\partial\varphi}{\partial\theta^j}\right)\right\rangle\right)\right| \sigma(\xi_1, \ldots, \xi_n) = |dx\, d\theta|(\xi_1, \ldots, \xi_n, \boldsymbol{\eta}_1, \ldots, \boldsymbol{\eta}_k) \tag{7.2}$$

where $\boldsymbol{\xi}_1, \ldots, \boldsymbol{\xi}_n$ are tangent to C_φ and $\boldsymbol{\eta}_1, \ldots, \boldsymbol{\eta}_k$ are independent of $T(C_\varphi)$. Now let the function a be defined on C_φ by

$$a\sigma^{1/2} = \mu^* v,$$

where v is our half density on Λ. Here a is a smooth function on C_φ. Extend it to be a C^∞ function of (x, θ) compactly supported in θ. Now consider the half density

$$\left[(\tau/2\pi)^{k/2} \int e^{i\tau\varphi(x,\theta)} a(x, \theta)\, d\theta\right] |dx|^{1/2} \tag{7.3}$$

where k is the number of θ variables. This half-density is clearly continuous in x.

We will show that at regular values of $\pi \circ \iota$ this half-density differs from a solution of the type (7.1) by terms of order τ^{-1}. This will then determine the choice of phases in (7.1). The proof of this fact is a straightforward application of the method of stationary phase.

We shall give a more invariant interpretation of this procedure in Chapter VII. For the moment let us simply apply the method of stationary phase to the integral in (7.3) at a regular value x. Let (x, θ) be a point of C_φ lying above the regular value x. Then no $\partial/\partial\theta^i$ is tangent to C_φ at (x, θ) and so we can take the η_i in (7.2) to be $\partial/\partial\theta^i$. Then (7.2) shows that

$$\sigma = \frac{\pi^* |dx|}{\left|\det \dfrac{\partial^2 \varphi}{\partial\theta^i \partial\theta^j}\right|}$$

where $\pi : C_\varphi \to M$. By abuse of notation we shall write $|dx|$ instead of $\pi^* |dx|$, and then (7.2) shows that

$$\sigma = \frac{|dx|}{\left|\det\left(\dfrac{\partial^2 \varphi}{\partial\theta^i \partial\theta^j}\right)\right|} = \frac{|dx|}{|\det H_\theta(x, \theta)|}$$

where H_θ is the fiber Hessian of φ. Now $v = (\pi \circ \iota)^* u$. Suppose that $u = A|dx|^{1/2}$. Then the definition of a shows that

$$a(x, \theta) = A(x)|\det H_\theta(x, \theta)|^{1/2}$$

for $(x, \theta) \in C_\varphi$. We have extended a to be defined on a whole neighborhood of C_φ. Near (x,θ) we can solve for θ as a function of x on C_φ, say $\theta = G(x)$. Let

$$F(x) = \varphi(x, G(x)).$$

Then

$$dF(x) = (d_x \varphi)(x, G(x)) + (d_\theta \varphi)(x, G(x)) \circ dG = (d_x \varphi)(x, G(x)) \in \Lambda$$

since $(x, G(x)) \in C_\varphi$ so that $d_\theta \varphi = 0$ at $(x, G(x))$. Thus, F is a suitable phase function for Λ. This applies to each sheet of C_φ lying over x. If we now apply the method of stationary phase we see that

$$\begin{aligned}\left(\frac{\tau}{2\pi}\right)^{k/2} \int e^{i\tau\varphi} a(x, \theta)\, d\theta &= \sum A_j(x) e^{i[\tau F_j + \sigma_j \pi/4]} + O(\tau^{-1}) \\ &= \sum A_j(x) e^{i\sigma_j \pi/4} \cdot e^{i\tau F_j} + O(\tau^{-1})\end{aligned} \tag{7.4}$$

where $\sigma_j = \operatorname{sign} H_\theta(x, \theta_j)$.

We draw two conclusions from (7.4). In the first place, we see that A_j should be replaced by $A_j e^{i\sigma_j \pi/4}$. Thus the correct coefficients for the oscillatory terms $e^{i\tau\varphi_j}$ in (7.1) are not quite the projections of the solution v of the half density transport equation, but differ from it by constant phase factors, differing from one another by powers of i. It will turn out that we can account for the correct coefficients by regarding the transport equation as an equation on Λ for "half-forms" with values in a certain locally constant line bundle instead of for half-densities. The necessary machinery for these notions will be developed in Chapter V.

Suppose we make a number of simplifying assumptions about the nature of the singular set of Λ. Assume that the map $\pi \circ \iota$ has rank $n - 1$ on a submanifold, Z, of codimension one in Λ and has rank less that $(n - 1)$ on a subset of dimension at most $n - 3$. We shall see in Chapter VII that this is the situation for a "generic" Lagrangian manifold in $T^* M$. Thus, any smooth curve which intersects the singular set can be deformed so as to avoid the subset of codimension 3, along with any homotopy of curves, and can be assumed to intersect Z transversally. A simple computation shows that σ_j changes by ± 2 as we cross Z. This means that Z has a preferred relative orientation in Λ (a curve crosses Z in the "positive" direction if σ_j increases). The submanifold Z with this orientation is called the Maslov cycle. We shall discuss it in detail (together with the above mentioned computation) in Chapter IV.

Notice that we have now given a generalization of the optical phenomenon described in Chapter I, namely that light undergoes a phase shift of $\pi/2$ as we pass through a caustic.

The above procedure gives us a method for fixing the phases φ_j for all regular x lying under a connected open set $U \subset \Lambda$ if $U = \mu(C_\varphi)$ for a suitable phase function $\varphi = \varphi(x, \theta)$. Indeed, we have set $F_j(x) = \varphi \circ \mu^{-1}(\lambda_j)$, where $\pi(\lambda_j) = x$. Now we claim that the function $\varphi \circ \mu^{-1}$ satisfies

$$d(\varphi \circ \mu^{-1}) = \alpha_{|U}, \tag{7.5}$$

where $\alpha = \sum \xi_i \, dx^i$ is the action form. If we establish (7.5), then we will have shown that the function $\varphi \circ \mu^{-1}$ is, up to an additive constant, determined on U independently of the choice of φ. This fixes the relative phases of the F_j, i.e., fixes the F_j up to a single additive constant. To establish (7.5), let $\boldsymbol{\eta}$ be any vector tangent to Λ at $z = \mu(x, \theta)$. Then, by definition,

$$\langle \boldsymbol{\eta}, \alpha \rangle = \langle d\pi\boldsymbol{\eta}, z \rangle = \langle d\pi\boldsymbol{\eta}, \mu(x, \theta) \rangle.$$

On the other hand $\pi \circ \mu(x, \theta) = x$ so that, if $\boldsymbol{\xi} = d\mu^{-1}\boldsymbol{\eta}$,

$$\begin{aligned} \langle \boldsymbol{\eta}, d(\varphi \circ \mu^{-1}) \rangle &= \langle \boldsymbol{\xi}, d\varphi \rangle = \langle \boldsymbol{\xi}, d_x\varphi \rangle \text{ since } d_\theta \varphi = 0 \text{ at } (x, \theta) \\ &= \langle d\pi\boldsymbol{\xi}, \mu(x, \theta) \rangle = \langle d\pi\boldsymbol{\eta}, \mu(x, \theta) \rangle. \end{aligned}$$

We have so far discussed the problem of solving the transport equation locally. There remains the question of piecing together the local solutions to get a global solution. As we shall prove in Chapter IV, we can always cover Λ by contractible open sets U_j such that each U_j is defined by a φ_j. We can also arrange that $U_i \cap U_j$ are contractible. Now φ_j at each regular $\lambda \in U_j$ gives rise to a contribution to our global half density of

$$e^{i[\tau\varphi_j(\lambda)+(\pi/4)\operatorname{sign} H_j(\lambda)]}v(\lambda),$$

where H_j is the Hessian of φ_j with respect to θ at $\mu^{-1}(\lambda)$. If the contributions due to φ_j and φ_k are to match up, then $\tau\varphi_j + (\pi/4)\operatorname{sgn} H_j$ must equal $\tau\varphi_k + (\pi/4)\operatorname{sgn} H_k (\operatorname{mod} 2\pi\mathbf{Z})$ on the regular points of $U_j \cap U_k$. Set $g_{jk} = \varphi_j - \varphi_k$ and $h_{jk} = \frac{1}{2}(\operatorname{sgn} H_j - \operatorname{sgn} H_k)$. It follows from (7.5) that g_{jk} is constant on the regular points of $U_j \cap U_k$ and hence extends to a constant on all of $U_j \cap U_k$. It turns out that $h_{jk} \in Z$. Since h_{jk} is plainly a cocycle, we obtain a class $\mathfrak{M} \in H^1(\Lambda, Z)$, the Maslov class of Λ. (The construction of $\mathfrak{M}$ was first suggested by Keller [6] and defined by Maslov [7].) It is the Čech definition giving the dual to the geometrical cycle described above.

In other words, if we make the genericity assumptions described above, then the value of the class $\mathfrak{M}$ on the class of any smooth curve C intersecting Z transversely is given by the number of intersections of C with Z, each counted with multiplicity.

On the other hand the cocycle g_{jk} defines a class β in $H^1(\Lambda, S)$, where S is the sheaf of germs of locally constant C^∞ functions on Λ. Since $d\varphi_j|_\Lambda = \alpha|_\Lambda$ on U_j, β is just the image of the deRham class $[\alpha|_\Lambda]$ under the canonical isomorphism of deRham cohomology with Čech cohomology. (Note that $d(\alpha|_\Lambda) = 0$ since Λ is Lagrangian.) We write $\beta = [\alpha|_\Lambda]$. Letting γ be the canonical map of $H^1(\Lambda, Z)$ into $H^1(\Lambda, R)$, we obtain

$$\frac{\tau}{2\pi}\beta + \tfrac{1}{4}\mathfrak{M} \in \gamma H^1(\Lambda, Z) \tag{7.6}$$

as a necessary and sufficient condition for finding a global half-density on $\mathfrak{M}$ corresponding to v. This condition on τ and Λ is independent of v and is known as the *Bohr-Sommerfeld quantization condition.* (The above argument was first suggested by Brillouin in the case of the Schrödinger equation in order to explain the old quantum theory in terms of quantum mechanics.) We shall give a rather different interpretation of the Bohr-Sommerfeld quantization condition, using the notion of half-form and "discrete" asymptotics in Chapter VII.

In terms of the geometric interpretation of the classes β and $\mathfrak{M}$, what we are requiring is that the total change in phase around any closed curve is an integer multiple of 2π. Here we get one contribution from α and the other from each intersection with Z. That this total change be an integer is obviously necessary and sufficient in order that we be able to give a global meaning to $e^{i\tau\varphi}v$ at regular points of Λ.

To summarize: If Λ satisfies the Bohr-Sommerfeld quantization conditions then we have a way of associating an asymptotic half density, u, on M to an asymptotic half density, ρ, on Λ, where u is determined up to term of order τ^{-1}. If ρ satisfies the transport equation for the operator L then $Lu = O(\tau^{-2})$.

The map assigning (the equivalence class of) u to ρ is the Maslov canonical "operator". It is unitary up to $O(\tau^{-1})$, i.e.,

$$\int_\Lambda |\rho|^2 = \int_M |u|^2 + O(\tau^{-1})$$

if ρ (and hence u) has compact support, for instance if Λ is compact.

Notice that if ρ vanishes near the Maslov cycle, so that we have, globally,

$$u = \sum e^{i\tau\varphi_j}\rho_j,$$

the sum being over the various sheets of Λ, then

$$u\bar{u} = \sum e^{i\tau(\varphi_j - \varphi_k)}\rho_j\rho_k \quad \text{and} \quad \int e^{i\tau(\varphi_j-\varphi_k)}\rho_j\rho_k = O(\tau^{-N}) \text{ for any } N \text{ if } j \neq k$$

by stationary phase. Thus $\int |u|^2 \sim \int |\rho|^2$ to all orders. The general argument, due to Duistermaat [8], is just a more delicate application of stationary phase and is quite straightforward.

We can now relate the Bohr-Sommerfeld conditions to the spectrum of L following a beautiful argument also due to Duistermaat. Suppose that L is a self-adjoint operator. Then given a ρ with $\|\rho\| = c > 0$ satisfying the transport equation on Λ, and given a τ satisfying the Bohr-Sommerfeld conditions we have produced a u such that

$$\|u\|_{L^2} = c + O(\tau^{-1}) \quad \text{and} \quad \|Lu\| < C\tau^{-2}\|u\| \qquad \text{for some } C.$$

This implies that there is some constant C such that the spectrum of L intersects the interval $(-C\tau^{-2}, C\tau^{-2})$, for otherwise $\|Lu\| \geqslant C\tau^{-2}\|u\|$. Applied to the Schrödinger operator

$$-h^2\Delta + (V - E)$$

this says that some spectrum intersects an interval of diameter h^2 about E, if the Bohr-Sommerfeld conditions can be satisfied for a given value of E.

Let us examine the meaning of the Bohr-Sommerfeld quantization condition in the classical case of the reduced (i.e, time independent) one dimensional Schrödinger equation

$$\frac{d^2\psi}{dx^2} + \frac{1}{h^2}(E - V(x))\psi = 0$$

where h^{-1} plays the role of our asymptotic parameter. Here the symbol is

$$\sigma = -\xi^2 + (E - V(x)) \quad \text{and} \quad \xi_\sigma = 2\xi\frac{\partial}{\partial x} - V'(x)\frac{\partial}{\partial \xi}.$$

The characteristic variety $\sigma = 0$ is thus given by

$$\xi^2 + V(x) = E$$

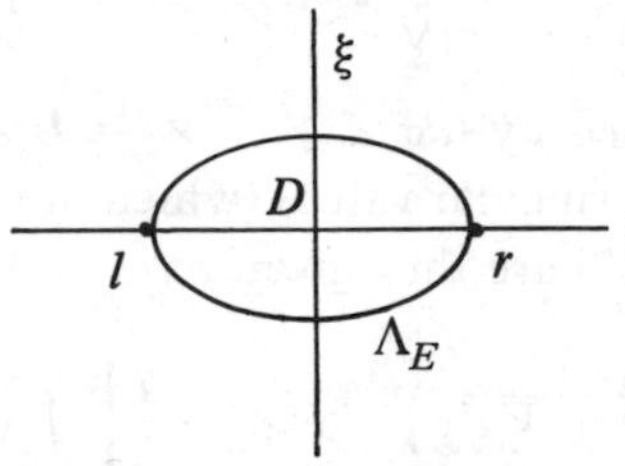

In the figure we are assuming that the function V looks something like

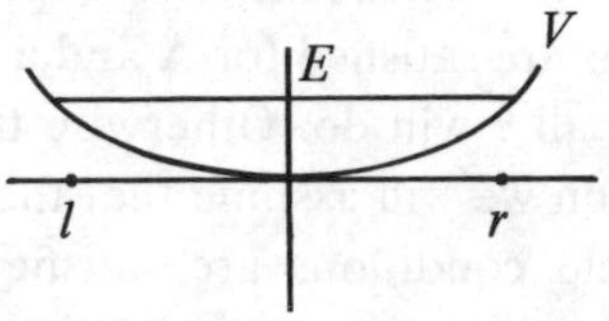

The characteristic variety in this case is one dimensional and thus coincides with a Lagrangian manifold. The Maslov cycle, Z, in this case clearly coincides with the intersection of the curve $\sigma = 0$ with the x-axis and the value of the Maslov class on the fundamental cycle (consisting of going around the curve once in the counterclockwise direction) is $+2$. Thus (7.6) becomes

$$\frac{1}{2\pi h}\int_{\Lambda_E} \alpha + \frac{1}{2} = n.$$

Now $\alpha = \xi dx$ and $\int_{\Lambda_E} \alpha = \iint_{D_E} d\xi \wedge dx$ is just the area, $A(E)$, enclosed by the curve Λ_E. Thus the Bohr-Sommerfeld conditions in this case reduce to

$$A(E) = 2\pi h(n + \tfrac{1}{2}), \qquad n = 0, 1, \ldots.$$

We can also explicitly write down the solution to the transport equation in this case. We are looking for a half density on Λ_E which is invariant under the vector field ξ_σ. It is easy to write down a one form (defined off Z) which is invariant under ξ_σ, namely $\iota^* dx/\xi$. In fact

$$D_{\xi_\sigma}\left(\frac{dx}{\xi}\right) = d\left(\xi_\sigma \lrcorner \frac{dx}{\xi}\right) + \xi_\sigma \lrcorner\, d\left(\frac{dx}{\xi}\right)$$
$$= d(2) - \xi_\sigma \lrcorner \frac{d\alpha}{\xi^2} = \frac{d\sigma}{\xi^2}$$

and $\iota^* d\sigma = 0$ since σ is constant on Λ_E. Thus, since $\xi = \sqrt{E - V}$ on Λ_E, we see that the half density is given by

$$\frac{|dx|^{1/2}}{\sqrt[4]{E-V}}.$$

The function φ is determined by $(d\varphi/dx)^2 + V = E$ and so $\varphi = \pm\int\sqrt{E-V}$. The solution away from the singular values (which in this case are called "turning points" instead of "caustics") are thus given by

$$\frac{c}{\sqrt[4]{E-V}}\left[\exp\left(\frac{i}{h}\int\sqrt{E-V}\,dx\right) + \exp\left(-\frac{i}{h}\left[\int\sqrt{E-V}\,dx - 1/4\right]\right)\right].$$

This is the standard JWKB approximate solution to the Schrödinger equation.

Let us return to more general considerations. Suppose that the Bohr-Sommerfeld quantization conditions are satisfied for Λ and a given set of τ. (For example if $\beta = 0$ and $\mathfrak{M} = 0$ then all τ will do. Otherwise this imposes a restriction on τ.) For the rest of the section we will assume that the τ range through the set for which the Bohr-Sommerfeld conditions are satisfied. Then the procedure described above gives us a method for passing from half densities on Λ to half densities at regular points on M. We can formulate this method as follows: Choose a regular point on Λ and let $y_1, \ldots, y_k$ be the preimages of x. Then (assuming that Λ is connected) since $d\varphi = \alpha$, the explicit determination of the phases in (7.1) is given by

$$\sum_j u_j \exp\left[i\left(\tau\int_{\gamma_j}\alpha + \#_j\,\pi/4 + \tau\varphi(z_0)\right)\right] \tag{7.7}$$

where γ_j is any curve joining z_0 to y_j which intersects Z transversally and $\#_j$ is the intersection number of γ_j with Z and where $\varphi(z_0)$ is an arbitrary constant.

Notice that the fact that the expression (7.7) is independent of the choice of the γ_i is exactly the assertion of the Bohr-Sommerfeld quantization conditions.

There might be circumstances in which there is a preferred choice of (class of) curves γ_j so that (7.7) makes sense even though the Bohr-Sommerfeld condition is not satisfied. For example, suppose that Λ is obtained by sweeping out a submanifold Λ_0 under the flow generated by a vector field, ξ, where we assume that each integral curve intersects Λ_0 at exactly one point. Suppose also that $\alpha_{|\Lambda_0}$ is exact. Then (up to an arbitrary constant) there is a well defined choice of phase function on Λ_0, and we can take the γ_j in (7.7) to be the trajectories of ξ joining y_j to Λ_0. For example, consider the (time dependent) Schrödinger equation

$$\frac{\hbar}{i}\frac{\partial\psi}{\partial t} = -\frac{\hbar^2}{2}\sum\frac{\partial^2\psi}{\partial x_i^2} + V(x)\psi$$

on $\mathbf{R}^n \times \mathbf{R}$. Here $1/\hbar$ plays the role of the parameter τ. If ξ_0 denotes the dual variable to t and ξ_i to x_i then the symbol of the operator is

$$\xi_0 - \frac{1}{2}\sum_1^n \xi_i^2 - V(x) = \sigma$$

and the associated vector field is

$$\xi = \frac{\partial}{\partial t} + \sum \frac{\partial V}{\partial x_i}\frac{\partial}{\partial \xi_i} - \sum \xi_i \frac{\partial}{\partial x_i}$$

while

$$\langle \xi, \alpha \rangle = \xi_0 - \sum \xi_i^2$$

so that on the characteristic variety, $\sigma = 0$, we have

$$\langle \xi, \alpha \rangle = V(x) - \tfrac{1}{2}\sum \xi_i^2 = -L$$

where L is the classical Lagrangian. Since all the terms in the Schrödinger operator are self-adjoint the subprincipal symbol vanishes and so the transport equation becomes

$$D_\xi \rho = 0$$

for the half density ρ.

Let $\Lambda \subset T^*(\mathbf{R}^n \times \mathbf{R})$ be the Lagrangian manifold swept out by $\Lambda_0 \times \{0\}$ under the flow, f_t, generated by ξ, where $\Lambda_0 = \text{graph}\, d\varphi_0$ is a Lagrangian submanifold of $T^*\mathbf{R}^n$. Let the map $F_t : \mathbf{R}^n \to \mathbf{R}^n$ be defined by

$$F_t(x) = \pi f_t(d\varphi_0(x)).$$

Now if v_0 is an initial half density on Λ_0 then $f_{-t}^* v_0$ is a half density on $f_t \Lambda_0$ and

$$v = |f_{-t}^* v_0| \otimes |dt|^{1/2}$$

is clearly invariant under the flow, i.e., satisfies the transport eqation. If $v_0 = \pi^* u_0$, then the corresponding half density, u, on $\mathbf{R}^n \times \mathbf{R}$ is given at a point z by

$$u(z)\left[\sum \left|\det \frac{\partial F_t}{\partial x}\right|^{-1/2} (x_i) u_0(x_i)\right] \otimes |dt|^{1/2},$$

the sum ranging over the inverse image of z (where we assume that z is a regular value of F_t). Thus if $u_0 = a|dx|^{1/2}$ then the full first order term in the asymptotic solution of the Schrödinger equation is given by

$$\sum_{x_j \in F_t^{-1}(x)} \left|\det \frac{\partial F_t}{dx}(x_i)\right|^{-1/2} a(x_i) \exp i\left[-\hbar^{-1} \int_{\gamma_j} L + H_j + \hbar^{-1}\phi_j\right]$$

where γ_j is the classical path and H_j the intersection number of the corresponding curve on Λ with the Maslov cycle, and where ϕ_j are initial phases.

We can use the above expression, together with the method of stationary phase, to derive an asymptotic expression for the fundamental solution of Schrödinger's equation. Let us write

$$\delta(x - x_0) = a(x) \int e^{i\eta\cdot(x-x_0)}\, d\eta$$

where $a(x)$ is a function of compact support in a neighborhood of x_0 and takes on the value one at x_0. Setting $\xi = \hbar\eta$ we can write this as

$$\delta(x - x_0) = \frac{a(x)}{\hbar^n} \int \exp i\left[\frac{\xi}{\hbar}(x - x_0)\right] d\xi.$$

Suppose that there are only a finite number of classical trajectories joining x_0 to a given point y at time t, and that y is not a conjugate point of x_0, i.e., that y is a regular value for the projection of the manifold of all trajectories emanating from x_0. The same will be true for all trajectories joining x to y if x is close enough to x_0. We thus may assume that this holds for all x in supp a. We now consider the initial value problem for each frequency, i.e., the Schrödinger equation with the initial condition

$$\frac{a(x)}{\hbar^n} \exp i\frac{\xi}{\hbar}(x - x_0).$$

The corresponding initial isotropic manifold is graph $d(\xi \cdot (x - x_0))$ which is the set of points (x, ξ) (with ξ fixed and x varying over $\mathbf{R}^n$). Then we obtain the corresponding Λ, and our assumption implies that there are only finitely many trajectories $\gamma_{j,\xi}$ lying in Λ and joining a point in supp a to y. Let $(x_j(\xi), \xi)$ be the initial point of $\gamma_{j,\xi}$. Then the asymptotic solution for the initial phase $\xi \cdot (x - x_0)$ is given by

$$\sum a(x_j(\xi)) \left|\frac{\partial y}{\partial x}(x_j(\xi))\right|^{-1/2} \exp\left(\frac{i}{\hbar}\left[\int_{\gamma_{j,\xi}} \alpha + \xi \cdot (x_j(\xi) - x_0)\right] + \frac{\pi i}{4}\,\#_j\right).$$

Integrating this expression over ξ gives

$$G(y, t, x_0) = \sum_j \frac{1}{\hbar^n} \int a(x_j(\xi)) \left|\frac{\partial y}{\partial x}(x_j(\xi))\right|^{-1/2}$$

$$\cdot \exp\left(\frac{i}{\hbar}\left[\int \alpha + \xi \cdot (x_j(\xi) - x_0)\right] + \frac{\pi i}{4}\,\#_j\right) d\xi.$$

To evaluate this integral by the method of stationary phase we must first find the critical points of the phase function

$$\phi(\xi) = \int_{\gamma_{j,\xi}} \alpha + \xi \cdot (x(\xi) - x_0)$$

as a function of ξ. To do so we will need some general facts about symplectic geometry: *Let M be a manifold, and Λ a Lagrangian submanifold of $T^* M$. Let α be the action form. For each $s \in \mathbf{R}$ let γ_s be a smooth curve on Λ and suppose γ_s depends smoothly on s. Let $a(s)$ and $b(s)$ be the initial and terminal points of γ_s. Then*

$$\frac{d}{ds}\int_{\gamma_s} \alpha = \alpha\left(\frac{db}{ds}\right) - \alpha\left(\frac{da}{ds}\right).$$

PROOF. There is a tubular neighborhood U of γ_s in Λ on which α is exact; i.e., $\alpha = df$ on U. Thus

$$\int_{\gamma_s} \alpha = f(\beta(s)) - f(\alpha(s)).$$

Differentiating with respect to s we get the assertion above, proving the formula. We will meet this formula several times again. For example in Chapter III we shall see that this formula gives the Abbe sine condition in microscopy.

Now let us compute

$$\frac{\partial}{\partial \xi_i}\int_{\gamma_{j,\xi}} \alpha.$$

Note that the curves $\gamma_{j,\xi}$ all lie on a fixed Lagrangian manifold in $T^*(\mathbf{R}^n \times \mathbf{R})$, namely the set of all trajectories that at time t lie above the point (y,t). Let $\boldsymbol{\eta}_i$ and ζ_i be the tangent vectors to the initial and terminal curves of $\gamma_{j,\xi}$ obtained by varying ξ_i and leaving the other coordinates of ξ fixed. By (7.8),

$$\frac{\partial}{\partial \xi_i}\int_{\gamma_{\alpha,\xi}} \alpha = \langle \boldsymbol{\eta}_i, \alpha\rangle - \langle \zeta_i, \alpha\rangle.$$

The end point of $\gamma_{j,\xi}$ projects onto the fixed point (y,t) in the base for all ξ, so $(d\pi)\boldsymbol{\eta}_i = 0$ and hence $\langle \boldsymbol{\eta}_i, \alpha\rangle = 0$. On the other hand,

$$(d\pi)\zeta_i = \frac{\partial x_j(\xi)}{\partial \xi_i},$$

so

$$\langle \zeta_i, \alpha\rangle = \xi \cdot \frac{\partial x_j}{\partial \xi_i}$$

at $(x(\xi), \xi)$. Therefore, we get

$$\frac{\partial}{\partial \xi_i}\int_{\gamma_{j,\xi}} \alpha = -\xi \cdot \frac{\partial x_j}{\partial \xi_i} \quad \text{and} \quad \frac{\partial}{\partial \xi_i}\phi(\xi) = (x(\xi) - x_0)_i.$$

This proves that *the critical points of the phase function in the integral expression for G are precisely those* ξ *for which* $x(\xi) = x_0$, i.e., for which the integral curve $\gamma_{j,\xi}$ joins x_0 to y.

If we apply stationary phase and use the fact that $a(x_0) = 1$ we get the following asymptotic formula for G:

$$G(y,t,x_0) \sim \sum_\alpha \frac{1}{(2\pi h)^{n/2}} \left| \frac{\partial y}{\partial \xi} \right|^{-1/2} (x_0, \xi_\alpha) \exp\left(\frac{i}{\hbar} \int_0^t L(x_\alpha, \dot{x}_\alpha, s)\, ds + \frac{\pi i}{4} \#_j \right)$$

where $x_\alpha(\tau)$, $0 \leqslant \tau \leqslant t$, is a classical trajectory going from x_0 to y. This is Maslov's asymptotic approximation to the fundamental solution of the Schrödinger equation.

Maslov gives an alternative "proof" of the formula using Feynman integrals: This starts with Feynman's representation of the fundamental solution of the Schrödinger operator:

$$G(y,t,x_0) = \int \exp\left(\frac{i}{\hbar} \int_0^t L(x, \dot{x}, \tau)\, d\tau \right) d\mu$$

where $x(\tau)$ is *path joining* x_0 *to* y and μ is "Feynman measure" on path space; see [9].

Let us apply stationary phase to the above (ignoring the fact that the integral is not over a finite dimensional region). The critical points of the phase function are just those paths for which the first variation $\delta \int L = 0$, which by the principal of least action are just the classical trajectories, that is, the $x_\alpha(\tau)$ above. The signature of $\delta^2 \int L$ at each of these trajectories is formally computed to be the value $\#_j$ along the trajectory. Therefore we obtain an asymptotic formula for the right hand side of the Feynman integral as

$$G(x,y,t) \sim \sum K_j \exp\left(\frac{i}{\hbar} \int_0^t L(x_\alpha, \dot{x}_\alpha, s)\, ds + i\frac{\pi}{4} \#_j \right).$$

Here K_α is the quotient of two infinite quantities, namely $(2\pi h)^{\infty/2}$, and $|\det \delta^2 L|$, but apparently these cancel each other out and give the finite answer computed above. It would be interesting to study this formula further.

References, Chapter II

1. R. K. Luneburg, *Mathematical theory of optics*, Univ. of California Press, Berkeley, Calif., 1964. MR **30** #2808.
2. P. Lax, Duke Math. J. **24** (1957), 627–646. MR **20** #4096.
3. D. Ludwig, Comm. Pure Appl. Math. **13** (1960), 473–508. MR **22** #5816.
4. L. H. Loomis and S. Sternberg, *Advanced calculus*, Addison-Wesley, Reading, Mass., 1968. MR **37** #2912.
5. V. Guillemin, D. Quillen and S. Sternberg, Comm. Pure Appl. Math. **23** (1970).
6. J. B. Keller, Ann. Physics **4** (1958), 180–188. MR **20** #5650.
7. V. P. Maslov, *Theory of perturbations and asymptotic methods*, Izdat. Moskov. Gos. Univ., Moscow, 1965; French transl., Dunod, Paris, 1972.
8. J. J. Duistermaat, Comm. Pure Appl. Math. **27** (1974), 207–281.
9. R. P. Feynman and A. R. Hibbs, *Quantum mechanics and path integrals*, McGraw-Hill, New York, 1965.

Chapter III. Geometrical Optics

In this chapter we illustrate some of the applications of the methods of the preceding chapter to optics. We shall present only a sketch of the subject, referring the reader to excellent existing texts such as Born and Wolf [1], Carathéodory [2], Luneburg [3], Sommerfeld [4], and Synge [5].

§1. The laws of refraction and reflection.

If we accept Maxwell's equations as describing the propagation of electromagnetic waves, then geometrical optics can be regarded as the study of the associated characteristic equation, the "light rays" being the corresponding bicharacteristic curves—that is the projection onto the base of the bicharacteristics. Thus geometrical optics deals with the "zeroth order approximation" in the asymptotic expansion, when we let the large parameter, τ, represent the frequency—or, as an alternative interpretation, as the equation describing the propagation of discontinuities of the electromagnetic field.

Let x, y, z denote rectilinear coordinates on $\mathbf{R}^3$. Then the characteristic equation for Maxwell's equations in an isotropic medium takes the form

$$\varphi_x^2 + \varphi_y^2 + \varphi_z^2 - \frac{\varepsilon\mu}{c^2}\varphi_t^2 = 0 \tag{1.1}$$

where c is the velocity of light (a constant, $c = 3 \cdot 10^{10}$ cm/sec approximately) and $\varepsilon = \varepsilon(x,y,z)$ is the dielectric constant and $\mu = \mu(x,y,z)$ is the magnetic permeability, both functions of the medium. See §6 for a brief discussion of Maxwell's equations and a derivation of (1.1). We seek a solution of the form

$$\varphi = \psi - ct$$

where $\psi = \psi(x,y,z)$ and obtain the equation

$$\psi_x^2 + \psi_y^2 + \psi_z^2 = n^2 \tag{1.2}$$

where $n = \sqrt{\varepsilon\mu}$ is called the refractive index. Equation (1.2) is called the *eikonal equation*.

Let p, q, and r denote the coordinates dual to x, y, z so that (x,y,z,p,q,r) are coordinates on $T^*\mathbf{R}^3$. We can rewrite (1.2) as

$$H(d\psi) = \tfrac{1}{2} \quad \text{where} \quad H = \frac{1}{2n^2}(p^2 + q^2 + r^2).$$

The bicharacteristic vector field ξ_H is then given by

$$\begin{aligned} \xi_H = {} & \frac{1}{n^2}\left(p\frac{\partial}{\partial x} + q\frac{\partial}{\partial y} + r\frac{\partial}{\partial z}\right) \\ & + \frac{(p^2 + q^2 + r^2)}{n^3}\left(\frac{\partial n}{\partial x}\frac{\partial}{\partial p} + \frac{\partial n}{\partial y}\frac{\partial}{\partial q} + \frac{\partial n}{\partial z}\frac{\partial}{\partial r}\right). \end{aligned} \tag{1.3}$$

Now H is the Hamiltonian corresponding to the Lagrangian

$$L = \tfrac{1}{2}n^2(\dot{x}^2 + \dot{y}^2 + \dot{z}^2),$$

and represents the kinetic energy associated to the Riemann metric

$$n\sqrt{dx^2 + dy^2 + dz^2} = nds.$$

Thus the differential equations given by the vector field (1.3) are exactly the differential equations for the geodesics of the corresponding Riemann metric, i.e., the Euler-Lagrange equations (in Hamiltoniam form; cf Loomis-Sternberg [6] p. 535). Now the geodesics are precisely the curves which extremize the arc length relative to this metric. This arc length is called the *optical length* of the path. Thus we obtain *Fermat's principle* that the light rays can be characterized as those curves which extremize the optical length, $\int n\,ds$.

If we consider the characteristic equation as describing the propagation of discontinuities, then the surfaces

$$\psi - ct = 0$$

are the surfaces of discontinuity. Relative to the *Euclidean* metric it is clear that the larger $|d\psi|^2 = (\psi_x^2 + \psi_y^2 + \psi_z^2)$ is, the closer the surfaces $\psi = \text{const}$ are together. Thus, in view of (1.2), in *regions of higher refractive index the discontinuities propagate more slowly*. The function ψ gives a section, $d\psi$, of the cotangent bundle. If we use the Euclidean metric we can identify covectors with vectors and consider $d\psi$ as a vector field. A particle moving according to this vector field will have velocity $|d\psi|$. With this interpretation, a particle will move *faster* in a region of higher refractive index.

Let us now derive the elementary laws of refraction and reflection. Suppose that the index of refraction takes on the values n_0 and n_1 in each of two regions, 0 and 1, except for a narrow band near the surface S bounding the two regions, where n changes smoothly from n_1 to n_0. In each of the regions we shall write

$$d\psi = nu$$

where u is the unit vector field in the direction of $d\psi$. Again we are using the Euclidean metric and identifying vector fields with covector fields.

A solution ψ of (1.2) corresponds to a Lagrangian manifold Λ lying in the characteristic variety, $H = \frac{1}{2}$. Any curve in Λ projects onto a curve in $\mathbf{R}^3$. If $\alpha = pdx + qdy + rdz$ is the fundamental one form on $T^*\mathbf{R}^3$ the integral of α around a path bounding any region in Λ is zero, by Stokes' theorem. Let us consider a solution ψ which extends across the boundary of the regions and integrate $d\psi$ around the closed path indicated in the figure, the major portion being curves parallel to the surface S , but lying in regions where n is constant.

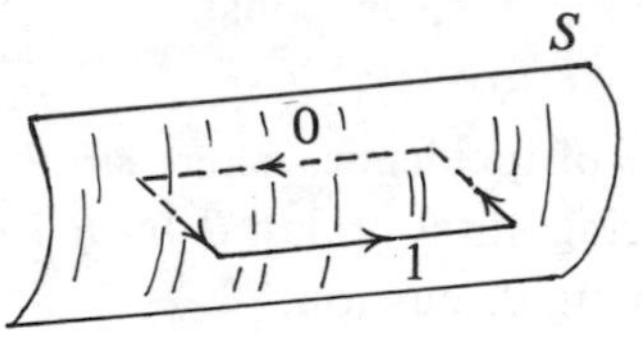

The values of $d\psi$ along the curves parallel to S are $n_1 u_1$ and $-n_0 u_0$ whereas the contribution from the short transversal curves is negligible. The total integral must vanish. We consider the limiting case where the n changes abruptly across S. We conclude that $n_1 u_1 - n_0 u_0$ has zero scalar product with any tangent vector to the surface. Thus $n_1 u_1 - n_0 u_0$ is normal to the surface. Let e be a unit normal vector to S. We obtain

$$e \wedge (n_1 u_1 - n_0 u_0) = 0.$$

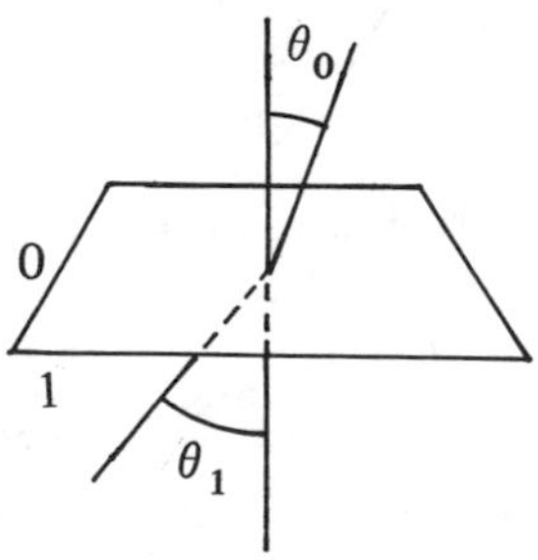

We conclude that e, u_1 and u_1 lie in the same plane. Let θ_0 and θ_1 be the angles between u_0 and u_0 with e. We conclude

Snell's law: *The incident and refracted ray lie in the same plane with the normal and*

$$n_1 \sin \theta_1 = n_0 \sin \theta_0 . \tag{1.4}$$

Actually, Snell (1591-1626) did not have at his disposal any independent measurement of n_0 or n_1 but rather was able to show by experiment that

$$\frac{\sin \theta_1}{\sin \theta_0} = \text{const.} \tag{1.5}$$

We refer the reader to Mach [7, Chapters III and V] for an extremely enlightening discussion of the early history of the laws of refraction and reflection. Snell's law was an improvement on the law proposed by Kepler who, in turn, by many ingenious experiments improved on the earliest known quantitative statement of a law of refraction by Ptolemy (2nd century C. E.) which was

$$\frac{\theta_1}{\theta_0} = \text{const.} \tag{1.6}$$

Notice that for small angles of incidence, where $\sin \theta \sim \theta$, (1.5) reduces to (1.6). This approximation, replacing $\sin \theta$ or $\tan \theta$ by θ for small angles, will occur again and again in the ensuing discussion.

For example, suppose that the region 1 is (in part) bounded by a portion of a

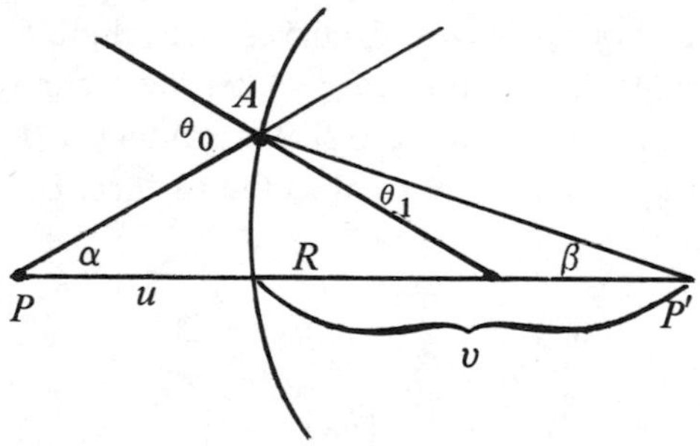

sphere of radius R. Then from the triangles we have

$$\frac{\sin(\pi - \theta_0)}{\sin \alpha} = \frac{R + u}{R} \quad \text{and} \quad \frac{\sin \theta_1}{\sin \beta} = \frac{v - R}{R}$$

while Snell's law yields

$$\frac{\sin \theta_1}{\sin \theta_0} = \frac{n_0}{n_1}.$$

Also, from the triangle PAP' we see that

$$\alpha + [(\pi - \theta_0) + \theta_1] + \beta = \pi \quad \text{or} \quad \beta = \theta_0 - \theta_1 - \alpha.$$

If α is small, then θ_0, θ_1 and β will also be small. If we replace the sine of each angle by the angle itself in each of the preceding equations we can eliminate the angles to obtain (after a little algebraic manipulation)

$$\frac{n_0}{u} + \frac{n_1}{v} = \frac{n_1 - n_0}{R}. \tag{1.7}$$

Notice that in this expression v does not depend on the angle α. In other words all rays of small angle leaving P converge to P'. In other words P' is the "image" of P. Of course, this is only true in the approximation described above. If we take $u = \infty$ then

$$v_\infty = \left(\frac{n_1}{n_1 - n_0}\right) R$$

is called the focal length of the system and represents the distance behind the surface S of the image of a very distant point. For example, in the eye the region behind the cornea is filled with a liquid called the aqueous humor and the region behind the lens is filled with a liquid called the vitreous humor. Both of these have an index of refraction about equal to that of water, namely 1.336. Thus, ignoring the lens, we would obtain $v_\infty = 3.98R$. The fact that the cornea has larger curvature (hence smaller R) than the rest of the eyeball allow the focal length to be about the diameter of the eye so that the image point lies on the retina. Thus most of the focusing of the eye is done without the help of the lens. The lens provides corrections and, by adjusting its radii of curvature, accomodation—that is, the focusing of points not at infinity on the retina. The laws for

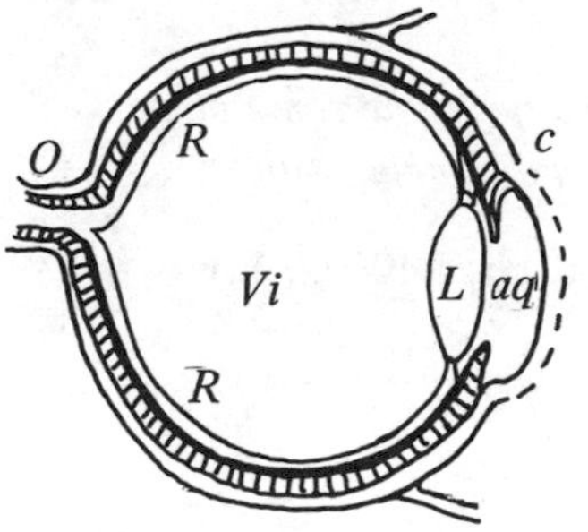

c cornea
aq aqueous humor
L lens
Vi Vitreous humor
R retina
O optic nerve

determining the image points for lenses made up of spherical surfaces can be obtained by repeated application of (1.7). We refer the reader to any elementary text. We shall also return to this point later on.

Now to reflection. Again suppose there is a small boundary layer around the surface S across which n changes for n_0 to n_1. This time, however let us suppose that there is a solution Lagrangian manifold which folds back on itself near S, (just as occured in some of the examples of the preceding chapter.) Thus we have a solution Λ and consider a closed curve on Λ whose main components consist

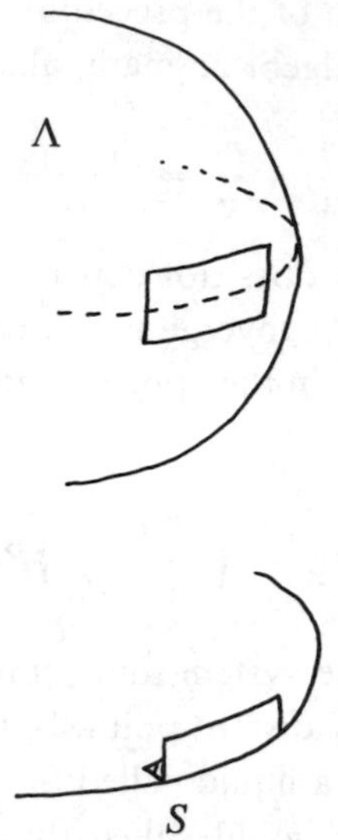

of arcs whose projection to $\mathbf{R}^3$ are curves parallel to S. Since Λ is Lagrangian, the integral of α around this curve is zero. Now Λ splits into two branches away from S, the one corresponding to the "incident rays" and given by $d\psi_i = n_0 u_i$ and the second given by $d\psi_r = n_0 u_r$. As before we conclude that

$$e \wedge (u_i - u_r) = 0.$$

Thus we conclude that

the incident and reflected rays lie in the same plane with the normal, and make equal angles with it. (1.8)

This law was known to Hero of Alexandria (circa 100 C. E.).

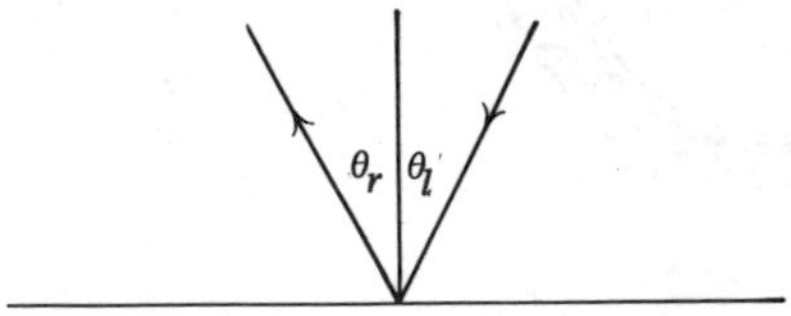

(Notice that in the limit, both for reflection and for refraction we obtain discontinuous Lagrangian manifolds).

One can apply the same arguments as above to the case of a spherical mirror to obtain

$$\frac{1}{u} - \frac{1}{v} = \frac{2}{R}. \tag{1.9}$$

Here, and in what follows, when we reduce to one dimensional results for spherical surfaces we adopt the convention that light will be drawn as incident from the left, use the same orientation in the object space and image space and assign a signature to a radius of curvature, positive or negative according to the relative location of the center. Thus, for example, if we take $u = \infty$ in (1.9) we get the two cases of a spherical mirror.

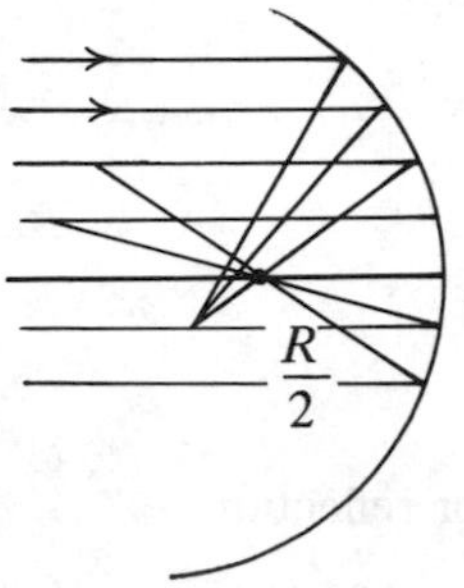

concave mirror

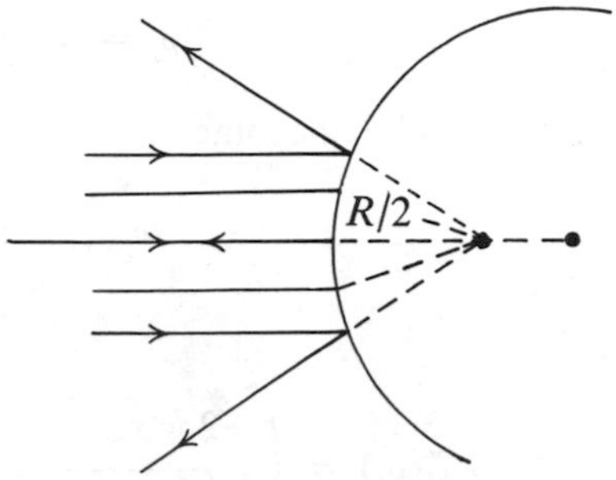

convex mirror

Notice that even in the case of reflection in a spherical mirror where $\theta_i = \theta_r$ the focusing is only approximately correct, because of the replacement of sin α by α, etc. in dealing with the equations coming from the triangles. A more accurate picture of what happens for reflection from a concave spherical mirror is drawn below. The envelope of the rays being the caustic.If we had a parabolic

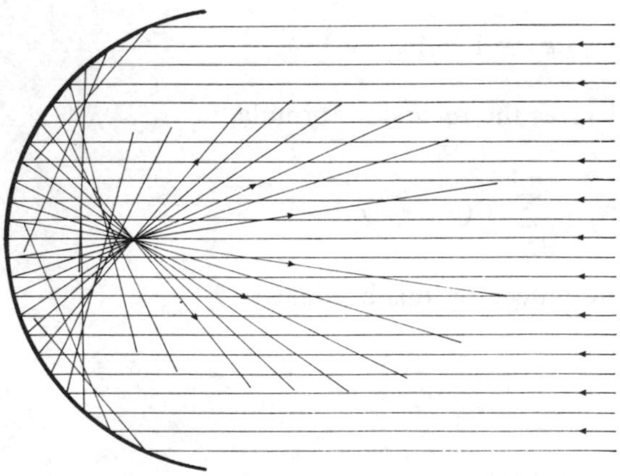
spherical mirror

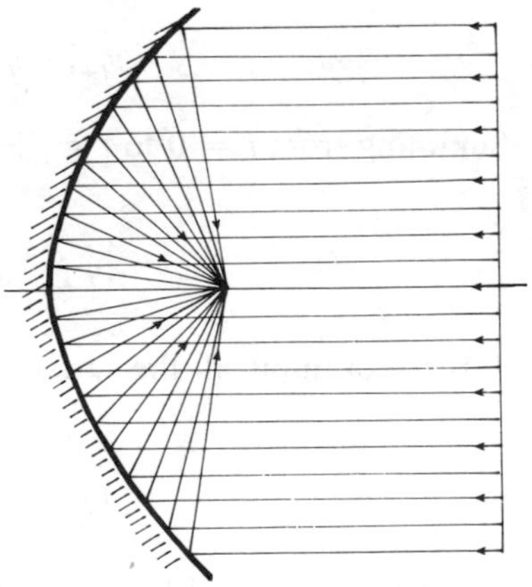
parabolic mirror

mirror, and the light was incident from infinity parallel to the axis, then we would obtain perfect focusing. The deviation of the spherical mirror from perfect focusing is called *spherical aberation.* We will discuss this and other aberations later on.

We close this section by describing Luneberg's algorithm for the tracing of rays through a system of plane reflecting or refracting surfaces. Let $v_i = d\psi(s_i) = n_i u_i$ be the ray vectors in the ith region. Let e_i be the unit normal pointing into the $(i + 1)$st region. Let θ_i be the ith angle of incidence and set

$$\rho_i = n_i \cos \theta_i = v \cdot e_i .$$

Then (1.4) and (1.7) become

$$v_{i+1} = v_i + \Gamma_i(\rho_i)e_i \tag{1.10}$$

where

$$\Gamma_i(\rho_i) = \begin{cases} -2\rho_i & \text{for reflection,} \\ \sqrt{n_{i+1}^2 - n_i^2 + \rho_i^2} - \rho_i & \text{for refraction.} \end{cases}$$

Thus the final (emerging) ray is given by

$$v_R = v_0 + \sum_{i=0}^{k-1} \Gamma_i(\rho_i)e_i . \tag{1.11}$$

The e_i are given, and we wish to find a simple recursive formula for the ρ_i. Take the scalar product of (1.10) with e_j. This gives

$$v_{i+1} \cdot e_j - v_i \cdot e_j = \Gamma_i(\rho_i)(e_i \cdot e_j).$$

Summing from $i = 0$ to $i = j - 1$ gives the recursion formula

$$\rho_j = v_0 \cdot e_j + \sum_{i=0}^{j-1} \Gamma_i(\rho_i)(e_i \cdot e_j).$$

If for example all the surfaces are reflecting, this becomes

$$\rho_j = (v_0 \cdot e_j) - 2 \sum_{i=1}^{j-1} (e_i \cdot e_j)\rho_i$$

which is a system of linear recursion formulae.

§2. Focusing and magnification.

Suppose we are given a curve, γ, on a Riemann manifold M. Then γ' is a curve on TM, assigning to each t the tangent vector γ'. Identifying TM with T^*M via the Riemann metric gives a curve, $\hat{\gamma}$, on T^*M. Now $\hat{\gamma} : I \to T^*M$ where I is an interval in **R**, and

$$\left\langle \frac{\partial}{\partial t}, \hat{\gamma}^* \alpha \right\rangle r = \left\langle d\hat{\gamma}\left(\frac{\partial}{\partial t}\right), \alpha \right\rangle = \left\langle d\pi \cdot d\hat{\gamma}\left(\frac{\partial}{\partial t}\right), \hat{\gamma}(t) \right\rangle$$
$$= \left\langle d\gamma\left(\frac{\partial}{\partial t}\right), \hat{\gamma}(t) \right\rangle = \left\langle \gamma'(t), \hat{\gamma}(t) \right\rangle = \|\gamma'(t)\|^2.$$

Suppose that γ is parametrized by arc length. Then $\|\gamma'(t)\| = 1$, and we conclude that

$$\text{arc length of } \gamma = \int_{\hat{\gamma}} \alpha. \tag{2.1}$$

If we start with a curve on T^*M, it does not necessarily arise from a curve on M. However half of the Hamiltonian differential equations (the equations $dx/dt = \partial H/\partial \xi$) imply that their solution curve arises from a curve in M. In particular, every trajectory of ξ_H arises from a curve on M. (Here H is the associated energy, $H(\xi) = \frac{1}{2}\|\xi\|^2$. In our optical application it is the H given in §1.) A curve γ on M is parametrized by arc length if and only if $\frac{1}{2}\|\hat{\gamma}(t)\|^2 = \frac{1}{2}$, i.e., if and only if $\hat{\gamma}$ lies in the characteristic variety $H = \frac{1}{2}$. Now a bicharacteristic is exactly a solution curve for ξ_H lying in $H = \frac{1}{2}$. We thus conclude:

Let $C : I \to T^*$ be a bicharacteristic and $\gamma = \pi \circ C$ the corresponding curve in M. Then (optical) length of $\gamma = \int_C \alpha$. (2.2)

Let us now consider the problem of focusing. Suppose we start with a point P and consider the sphere $\Lambda_{P,0}$ of all unit covectors over P. (This is the intersection of T_P^* with the characteristic variety.) The manifold $\Lambda_{P,0}$ is clearly isotropic (since $\alpha_{|\Lambda_{p,0}} = 0$) and transversal to ξ_H. It therefore sweeps out a Lagrangian manifold, Λ_P, consisting of all rays emanating from P. At $\pi^{-1}P$, the map π has the maximum possible singularity, of order $n - 1$ (i.e., of order 2 if we think of $\mathbf{R}^3$); when restricted to Λ_P, all the points of the sphere $\Lambda_{P,0}$ project onto P. We say that P has a perfect focus, P', if all the rays leaving P meet again at P'. In other words we say that P has a perfect focus at P' if

$$\Lambda_P \cap T_{P'}^* = \Lambda_{P',0}.$$

Since Λ_P is an immersed submanifold we should specify the above definition a little more precisely. We say that P has a *perfect focus* if there is a sphere $\Lambda' \subset \Lambda_P$ such that $\iota\Lambda' = \Lambda_{P',0}$ where $\iota : \Lambda_P \to T^*M$ is the immersion. We shall say that P has a *strong focus* at P if an open subset of rays from P meet at P', thus that there be a submanifold Λ' of dimension $n-1$ in Λ_P with $\iota\Lambda' \subset \Lambda_{P',0}$. We now come to an important observation:

PROPOSITION 2.1. *Let γ_t be a smooth one parameter family of rays joining P to P'. Then all the rays have the same optical length.*

Indeed we can write $\gamma_t = \pi C_t$ where $C_t : I_t \to \Lambda_P$ with $0 \leqslant t \leqslant a$, say, where

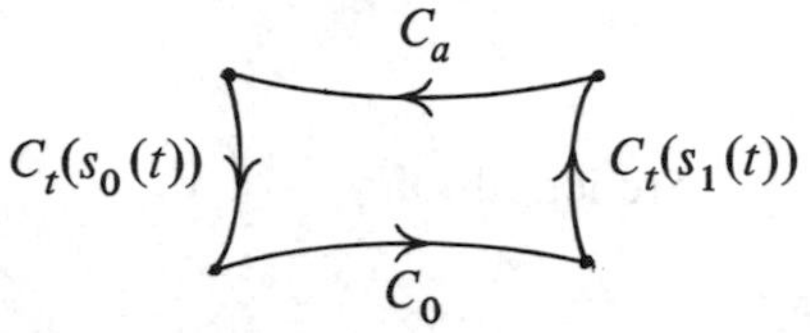

I_t is the interval $s_0(t) \leqslant s \leqslant s_1(t)$ and the arc length of γ_t is given by $\int_{C_t} \alpha$. We may apply Stokes' theorem to the closed curve made up of C_0, $C_t(s_1(t))$, C_a and $C_t(s_0(t))$ (these last two traversed in the reverse direction). Now $\pi \circ C_t(s_0(t)) = P$ and $\pi \circ C_t(s_1(t)) = P'$ so that α vanishes along the two "end" curves and we conclude, since $d\alpha = 0$ on Λ, that

$$\int_{C_0} \alpha = \int_{C_a} \alpha$$

and thus the arc length of γ_0 is the same as the arc length of γ_a. The same argument leads to the following formula: let $C_t : I_t \to \Lambda$ be a smooth family of curves on a Lagrangian manifold , Λ, where I_t is the interval $s_0(t) \leqslant s \leqslant s_1(t)$. Let

η_1 be the tangent to the curve $C_t(s_1(t))$
η_0 be the tangent to $C_t(s_0(t))$ and at $t = 0$.

Then

$$\frac{d}{dt}\int_{C_t} \alpha = \langle d\pi\eta_1, C_t(s_1(t))\rangle - \langle d\pi\eta_0, C_t(s_0(t))\rangle. \tag{2.3}$$

(This formula, as we shall see a little later on in this section, is the basis of the sine condition of Abbé. It was already proved, in a slightly different context, in §7 of Chapter II.)

In particular, in the case of perfect focusing, all the rays from P to P' (corresponding to the same Λ') have the same optical length. There are some interesting theoretical considerations concerning perfect focusing which will be of interest to us in later chapters. These considerations were first studied by Maxwell [1, pp. 143-150]. An *absolute instrument* is an optical system (i.e., for our purposes any Riemannian manifold) for which there are open sets U and U' such that each point $P \in U$ has a perfect focus, $P' \in U'$. We may assume also without loss of generality that the map from $U \to U'$ sending P to P' is continuous. (We let the reader provide the details or add this as an extra hypothesis.) Maxwell's theorem asserts:

THEOREM 2.1. *In an absolute instrument the map from U to U′ is an isometry, i.e., preserves optical length.*

(Thus, for example, if the refractive index of U' is the same as that of U no magnification is possible in an absolute instrument.)

PROOF. Notice that since every arc in U can be approximated by a geodesic polygon (i.e., by a curve made up of broken rays) it suffices to prove that a geodesic arc is carried into a geodesic arc of the same length. The proof that we present of this fact is due to Lenz [8]. It assumes that every point of U is perfectly focused (i.e., rays in all directions converge—a "360 degree lens"). The theorem is also true for the case of strong focusing and was proved in this more general case by Caratheodory [2]. We refer the reader to Caratheodory [2] for the proof in this more general case.

Let P and Q be two nearby points of U. Then (cf. Loomis-Sternberg [6, p.541]) there is a unique shortest geodesic joining P to Q. Extend this geodesic until it passes through P' and Q'. The geodesic from Q to P along the same arc but in the reverse sense must also pass, when extended, through Q' and P'. If P and Q

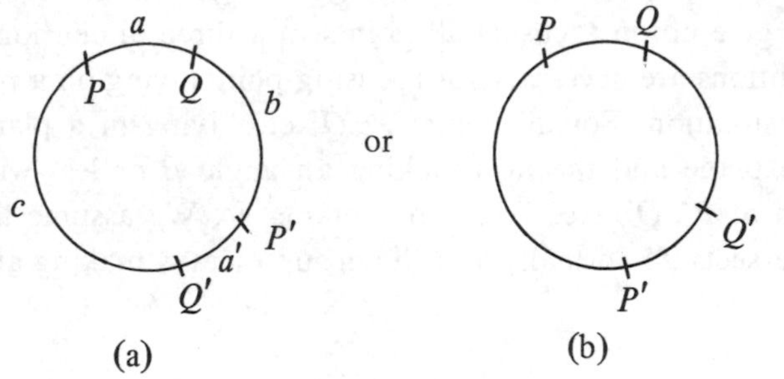

(a) (b)

are close enough together so that P' and Q' are joined by a unique geodesic, we conclude that the geodesic from Q to P must coincide with the geodesic from P to Q, i.e., that we have a closed geodesic. We thus have two configurations. We claim that the second alternative is impossible. Indeed by Proposition 2.1 the geodesic PP' has the same length as $PQQ'P'$. But, QQ' also has the same length as $QPP'Q$ and this is clearly impossible. In case (a) we have that the length of PQP' is the same as $PQ'P'$ or (with a, a', b and c denoting the lengths as indicated in the figure)

$$a + b = a' + c$$

and, similarly $QP'Q'$ has the same length as QPQ' so

$$a + c = a' + b.$$

These two equations yield

$$a = a',$$

which was what we wanted to prove.

Notice that this also proves that $a + b = b + a'$ or (if U is connected) *the length of the geodesic joining* P to P' does not depend on P.

There is one obvious example of an absolute instrument and that is the sphere. All geodesics leaving a point on the sphere converge at the antipodal point. Since stereographic projection maps the sphere (with one point removed) onto Euclidian space conformally, we can consider the (induced) spherical metric on $\mathbf{R}^3$. This system is called Maxwell's "fish eye"; cf. Born and Wolf [1, pp. 147-149]. It is an absolute instrument (if we do not worry about the rays going off to infinity).

It is a conjecture of Blashke that this is the only example of an absolute instrument with U being the whole manifold. More precisely, let us call a Riemannian manifold M a *wiedersehn* manifold (terminology transmitted to us by Professor S.S. Chern) if M is connected and all geodesics leaving any point, P, of M meet again at the first conjugate point, P', of P. Blashke's conjecture is that the only *wiedersehn* manifolds are isometric to spheres.

Blashke's conjecture has been proved by L. Green [9] for the case of two dimensions. It is unproved as of this writing for higher dimensions.

Let us then give up on focusing all points in a three dimensional region, and see what conditions are necessary for focusing points lying on a plane. Consider the following situation. For all points P, Q, etc., lying in a plane Σ, the rays normal to the plane and the rays making an angle β or less with the normal intersect again at P', Q', etc., lying on a plane, Σ'. We assume the normal ray through P intersects Σ' normally and the group of rays making an angle β with

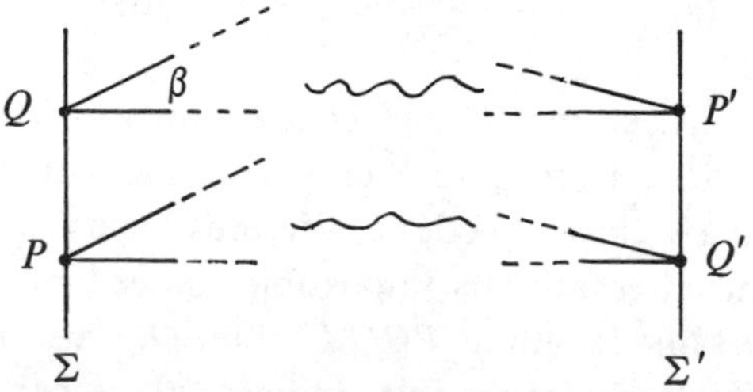

the normal to Σ make an angle β' with Σ'. For example, we might consider Σ as the objective plane and Σ' as the image plane of a microscope, with P centered in the objective.

For each unit vector u, let Λ_u denote the Lagrangian manifold consisting of those rays which leave Σ tangent to u. For u close enough to the normal these rays intersect Σ' in the direction u' (where u' may depend on the initial point as well as on u). If the refractive index at Σ is n then the covector of Λ_u sitting over P is nu, while the covector sitting over P' is $n'u'$ where n' is the refractive index at P'. Now let P_t be a curve of points on Σ with $P_0 = P$ and with tangent ξ at P and let P_t' be the corresponding family of curves on Σ' with tangent ξ' at P'. On Λ_u the rays γ_t joining P_t to P_t' correspond to curves C_t. The γ_t and C_t depend also on u, of course. Let L_t denote the optical length of γ_t. Then, by (2.1) and (2.3),

$$\frac{d}{dt}L_t = \langle \xi', n'u' \rangle - \langle \xi, nu \rangle.$$

On the other hand, by Proposition 2.1, L_t *does not depend on* u. If we take u to be normal both terms on the right vanish. Thus this gives the *Abbé sine condition*

$$n'\ell' \sin \beta' = n\ell \sin \beta, \tag{2.4}$$

where $\ell = \|\xi\|$ and $\ell' = \|\xi'\|$. This gives a formula for the magnification

$$m = \ell'/\ell.$$

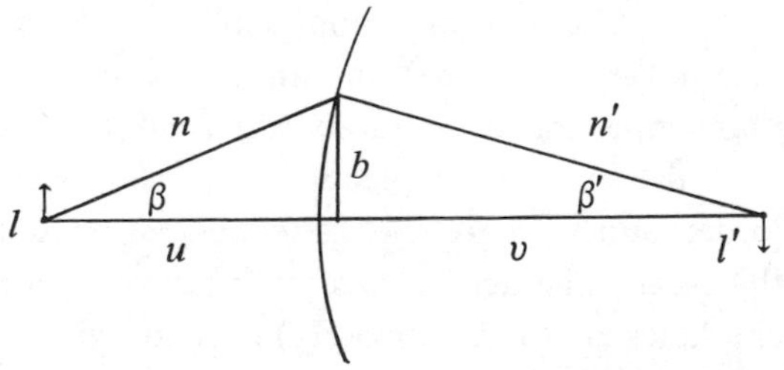

For example, consider the system given by two media with a spherical bounding surface as in the preceding section. For small angles we can replace sin by tan so that $\sin \beta \sim h/u$ and $\sin \beta' \sim h/v$. In this approximation, (2.4) becomes

$$m = \ell'/\ell = (v/u)(n/n'). \tag{2.5}$$

§3. Hamilton's method.

In this section we describe the method developed by Hamilton for dealing with problems in geometrical optics. It is interesting to observe that Hamilton developed his celebrated method precisely for the purpose of dealing with problems in optics. It was not until substantially later that he applied the same methods to mechanics, methods which were then extended by Jacobi and incorporated into the basic formulation of classical mechanics, finally leading to a key concept in quantum mechanics. It is also interesting to notice that Hamilton published his fundamental paper in 1828, a full ten years after Fresnel's conclusive experiments had shown that the wave theory of light was correct and thus invalidating geometrical optics as the precise physical theory. Our treatment will follow that given by Luneburg.

We will be primarily interested in the following situation. Light leaves a plane, passes through an optical system and again impinges on a plane. We will assume that the optical system lies in some cylinder about say, the z axis. We will assume that all the light rays of interest project diffeomorphically onto the z axis and thus can be described by functions $x(z)$ and $y(z)$. (We are thus excluding reflections from our system. It is not that difficult to deal with them but it complicates the presentation.) The initial and final plane will be given by values z_0 and z_1 (measured from an appropriate origin). The space of initial positions or directions is four dimensional (parametrized by $x_0, y_0, \dot{x}_0, \dot{y}_0$ where $\dot{x}_0 = (dx/dz)(z_0)$, etc.) as is the space of final positions and directions (parametrized by $x_1, y_1 \dot{x}_1, \dot{y}_1$). We obtain a transformation from the initial to the final conditions given by the light rays. *This transformation, F_{z_0,z_1} depends on z_0 and z_1.* We shall see that (after applying Legendre transformations, which amounts to the introduction of an appropriate symplectic structure) the transformations F_{z_0,z_1} is Hamiltonian, i.e., respects the symplectic structure. Thus, the theory of geometrical optics becomes the study of Hamiltonian (or "canonical") transformations. *First order optics* (wherein we make the approximations $\sin x \sim x$, etc.) consists in studying the corresponding tangential transformations (the Jacobian matrices) and hence becomes the study of the (linear) symplectic group. *Gaussian optics* deals with that case of first order optics where the system is assumed to be rotationally symmetric about the z-axis. The dependence on z_0 and z_1 is crucial for us, as we wish to know which planes are (approximately) in focus with each other, etc. We now proceed with the details.

Any curve γ given by $x(z)$, $y(z)$ has optical length

$$L(\gamma) = \int_{z_0}^{z_1} n\sqrt{1 + \dot{x}^2 + \dot{y}^2}\,dz = \int_{z_0}^{z_1} \mathfrak{L}dz$$

where

$$\mathfrak{L}(x, y, \dot{x}, \dot{y}, z) = n(x, y, z)\sqrt{1 + \dot{x}^2 + \dot{y}^2}.$$

We make the Legendre transformation

$$p = \partial\mathfrak{L}/\partial\dot{x} = \frac{\dot{x}n(x, y, z)}{\sqrt{1 + \dot{x}^2 + \dot{y}^2}},$$

$$q = \partial\mathfrak{L}/\partial\dot{y} = \frac{\dot{y}n(x, y, z)}{\sqrt{1 + \dot{x}^2 + \dot{y}^2}},$$

and use x, y, p, q as our coordinates. Notice that $np = a/c$ and $nq = b/c$ where a, b, and c are the direction cosines of the ray and so p and q have simple geometric interpretations. From the general theory we know that light rays are the solutions of the Hamiltonian equations

$$\frac{dx}{dz} = \frac{\partial H}{\partial p}, \qquad \frac{dy}{dz} = \frac{\partial H}{\partial q},$$
$$\frac{dp}{dz} = -\frac{\partial H}{\partial x}, \qquad \frac{dq}{dz} = -\frac{\partial H}{\partial y},$$

where

$$H = p\dot{x} + q\dot{y} - \mathfrak{L} = \frac{-n}{\sqrt{1 + \dot{x}^2 + \dot{y}^2}} = -\sqrt{n^2 - p^2 - q^2}$$

since

$$n^2 - p^2 - q^2 = \frac{n^2}{1 + \dot{x}^2 + \dot{y}^2}.$$

(The reader can verify directly that the differential equations listed above are indeed the Euler-Lagrange equations for optical length.) We also know from the general theory that the optical length is given by

$$\begin{aligned} L(\gamma) &= \int_C p\,dx + q\,dy - H\,dz \\ &= \int_C \alpha - H\,dz, \qquad \alpha = p\,dx + q\,dy, \end{aligned} \tag{3.1}$$

where $C(z) = (x(z), y(z), p(z), q(z))$ with

$$p(z) = \frac{\partial \mathfrak{L}}{\partial \dot{x}}(x(z), y(z), \dot{x}(z), \dot{y}(z)),$$

etc. Formula (3.1) can also be verified directly in the present case. We think of the space parametrized by x, y, p, q as $T^*\mathbf{R}^2$. We can then reformulate the above differential equations as saying that there is a (time dependent) Hamiltonian vector field ξ_H where

$$\xi_H \lrcorner\, \omega = -dH$$

where

$$\omega = dp \wedge dx + dq \wedge dy = d\alpha.$$

If we integrate this vector field from z_0 to z_1 we obtain a transformation, F_{z_0, z_1}, which we shall write as

$$(x_0 y_0 p_0 q_0 ; z_0) \to (x_1, y_1, p_1, q_1 ; z_1)$$

and conclude that

$$F^*_{z_0, z_1} \omega = \omega,$$

i.e., that F_{z_0, z_1} is Hamiltonian. We can also describe the situation as follows. Let us introduce the vector field

$$\boldsymbol{\eta} = \xi_H + \frac{\partial}{\partial z}$$

on $T^* \mathbf{R}^2 \times \mathbf{R}$. Then

$$\boldsymbol{\eta} \lrcorner (d\alpha - dH \wedge dz) = 0 \quad \text{and} \quad \langle \boldsymbol{\eta}, dz \rangle \equiv 1,$$

and it is easy to see that this characterizes $\boldsymbol{\eta}$ uniquely. The light rays are (the projections onto x, y, z space of) the solution curves of the vector field $\boldsymbol{\eta}$. Each curve intersects each of the hyperplanes $z = z_0$ and $z = z_1$ in a unique point and thus determines the transformation F_{z_0, z_1}.

Let γ_t be a one parameter family of light rays, so that $\gamma_t(s)$ goes from $x_0(t)$, $y_0(t)$, $z_0(t)$ to $x_1(t)$, $y_1(t)$, $z_1(t)$, and let C_t be the corresponding one parameter family of curves in $T^* \mathbf{R}^2 \times \mathbf{R}$. Let

$$\zeta_0 \text{ be the tangent to } C_t(\cdots, z_0(t))$$

and

$$\zeta_1 \text{ be the tangent to } C_t(\cdots, z_1(t)).$$

Then, applying Stokes' theorem as in the proof of (2.3), we obtain

$$\begin{aligned} \frac{d}{dt} L(\gamma_t) &= \frac{d}{dt} \int_{C_t} \alpha - H dz \\ &= \langle \zeta_1, \alpha - H dz \rangle - \langle \zeta_0, \alpha - H dz \rangle - \int \dot{C}_t \lrcorner (d\alpha - dnh \wedge dz) \\ &= \langle \zeta_1, \alpha - H dz \rangle - \langle \zeta_0, \alpha - H dz \rangle \end{aligned} \tag{3.2}$$

since $\dot{C}_t \lrcorner d(\alpha - H dz) = 0$ as C_t is an extremal. We can reformulate these last results as follows: Consider two copies of $T^* \mathbf{R}^2$ with forms ω_0 and ω_1. We are given a submanifold Λ of $T^* \mathbf{R}^2 \times T^* \mathbf{R} \times \mathbf{R} \times \mathbf{R}$ whose intersection, Λ_{z_0, z_1}, with $T^* \mathbf{R}^2 \times T^* \mathbf{R}^2 \times \{z_0\} \times \{z_1\}$ is the graph of F_{z_0, z_1}. On $T^* \mathbf{R}^2 \times T^* \mathbf{R}^2$ we put the symplectic structure given by the form

$$\omega_1 - \omega_0 = dp_1 \wedge dx_1 + dq_1 \wedge dy_1 - dp_0 \wedge dx_0 - dq_0 \wedge dy_0$$

so that Λ_{z_0,z_1} is a Lagrangian manifold. Each point of Λ corresponds to a unique light ray. Therefore the length of the ray becomes a function, L, on Λ. We can thus reformulate (3.2) as

$$\begin{aligned} dL &= \alpha_1 - Hdz_1 - \alpha_0 + Hdz_0 \\ &= p_1\,dx_1 + q_1\,dy_1 - H(x_1,y_1,p_1,q_1;z_1)dz_1 - p_0\,dx_0 - q_0\,dy_0 \qquad (3.3)\\ &\quad + H(x_0,y_0,p_0,q_0;z_0)dz_0\,. \end{aligned}$$

This equation is the fundamental equation of Hamiltonian optics. It is applied as follows: We have a projection $\pi : T^*\mathbf{R}^2 \to \mathbf{R}^2$ which induces a projection

$$\pi_0 \times \pi_1 : T^*\mathbf{R}^2 \times \mathbf{R} \times T^*\mathbf{R}^2 \times \mathbf{R} \to \mathbf{R}^2 \times \mathbf{R} \times \mathbf{R}^2 \times \mathbf{R},$$

where we have written $\pi_0 \times \pi_1$ instead of $\pi_0 \times \mathrm{id} \times \pi_1 \times \mathrm{id}$.

Suppose that $\pi_0 \times \pi_1$ when restricted to an open neighborhood in Λ is a diffeomorphism. (Thus we assume that there is a unique ray joining (x_0,y_0,z_0) to (x_1,y_1,z_1).) Then we can define

$$V = L \circ (\pi_0 \times \pi_1)^{-1}_{|\Lambda}\,. \qquad (3.4)$$

Then it follows from (3.3) that

$$\begin{aligned} \frac{\partial V}{\partial x_0} &= -p_0, & \frac{\partial V}{\partial y_0} &= -q_0, \\ \frac{\partial V}{\partial x_1} &= p_1, & \frac{\partial V}{\partial y_1} &= q_1\,. \end{aligned} \qquad (3.5)$$

If we know $V = V(x_0,y_0,z_0,x_1y_1,z_1)$ then we can determine the directions of the light rays at (x_0,y_0,z_0) and (x_1,y_1,z_1). Hamilton called the function V the *characteristic function*, or the *point characteristic* of the system.

We can identify $T^*(\mathbf{R}^2)$ with $\mathbf{R}^2 + \mathbf{R}^{2*}$ and thus have a projection $\rho : T^*(\mathbf{R}^2) \to \mathbf{R}^{2*}$, $\rho(x,y,p,q) = (p,q)$. Suppose that $\rho_1 \times \rho_2 : T^*\mathbf{R}^2 \times \mathbf{R} \times T^*\mathbf{R}^2 \times \mathbf{R}$ maps (an open subset of) Λ diffeomorphically onto a region of $\mathbf{R}^{2*} \times \mathbf{R} \times \mathbf{R}^{2*} \times \mathbf{R}$. (Thus we assume that there is a unique ray meeting the z_0 and z_1 planes at given angles.) Define the function T by

$$T = (L - (p_1x_1 + q_1y_1) + (p_0x_0 + q_0y_0)) \circ (\rho_0 \times \rho_1)^{-1}_{|\Lambda}\,. \qquad (3.6)$$

Then

$$T = T(p_0,q_0,z_0;p_1,q_1,z_1)$$

and it follows from (3.3) that

$$\frac{\partial T}{\partial p_0} = x_0, \qquad \frac{\partial T}{\partial p_1} = -x_1,$$
$$\frac{\partial T}{\partial q_0} = y_0, \qquad \frac{\partial T}{\partial q_1} = -y_1. \tag{3.7}$$

The function T is called the *angle characteristic*. It allows us to determine the initial or final positions from a knowledge of the initial or final directions of the rays, once we have the explicit form of T.

Finally, if $\pi_0 \times \rho_1$ restricts to a diffeomorphism on Λ we can define

$$W = (L - (p_1 x_1 + q_1 y_1)) \circ (\pi_0 \times \rho_1)_{|\Lambda}^{-1} \tag{3.8}$$

where

$$W = W(x_0, y_0, z_0; p_1, q_1, z_1)$$

and

$$\frac{\partial W}{\partial x_0} = -p_0, \qquad \frac{\partial W}{\partial y_0} = -q_0,$$
$$\frac{\partial W}{\partial p_1} = -x_1, \qquad \frac{\partial W}{\partial q_1} = -y_1. \tag{3.9}$$

The function W is called the *mixed characteristic*. If we know W then we can determine the final position and the initial angle from a knowledge of the initial position and final angle.

Suppose that $(x_0, y_0, p_0, q_0, z_0; x_1, y_1, p_1, q_1, z_1)$ and $(x_1, y_1, p_1, q_1, z_1; x_2, y_2, p_2, q_2, z_2)$ are in Λ. Then clearly so is $(x_0, y_0, p_0, q_0, z_0; x_2, y_2, p_2, q_2, z_2)$ and it follows directly from the definition that

$$L(x_0, y_0, p_0, q_0, z_0; x_2, y_2, p_2, q_2, z_2) = L(x_0, y_0, p_0, q_0, z_0; x_1, y_1, p_1, q_1, z_1) + L(x_1, y_1, p_1, q_1, z_1; x_2, y_2, p_2, q_2, z_2).$$

Therefore it follows that

$$V(x_0, y_0, z_0; x_2, y_2, z_2) = V(x_0, y_0, z_0; x_1, y_1, z_1) + V(x_1, y_1, z_1; x_2, y_2, z_2) \tag{3.10}$$

and

$$T(p_0, q_0, z_0; p_2, q_2, z_2) = T(p_0, q_0, z_0; p_1, q_1, z_1) + T(p_1, q_1, z_1; p_2, q_2, z_2), \tag{3.11}$$

whenever both sides of this equation make sense.

We will frequently apply these functions in a situation where z_0 and z_1 both lie in regions where n is constant. Now, from (3.3),

$$\frac{\partial L}{\partial z_0} = H = -\sqrt{n^2 - p_0^2 - q_0^2}$$

and

$$\frac{\partial L}{\partial z_1} = -H = \sqrt{n^2 - p_1^2 - q_1^2}$$

and thus *if n is independent of z for z near z_0 and z_1 then*

$$G = G_0 + z_0\sqrt{n^2 - p_0^2 - q_0^2} - z_1\sqrt{n^2 - p_1^2 - q_1^2} \tag{3.12}$$

for $G = V$, W or T. (In the case of V or *W we substitute the p, q from* (3.5) *or* (3.7).) *Here G_0 is independent of z_0 and z_1.*

§4. First order optics.

Let us apply the methods of the preceding section to the study of first order optics. We will derive the basic results under the assumption that we can apply the angle characteristic, i.e., that $\rho_0 \times \rho_1$ is a diffeomorphism on Λ. (This will exclude a system in which a bundle of parallel rays enters and leaves the system, for example a telescope.) The general case can then be obtained by a limiting argument.

We assume that the ray $x(z) \equiv 0 \equiv y(z)$ is in our system so that $(0,0,0,0,z_0;\ 0,0,0,0,z_1) \in \Lambda$ for all z_0 and z_1. Thus by (3.7) we know that the first derivatives of T vanish at $(0,0,z_0;0,0,z_1)$. We assume that n is constant in the regions of interest. Then by (3.11) we know that

$$T = C + T_0(p_0,q_0;q_1,p_1) + z_1\sqrt{n_1 - p_1^2 - q_1^2} - z_0\sqrt{n_0^2 - p_0^2 - q_0^2}$$

where C is a constant and T_0 starts with quadratic terms. If we are interested in the linear approximation to the maps F_{z_0,z_1} determined by T we must only retain those terms which are quadratic in the p's and q's. This is the approximation of first order optics.

The terms involving z will come from the approximation

$$z\sqrt{n^2 - p^2 - q^2} = nz - \frac{1}{2}\frac{z}{n}(p^2 + q^2) + \cdots.$$

Thus, dropping the irrelevant constant, we get, for our "linearized" T_0,

$$T_0 = Q_A(p_0,q_0) + Q_C(p_1,q_1) + B_F(q_0,p_0;p_1,q_1)$$

where Q_A, Q_C are quadratic forms while B_F is bilinear:

$$Q_A(p_0, q_0) = \frac{1}{2}(A_{11}p_0^2 + 2A_{12}p_0q_0 + A_{22}q_0^2)$$
$$= \frac{1}{2}(p_0, q_0)\begin{bmatrix} A_{11} & A_{12} \\ A_{21} & A_{22} \end{bmatrix}\begin{pmatrix} p_0 \\ q_0 \end{pmatrix},$$
$$Q_C(p_1, q_1) = \frac{1}{2}(p_1, q_1)\begin{bmatrix} C_{11} & C_{12} \\ C_{21} & C_{22} \end{bmatrix}\begin{pmatrix} p_1 \\ q_1 \end{pmatrix}$$

and

$$B_F(p_0, q_0, p_1, q_1) = \frac{1}{2}(p_0, q_0)\begin{bmatrix} F_{11} & F_{12} \\ F_{21} & F_{22} \end{bmatrix}\begin{pmatrix} p_1 \\ q_1 \end{pmatrix}.$$

Then (3.7) becomes

$$\begin{pmatrix} x_0 \\ y_0 \end{pmatrix} = \left[\begin{bmatrix} A_{11} & A_{12} \\ A_{21} & A_{22} \end{bmatrix} + \left(\frac{z_0}{n_0}\right)\begin{bmatrix} 1 & 0 \\ 0 & 1 \end{bmatrix}\right]\begin{pmatrix} p_0 \\ q_0 \end{pmatrix} - \begin{bmatrix} F_{11} & F_{12} \\ F_{21} & F_{22} \end{bmatrix}\begin{pmatrix} p_1 \\ q_1 \end{pmatrix},$$
$$-\begin{pmatrix} x_1 \\ y_1 \end{pmatrix} = -\begin{bmatrix} F_{11} & F_{21} \\ F_{12} & F_{22} \end{bmatrix}\begin{pmatrix} p_0 \\ q_0 \end{pmatrix} + \left[\begin{bmatrix} C_{11} & C_{12} \\ C_{21} & C_{22} \end{bmatrix} - \left(\frac{z_1}{n_1}\right)\begin{bmatrix} 1 & 0 \\ 0 & 1 \end{bmatrix}\right]\begin{pmatrix} p_1 \\ q_1 \end{pmatrix}.$$

There are the equations of first order optics. In *Gaussian optics* we assume that the system is invariant under rotation about the z-axis. Any quadratic form that is invariant under rotation is some multiple of the length squared, i.e., $Q_A(p_0, q_0) = \frac{1}{2}a(p_0^2 + q_0^2)$ and similarly for Q_C. Also, any invariant bilinear form is some multiple of the scalar product so $B_F(p_0, q_0, p_1, q_1) = f(p_0p_1 + q_0q_1)$. Thus, in the above equation all the matrices become scalar multiples of the identity. The equations for x and y decouple and are the same, so we need only treat the equations in x, say, which become

$$x_0 = \left(a + \frac{z_0}{n_0}\right)p_0 - fp_1,$$
$$-x_1 = -fp_0 + \left(c - \frac{z_1}{n_1}\right)p_1 \tag{4.1}$$

Solving for x_1 and p_1 in terms of x_0 and p_0 gives

$$x_1 = \frac{1}{f}\left(\left(c - \frac{z_1}{n_1}\right)x_0 - \left[\left(a + \frac{z_0}{n_0}\right)\left(c - \frac{z_1}{n_1}\right) - f^2\right]p_0\right),$$
$$p_1 = -\frac{1}{f}\left(x_0 - \left(a + \frac{z_0}{n_0}\right)p_0\right). \tag{4.2}$$

The condition that the planes z_0 and z_1 be in focus is that all rays leaving x_0 end at x_1, i.e., that x_1 not depend on p_0. Setting the coefficient of p_0 equal to zero in the above equation for x_1 yields the condition

$$\left(a + \frac{z_0}{n_0}\right)\left(c - \frac{z_1}{n_1}\right) = f^2. \tag{4.3}$$

When (4.3) is satisfied, (4.2) becomes

$$x_1 = \frac{c - \dfrac{z_1}{n_1}}{f} x_0 \tag{4.4}$$

where $m = (c - z_1/n_1)/f$ is the magnification of our system. Notice that we are free to adjust our choice of origin for z_0 and for z_1 (which would entail corresponding changes in a and c). Suppose we have chosen our origins such that $z_0 = 0$ and $z_1 = 0$ are conjugate. (This fixes the origin of z_1 relative to the origin of z_0; we still have freedom in the choice of the origin of z_0.) This means that (4.3) holds with $z_0 = 0 = z_1$ so that

$$ac = f^2. \tag{4.3$'$}$$

Let

$$m_0 = \frac{c}{f} = \frac{f}{a}$$

denote the magnification for this pair of planes. Then for any other pair of conjugate planes (i.e., those satisfying (4.3)) we can rewrite (4.3) as

$$\left(\frac{f}{m_0} + \frac{z_0}{n_0}\right)\left(fm_0 - \frac{z_1}{n_1}\right) = f^2 \tag{4.5}$$

or

$$m_0 \frac{n_1}{z_1} - \frac{1}{m_0}\frac{n_0}{z_0} = \frac{1}{f} \tag{4.6}$$

while (4.4) becomes

$$x_1 = mx_0$$

where

$$m = m_0 - \frac{1}{f}\frac{z_1}{n_1} = \frac{1}{\dfrac{1}{m_0} + \dfrac{1}{f}\dfrac{z_0}{n_0}}$$

giving

$$mm_0 = \frac{z_1/n_1}{z_0/n_0}. \tag{4.7}$$

The *unit planes* are those conjugate planes for which the magnification is one. For these (4.6) and (4.7) simplify to

$$\frac{n_1}{z_1} - \frac{n_0}{z_0} = \frac{1}{f} \tag{4.8}$$

and

$$m = \frac{z_1}{z_0}\frac{n_0}{n_1}. \tag{4.9}$$

Recall that we have already obtained these equations for a single refracting spherical surface (1.4) and (2.5) by more elementary means. We now see that these equations are consequences of the rotational symmetry and the first order approximation. Of course for the case of the spherical refracting surface we also obtained one more piece of information, namely the formula for f:

$$f = R/(n_1 - n_0). \tag{4.10}$$

The focal points of the system are obtained as $z_0 = F_0$ when $z_1 = \infty$ and $z_1 = F_1$ when $z_0 = \infty$ or

$$F_0 = -n_0 f, \qquad F_1 = n_1 f,$$

if we measure from the unit plane. We can then rewrite (4.5) (with $m_0 = 1$) as

$$(z_0 - F_0)(z_1 - F_1) = -n_0 n_1 f^2,$$

or setting

$$Z_0 = z_0 - F_0, \qquad Z_1 = z_1 - F_1.$$

This takes the form given by Newton:

$$Z_0 Z_1 = -n_0 n_1 f^2. \tag{4.11}$$

From the expression $x_1 = mx_0$ and

$$m = \frac{n_0 f}{n_0 f + z} = \frac{n_0 f}{Z_0}$$

we see that

$$x_1 = \frac{n_0 f}{Z_0} \quad \text{and} \quad Z_1 = \frac{-n_0 n f^2}{Z_0}.$$

Thus the transformation from x_0, y_0, z_0 space to x_1, y_1, z_1 space is a *projective* transformation.

Getting to equations (4.1) we see that for a pair of conjugate planes they can be written as

$$x_0 = \frac{f}{m}p_0 - fp_1 \qquad -x_1 = -fp_0 + mfp_1. \tag{4.12}$$

In particular:

$$\text{If } p_0 = 0 \text{ then } -x_0/p_1 = f. \tag{4.13}$$

$$\text{If } x_0 = 0 \text{ then } m = p_0/p_1\,. \tag{4.14}$$

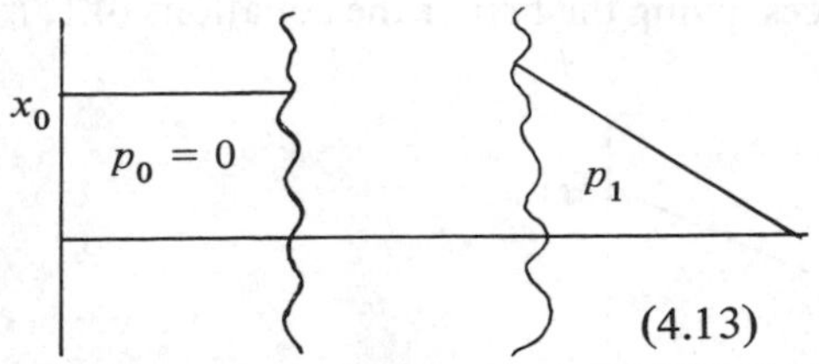

(4.13)

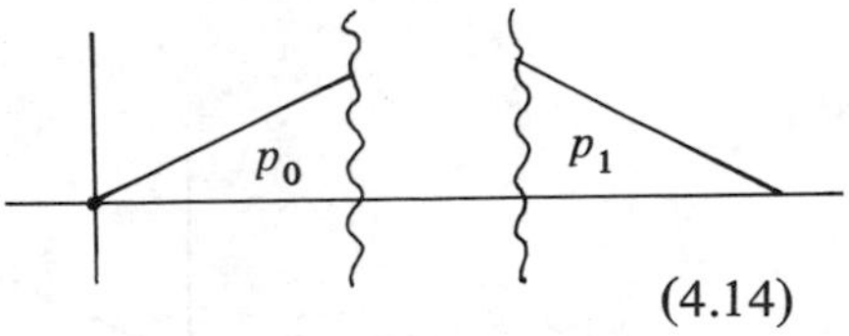

(4.14)

In principle, we have completely solved the problem of first order optics. If we are given, for example, a sequence of centered spherical refracting surfaces then a repeated application of (4.8), (4.9) and (4.10) will yield the relevant data (f, m and the location of the unit planes) for the system as a whole. It is amusing to observe (following Luneburg) that the procedure can be cast as the problem of solving a system of linear difference equations whose form is similar to Hamiltonian differential equations. We introduce some slightly different notation from the one we used before. Let the k refracting surfaces intersect the z-axis at the position $z_2, z_4, \ldots, z_{2k}$ and have radii of curvature $R_2, R_4, \ldots, R_{2k}$. Let $t_3, t_5, \ldots, t_{2k-1}$ be the distances between these surfaces and $n_1, n_3, \ldots, n_{2k+l}$ the indices of refraction of the regions separated by the surfaces. Let $z = z_0$ be the initial plane and $z_{2k+2} = z'$ be the final plane. A ray passing through the system will take on the values $x = x_0, x_2, \ldots, x_{2k}, x_{2k+2} = x'$ at these planes and its direction in the region between the planes is determined by the values $p_1, p_3, \ldots, p_{2k+1}$. We let t_1 be the (variable) distance between z_0 and z_1 and

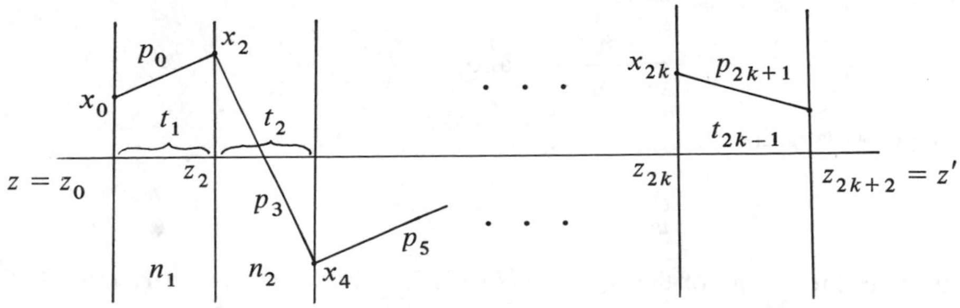

between z_{2k} and z_{2k+2}. We claim that the following difference equations hold:

$$\begin{aligned} x_{i+1} - x_{i-1} &= \delta_i p_i \qquad \text{where } \delta_i = \frac{t_i}{n_i},\ i = 1, 3, 5, \ldots, \\ p_{i+1} - p_{i-1} &= D_i x_i \qquad \text{where } D_i = \frac{n_{i+1} - n_{i-1}}{R_i},\ i = 2, 4, 6, \ldots, 2k. \end{aligned} \tag{4.15}$$

Indeed

$$\frac{x_{i+1} - x_{i-1}}{t_i} = \frac{\sin \theta_i}{\cos \theta_i} = \frac{a_i}{c_i} = n_i p_i$$

where a and c are the direction covariances giving the first of the equations (4.15).

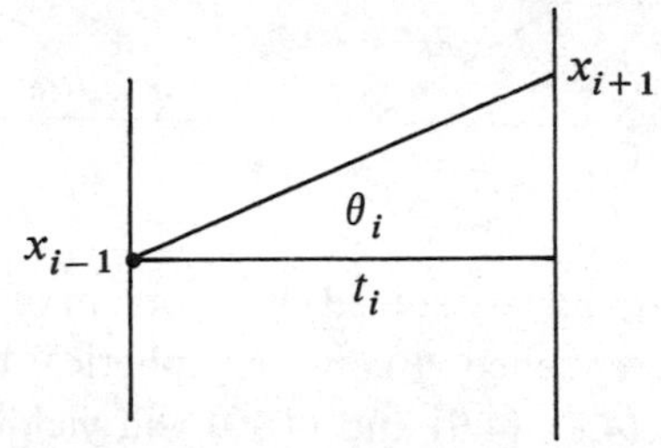

To obtain the second we apply (1.4) to the ith surface: Let u_i and v_i be the distances from the ith plane of the points of intersection of the $(i-1)$st and $(i+1)$st ray with the z axis.

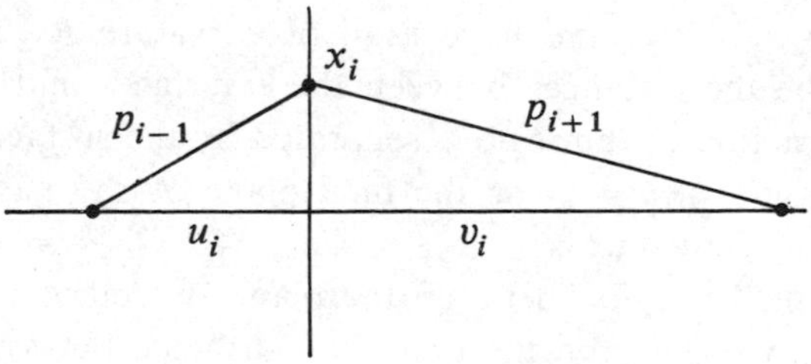

Then $x_i/u_i = -p_{i-1}/n_{i-1}$ so that

$$\frac{n_{i-1}}{u_i} = \frac{-p_{i-1}}{x_i} \quad \text{and} \quad \frac{n_{i+1}}{v_i} = \frac{-p_{i+1}}{x_i}$$

while (1.4) says that

$$\frac{n_{i+1}}{u_i} - \frac{n_{i-1}}{v_i} = \frac{n_{i+1} - n_{i-1}}{R_i} = D_i.$$

This gives the second of the equations (4.15). We let the reader verify that (4.15) constitute the "Euler equations" of finding an extremum for the "Lagrangian"

$$I = \frac{1}{2} \sum_{i=1,3,\ldots} \frac{1}{\delta_i}(x_{i+1} - x_i)^2 - \frac{1}{2} \sum D_i x_i^2$$

or, after making the "Legendre transformation",

$$p_i = \frac{x_{i+1} - x_{i-1}}{\delta_i},$$

of extremizing the analogue of $\alpha - Hdt$ which is

$$\Sigma\,(x_{i+1} - x_{i-1})p_i - \frac{1}{2}\Sigma\,\delta_i p_i^2 - \frac{1}{2}\Sigma\,D_i x_i^2 .$$

Equations (4.15) have two linearly independent solutions which can be specified by given the initial conditions. The *axial ray*, x^A, p^A is specified by the initial conditions

$$x_0^A = 0, \quad p_1^A = 1$$

while the *field ray* x^F, p^F is specified by the initial conditions

$$x_0^F = 1, \quad p_1^F = 0.$$

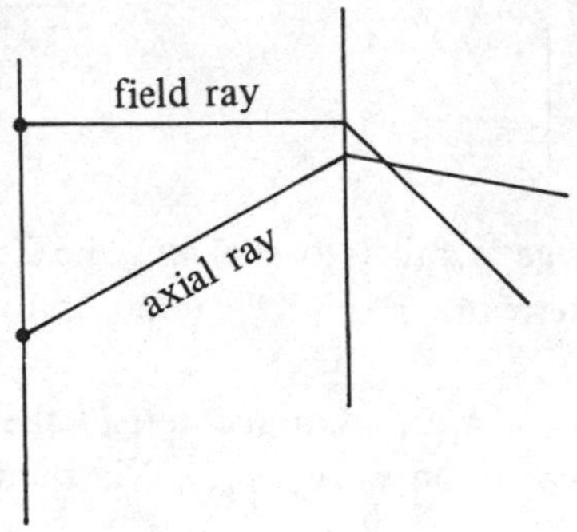

By definition, the planes z_0 and z_{2k+2} are conjugate if $x_{2k+2}^A = 0$. It then follows from (4.13) and (4.14) (or directly from (4.15)) that the magnification and the focal length of the system are given by

$$m = x_{2k+2}^F = 1/p_{2k-1}^A \quad \text{and} \quad f = -1/p_{2k-1}^F .$$

Starting with any solution x, p of (4.15) we can determine δ_i and D_i. If we also assign n_i then this determines t_i and R_i. Thus the knowledge of a ray and the n_i specifies the optical system. The ray values x_i, p_{i+1} together with the n_i are frequently useful parameters in actual optical design.

§5. The Seidel aberrations.

In this section we use Hamilton's method to describe the next term in the expansion of F_{z_0,z_1} for a system of rotational symmetry. We will assume that the planes z_0 and z_1 are conjugate with respect to the linear approximation. The deviations

$$\Delta x_1 = x_1 - mx_0,$$
$$\Delta y_1 = y_1 - my_0$$

are called the *aberrations* of the system. To study these aberrations it is convenient to use the mixed characteristic, W. Thus, the equations (4.9),

$$x_1 = -\frac{\partial W}{\partial p_1}, \qquad y_1 = -\frac{\partial W}{\partial q_1},$$

express x_1, y_1 in terms of x_0, y_0, p_1 and q_1, i.e., the final positions in terms of the initial position and the final directions. A geometric interpretation of this procedure is the following: Imagine an optical instrument with a diaphragm placed at the first focal point, F_0, of the system. Any ray which passes through

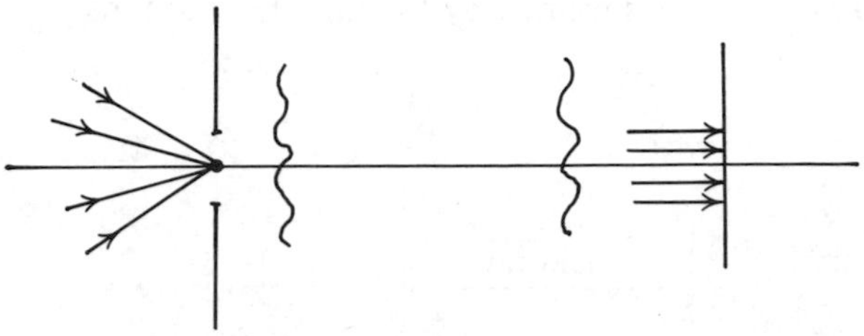

the focal point will emerge parallel to the axis, i.e., with $p_1 = q_1 = 0$. Thus expanding x_1 and y_1 in terms of p_1 and q_1 amounts to observing the effect of opening the diaphragm.

Now $W = W(x_0, y_0, z_0; p_1, q_1, z_1)$. On account of the rotational symmetry of the system W can only depend on x_0, y_0; p_1, q_1 via the functions

$$u = x_0^2 + y_0^2, \quad v = p_1^2 + q_1^2 \quad \text{and} \quad w = 2(x_0 p_1 + y_0 q_1),$$

i.e.,

$$W = W(u, v, w; z_0, z_1).$$

From this we see that there are no third order terms in the expansion of W about $x_0 = 0$, $y_0 = 0$, $p_1 = 0$, $q_1 = 0$ and thus the aberrations Δx_1 and Δy_1 start with third order terms, coming from the quadratic terms in W when considered as a function of u, v, and w. These third order terms are known as the Seidel aberrations, and it is our purpose in this section to describe them. We thus assume that we can ignore the higher terms in W so

$$W = W_0 + W_1 + W_2,$$

$$W_2 = W_2(u, v, w);$$

thus (2.7) gives

$$\begin{aligned} \Delta x_1 &= -2\left(\frac{\partial W_2}{\partial w} x_0 + \frac{\partial W_2}{\partial v} p_1\right), & \Delta y_1 &= -2\left(\frac{\partial W_2}{\partial w} y_0 + \frac{\partial W_2}{\partial v} q_1\right), \\ \Delta p_0 &= -2\left(\frac{\partial W_2}{\partial u} y_0 + \frac{\partial W_2}{\partial w} p_1\right), & \Delta q_0 &= -2\left(\frac{\partial W_2}{\partial u} y_0 + \frac{\partial W_2}{\partial w} q_1\right). \end{aligned} \tag{5.1}$$

The most general quadratic form in three variables depends on six parameters. With a view to later interpretation let us set

$$W_2 = -\left[\frac{1}{4}Fu^2 + \frac{1}{4}Av^2 + \frac{(C-D)}{8}w^2 + \frac{1}{2}Duv + \frac{1}{2}Euw + \frac{1}{6}Bvw\right].$$

Then (5.1) becomes

$$\begin{aligned}\Delta x_1 &= \left[Eu + \tfrac{1}{3}Bv + \tfrac{1}{2}(C-D)w\right]x_0 + \left[Av + Du + \tfrac{1}{3}Bw\right]p_1,\\ \Delta y_1 &= \left[Eu + \tfrac{1}{3}Bv + \tfrac{1}{2}(C-D)w\right]y_0 + \left[Av + Du + \tfrac{1}{3}Bw\right]q_1.\end{aligned} \tag{5.2}$$

To investigate the meaning of these equations let us consider incoming rays with

$$y_0 = 0$$

and introduce polar coordinates

$$\begin{aligned} p_1 &= \rho\cos\varphi,\\ q_1 &= \rho\sin\varphi,\end{aligned}$$

so that

$$u = x_0^2, \quad v = \rho^2 \quad \text{and} \quad w = 2x_0\rho\cos\varphi$$

and (5.2) becomes

$$\begin{aligned}\Delta x_1 &= A\rho^3\cos\varphi + \tfrac{1}{3}B\rho^2(3\cos^2\varphi + \sin^2\varphi)x_0 + C(\rho\cos\varphi_x)_0^2 + Ex_0^3,\\ \Delta y_1 &= A\rho^3\sin\varphi + \tfrac{2}{3}B\rho^2(\sin\varphi\cos\varphi)x_0 + D(\rho\sin\varphi)x_0^2,\end{aligned} \tag{5.3}$$

so that A, B, C and D, and E are the coefficients of terms involving increasing powers of x_0. For $x_0 = 0$ only the first term appears. It represents the fact that the point is not perfectly focused, but is imaged on a disk of radius $A\rho^3$, where $\rho^2 = p_1^2 + q_1^2$ is determined, as we remarked above, by the opening of the diaphragm. This is spherical aberration.

Suppose we have corrected for spherical aberration (or are willing to ignore it), and let us consider the contribution of the second term. For fixed ρ and x the curve

$$\begin{aligned}\Delta_B x_1 &= \tfrac{1}{3}B\rho^2 x_0(3\cos^2\varphi + \sin^2\varphi) = \tfrac{1}{2}B\rho^2 x_0(2 + \cos 2\varphi),\\ \Delta_B y_1 &= \tfrac{2}{3}B\rho^2 x_0 \sin\varphi\cos\varphi = \tfrac{1}{3}B\rho^2 x_0\sin 2\varphi\end{aligned}$$

describes the circle of

$$\text{center} = \tfrac{2}{3}B\rho^2 x_0 \quad \text{and} \quad \text{radius} = \tfrac{1}{3}B\rho^2 x_0 \tag{5.4}$$

twice as φ goes from 0 to 2π. Now if we put a ring diaphragm (i.e., one which stops all rays except those passing through a circular ring) at the focal plane, then rotational symmetry guarantees that the corresponding rays all have $p_1^2 + q_1^2$ constant, and thus they image on one of the above circles. The circles have the appearance of the figure below: The effect is that rays emanating from x form a

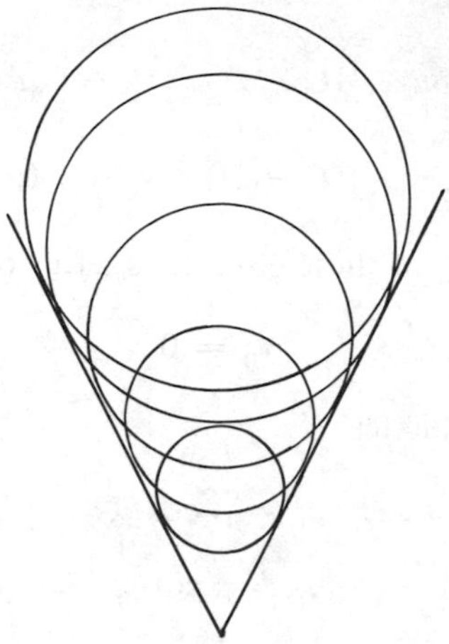

plane-like image reminiscent of a comet, hence this aberration is known as *coma*.

Suppose the system is corrected for spherical aberration and coma so that $A = B = 0$, and let us consider the effects of C and D:

$$\begin{aligned} \Delta_{C,D} x_1 &= C x_0^2 \rho \cos\varphi, \\ \Delta_{C,D} y_1 &= D x_0^2 \rho \sin\varphi. \end{aligned} \tag{5.5}$$

Thus rays passing through the ring diaphragm form an ellipse on the image plane. In order to understand this effect see what happens if we try to improve the focus by moving the plane $z = z_1$ by an amount s. Thus we wish to consider φ_{z_0, z_1+s}. We assume that n_1 is constant. Then the ray given by $(x_0, 0; p_1, q_1)$ will intersect the plane $z = z_1 + s$ at $x_1(s)$, $y_1(s)$ where

$$x_1(s) = x_1 + s\frac{p_1}{n_1} = mx_0 + \left(Cx_0^2 + \frac{s}{n_1}\right)\rho \cos\varphi,$$

$$y_1(s) = y_1 + s\frac{q_1}{n_1} = \left(Dx_0^2 + \frac{s}{n_1}\right)\rho \sin\varphi,$$

thus describing an ellipse about mx_0 with axes

$$Cx_0^2 + \frac{s}{n_1} \quad \text{and} \quad Dx_0^2 + \frac{s}{n_1}.$$

We can make the first axis vanish by choosing

$$s_p = -n_1 Cx_0^2 = -\frac{n_1}{m^2} Cx_1^2 \qquad \text{up to higher order terms}$$

and the second axis vanish by choosing

$$s_q = -n_1 Dx_0^2 = -\frac{n_1}{m^2} Dx_1^2 + \cdots.$$

Thus s_p will not coincide with s_q for $x_0 \neq 0$ unless $C = D$, i.e., the rays emanating from $(x_0, 0)$ do not focus, but we have two focal lines instead of a focal point. This effect is known as *astigmatism.* The fact that the focusing does not occur on a plane but rather along the surfaces of rotation swept out by the

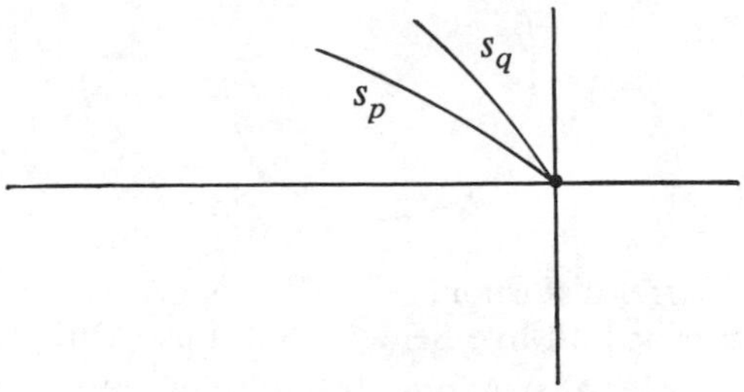

curves s_p or s_q is known as *curvature of the field.* The difference $s_q - s_p - (n_1/m^2)(D - C)x_1^2$ is called the *astigmatism*, and the average

$$\frac{1}{2}(s_p + s_q) = -\frac{1}{2}\frac{n_1}{m^2}(D + C)x_1^2$$

is called the *curvature of field.*

Suppose that A, B, C and D all vanish. Then

$$\Delta_E x_1 = Ex_0^3,$$

$$\Delta_E y_1 = 0.$$

In this case (to this order of approximation) all the points x_0, y_0 are focused perfectly. But the map from (x_0, y_0) to (x_1, y_1) is not linear. Indeed we have $(x_0, 0)$ going into $mx_0 + Ex_0^3$ and so, by symmetry,

$$x_1 = mx_0 + Ex_0(x_0^2 + y_0^2),$$

$$y_1 = my_0 + Ey_0(x_0^2 + y_0^2).$$

This effect is known as *distortion*. If E and m have the same sign, then a point is moved further away from the origin depending on its distance, thus the image of a centered square appears as:

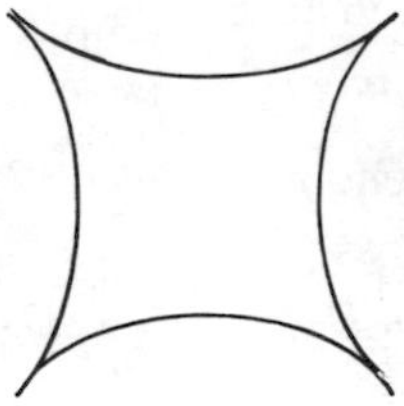

This is known as pincushion distortion. If E and m have opposite signs then the image of a centered square is:

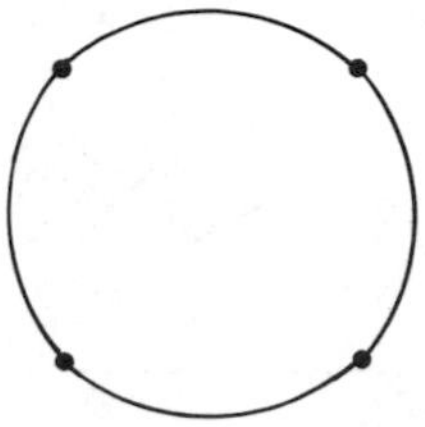

This is known as barrel distortion.

We have thus described all five Seidel aberrations. One might go further and study fifth or higher order aberrations. In addition there is *chromatic aberration* which arises from the fact that the refractive index n depends, in general, on the wave length, and so light of different frequency is focused differently.

For the actual determination of the Seidel aberrations in terms of the data of the optical system (the Seidel formulas) we refer the reader to Born and Wolf [1, Chapter V] or Luneburg [3, Chapter V]. They also describe methods of eliminating these various aberrations by appropriate use of stops and lens combinations.

All of the above was based on the zeroth order approximation given by geometrical optics. One also must take into account interference and diffraction effects which arise from the next order approximations to the Maxwell equations.

Let us now sketch the method of Debye and Luneburg for dealing with the problem of diffraction theory of an optical instrument. The basic philosophy is to impose a "physical condition" which, roughly speaking, says that we are interested in that solution of Maxwell's equations, of a given frequency, which behaves, at infinity, as if the only source of light impinging on the image comes from the instrument, all the rest representing "outgoing radiation". In a sense, the condition imposed is a generalization of the Sommerfeld radiation condition. We consider the solution of geometrical optics in the image space extending out to the ideal infinity in all directions from data on some fixed phase surface $\psi =$ const. by solving the transport equation in $\mathbf{R}^3$ with constant n_1 (the ideal extension of the image space). This gives a definite function u_0, and we seek that solution of Maxwell's equations which satisfies $\lim_{R\to\infty} R(u - u_0) = 0$. To solve this problem one applies the methods of Chapter I and passes to the limit over more and more distant surfaces. For the detailed hypotheses and conclusions we refer the reader to Sommerfeld [4, pp. 318–320], Luneburg [3, pp. 304–353] or Kline-Kay [10, Chapter XI]. As of yet the nature of the hypothesis has not been clearly elucidated or generally formulated as the Sommerfeld radiation condition in Lax and Phillips [11]. What is of interest to us is the form of the final answer, which asserts that the diffraction pattern for light from source point x_0, y_0, z_0 is given by an integral of the form

$$\frac{ik}{2\pi}\int ue^{ik[W+xp+yq+zr]}dpdq$$

where W is the mixed characteristic of the system, and where u is to be interpreted as a (vector valued) half density. Now W determines a Lagrangian manifold, parametrized by the coordinates p, q, where in fact Λ is the Lagrangian manifold of geometrical optics and, in a certain sense (see Chapter IV, §5), every Lagrangian submanifold of T^*M has such a parametrization. It is also not difficult to show that the method of stationary phase implies that the above integral determines a half density on Λ, and two integrals which determine the same have density different by terms of order $1/k$. We can then make use of the following procedure. Choose a half density on Λ which is a solution of the transport equation. Then choose any parametrization Φ of Λ (as sketched in Chapter II and described in detail in Chapter IV) and any integral of the form

$$\frac{ik}{2\pi}\int ue^{ik\Phi},$$

determining the prescribed half density on Λ. Then the highest order terms in the asymptotic expansion are determined purely by the geometry of Λ and the half density on Λ. This is Maslov's prescription. In its highest order term it does not depend on the choice of extensions and is thus independent of "radiation condition" type requirements. Near regular points we pointed this out already in Chapter II by a straightfoward application of stationary phase. Near caustics the behavior of the integral will be studied by applying methods from the theory of singularities, in Chapter VII.

§6. The asymptotic solution of Maxwell's equations.

To describe Maxwell's equations we introduce coordinates x^1, x^2, x^3 and t on $\mathbf{R}^4$ which are orthogonal with respect to the Lorentz metric, so that

$$(dx^i, dx^j) = \delta_{ij},$$

$$(dx^i, dt) = 0,$$

$$(cdt, cdt) = -1.$$

We have the four dimensional star operater $*\colon \wedge^k(\mathbf{R}^4) \to \wedge^{4-k}(\mathbf{R}^4)$ defined by

$$\omega_1 \wedge \omega_2 = (*\,\omega_1, \omega_2)(dx^1 \wedge dx^2 \wedge dx^3 \wedge c\,dt)$$

for $\omega_1 \in \wedge^k$ and $\omega_2 \in \wedge^{4-k}$, and where $(\ ,\ )$ is the induced metric on $\wedge^{4-k}$. Thus, for example,

$$* \, dx^1 = -dx^2 \wedge dx^3 \wedge c\,dt,$$

$$* \, dx^2 = dx^1 \wedge dx^3 \wedge c\,dt,$$

$$* \, dx^3 = -dx^1 \wedge dx^2 \wedge c\,dt,$$

$$* \, c\,dt = dx^1 \wedge dx^2 \wedge dx^3,$$

and

$$* \, (dx^1 \wedge dx^2) = -dx^3 \wedge c\,dt,$$

$$* \, (dx^1 \wedge c\,dt) = dx^2 \wedge dx^3, \qquad \text{etc.}$$

while

$$* * \omega = -(-1)^{k(4-k)}\omega \qquad \text{for } \omega \in \wedge^k.$$

Let us also introduce the three dimensional star operator $\circledast\ \wedge^k(\mathbf{R}^3) \to \wedge^{3-k}(\mathbf{R}^3)$ by

$$(\circledast\, \omega_1, \omega_2) dx^1 \wedge dx^2 \wedge dx^3 = \omega_1 \wedge \omega_2$$

for $\omega_1 \in \wedge^k(\mathbf{R}^3)$ and $\omega_2 \in \wedge^{3-k}(\mathbf{R}^3)$ so that

$$\circledast\, dx^1 = dx^2 \wedge dx^3 \qquad \text{etc.}$$

and

$$\circledast\circledast = \text{id}.$$

We let d denote the usual exterior derivative on $\mathbf{R}^4$ and d_x the exterior derivative with respect to the x variables only. The fundamental objects of Maxwell's theory are the following.

ρ— The charge density, a time dependent function on $\mathbf{R}^3$. The three form $\circledast\,\rho = \rho dx^1 \wedge dx^2 \wedge dx^3$ is the three form on $\mathbf{R}^3$ whose integral over any given three dimensional region gives the total charge contained in that region. (We assume that a unit, say the coulomb, has been chosen for charge. Otherwise, ρ should be considered as a section of a line bundle.)

J— The current density, a time dependent vector field on $\mathbf{R}^3$ such that $\circledast J$ is the two form whose integral over any surface fixed in space measures the rate of flow of charge through the surface.

D— The dielectric displacement, a time dependent vector field on $\mathbf{R}^3$ such that

$$d_x \circledast D = 4\pi\rho dx^1 \wedge dx^2 \wedge dx^3.$$

E— The electric field, a time dependent vector field on $\mathbf{R}^3$, the force field exerted on an extremely small test charge, divided by the charge.

B— The magnetic induction, a time dependent vector field on $\mathbf{R}^3$, the force field exerted on an extremely small magnetic pole, divided by the pole strength.

H— The magnetic excitation (Sommerfeld's notation), a time dependent vector field on $\mathbf{R}^3$.

Let Σ be a surface in $\mathbf{R}^3$ bounded by a curve, γ. We identify forms and vector fields in $\mathbf{R}^3$ via the Euclidean metric. Then $\int_\gamma E$ is called the electric loop tension while $\int_\gamma H$ is called the magnetic loop tension. Faraday's law of induction says that

$$\frac{1}{c}\frac{d}{dt}\int_\Sigma \circledast B = -\int_\gamma E,$$

while Ampere's law says that

$$\int_\Sigma \left(\frac{1}{c} \circledast \frac{\partial D}{\partial t} + \frac{4\pi}{c} \circledast J\right) = \int_\gamma H.$$

In differential form these become

$$\frac{1}{c}\frac{d}{dt} \circledast B = -d_x E, \tag{6.1}$$

$$\circledast \left[\frac{1}{c}\frac{\partial D}{\partial t} + \frac{4\pi}{c} J\right] = d_x H. \tag{6.2}$$

It follows from (6.1) that $(d/dt)d_x \circledast B = 0$. This is strengthened by the law

$$d_x \circledast B = 0 \tag{6.3}$$

and we repeat that one of the defining properties of D is

$$d_x \circledast D = 4\pi\rho dx^1 \wedge dx^2 \wedge dx^3. \tag{6.4}$$

If we set, as forms on $\mathbf{R}^4$,

$$\beta = E \wedge c\,dt + \circledast B,$$

$$\delta = -H \wedge c\,dt + \circledast D,$$

and

$$\Omega = \circledast J \wedge dt - \rho dx^1 \wedge dx^2 \wedge dx^3,$$

then (6.1) and (6.3) become

$$d\beta = 0 \tag{6.5}$$

while (6.2) and (6.4) become

$$d\delta + 4\pi\Omega = 0. \tag{6.6}$$

There are many advantages to writing the Maxwell equations in the form (6.5) and (6.6). One is that in vacuo (where, in the appropriate units, $\epsilon = \mu = 1$ and where $\Omega = 0$) the equations reduce to $d\beta = 0$ and $d * \beta = 0$ which are invariant under any conformal transformation of the Lorentz metric (and thus in particular under the inhomogeneous Lorentz group). Indeed, we claim that $*$: $\wedge^2(\mathbf{R}^4) \to \wedge^2(\mathbf{R}^4)$ is a conformal invariant: Let S_λ be scalar multiplication by λ on $\mathbf{R}^4$. Then the induced linear transformation, S_λ, on $\wedge^k$ is multiplication by λ^k. Thus, as $*$ maps $\wedge^k \to \wedge^{n-k}$ we see that $S_\lambda * S_\lambda^{-1}$ is multiplication by λ^{n-2k}. If $k = n/2$ we see that $*$ is invariant under change of scale. Thus, if f is a conformal transformation we see that $d * f * \beta = f * d * \beta = 0$, if $d * \beta = 0$. Secondly, the laws themselves take a simpler and more general form.

We can also read off from (6.5.)—(6.6) the existence of the "vector potential" for simply connected regions, in fact $d\beta = 0$ implies $\beta = dA$ for some $A \in \wedge^1(\mathbf{R}^4)$ and (6.6) says that $d * dA + 4\pi\Omega = 0$. In vacuo, this equation becomes $d * dA = 0$. The vector potential A is not determined, of course, by $dA = \beta$. It is common practice to impose on A the additional constraint $d * A = 0$. It is interesting to observe that the pair of equations

$$d * A = 0, \qquad d * dA = 0$$

can "almost" be derived from a simple variational principle. (This remark is not needed for what follows and can be skipped if the reader is uninterested.) Indeed, consider the (four dimensional) Lagrangian

$$\mathbf{L}(A) = \frac{1}{2}\langle dA, dA\rangle + \frac{m^2}{2}\langle A, A\rangle.$$

(This amounts to assigning a "mass", m, to the electromagnetic field.) It is easy to check that the Euler-Lagrange equations become

$$d * dA - mA = 0$$

where $d *$ is the adjoint of d, given up to sign by $* \, d \, *$. Then

$$d * A = \frac{1}{m} d * d * (dA) = 0$$

so

$$d * A = 0 \qquad \text{and} \qquad d * dA = mA.$$

Letting $m \to 0$ gives the desired equations

$$d * A = 0 \qquad \text{and} \qquad d * dA = 0.$$

Despite the advantages of the relativistic formulation, we shall return to the non-relativistic notation to continue the discussion of the Maxwell equation in a material medium.

In a material medium which is isotropic, there are relations between the various vector fields: We have

$$J = \sigma E, \qquad \text{Ohm's law,} \tag{6.7}$$

$$D = \epsilon E, \tag{6.8}$$

$$B = \mu H, \tag{6.9}$$

where σ is called the *specific conductivity*, ϵ is the *dielectric constant*, and μ is the *magnetic permeability*. If $\psi = \psi(x^1, x^2, x^3, t)$ is a phase function then the symbols of the equations (6.1)—(6.4) become

$$\frac{1}{c}\frac{\partial \psi}{\partial t} \circledast B + d_x\psi \wedge E, \tag{6.10}$$

$$\frac{1}{c}\frac{\partial \psi}{\partial t} \circledast D - d_x\psi \wedge H, \tag{6.11}$$

$$d_x\psi \wedge \circledast B, \tag{6.12}$$

$$d_x\psi \wedge \circledast D. \tag{6.13}$$

If we substitute (6.8) and (6.9) into (6.10)—(6.13) we obtain

$$\frac{\mu}{c}\frac{\partial \psi}{\partial t} \circledast H + d_x\psi \wedge E,$$

$$\frac{\epsilon}{c}\frac{\partial \psi}{\partial t} \circledast E - d_x\psi \wedge H,$$

$$\mu d_x\psi \wedge \circledast H,$$

$$\epsilon d_x\psi \wedge \circledast E$$

as the symbol of the operator applied to E and H. If E and H are in the kernel of the symbol then it follows from $d_x\psi \wedge \circledast H = 0$ that $d_x\psi$ is orthogonal to H, and similarly that $d_x\psi$ is orthogonal to E. (Thus E and H are *transverse*, e.g., if $\psi = ct - \varphi$ then E and H are perpendicular to the direction of propagation given by $d\varphi$.)

Since $d_x\psi$ is perpendicular to H it follows that

$$d_x\psi \wedge \circledast (d_x\psi \wedge H) = |d_x\psi|^2 \circledast H.$$

(Indeed since both sides are quadratic in $d_x\psi$ and linear in H, it suffices to show that $i \wedge \circledast (i \wedge j) = -\circledast j$ for any oriented orthogonal basis i, j, k. But $\circledast (i \wedge j) = k$ and $i \wedge k = -\circledast j$.)

Applying $\circledast$ to the second of the four lines above gives

$$\frac{\epsilon}{c}\frac{\partial\psi}{\partial t}E - \circledast (d_x\psi \wedge H)$$

and wedging with $d_x\psi$ gives

$$\frac{\epsilon}{c}\frac{\partial\psi}{\partial t}d_x\psi \wedge E + |d_x\psi|^2 \circledast H = 0.$$

Multiply the first line by $(\epsilon/c)(\partial\psi/\partial t)$ and substituting gives

$$\left[\frac{\mu\epsilon}{c^2}\left(\frac{\partial\psi}{\partial t}\right)^2 - |d_x\psi|^2\right] \circledast H = 0$$

and we similarly obtain

$$\left[\frac{\mu\epsilon}{c^2}\left(\frac{\partial\psi}{\partial t}\right)^2 - |d_x\psi|^2\right] \circledast E = 0.$$

Thus ψ must satify the characteristic equation

$$\frac{\mu\epsilon}{c^2}\left(\frac{\partial\psi}{\partial t}\right)^2 - \left[\left(\frac{\partial\psi}{\partial x^1}\right)^2 + \left(\frac{\partial\psi}{\partial x^2}\right)^2 + \left(\frac{\partial\psi}{\partial x^3}\right)^2\right] = 0. \tag{6.14}$$

We can apply the method discussed at the end of §3 of Chapter II to discuss the singularities of the electromagnetic field. Suppose that E and H have jump discontinuities along $\psi = 0$ and that ρ and J have (possible) δ function discontinuities on $\psi = 0$ given by $\hat{\rho}|d\psi|\delta(\psi)$ and $\hat{J}|d\psi|\delta(\psi)$. Let ΔE denote the jump in E along $\psi = 0$, etc. Then (3.8) of Chapter II and (6.1)—(6.4) give

$$d_x\psi \wedge \Delta E + \frac{1}{c}\frac{\partial \psi}{\partial t} \circledast \Delta B = 0, \tag{6.15}$$

$$d_x\psi \wedge \Delta H - \frac{1}{c}\frac{\partial \psi}{\partial t} \circledast \Delta D = \frac{4\pi}{c}|d\psi| \circledast \hat{J}, \tag{6.16}$$

$$d_x\psi \wedge \circledast \Delta B = 0, \tag{6.17}$$

$$d_x\psi \wedge \circledast \Delta D = 4\pi|d\psi| \circledast \hat{\rho} \tag{6.18}$$

along $\psi = 0$. There are a number of important consequences of (6.15)—(6.18) of which we shall mention two. If we take ψ to be a function of x^1, x^2, x^3 alone then (6.15)—(6.18) give the boundary conditions of the electromagnetic field at a surface of discontinuity of the medium. The terms involving $\partial\psi/\partial t$ disappear. Equation (6.15) says that the jump in E must be entirely normal, i.e., the tangential components of E must be continuous across $\psi = 0$ while (6.17) says that the normal component of B must be continuous. Equations (6.16) and (6.18) give the jump in the tangential components of H and normal components of D in terms of the surface distributions of current and charge.

Let us consider the case where $\psi = \varphi - ct$ (with $\hat{J} = 0$ and $\hat{\rho} = 0$). We will use (6.8) and (6.9) and set $\mathbf{e} = \Delta E$ and $\mathbf{h} = \Delta H$. Then

$$\begin{aligned} d_x\varphi \wedge \mathbf{e} - \mu \circledast \mathbf{h} &= 0, \\ d_x\varphi \wedge \mathbf{h} + \epsilon \circledast \mathbf{e} &= 0, \\ d_x\varphi \wedge \circledast \mathbf{e} &= 0, \\ d_x\varphi \wedge \circledast \mathbf{h} &= 0. \end{aligned} \tag{6.19}$$

These are the analogues of (3.3) of Chapter II. If $\mathbf{e}$, $\mathbf{h}$ are actual discontinuities of the electromagnetic field they must satisfy the transport equation in addition to (6.19); we refer the reader to Luneberg [3] or Kline and Kay [10] for details. As before, the eikonal equation $\|d\varphi\|^2 = n^2$ follows from (6.19).

References, Chapter III

1. M. Born and E. Wolf, *Principles of optics*, 4th ed., Pergamon Press, Oxford and New York, 1970.
2. C. Carathéodory, *Geometrische Optik*, Springer, Berlin, 1937.
3. R. K. Luneburg, *Mathematical theory of optics*, Univ. of California Press, Berkeley, Calif., 1964. MR **30** #2808.
4. A. Sommerfeld, *Optics*, Lectures on Theoretical Physics, vol. 4, Academic Press, New York, 1954. MR **16**, 1074.
5. J. L. Synge, *Geometrical optics*, Cambridge Univ. Press, New York, 1937.

6. L. H. Loomis and S. Sternberg, *Advanced calculus*, Addison-Wesley, Reading, Mass., 1968. MR **37** #2912.
7. E. Mach, *The principles of physical optics*, Dover, New York, 1949.
8. W. Lenz, *Probleme der Modernen Physik*, Herzel, Leipzig, 1929.
9. L. W. Green, Ann. of Math. (2) **78** (1963), 289–299. MR **27** #5206.
10. M. Kline and I. W. Kay, *Electromagnetic theory and geometric optics*, Pure and Appl. Math., vol. 12, Interscience, New York and London, 1965. MR **31** #4330.
11. P. D. Lax and R. S. Phillips, *Scattering theory*, Pure and Appl. Math., vol. 26, Academic Press, New York and London, 1967. MR **36** #530.
12. R. P. Feynman, R. B. Leighton and M. Sands, *The Feynman lectures on physics*. Vol. 2: *Mainly electromagnetism and matter*, Addison-Wesley, Reading, Mass., 1964. MR **35** #3943.

Chapter IV. Symplectic Geometry

§1. The Darboux-Weinstein theorem.

In this chapter we shall collect various facts about the geometry of symplectic manifolds and of their Lagrangian submanifolds which will be of use to us later. Recall that a symplectic manifold is a manifold X together with a non-degenerate closed two form, ω. The first basic fact about symplectic manifolds is that, locally, all symplectic manifolds of the same finite dimension, n, look the same. A beautiful proof of this theorem, together with a strong generalization of it, has been recently given by Weinstein [8]. The method of proof is quite similar to the proof we gave for Morse's lemma and is one that we shall have occasion to use again several times.

Let X be a manifold and Y an embedded submanifold. If σ is a differential form on X, we shall let $\sigma_{|Y}$ denote the restriction of σ to $(\wedge\ TX)_{|Y}$. (Thus $\sigma_{|Y}$ can be evaluated on vectors which are not necessarily tangent to Y.)

THEOREM 1.1 (*Darboux-Weinstein*). *Let Y be a submanifold of X and let ω_0 and ω_1 be two non-singular closed two forms on X such that $\omega_{0|Y} = \omega_{1|Y}$. Then there exists a neighborhood, U, of Y and a diffeomorphism $f\colon U \to X$ such that*

(i) $f(y) = y$ *for all* $y \in Y$,

(ii) $f^*\omega_1 = \omega_0$.

If we take Y to be a point, the theorem asserts that if the two forms agree on the tangent space at a point, then, up to a diffeomorphism, f, they agree in a neighborhood of the point. For finite dimensional vector spaces, all non-degenerate anti-symmetric forms are equivalent up to linear transformation (see, for example, [2, Chapter 1]). Extend this linear transformation to some neighborhood, (by some exponential map, say, i.e., by using the linear coordinates in some neighborhood). Thus, the theorem implies that given any pair ω_0 and ω_1 defining a symplectic structure on X then near any point p there is a diffeomorphism g with $g(p) = p$ and $g^* \omega_1 = \omega_0$. This is the content of the Darboux theorem.

For the proof of Theorem 1.1 we need a basic formula of differential calculus which we now recall. Let W and Z be differentiable manifolds and let $\varphi_t: W \to Z$ be a smooth one parameter family of maps of W into Z. In other words the map $\varphi: W \times I \longrightarrow Z$ given by $\varphi_t(w) = \varphi(w, t)$ is smooth. Then we let ξ_t denote the tangent field along φ_t, i.e., $\xi_t: W \longrightarrow TZ$ is defined by letting $\xi_t(w)$ be the tangent vector to the curve $\varphi(w, \cdot)$ at t. If σ is a differential $k+1$ form on Z, then $\varphi_t^*(\xi_t \lrcorner \sigma)$ is a well defined differential k form on W given by

$$\varphi_t^*(\xi_t \lrcorner \sigma)(\eta_1, \ldots, \eta_k) = (\xi_t(w) \lrcorner \sigma)(d\varphi_t \eta_1, \ldots, d\varphi_t \eta_k).$$

(Notice that since ξ_t is not a vector field on Z the expression $\xi_t \lrcorner \sigma$ does not define a differential form on Z.)

Let σ_t be smooth one parameter family of forms on Z. Then $\varphi_t^* \sigma_t$ is a smooth family of forms on W and the basic formula of the differential calculus of forms asserts that

$$\frac{d}{dt} \varphi_t^* \sigma_t = \varphi_t^* \frac{d\sigma_t}{dt} + \varphi_t^*(\xi_t \lrcorner d\sigma_t) + d\varphi_t^*(\xi_t \lrcorner \sigma_t). \tag{1.1}$$

For the sake of completeness we shall present a proof of this formula at the end of this section.

Let $Y \subset X$ be an embedded submanifold and suppose that there exists a smooth retraction, φ_t, of X onto Y. Thus we assume that φ_t is a smooth family of maps of $X \to X$ such that

$$\varphi_0: X \to Y, \qquad \varphi_1 = \mathrm{id}$$

and

$$\varphi_t y = y \qquad \text{for all } y \in Y \text{ and all } t.$$

(Notice that if X were a vector bundle and Y were the zero section then multiplication by t would provide such a retraction. Also if X were a convex open neighborhood of the zero section. By choosing a Riemann metric and using the exponential map on the normal bundle of Y, we can thus arrange that some neighborhood of Y has a differentiable retraction onto Y.) Then, for any form σ on X we have (in some neighborhood of Y)

$$\begin{aligned} \sigma - \varphi_0^* \sigma &= \int_0^1 \frac{d}{dt}(\varphi_t^* \sigma)\, dt = \int_0^1 (\varphi_t^*(\xi_t \lrcorner d\sigma))\, dt + d \int_0^1 (\varphi_t^*(\xi_t \lrcorner \sigma))\, dt \\ &= I d\sigma + dI\sigma \end{aligned}$$

where we have set

$$I\beta = \int_0^1 [\varphi_t^*(\xi_t \lrcorner \beta)]\, dt$$

for any form β on X. In other words I: $\wedge^k(X) \to \wedge^{k-1}(X)$ and

$$\sigma - \varphi_0^* \sigma = dI\sigma + Id\sigma. \tag{1.2}$$

PROOF OF THE DARBOUX-WEINSTEIN THEOREM. Set

$$\omega_t = (1 - t)\omega_0 + t\omega_1 = \omega_0 + t\sigma \quad \text{where} \quad \sigma = \omega_1 - \omega_0 .$$

Notice that

$$\sigma_{|Y} = 0$$

so that, in particular,

$$\varphi_0^* \sigma = 0 \qquad \text{and} \qquad d\sigma = 0.$$

Hence, by (1.2),

$$\sigma = d\beta \qquad \text{where } \beta = I\sigma.$$

Notice that

$$\beta_{|Y} = 0.$$

Now

$$\omega_{t|Y} = \omega_{0|Y} = \omega_{1|Y}$$

and so $\omega_{t|Y}$ is non-degenerate for all $0 \leqslant t \leqslant 1$. We can therefore find some neighborhood of Y on which ω_t is non-degenerate for all $0 \leqslant t \leqslant 1$. We can therefore find a vector field $\boldsymbol{\eta}_t$ such that

$$\boldsymbol{\eta}_t \lrcorner \omega_t = -\beta. \tag{1.3}$$

We can integrate the vector field $\boldsymbol{\eta}_t$ to obtain a one parameter family of maps, f_t, whose tangent vector is $\boldsymbol{\eta}_t$. Notice that $f_{t|Y} = \text{id}$. By restricting to a smaller neighborhood of Y we may assume that f_t is also defined for all $0 \leqslant t \leqslant 1$. (Strictly speaking, in proving this fact, we may want $\boldsymbol{\eta}_t$, etc. to be defined for some range of $t > 1$.) Then $f_0 = \text{id}$ and, by (1.1) and the fact that

$$\frac{d}{dt}\omega_t = \sigma,$$

we see that

$$f_1^* \omega_1 - \omega_0 = \int_0^1 \frac{d}{dt}(f_t^* \omega_t)\, dt = \int_0^1 f_t^* (\sigma + d(\eta_t \lrcorner \omega_t))\, dt = 0$$

since $d\omega_t = 0$. Thus f_1 provides the desired diffeomorphism, proving the theorem.

Let Λ be an embedded Lagrangian submanifold of a symplectic manifold X. As an example, let $X = T^* M$ and let Λ be the zero section of $T^* M$. It is an observation due to Kostant, that, locally, this is the only example.

Proposition 1.1 (*Kostant*). *Let Λ be an embedded Lagrangian submanifold of a symplectic manifold X, whose sympectic form is ω. Let Λ also be regarded as the zero section of $T^* \Lambda$ and let ω' be the symplectic form on $T^* \Lambda$. Then there exists a neighborhood, U, of Λ in X and a diffeomorphism h of U into $T^* \Lambda$ such that $h_{|\Lambda} = \mathrm{id}$ and $h^* \omega' = \omega$.*

The proof of the proposition will use Theorem 1.1 and an algebraic fact concerning Lagrangian subspaces of a symplectic vector space which we shall prove in the next section. The algebraic fact is as follows: Let V be a symplectic vector space, and let Z be a Lagrangian subspace of V. Then the set of all Lagrangian subspaces, W, such that $W \cap Z = \{0\}$ is an affine space. For the precise statement, see Proposition 2.3 below. For us this fact has the following consequence.

We can find a smooth bundle, E, of Lagrangian subspaces of $TX_{|\Lambda}$ such that $E_\lambda \cap T\Lambda_\lambda = \{0\}$ for all $\lambda \in \Lambda$.

In fact the bundle of all Lagrangian subspaces of $TX_{|\Lambda}$ which have zero intersection with $T\Lambda$ is an affine bundle by the above algebraic fact, and hence has a smooth section. (Just choose sections locally and patch together by averaging, using a partition of unity, i.e., give s_i locally and let $s = \sum \phi_i s_i$ where (ϕ_i) is a suitable partition of unity. Averaging makes sense in an affine space.) Now once we have fixed E, this determines an isomorphism, for each $\lambda \in \Lambda$, of TX_λ with $T\Lambda_\lambda \oplus T^* \Lambda_\lambda$, since E_λ is naturally dual to $T\Lambda_\lambda$. Now if we regard Λ as the zero section of $T^* \Lambda$ then the tangent space to $T^* \Lambda$ at λ splits into a direct sum of the tangent to the fiber and the tangent to the zero section. Now the fiber is a vector space, so we may identify the tangent to the fiber with $T^* \Lambda_\lambda$. In this way we have an identification

$$T(T^* \Lambda)_\lambda = T\Lambda_\lambda \oplus T^* \Lambda_\lambda .$$

We have thus an isomorphism of TX_λ with $T(T^* \Lambda)_\lambda$ which clearly preserves the symplectic structure and varies smoothly with λ. In other words we have a map of vector bundles $TX_{|\Lambda} \to T(T^* \Lambda)_{|\Lambda}$ which is an isomorphism of symplectic structures. We now choose some diffeomorphism, g, of some neighborhood of Λ in X into $T^* \Lambda$ such that $g_{|\Lambda} = \mathrm{id}$ and dg is the isomorphism constructed above on $TX_{|\Lambda}$. (This is always possible by using some exponential map. Notice that we do not yet require g to have any properties relative to the symplectic structure.) Then let $\omega_1 = g^*\omega'$. By construction

$$\omega_{1|\Lambda} = \omega_{|\Lambda}$$

and therefore by Theorem 1.1 there exists an f mapping some neighborhood of Λ in X into X with $f^*\omega_1 = \omega$. Thus $f^*g^*\omega' = \omega$ and $h = g \circ f$ is the desired diffeomorphism.

Let us now give a proof of (1.1). We first prove the formula in the special case where $W = Z = M \times I$ and φ_t is the map $\psi_t \colon M \times I \to M \times I$ given by

$$\psi_t(x,s) = (x, s+t).$$

The most general differential form on $M \times I$ can be written as

$$ds \wedge a + b$$

where a and b are forms on M which may depend on t and s. (In terms of local coordinates, $s, x^1, \ldots, x^n$, these forms are sums of terms which look like

$$c\,dx^{i_1} \wedge \cdots \wedge dx^{i_k}$$

where c is a function of t, s and x.) To show the dependence on x and s we shall rewrite the above expression as

$$\sigma_t = ds \wedge a(x,s,t)dx + b(x,s,t)dx.$$

With this notation it is clear that

$$\psi_t^*\sigma_t = ds \wedge a(x,s+t,t)dx + b(x,s+t,t)dx$$

and therefore

$$\begin{aligned}\frac{d\psi_t^*\sigma_t}{dt} &= ds \wedge \frac{\partial a}{\partial s}(x,s+t,t)dx + \frac{\partial b}{\partial s}(x,s+t,t)dx\\ &\quad + ds \wedge \frac{\partial a}{\partial t}(x,s+t,t)dx + \frac{\partial b}{\partial t}(x,s+t,t)dx,\end{aligned}$$

so that

$$\frac{d\psi_t^*\sigma_t}{dt} - \psi_t^*\left(\frac{d\sigma_t}{dt}\right) = ds \wedge \frac{\partial a}{\partial s}(x,s+t,t)dx + \frac{\partial b}{\partial s}(s,s+t,t)dx. \qquad \text{(a)}$$

It is also clear that in this case the tangent to $\psi_t(x,s)$ is $\partial/\partial s$ evaluated at $(x, s+t)$.

In this case $\partial/\partial s$ is a vector field and

$$\frac{\partial}{\partial s} \lrcorner \sigma_t = a\,dx$$

so

$$\psi_t^*\left(\frac{\partial}{\partial s} \lrcorner \sigma_t\right) = a(x, s + t, t)dx$$

and therefore

$$d\psi_t^*\left(\frac{\partial}{\partial s} \lrcorner \sigma_t\right) = \frac{\partial a}{\partial s}(x, s + t, t)ds \wedge dx + d_x a(x, s + t, t)dx \tag{b}$$

(where d_x denotes the exterior derivative of the form $a(x,s + t,t)dx$ on the manifold M, holding s fixed). Similarly,

$$d\sigma_t = -ds \wedge d_x adx + \frac{\partial b}{\partial s}ds \wedge dx + d_x bdx$$

so

$$\frac{\partial}{\partial s} \lrcorner d\sigma_t = -d_x adx + \frac{\partial b}{\partial s}dx$$

and

$$\psi_t^* \frac{\partial}{\partial s} \lrcorner d\sigma_t = -d_x a(x, s + t, t)dx + \frac{\partial b}{\partial s}(x, s + t, t)dx. \tag{c}$$

Adding (a), (b), (c) proves (1.2) for ψ_t.

Now let $\varphi\colon W \times I \to Z$ be given by

$$\varphi(w, s) = \varphi_s(w).$$

Then the image under φ of the lines parallel to I through w in $W \times I$ is just the curves $\varphi_s(w)$ in Z. In other words

$$d\varphi\left(\frac{\partial}{\partial s}\right)_{(w,t)} = \xi_t(w).$$

If we let $\iota\colon W \to W \times I$ be given by

$$\iota(w) = (w, 0)$$

then we can write the map φ_t as

$$\varphi \circ \psi_t \circ \iota.$$

Thus

$$\varphi_t^* \sigma_t = \iota^* \psi_t^* \varphi^* \sigma_t$$

and, since ι and φ do not vary with t,

$$\frac{d}{dt}\varphi_t^* \sigma_t = \iota^* \frac{d}{dt}\psi_t^* (\varphi^* \sigma_t).$$

At the point w, t of $W \times I$, we have

$$\frac{\partial}{\partial s} \lrcorner \varphi^* \sigma_t = (d\varphi)^* \left\{ \left(d\varphi \frac{\partial}{\partial s} \lrcorner \sigma_t \right) \right\} = (d\varphi)^* \xi_t \lrcorner \sigma_t$$

and thus

$$\iota^* \psi_t^* \left(\frac{\partial}{\partial s} \lrcorner \varphi^* \sigma_t \right) = \iota^* \psi_t^* \varphi^* (\xi_t \lrcorner \sigma_t) = \varphi_t^* (\xi_t \lrcorner \sigma_t).$$

Substituting into the formula for

$$\frac{d\psi_t^*}{dt} \varphi^* \sigma_t$$

yields (1.1).

§2. Symplectic vector spaces.

In this section we list various facts concerning the geometry of symplectic vector spaces. Let V be a vector space and $(\,,)$ an antisymmetric bilinear form on V. If the form $(\,,)$ is non-singular, then V, together with $(\,,)$, is called a *symplectic vector space*. If $(\,,)$ is singular, then we set

$$\begin{aligned} V^\perp &= \{v \mid (v, w) = 0 \quad \text{all } w \in V\} \\ &= \{v \mid (w, v) = 0 \quad \text{all } w \in V\} \end{aligned}$$

and it is clear that we get an induced bilinear form on $V/V^\perp$ and that $V/V^\perp$ is a symplectic vector space.

Let V be a symplectic vector space. The symplectic group $Sp(V)$ consists of all non-singular linear transformations, B, such that

$$(Bu, Bv) = (u, v)$$

for all $u, v \in V$. The conformal symplectic group $CSp(W)$ consists of those non-singular linear transformations satisfying

$$(Bu, Bv) = \mu_B(u, v) \qquad \forall u, v \in V,$$

where μ_B is some scalar depending on B. The corresponding Lie algebras are the symplectic algebra, $sp(V)$, consisting of those $A \in \operatorname{Hom}(V, V)$ satisfying

$$(Au, v) + (u, Av) = 0 \qquad \forall u, v \in V$$

and the conformal symplectic algebra, $csp(V)$, consisting of those A which satisfy

$$(Au, v) + (u, Av) = \mu_A(u, v) \qquad \forall u, v \in V. \tag{2.1}$$

If $A \in csp(V)$ then $A - \frac{1}{2}\mu_A I \in sp(V)$, and so $csp(V) = sp(V) + Z$, where the center, Z, consists of all multiples of the identity transformation.

If V is a real symplectic vector space then its complexification, $V^{\mathbf{C}} = V \otimes \mathbf{C}$ is easily seen to be a complex symplectic vector space with the obvious bilinear form:

$$(x + iy, u + iv) = (x, u) - (y, v) + i\{(x, v) + (y, u)\}.$$

Let $A \in sp(V)$, where V is a finite dimensional symplectic vector space. Then for any scalar λ we have

$$([A - \lambda]u, v) = -(u, [A + \lambda]v) \qquad \forall u, v \in V$$

and therefore

$$([A - \lambda]^k u, v) = (-1)^k (u, [A + \lambda]^k v) \qquad \forall u, v \in V.$$

Let V_λ denote the generalized eigenspace of A corresponding to the eigenvalue λ. Thus V_λ consists of those $u \in V$ such that

$$[A - \lambda]^k u = 0$$

for sufficiently large k. Thus $u \in V_\lambda$ if and only if $u \in ([A + \lambda]^k V)^\perp$. In particular $\dim V_\lambda = \dim([A + \lambda]^k V)^\perp = \dim V_{-\lambda}$. Also, if $0 \neq u$ is an eigenvector:

$$(A - \lambda)u = 0$$

then $u \in ((A + \lambda)V)^\perp$, and the set of eigenvectors with eigenvalue λ is paired, under $(\,,)$, with the eigenvectors corresponding to the eigenvalues $-\lambda$. If λ is complex and V is real, we can apply the same results to A acting on $V^{\mathbf{C}}$. If $A \in csp(V)$ we can apply the above to $A - \frac{1}{2}\mu_A I \in sp(V)$. We thus obtain:

PROPOSITION 2.1. *Let $A \in csp(V)$. Then the eigenvalues of A are symmetric about $\frac{1}{2}\mu_A$. That is, if λ is an eigenvalue of A then so is $\mu_A - \lambda$ and*

$$\dim(V_\lambda^{\mathbf{C}}) = \dim(V_{\mu_A - \lambda}^{\mathbf{C}}).$$

In fact V_λ and $V_{\mu_A - \lambda}$ are non-singularly paired under $(\,,)$. Also the eigenspaces corresponding to λ and $\mu_A - \lambda$ are non-singularly paired under $(\,,)$.

A subspace $X \subset V$ is called *Lagrangian* if it is maximally isotropic. Thus $(u_1, u_2) = 0$ if $u_i \in X$ and X is maximal with respect to this property.

Let X be a fixed Lagrangian subspace and let Y be a second Lagrangian subspace such that $X \cap Y = \{0\}$. Then X and Y are non-singularly paired by $(\,,)$. Let P denote the projection of V onto Y with kernel X so that

$$0 \to X \to V \xrightarrow{P} Y \to 0.$$

Then $P \in csp(V)$ and $\mu_P = 1$. Indeed, we must show that

$$(Pu, v) + (u, Pv) = (u, v).$$

As X and Y span V we need only consider three cases: $u, v \in X$; $u \in X, v \in Y$; and $u, v \in Y$. If $u, v \in X$ then $Pu = Pv = 0$ and the right hand side vanishes since X is isotropic. If $u \in X$ and $v \in Y$ the equation becomes $(u, v) = (u, v)$ while if u and v both lie in Y both sides are zero. Conversely, let $P \in csp(V)$ satisfy $\mu_P = 1$ and $P_{|X} \equiv 0$. Then according to Proposition 2.1, P must have an eigenspace Y corresponding to the eigenvalue $\lambda = 1$ whose dimension is equal to $\dim X$. Obviously $X \cap Y = \{0\}$ and, if $u, v \in Y$ we have $(u, v) = (Pu, v) + (u, Pv) = 2(u, v)$ so $(u, v) = 0$; thus Y is Lagrangian. Thus, the set of Lagrangian subspaces Y with $Y \cap X = \{0\}$ is in one to one correspondence with the set of $P \in csp(V)$ such that $\mu_P = 1$ and $P_{|X} = 0$.

Given any element, P, of $csp(V)$ we obtain a symmetric bilinear form Q_P on V by setting

$$Q_P(x, y) = (Px, y) - \tfrac{1}{2}\mu_P(x, y). \tag{2.2}$$

Indeed

$$\begin{aligned} Q_P(y, x) &= (Py, x) - \tfrac{1}{2}\mu_P(y, x) = -(x, Py) + \tfrac{1}{2}\mu_P(x, y) \\ &= (Px, y) - \tfrac{1}{2}\mu_P(x, y) = Q_P(x, y). \end{aligned}$$

Conversely, given Q and μ_P the equation defines $P \in csp(V)$.

If $Px = 0$ for $x \in X$ then $Q_P(x, x) = 0$ for $x \in X$, while $Q_P(x, y) = -\frac{1}{2}(x, y)$ is a non-singular pairing between X and Y if $X \cap Y = \{0\}$ and $\mu_P = 1$. Thus Q_P has rank n. Conversely, let Q_P be any symmetric quadratic form such that

$$Q_P(x, v) = -\tfrac{1}{2}(x, v) \qquad \forall x \in X, v \in V.$$

Then we get a $P \in csp(V)$ with $\mu_P = 1$ and $(Px, y) \equiv 0$ for $x \in X$ and y arbitrary so that $Px = 0$ for $x \in X$. We have thus established

PROPOSITION 2.2. *Let X be a fixed Lagrangian subspace. Then the following sets are in one to one correspondence*:

(i) *The set of all Lagrangian subspaces Y such that $Y \cap X = \{0\}$.*

(ii) *The set of all $P \in csp(V)$ such that $\mu_P = 1$ and $P_{|X} \equiv 0$.*

(iii) *The set of all symmetric quadratic forms, Q, on V, such that*

$$Q(x,v) = -\tfrac{1}{2}(x,v) \qquad \forall x \in X,\, v \in V. \tag{2.3}$$

Here $Y = \ker(P - I)$ *while* P *and* Q *are related by* (2.2). *We shall denote the space* (i) *by* $\mathfrak{L}_X$.

The third description shows that the space in question has the structure of an affine space whose associated vector space is $S^2(V/X)$. Indeed, let Q_1 and Q_2 be two symmetric forms on V which satisfy (2.3). Then $Q_1 - Q_2 = H$ is a symmetric form on V such that $H(x,v) \equiv 0$ for $x \in X$ and all v. Thus H defines a symmetric bilinear form on V/X. Conversely, $S^2(V/X)$ can be considered as the space of symmetric bilinear forms H on V such that $H(x,v) \equiv 0$ for $x \in X$. Then $Q + H$ satisfies (2.3) if Q does. Thus we have proved:

PROPOSITION 2.3. *Let* X *be a fixed Lagrangian subspace. Then the space of all Lagrangian subspaces transversal to* X *is an affine space whose associated linear space is* $S^2(V/X)$. *In particular, if we fix a transversal Lagrangian subspace* Y *then* $\mathfrak{L}_X$ *becomes identified with* $S^2(Y)$, *since we may identify* V/X *with* Y. *If* W *is some other element of* $\mathfrak{L}_X$ *then the quadratic form associated with* W *on* Y *is given by*

$$H(y_1, y_2) = (P_W y_1, y_2) \tag{2.4}$$

where P_W *is the projection of* V *onto* W *along* X.

The element P_W described in Proposition 2.3 is the projection described by the exact sequence

$$0 \to X \to V \xrightarrow{P_W} W \to 0.$$

The quadratic form, Q_Y, associated to Y by Proposition 2.2 vanishes on Y so that the H defined by (2.4) does indeed satisfy

$$H = (Q_W - Q_Y)_{|Y}.$$

Notice that

$$H \text{ is non-singular if and only if } Y \cap W = \{0\}. \tag{2.5}$$

Indeed if $y \in W \cap Y$ then $P_W y = P_Y y = y$ and thus $Q_W(y,v) = Q_Y(y,v)$ for all $v \in V$ and hence $H(y,v) \equiv 0$. On the other hand, since $X \cap Y = \{0\}$ we know that P: $Y \to W$ is a isomorphism. If $W \cap Y = \{0\}$ then W and Y are non-singularly paired under $(\,,)$. Thus (2.4) defines a non-singular pairing.

Since $S^2(Y)$ has plenty of non-singular elements we conclude that there always is a Lagrangian subspace W transversal to two given Lagrangian subspaces X and Y, at least if X and Y are transversal to each other. Of course, if X and Y are not transversal to each other it should be even "easier" to pick a W transversal to both. Indeed consider the subspace $X + Y$ and the restriction of $(\,,)$ to it. The only elements of V which are orthogonal to X and to Y must be in $X \cap Y$ so that $(X + Y)/X \cap Y$ is a symplectic vector space. We can therefore find a subspace $W_0 \subset X + Y$ whose dimension equals that of $X/(X \cap Y)$ such that W_0 is totally isotropic and $W_0 \cap X = W_0 \cap Y = \{0\}$. If we choose a basis, $q_1, \ldots, q_r$ of $X \cap Y$ and a basis $p_{r+1}, \ldots, p_n$ of W_0 then the $q_1, \ldots, q_r, p_{r+1}, \ldots, p_n$ span a Lagrangian subspace. We can choose a dual basis $p_1, \ldots, p_r, q_{r+1}, \ldots, q_n$. Then the space W spanned by $p_1, \ldots, p_n$ clearly has the desired properties. We have thus proved:

PROPOSITION 2.4. *Given any pair X and Y of Lagrangian subspaces it is always possible to find a third Lagrangian subspace transversal to both.*

Let us return to the situation described by the pair of transversal Lagrangian subspaces X and W. If Y is a second Lagrangian subspace transversal to X then

$$P_Y - P_W \in sp(V)$$

since both P_Y and $P_W \in csp(V)$ and $\mu_{P_Y} = \mu_{P_W} = 1$. Furthermore, we claim that

$$(P_Y - P_W)^2 = 0. \tag{2.6}$$

Indeed, for $x \in X$ we have $P_Y x = P_W x = 0$; while for $w \in W$ we have

$$(P_Y - P_W)w = P_Y w - w \in X.$$

Since X and W span V this proves (2.6). Thus

$$\exp(P_Y - P_W) = 1 + (P_Y - P_W) \in Sp(V) \tag{2.7}$$

is the transformation which is the identity on X and maps W into Y. In (2.7) we have used the exponential map in the symplectic group. For reasons that will become clear later on, we will want to consider covering groups of the symplectic group (in particular the double covering). The exponential map again is well defined (but not given by the right hand side of (2.7)). We record these results as

PROPOSITION 2.5. *Let X be a Lagrangian subspace. If W and Y are two Lagrangian subspaces transversal to X then $P_Y - P_W \in sp(V)$ and $(P_Y - P_W)^2 = 0$ where P_Z denotes projection onto Z along X (with $Z = Y$ or W). The map $1 + P_Y - P_W$ is the identity on X, it carries W into Y, and lies in $Sp(V)$. If G is any covering group of $Sp(V)$ then* Exp $(P_Y - P_W)$ *lies in G and covers $1 + P_Y - P_W$, where* Exp *is the exponential map: $sp(V) \to G$ for the group G. In this way we have associated an element of G to each pair of Lagrangian subspaces transversal to X.*

Let us now examine the structure of the space of all Lagrangian subspaces of a real symplectic space, V. We will denote the set of all Lagrangian subspaces by $L(V)$. If $V = 2$, then any one dimensional subspace is Lagrangian. Thus $L(V)$ is just the one dimensional projective space, which is, topologically, a circle. In particular, $H^1(L(V)) = \mathbf{Z}$.

Notice that if we pick one Lagrangian subspace, X, then $L(V) - \{X\} = \mathfrak{L}_X$ consists of the projective line with a point omitted: If we consider X as the "point at infinity" $\mathfrak{L}_X$ becomes the affine line.

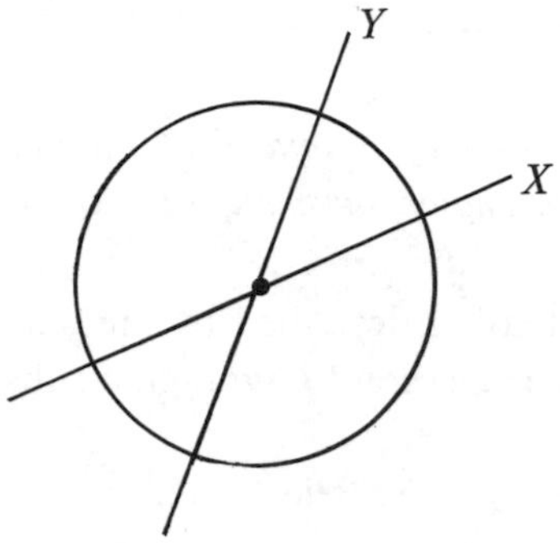

Fixing another Y determines the origin of the affine line, and hence a linear structure. To actually visualize $L(V)$ as a circle, we put a Riemann metric and an orientation on V. This makes V into a one dimensional complex vector space. Then each line will determine two points on the unit circle. If we let $U(1)$, the one dimensional unitary group, decribing the unit circle, and $O(1) = \{+1, -1\}$ be the subgroup of $U(1)$ consisting of the orthogonal group of the line (i.e., the real unitary group) then

$$L(\mathbf{R}^2) \sim U(1)/O(1).$$

The identification, of course, depends on the choice of Riemann metric.

We can perform the same construction in general: Suppose that V is a complex vector space with a Hermitian scalar product $\langle\, , \rangle$. Let

$$\langle v, w\rangle_R = \operatorname{Re} \langle v, w\rangle$$

and

$$\langle v, w\rangle_I = \operatorname{Im} \langle v, w\rangle.$$

Then V is a $2n$ dimensional real space and $\langle v, w\rangle_I$ is clearly an anti-symmetric form which is non-singular, hence a symplectic form. If Z is a Lagrangian subspace of V then we claim that Z is orthogonal to iZ with respect to $\langle\, , \rangle_R$. Indeed, for v and $w \in Z$ we have

$$\mathrm{Re}\langle v, iw\rangle = -\mathrm{Re}\langle iv, w\rangle = -\langle v, w\rangle_I = 0.$$

If U is any unitary transformation, then, by definition, it preserves $\langle\,,\rangle$ and hence $\langle\,,\rangle_I$ and so defines a symplectic transformation. Thus U acts on $L(V)$. We claim that it acts transitively on $L(V)$. Indeed, let X and X' be Lagrangian subspaces and let $\{e_1, \ldots, e_n\}$ and $\{e'_1, \ldots, e'_n\}$ be orthonormal bases of X and X' with respect to $\langle\,,\rangle_R$. Since $\langle e_i, e_j\rangle_I = 0$ we conclude that $\{e_1, \ldots, e_n\}$ is an orthonormal basis for the Hermitian structure as well, and the same for $\{e'_1, \ldots, e'_n\}$. Hence there is a unitary transformation with $U(e_i) = e'_i$, for all i, and therefore $UX = X'$. The set of unitary transformations keeping X fixed will be those unitaries with $Ue_i = \sum a_{ij} e_j$ where the a_{ij} are real; thus $U \in O(n)$. Thus

$$L(V) = U(n)/O(n). \tag{2.8}$$

We have derived this result starting from a Hermitian structure on V. Let us now show how to put a Hermitian structure on any symplectic vector space V (whose imaginary part gives the symplectic form) and, indeed, parametrize all such Hermitian structures.

Fix a Lagrangian subspace X. Suppose we are given a transversal Lagrangian subspace Y and a positive definite scalar product $\langle\,,\rangle_R$ on X. Then $\langle\,,\rangle_R$ determines an isomorphism of $X \to X^*$ and Y can be identified with X^* via the symplectic form. Thus we are given a map

$$X \to Y.$$

Call this map multiplication by i. Then $i(ix) = -x$ determines multiplication by i on all of V, making V into a complex vector space. Also set

$$\begin{aligned} \langle u, v\rangle &= \langle u, v\rangle_R, && u, v \in X,\\ \langle u, y\rangle &= i(u, y), && u \in X, y \in Y,\\ \langle y, z\rangle &= \langle iy, iz\rangle_R, && y, z \in Y, \end{aligned}$$

and extending by linearity defines a Hermitian form, $\langle\,,\rangle$, with

$$\langle\,,\rangle_I = (\,,).$$

Conversely, starting with a Hermitian form $\langle\,,\rangle$ on V then we have already observed that iX is a Lagrangian subspace transversal to X and $\langle\,,\rangle_R$ restricted to X is positive definite. We have thus proved:

PROPOSITION 2.6. *Let V be a finite dimensional real symplectic space with form $(\,,)$. Let H be the space of Hermitian forms $\langle\,,\rangle$ (and complex structures) with $\langle\,,\rangle_I = (\,,)$. Then*

$$H \cong \mathfrak{L}_X \times P_X$$

where X is a Lagrangian subspace of X and P_X denotes the space of positive definite quadratic forms on X. In particular, since $\mathfrak{L}_X$ and P_X are both diffeomorphic to cells, we conclude that H is a cell.

Let det be the determinant function mapping $U(n) \to S^1$. It maps $O(n) \to \{\pm 1\}$ which we shall denote by S^0. We then get a well defined function $\det^2$: $U(n)/O(n) \to S^1$. Let $S[U(n)/O(n)] = (\det^2)^{-1}(1)$. Now if $\det U = \pm 1$ we can find an $O \in O(n)$ such that $\det UO = 1$. Thus $SU(n)$ acts transitively on $S[U(n)/O(n)]$ and the isotropy group is $SO(n)$. Thus we have a fibration of

$$U(n)/O(n) \xrightarrow{\det^2} S^1$$

where the fiber over each point is diffeomorphic to $SU(n)/SO(n)$. Now $SU(n)$ is simply connected and $SO(n)$ is connected. Therefore $SU(n)/SO(n)$ is also simply connected. (Indeed, any curve starting and ending at $SO(n)$ in $SU(n)/SO(n)$ is homotopic to the image of a curve starting and ending at 1 in $SU(n)$, since $SO(n)$ is connected. But any closed curve in $SU(n)$ is homotopic to the trivial curve, and thus so is its image.) Thus,

$$\pi_1(L(V)) = \mathbf{Z} \tag{2.9}$$

and, in particular,

$$H^1(L(V), \mathbf{Z}) = \mathbf{Z}. \tag{2.10}$$

We shall present an independent proof of these facts in the next section. Now $dz/2\pi iz$ is a form on S^1 which generates $H^1(S^1)$. Hence

$$(\det^2)^* \frac{dz}{2\pi iz}$$

defines a form on $L(V)$, generating $H^1(L(V))$. Now $\det^2$ is defined on $U(n)/O(n)$ and hence becomes a map on $L(V)$ only after we have identified $L(V)$ with $U(n)/O(n)$ by a choice of Hermitian metric (and Lagrangian subspace). However, by Proposition 2.6, all such choices can be smoothly deformed into one another and hence the cohomology class is independent of the choice. This class is called the Maslov class.

We now give an explicit description of the universal covering space, $\tilde{L}(V)$, of the space of all Lagrangian subspaces, $L(V)$, of a symplectic vector space, V. We shall show, following Leray, that there exists an invariant, $m(u, u')$, for any two transversal elements, u and u' of $\tilde{L}(V)$. (Here u and u' are called transversal if πu and $\pi u'$ are transversal Lagrangian subspaces, where π denotes the projection of $\tilde{L}(V)$ onto $L(V)$.) We shall relate the invariants of three transversal elements, u, u' and u'' to the signature of the quadratic form associated to the three Lagrangian subspaces, πu, $\pi u'$ and $\pi u''$, and use the invariant, m, to give an alternative description of the Maslov class. For the purposes of obtaining these results, we continue in our choice of some Hermitian metric on V (and a choice of an orthonormal basis) which allows us to identify V with $\mathbf{C}^n$. Any Lagrangian subspace, X, is of the form $X = A\mathbf{R}^n$ for some $A \in U(n)$ and

$$A\mathbf{R}^n = B\mathbf{R}^n \quad \text{if and only if } A\overline{A}^{-1} = B\overline{B}^{-1},$$

where $\overline{A}$ denotes the matrix whose entries are the complex conjugates of the entries of A. We can thus define a map $\upsilon\colon L(V) \Rightarrow U(n)$ by

$$\upsilon(X) = A\overline{A}^{-1} \quad \text{if } X = A\mathbf{R}^n. \tag{2.11}$$

Notice that if $B \in U(n)$ then

$$\upsilon(BX) = B\upsilon(X)\overline{B}^{-1}. \tag{2.12}$$

If $z \in \mathbf{C}^n$ then $z = Ar$ for $r \in \mathbf{R}^n$ if and only if $z = A\overline{A}^{-1}\bar{z}$. Thus

$$z \in X \quad \textit{if and only if } z = \upsilon(X)\bar{z}. \tag{2.13}$$

If $X \cap Y \neq \{0\}$ the pair of equations $z = \upsilon(X)\bar{z}$ and $z = \upsilon(Y)\bar{z}$ has a nontrivial solution, so $\upsilon(X) - \upsilon(Y)$ is not invertible. Conversely: by applying a suitable element of $U(n)$ we may assume that $X = \mathbf{R}^n$ and $Y = A\mathbf{R}^n$. If $u = A\overline{A}^{-1}\bar{u}$ then $\bar{u} = A\overline{A}^{-1}u$ so we can find a $v \in \mathbf{R}^n$ with $v = A\overline{A}^{-1}\bar{v}$ so $v \in X \cap Y$. Thus,

$$X \textit{ and } Y \textit{ are transversal if and only if } \upsilon(X) - \upsilon(Y) \textit{ is invertible.} \tag{2.14}$$

Let $\tilde{U}(n)$ denote the space of all pairs

$$(A, \varphi),\ A \in U(n),\ \varphi \in \mathbf{R} \quad \textit{satisfying } \det A = e^{i\varphi}. \tag{2.15}$$

The multiplication $(A, \varphi) \cdot (A', \varphi') = (AA', \varphi + \varphi')$ makes $\tilde{U}(n)$ into a group and the map $\tilde{U}(n) \to U(n)$ sending (A, φ) into A makes $\tilde{U}(n)$ into a covering group of $U(n)$. The map

$$SU(n) \times \mathbf{R} \to \widetilde{U}(n) \text{ sending } (B, \psi) \text{ into } (Be^{i\psi}, n\psi)$$

is easily seen to be an isomorphism. Since $SU(n)$ is simply connected, it follows that $\tilde{U}(n)$ is the universal covering group of $U(n)$. This shows that the fundamental group of $U(n)$ is $\mathbf{Z}$. This also implies that the fundamental group of $Sp(V)$ is $\mathbf{Z}$; we shall sketch a proof of this fact here, referring the reader ahead to Chapter V, §5, for the proof of some of the group theoretical facts that we will use. We first observe the following fact about the algebra $sp(V)$: Let J denote multiplication by i (in the complex structure we have introduced on V).

We have the vector space direct sum decomposition $sp(V) = u(n) \oplus \mathcal{P}$ *where,* $u(n) = \{A \in sp(V) | JAJ^{-1} = A\}$ and $\mathcal{P} = \{B \in sp(V) | JBJ^{-1} = -B\}$. *Every element of* $\mathcal{P}$ *is of the form* $B = SC$, *where* C *denotes complex conjugation and* S *is a symmetric complex* $n \times n$ *matrix, i.e.* $Bz = S\bar{z}$ *for all* $z \in V \sim \mathbf{C}^n$.

PROOF. Since $J^2 = -1$, the operator, Θ on $sp(V)$ consisting of conjugation by J, i.e. the operator $\Theta(A) = JAJ^{-1}$ satisfies $\Theta^2 = 1$. Thus the direct sum in question is the decomposition of $sp(V)$ into $+1$ and -1 eigenspaces for Θ. The fact that $u(n)$ consists of the $+1$ eigenspace is just the characterization of $U(n)$ as the subgroup of the symplectic group preserving the complex (and hence the Hermitian) structure. The complex conjugation, C, clearly satisfies $JCJ^{-1} = -C$. Since S is complex linear, $JSJ^{-1} = S$ and hence $JBJ^{-1} = -B$ if $B = SC$. Let us show that all such B belong to $sp(V)$. We must show that

$$\operatorname{Im}(\langle Bu, v\rangle + \langle u, Bv\rangle) = 0.$$

Now $\langle Bu, v\rangle = \langle S\bar{u}, v\rangle$ and $\langle u, Bv\rangle = \langle u, S\bar{v}\rangle = \langle S^* u, \bar{v}\rangle = \langle \bar{S}u, \bar{v}\rangle$ since S is symmetric. But $\langle \bar{S}u, \bar{v}\rangle$ is the complex conjugate of $\langle S\bar{u}, v\rangle$ so that the above equality holds. Now the dimension, over the real numbers, of the space of complex symmetric matrices is $n(n+1)$, while $\dim u(n) = n^2$ and $\dim sp(V) = n(2n+1)$. Thus, by dimension count, we see that all elements of $\mathcal{P}$ have the desired form.

It now follows (cf. §5 of Chapter V) that we have the polar decomposition: every element of $Sp(V)$ can be uniquely written as the product

$$a = u \cdot \exp B \quad u \in U(n),\ B \in \mathcal{P},$$

where exp: $sp(V) \to Sp(V)$ is the exponential map. Since $\mathcal{P}$ is contractible, this shows that $Sp(V)$ and $U(n)$ have the same fundamental group. In fact, if $\tilde{S}p(V)$ denotes the universal covering group of $Sp(V)$ the polar decomposition theorem for $\tilde{S}p(V)$ implies that every element of $\tilde{S}p(V)$ can be written in the form

$$\tilde{a} = \tilde{u} \cdot \operatorname{Exp} B \quad \tilde{u} \in \tilde{U}(n),\ B \in \mathcal{P}$$

where Exp: $sp(V) \Rightarrow \tilde{S}p(V)$ is the exponential map for $\tilde{S}p(V)$. We shall identify the element

$$g = (I, 2\pi) \in \tilde{U}(n)$$

with the generator of the fundamental group of $Sp(V)$.

Let $\tilde{L}(V)$ denote the set of all pairs (X, θ), $X \in L(V)$, $\theta \in \mathbf{R}$ satisfying

$$\det v(X) = e^{i\theta}. \tag{2.16}$$

The group $\mathbf{Z}$ acts on $\tilde{L}(V)$ by

$$k(X, \theta) = (X, \theta + 2\pi k) \quad k \in \mathbf{Z},$$

and

$$\tilde{L}(V)/\mathbf{Z} = L(V),$$

making $\tilde{L}(V)$ into a covering space of $L(V)$. The group $\tilde{U}(n)$ acts transitively on $\tilde{L}(V)$ by

$$(A,\varphi)\cdot(X,\theta) = (AX, \theta + 2\varphi). \tag{2.17}$$

In view of (2.12) this is well defined. The subgroup which leaves the element $(\mathbf{R}^n, 0) \in \tilde{L}(V)$ fixed consists of all $(A,0)$ where

$$A = \bar{A} \quad \text{and} \quad \det A = 1$$

i.e. $A \in SO(n)$. Thus,

$$\tilde{L}(V) = \tilde{U}(n)/SO(n).$$

Since $\tilde{U}(n)$ is simply connected and $SO(n)$ is connected, this implies that $\tilde{L}(V)$ is simply connected. Thus

$\tilde{L}(V)$ is the universal covering space of $L(V)$.

Let $\tilde{\gamma}$ be a path in $\tilde{L}(V)$ such that $\tilde{\gamma}(0) = (X,\theta)$ and $\tilde{\gamma}(1) = (X, \theta + 2\pi)$. Let γ be the corresponding curve on $L(V)$, so that γ is a closed path starting and ending at X. It follows from (2.13) and (2.15) that

$$\int_\gamma (\det{}^2)^* \frac{dz}{2\pi i z} = 1. \tag{2.18}$$

Let $u = (X,\theta)$ and $u' = (X',\theta')$ be two elements of $\tilde{L}(V)$. We say that u and u' are transverse if X and X' are transverse Lagrangian subspaces. We now wish to define the Maslov index, $m(u,u')$, associated to a pair, u and u' of transverse elements of $\tilde{L}(V)$. For this purpose, we define the logarithm of an element A of Gl $(n,\mathbf{C})$ by the formula

$$\operatorname{Log} A = \int_{-\infty}^{0} \{(sI - A)^{-1} - (s-1)^{-1} I\}\, ds$$

where I is the unit matrix. This definition is valid for any: $A \in \mathrm{GL}\,(n,\mathbf{C})$ which does not have any eigenvalue on the negative real axis. It is easy to check that

$$\exp(\operatorname{Log} A) = A \quad \text{wherever } \operatorname{Log} A \text{ is defined,}$$

$$e^{tr(\operatorname{Log} A)} = \det A,$$

and

$$\operatorname{Log} A^{-1} = -\operatorname{Log} A.$$

If $u = (X,\theta)$ and $u' = (X',\theta')$ are transversal, then following Souriau [24], we define

$$m(u,u') = \frac{1}{2\pi}\{\theta - \theta' + itr\,\mathrm{Log}\,(-v(X)v(X')^{-1})\}. \tag{2.19}$$

By (2.14), $v(X) - v(X')$ is invertible, thus $-v(X)v(X')^{-1}$ does not have -1 as an eigenvalue, and hence does not have any negative real eigenvalues since it is unitary. Thus (2.19) is defined if u and u' are transverse. Since $e^{tr\,\mathrm{Log}\,A} = \det A$, it follows that

$$e^{2\pi i m(u,u')} = (-1)^n = e^{in\pi} \quad \text{where } \dim V = 2n.$$

Thus

$$m(u,u') \in \mathbf{Z} \quad \textit{if } n \textit{ is even and } m(u,u') \in \mathbf{Z} + \tfrac{1}{2} \quad \textit{if } n \textit{ is odd.} \tag{2.20}$$

Notice that

$$m(k \cdot u, k' \cdot u') = k - k' + m(u,u') \quad \textit{for } k,\, k' \in \mathbf{Z} \tag{2.21}$$

so that all values permitted by (2.20) are in fact taken on.

The group $Sp(V)$ is connected and acts on $L(V)$, and hence this action is covered by a unique action of $\tilde{S}p(V)$ on $\tilde{L}(V)$. If u and u' are transverse, then so are $\tilde{a}u$ and $\tilde{a}u'$, for any $\tilde{a} \in \tilde{S}p(V)$. The map $\tilde{a} \to m(\tilde{a}u, \tilde{a}u')$ is well defined, continuous, and takes values in a discrete set, and hence is constant. Thus

$$m(\tilde{a}u, \tilde{a}u') = m(u,u') \quad \text{for all } a \in \tilde{S}p(V). \tag{2.22}$$

In other words, $m(u,u')$ does not depend on the choice of complex structure, but is an invariant of $\tilde{S}p(V)$. It is clear from the definition that

$$m(u,u') + m(u',u) = 0. \tag{2.23}$$

Let X, X' and X'' be three transversal Lagrangian subspaces. By Proposition 2.3, the spaces X, X'' determine a quadratic form on X' given by

$$H(y_1, y_2) = (P_{X''}y_1, y_2) \quad y_1, y_2 \in X'$$

where $P_{X''}$ denotes projection onto X'' along X. We define

$$i(X, X', X'') = \tfrac{1}{2}\,\mathrm{sig}\,H. \tag{2.24}$$

This is a symplectic invariant assigned to any triple of transverse Lagrangian subspaces, that is

$$i(aX, aX', aX'') = i(X, X', X'') \tag{2.25}$$

for any $a \in Sp(V)$. (Recall that there is *no* invariant for pairs of transverse Lagrangian subspaces; the group $Sp(V)$ acts transitively on the set of all pairs of transverse Lagrangian subpsaces.)

Let $u = (X, \theta)$, $u' = (X', \theta')$ and $u'' = (X'', \theta'')$ be points of $\tilde{L}(V)$ sitting over X, X', and X''. We claim that the following formula

$$m(u, u') + m(u', u'') + m(u'', u) = i(X, X', X''), \tag{2.26}$$

due to Leray [23], holds.

To prove this formula we may apply any $a \in \tilde{Sp}(V)$ to the u, u', and u'' on left hand sides with the corresponding $a \in Sp(V)$ to the X, X', X'' on the right. We may thus assume that $X = \mathbf{R}^n$ and that $X' = i\mathbf{R}^n$, and that $X'' = b\mathbf{R}^n$ where $b \in Sp(V)$ has the form

$$b = \begin{pmatrix} I & S \\ 0 & I \end{pmatrix}$$

relative to the basis determined by $V \sim \mathbf{R}^n \oplus i\mathbf{R}^n$, where S is a symmetric matrix. It follows from (2.4) and (2.24) that

$$i(X, X', X'') = \tfrac{1}{2} \operatorname{sig} S.$$

Finally, by applying an element $c \in Sp(n)$ of the form

$$c = \begin{pmatrix} A & 0 \\ 0 & A^{t-1} \end{pmatrix} \quad A \in \mathrm{Gl}\,(n)$$

we may assume that S is a diagonal matrix with $+1$'s and -1's on the diagonal, i.e.

$$S = \operatorname{diag}\,(+1, \ldots, +1, -1, \ldots, -1)$$

where there are k $+$'s and $(n-k)$ $-$'s. Thus $\operatorname{sig} S = 2k - n$ and

$$i(X, X', X'') = \tfrac{1}{2}(2k - n).$$

If $\delta_1, \ldots, \delta_n$ denotes the standard basis of $\mathbf{R}^n$, then $i\delta_1, \ldots, i\delta_n$ is a basis of X', and the vectors $(1 \pm i)\delta_j$ form a basis of X'', where the choice of sign is $+$ for the first k vectors and $-$ for the last $n-k$. Since $\sqrt{2} \cdot e^{\pm \pi i/4} = 1 \pm i$, we may, as well, take $e^{\pm \pi i/4}\delta_j$ as the basis vectors. Thus $X'' = A\mathbf{R}^n$ where A is the diagonal unitary matrix

$$A = \operatorname{diag}\,(e^{\pi i/4}, \ldots, e^{\pi i/4}, e^{-\pi i/4}, \ldots, e^{-\pi i/4}).$$

Therefore, by (2.11),

$$v(X'') = \operatorname{diag}\,(e^{\pi i/2}, \ldots, e^{\pi i/2}, e^{-\pi i/2}, \ldots, e^{-\pi i/2}),$$

where, as always, there are k $+$'s and $(n-k)$ $-$'s. It is obvious that

$$v(X) = I \quad \text{and} \quad v(X') = -I.$$

Then if $u'' = (X'', \theta'')$ we must have, by (2.16)

$$\theta'' = 2\pi[q'' + \tfrac{1}{4}(2k - n)]$$

for some integer q''. Similarly,

$$\theta = 2\pi q \quad \text{and} \quad \theta' = 2\pi[q' + n/2]$$

where q and q' are integers. Thus, by (2.19),

$$m(u, u') = q - q' - n/2.$$

Also

$$\begin{aligned}\operatorname{Log}\,(-v(X')v(X'')^{-1}) &= \operatorname{Log}\operatorname{diag}\,(e^{-\pi i/2}, \ldots, e^{-\pi i/2}, e^{\pi i/2}, \ldots, e^{\pi i/2}) \\ &= \operatorname{diag}\,(-\pi i/2, \ldots, -\pi i/2, \pi i/2, \ldots, \pi i/2)\end{aligned}$$

since our choice of the Log function is the one which is given by analytic continuation from the positive real axis in both the upper and lower half planes, i.e. $\operatorname{Log} e^{\pi i/2} = \pi i/2$ and $\operatorname{Log} e^{-\pi i/2} = -\pi i/2$. Thus

$$itr\,\operatorname{Log}\,(-v(X')v(X'')^{-1}) = \tfrac{1}{2}\pi(2k - n)$$

and

$$m(u', u'') = q' + n/2.$$

Similarly,

$$itr\,\operatorname{Log}\,(-v(X'')v(X)^{-1}) = \tfrac{1}{2}\pi(2k - n)$$

so that

$$m(u'', u) = q'' - q + \tfrac{1}{2}(2k - n).$$

Adding up the three expressions for m proves Leray's formula, (2.26). Notice that it follows from (2.26) that $i(X, X', X'')$ is an antisymmetric function of its three variables. It also follows that if we define the Hormander cross index of four transverse Lagrangian subspaces X, Y, Z, and W by

$$(X, Y, Z, W) = i(X, Y, Z) - i(X, Y, W)$$

that

$$(X, Y, Z, W) = -(Z, W, X, Y).$$

Neither of these facts is immediately obvious from the definition (2.24). We shall discuss them further in the next section.

We are now in a position to discuss the relation between the Maslov class as defined earlier in this section, and the Maslov cycle as introduced in §7 of Chapter II. Let Y be a fixed Lagrangian subspace. (Eventually, Y will play the role of the "tangent to the vertical" in $T^* M$ in a local coordinate system.) Let $X(t)$ be a curve of Lagrangian subspaces defined for $0 \leqslant t \leqslant 1$, and suppose that $X(0)$ and $X(1)$ are both transversal to Y. Let $\tilde{Y}$ be a point of $\tilde{L}(V)$ covering Y and let $u(t)$ be a curve in $\tilde{L}(V)$ covering $X(t)$. It follows from (2.21) that the integer

$$l = m(\tilde{Y}, u(1)) - m(\tilde{Y}, u(0))$$

is independent of our choices of liftings. The function $m(\tilde{Y}, u(\cdot))$ fails to be defined at precisely those values of t for which $X(t)$ is not transversal to Y. Suppose that s is an isolated point where this happens. That is, suppose $X(t')$ is transversal to Y for all $t' < s$ and sufficiently close to s and also that $X(t'')$ is transversal to Y for all $t'' > s$ and sufficiently close to s. We can use (2.26) to evaluate the jump in $m(\tilde{Y}, u(t))$ as we cross the value s in terms of the difference of signatures of quadratic forms defined downstairs on $L(V)$: In fact, let us choose some other Lagrangian subspace, Z, which is transverse to Y and to $X(t)$ for all $t' \leqslant t \leqslant t''$. This is clearly possible if t' and t'' are sufficiently close. Then, by (2.26)

$$\begin{aligned} i(Z, Y, X(t'')) - i(Z, Y, X(t')) &= m(\tilde{Y}, u(t'')) - m(\tilde{Y}, u(t')) \\ &\quad - (m(\tilde{Z}, u(t'')) - m(\tilde{Z}, u(t'))) \end{aligned}$$

for some choice of $\widetilde{Z}$ sitting over Z. But $m(\widetilde{Z}, u(t))$ is a continuous function on the interval $t' \leqslant t \leqslant t''$ with discrete values and hence $m(\widetilde{Z}, u(t'')) = m(\widetilde{Z}, u(t'))$. Hence

$$m(\tilde{Y}, u(t'')) - m(\tilde{Y}, u(t')) = i(Z, Y, X(t'')) - i(Z, Y, X(t')). \tag{2.27}$$

Now suppose that $X(1) = X(0)$, i.e. that the curve is closed. It follows from (2.21) that $l \cdot u(0) = u(1)$ and hence from (2.18) that

$$l = \int (\det{}^2)^* \frac{dz}{2\pi i z}. \tag{2.28}$$

Suppose that Λ is a Lagrangian submanifold of V. We may identify the tangent space, $T\Lambda_\lambda$, with a Lagrangian subspace of V, when we identify TV_λ with V, under the usual identification of the tangent space of a vector space with the vector space itself. Any curve, γ, on Λ then gives rise to a curve of Lagrangian subspaces defined by $X(t) = T\Lambda_{\gamma(t)}$. If γ is a closed curve then (2.28) defines an integer associated to γ; in fact, we have defined an element of $H^1(\Lambda, \mathbf{Z})$ which can be computed as an integral, (2.28) once a complex structure has been chosen, or as sum of "crossing numbers" (2.27) in terms of a fixed Lagrangian subspace,

Y. The actual class is independent of the choice of Y or of the complex structure. We shall call this the Leray class of Λ.

Similarly, suppose that M is a differentiable manifold. For each $z \in T^* M$, the tangent space $T(T^* M)_z$ is a symplectic vector space and is equipped with a preferred Lagrangian subspace, Y_z, where Y_z is tangent to the vertical, i.e. Y_z is the subspace consisting of those vectors, ζ, which satisfy $d\pi_z \zeta = 0$, where π denotes the standard projection of $T^* M$ onto M. A choice of a Riemann metric on M puts a positive definite scalar product on Y_z, giving an identification of $L(T(T^* M))$ with $U(n)/O(n)$ so that $(\det^2)^*(dz/2\pi iz)$ is a well-defined differential form on the bundle $L(T^* M)$, where $L(T^* M)$ denotes the bundle over $T^* M$ whose fiber over z consists of the set of all Lagrangian subspaces of $T(T^* M)_z$. If Λ is a Lagrangian submanifold of $T^* M$, then, once again, any curve, γ, on Λ determines a curve on $L(T^* M)$ and we can use (2.28) to define an integer, i.e. we have defined an element of $H^1(\Lambda, \mathbf{Z})$ which is the Maslov class of Λ. This class does not depend on the choice of Riemann metric, since we can continuously deform any two Riemann metrics into one another. In terms of local coordinates on M, we get a local identification of a neighborhood of $T^* M$ with an open subset of a symplectic vector space, V, in which all the Y_z are identified with a fixed Lagrangian subspace, Y. We can then use (2.27) for the computation of this class in terms of local crossing numbers. In the next section we shall describe the definition of the Maslov class due to Hörmander using Čech theory.

§3. The cross index and the Maslov class.*

Let us begin this section by giving a somewhat different presentation of the computation of $\pi_1(L(V))$ and of the class introduced in the preceding section. We will use induction on dim V rather than the introduction of a complex structure, and we will find some applications for this alternative approach in what follows.

Let R be an isotropic subspace of V. Then $R^\perp \supset R$ and since $R = (R^\perp)^\perp$, we see that $W = R^\perp/R$ is again a symplectic vector space with

$$\begin{aligned} \dim R^\perp/R &= \dim R^\perp - \dim R \\ &= \dim V - 2 \dim R \end{aligned}$$

since

$$\dim R + \dim R^\perp = \dim V.$$

Let X be any Lagrangian subspace of V. Then $X \cap R^\perp/X \cap R$ is clearly an isotropic subspace of W. We claim that it is Lagrangian, i.e., that

$$\dim(X \cap R^\perp/X \cap R) = \tfrac{1}{2} \dim W = \tfrac{1}{2} \dim V - \dim R.$$

To prove this, notice that

* This section should be omitted on first reading.

$$\dim X \cap R^{\perp} = \dim V - \dim(X \cap R^{\perp})^{\perp}$$
$$= \dim V - \dim(X + R)$$

and

$$\dim(X + R) + \dim(X \cap R) = \dim X + \dim R$$

so that

$$\dim X \cap R^{\perp}/X \cap R = \dim X \cap R^{\perp} - \dim X \cap R$$
$$= \dim V - [\dim(X + R) + \dim(X \cap R)]$$
$$= \dim V - [\dim X + \dim R]$$
$$= \tfrac{1}{2} \dim V - \dim R$$

as $\dim X = \frac{1}{2} \dim V$. We have thus proved:

PROPOSITION 3.1. *Let R be an isotropic subspace of V. Then $W = R^{\perp}/R$ is a symplectic vector space and the map ρ defined by*

$$\rho(X) = X \cap R^{\perp}/X \cap R$$

sends $L(V) \to L(W)$.

Unfortunately, the map ρ is not continuous. For example, let us examine the map ρ for the case $V = \mathbf{R}^4$ with basis $\{e_1, e_2, f_1, f_2\}$ where

$$(e_1, e_2) = (f_1, f_2) = 0 \qquad \text{and} \qquad (e_i, f_j) = \delta_{ij},$$

and where we take

$$R = \{e_1\}, \qquad \text{so that} \qquad R^{\perp} = \{e_1, e_2, f_2\}.$$

We will describe the map ρ locally, in the coordinate chart consisting of those

$$X \in \mathfrak{L}_{\{f_1, f_2\}}.$$

For such X the projection onto $\{e_1, e_2\}$ is non-singular and X determines a symmetric map of $\{e_1, e_2\}$ into $\{f_1, f_2\}$. In particular, X is spanned by vectors

$$e_1 + a_{11} f_1 + a_{12} f_2 \qquad \text{and} \qquad e_2 + a_{21} f_1 + a_{22} f_2$$

where $a_{12} = a_{21}$ and the matrix $A = (a_{ij})$ determines, and is determined by X. We shall therefore denote X by X_A. Now $X \cap R^{\perp}$ consists of combinations of the above vectors having 0 as the coefficient of f_1. Thus, there are two cases to consider:

(i) $a_{11} = a_{21} = 0$. This amounts to the assumption that $X_A \subset R^\perp$. In this case $a_{12} = 0$ and

$$\rho(X_A) = ([e_2] + a_{22}[f_2])$$

where $[e_2] = e_2/R$. We write this for short as

$$\rho(X_A) = (1, a_{22})$$

(ii) $(a_{11}, a_{21}) \neq (0,0)$. Then $\dim X_A \cap R^\perp = 1$ and

$$\rho(X_A) = (a_{11}, \det A).$$

If, in the above formulae we let

$$A = \begin{pmatrix} 0 & s \\ s & 0 \end{pmatrix}$$

we see that

$$\rho(X_A) = \begin{cases} (1,0) & \text{for } s = 0, \\ (0,s) = (0,1) \text{ projectively} & \text{for } s \neq 0, \end{cases}$$

so that ρ is not continuous.

From this example we see that we can expect trouble from ρ at those X satisfying $X \cap R \neq 0$. In fact, this is indeed the case, and indeed the only troublesome locus for ρ for general V and R. By choosing a basis of R, we may proceed inductively on the dimension of R. Let us therefore analyse, in some detail, the map ρ in the case where $R = \{e\}$ is one dimensional. Thus $\dim V = 2n$ and letting $W = R^\perp/R$, we get $\dim W = 2n - 2$. Recall that $\dim L(V) = \dim(S^2(\mathbf{R}^n)) = n(n+1)/2$. We will let

$$S_R = \{X \in L(V) \mid X \supset R\} = \{X \in L(V) \mid X \subset R^\perp\}$$

since $X = X^\perp$.

PROPOSITION 3.2. *The set S_R is a submanifold of codimension n in $L(V)$. The map ρ restricted to S_R is a diffeomorphism of S_R onto $L(W)$. The map ρ, when restricted to $L(V) - S_R$, is a smooth map making $L(V) - S_R$ into a fiber bundle over $L(W)$ with fiber $\mathbf{R}^n$.*

Notice that if $n > 2$, then the proposition implies that the inclusion of $L(V) - S_R$ into $L(V)$ induces an isomorphism on π_1. Indeed any smooth curve can be deformed so as to avoid S_R and so can any smooth homotopy. Since $L(V) - S_R$ is asserted to be a fiber bundle over $L(W)$ with homotopically trivial fiber, we conclude that ρ induces an isomorphism of $\pi_1(L(V))$ with $\pi_1(L(W))$. We shall see by direct calculation that the same is true for the case that dim $V = 4$. This will provide an alternative proof of the fact that $\pi_1(L(V)) = \mathbf{Z}$. Also, the generator in the plane will then determine the generator of $\pi_1(L(V))$. We will see in the course of a subsequent calculation, that this generator coincides with the one previously obtained from the Hermitian structure.

PROOF OF THE PROPOSITION. The fact that S_R has codimension n is pretty obvious. It suffices to check that $S_R \cap \mathfrak{L}_X$ is a submanifold of codimension n for each $X \in L(V)$. Now choose a complementary $Y \in \mathfrak{L}_X$, giving a direct sum decomposition $V = X \oplus Y$ with corresponding projections π_X and π_Y. Now if $S_R \cap \mathfrak{L}_X \neq \varnothing$ we conclude that $\pi_Y e \neq 0$, where $R = \{e\}$. Any $Z \in \mathfrak{L}_X$ corresponds to a symmetric map, A, from Y to X, and $Z \in S_R$ if and only if $A\pi_Y e = \pi_X e$. This clearly represents n linear conditions on A.

To see that $\rho_{|S_R}$ is a bijection notice that if $R \subset X \subset R^\perp$ then $\rho(X) = X/R$. If X' is a Lagrangian subspace of $R^\perp/R$ then its inverse image in $R^\perp$ is a Lagrangian subspace, the unique X with $\rho(X) = X'$.

Let us now examine $L(V) - S_R$. For $X \not\subset R^\perp$ it is clear that the map $X \to X \cap R^\perp$ is smooth, and $X \cap R^\perp$ does not contain R. The map $X \cap R^\perp \longrightarrow X \cap R^\perp / X \cap R$ is thus also smooth, proving that ρ is smooth on $L(V) - S_R$. Let us examine the inverse image. Let Z_1' and Z_2' be two $(n-1)$ dimensional isotropic spaces of $R^\perp$ with $Z_1'/R = Z_2'/R = Z''$. Then given $z \in Z''$ if we get $z_1 \in Z_1'$ and $z_2 \in Z_2'$ and so $z_1 - z_2 \in R$. In this way it is clear that the inverse image of Z'' is an affine space whose associated linear space is Hom (Z'', R). Now for a given $n-1$ dimensional isotropic Z' lying in $R^\perp$ we must determine all possible Z's in $L(V)$ with $Z \cap R^\perp = Z'$. Such a Z must lie in $(Z')^\perp$ which is $n+1$ dimensional. We are thus looking for all lines in $(Z')^\perp/Z'$, with the exclusion of the line $\{e + Z'\}$. Since $\dim(Z')^\perp/Z' = 2$ we are essentially adding the affine line. Thus the entire inverse image of Z'' in $L(V) - S_R$ will be diffeomorphic to $\mathbf{R}^n$.

Let us now examine what happens when dim $V = 4$. In this case $L(V)$ is diffeomorphic to $U(2)/O(2)$. Now we can write every unitary 2×2 matrix as $U = U'\Delta$ where Δ is diagonal and $U' \in SU(2)$. This decomposition is unique up to multiplying both factors by $-I$. Notice that $-I \in SO(2)$. Thus

$$U(2)/O(2) \sim SU(2)/SO(2) \times S^1/S^0$$

where $S^0 = \{\pm I\}$. Now $SU(2)/SO(2)$ is a compact simply connected two dimensional manifold and hence is S^2. Thus

$$L(V) = S^2 \times S^1.$$

Let us sketch this result directly. We may assume $V = \mathbf{R}^4$. If we pick a unit vector $v = (a, b, c, d)$ then the set of unit vectors in $v^\perp$ which are also orthogonal to v with respect to the scalar product forms a circle. Once we pick a point on this circle we determine a Lagrangian subspace, together with an orthogonal basis. We may rotate any one of these bases into any other so that we must divide by S^1. Thus we have $(S^3 \times S^1)/S^1$. Actually the second S^1 acts only on the S^3 to fiber S^3 over S^2 (the so called Hopf fibration) so we end up with $S^2 \times S^1$. In more detail: Define the operators E and F of $\mathbf{R}^4 \to \mathbf{R}^4$ by

$$E(a,b,c,d) = (-c,d,a,-b),$$
$$F(a,b,c,d) = (d,c,-b,-a).$$

Then $(v, Ev) = 0$ and $\langle v, Ev \rangle = 0$ where $\langle , \rangle$ denotes scalar product and similarly for F. Also $E^2 = -I = F^2$ and $EF = -FE$. Given any θ, let

$$E_\theta = \cos\theta E + \sin\theta F \qquad \text{so that} \qquad E_\theta^2 = -I.$$

Then v, $E_\theta v$ spans all Lagrangian planes as v runs over S^3 and θ over S^1. On the other hand, any two dimensional orthogonal transformation acting on v, $E_\theta v$ spans the same space. Let

$$\begin{pmatrix} c & s \\ -s & c \end{pmatrix}, \qquad c^2 + s^2 = 1$$

be any rotation. Then

$$(cv + sE_\theta v, -sv + cE_\theta v) = (cv + sE_\theta v, E_\theta(cv + sE_\theta v)).$$

(Reflections we can eliminate by identifying v, $E_\theta v$ with v, $-E_\theta v$.) Thus each choice of θ determines a different fibration of S^3 by great circles over S^1. In terms of this picture it is clear that for any R, the set S_R consists of $S^1 \times p$ where p is a point of S^2.

Since removing a point from a sphere has no effect on the fundamental group we see that $\pi_1(L(V) - S_R) \cong \pi_1(L(V))$. (Incidently in this case we also see that $L(V) - S_R = S^1 \times \mathbf{R}^2$, as we have already proved.)

We now wish to show how to associate an integer (A, B, C, D) to a quadruplet of Lagrangian subspaces where

$$C \cap A = \{0\} = C \cap B$$

and

$$D \cap A = \{0\} = D \cap B.$$

It is defined as follows: Let

$$R = A \cap B.$$

Then $\rho(A)$ and $\rho(B)$ are transversal Lagrangian subspaces of $R^\perp/R$. By Proposition 3.4, any third Lagrangian subspace of $R^\perp/R$ which is transversal to $\rho(A)$ and

$\rho(B)$ corresponds to a quadratic form on $\rho(B)$. Since C and D are both transversal to A and B we conclude that $\rho(C)$ and $\rho(D)$ each determine non-singular quadratic forms, Q_C and Q_D, on $\rho(B)$. The quadratic form, Q_C, on $\rho(B)$ is given by $Q_C(b) = (P_A^C b, b)$ where P_A^C is the projection of $\rho(B)$ onto $\rho(A)$ along $\rho(C)$. Let

$$(A, B, C, D) = \tfrac{1}{2}[\operatorname{sig} Q_C - \operatorname{sig} Q_D] = \operatorname{ind} Q_D - \operatorname{ind} Q_C.$$

We shall write $i(A, B, C) = \frac{1}{2} \operatorname{sig} Q_C$ so that

$$(A, B, C, D) = i(A, B, C) - i(A, B, D).$$

It is clear that

$$(A, B, C, D) = -(A, B, D, C) \tag{3.1}$$

and

$$(A, B, C, D) + (A, B, D, E) + (A, B, E, C) = 0. \tag{3.2}$$

If $A \cap B = \{0\}$ then for all nearby A, B, C and D it is clear that the transversality conditions will still be satisfied and, since the signatures of Q_C and Q_D will not change, we conclude that (A, B, C, D) is locally constant. We wish to prove that this remains true even if $A \cap B \neq \{0\}$, provided that C and D each remain transversal to A and B. For this purpose, we will give an alternative definition of (A, B, C, D). Recall that $\mathfrak{L}_A$ is a cell. Since C and D both belong to $\mathfrak{L}_A$ there is a curve, γ_{CD}, joining C to D in $\mathfrak{L}_A$, and two such curves are homotopic. Similarly there is a unique curve (up to homotopy), γ_{DC}, joining D to C in $\mathfrak{L}_B$. This then defines a closed curve γ_{CDC} in $L(V)$ up to homotopy, i.e., an element of $\pi_1(L(V))$. It is some multiple of the basic generator. Our claim is that this multiple is exactly (A, B, C, D). In other words that

$$(A, B, C, D) = \int_{\gamma_{CDC}} (\det{}^2)^* \frac{dz}{2\pi i z} \tag{3.3}$$

in terms of some choice of identification of $L(V)$ with $U(n)/O(n)$. Since the curve γ_{CDC} can be made to vary smoothly with A, B, C and D so long as C and D remain transversal to A and to B we see that the left hand side is indeed a continuous (and hence constant) function of its arguments. The proof of (3.3) that we present below is essentially due to Kostant.

Before presenting the proof let us look at the various possibilities for the curve γ_{CDC} when $\dim V = 2$. We begin with the geometric picture of $i(A, B, C)$. For $A = B$ we have $i(A, B, C) = 0$ by definition. For $A \neq B$ we have, for $b \in B$ and $P_A^C b \in A$, that the signature of Q_C will be ± 1 according as to whether the vectors $P_A^C b$, b form a positive or negative orientation of the plane.

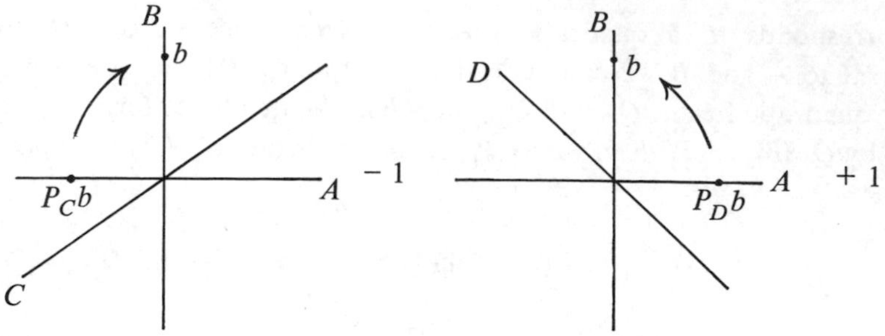

Now the various cases for (A, B, C, D):

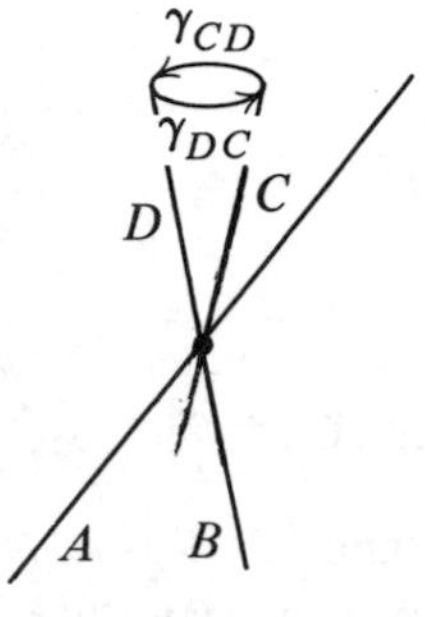

(a) $A = B$.

$$\int_{\gamma_{CDC}} (\det{}^2)^* \frac{dz}{2\pi i z} = 0$$

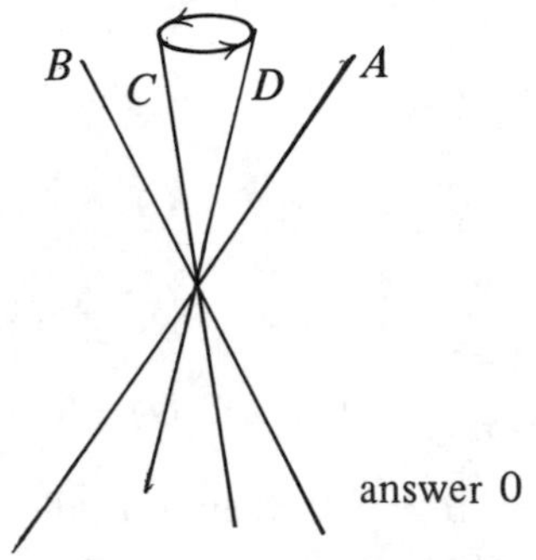

(b) $A \neq B$.

D, C lie in the same component of $\mathfrak{L}_A \cap \mathfrak{L}_B$.

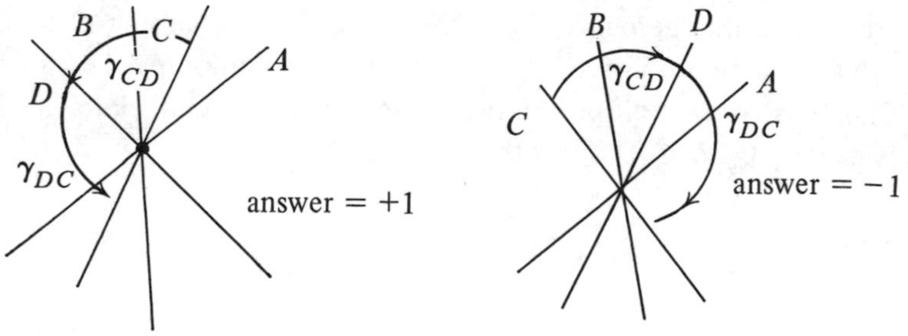

(c) $A \neq B$.

D, C lie in different components of $\mathfrak{L}_A \cap \mathfrak{L}_B$.

The formula is thus clear from the diagrams when $\dim V = 2$. (We shall give an analytical proof soon.) To prove the formula in general let us observe that it suffices to prove the formula for the case that $A \cap B = \{0\}$. Indeed, suppose that

$$A \cap B = R.$$

Since γ_{DC} lies in $\mathfrak{L}_A$ and γ_{CD} lies in $\mathfrak{L}_B$ we see that γ_{CDC} lies in $L(V) - S_R$. Now ρ: $L(V) - S_R \to L(R^\perp/R)$ determines an isomorphism between $\pi_1(L(V))$ and $\pi_1(L(R^\perp/R))$. Thus if the class of γ_{CDC} is k times the generator of $\pi_1(L(V))$ then $\rho(\gamma_{CDC})$ will determine the same multiple, k, of the generator of $\pi_1(L(R^\perp/R))$. But $\rho(A)$ and $\rho(B)$ are transversal in $R^\perp/R$. Hence we are reduced to the transversal case.

Now both sides of (3.3) don't change under deformations so long as transversality is maintained. Our object will be to deform the spaces until the formula becomes obvious.

First choose a Hermitian structure with $B = iA$, and choose an orthonormal basis, $e_1, \ldots, e_n$ in A. This determines a dual basis $f_1, \ldots, f_n$ in B. With respect to the basis $e_1, \ldots, e_n$, the subspace C determines a non-singular symmetric matrix $C = (C_{ij})$. Now we can find an orthogonal matrix, $O \in SO(n)$, such that $O(C)O^{-1}$ is diagonal. Since $SO(n)$ is connected we can find a curve $O(t)$ with $O(0) = \text{id}$ and $O(1) = O$. Then $O(t)CO(t)^{-1}$ deforms C into a diagonal matrix. By further deformation we may assume that the entries of C are $\pm$ 1, and similarly for D. It is clear from (3.2) and from the fact that we can choose $\gamma_{CE} = \gamma_{CD} \circ \gamma_{DE}$, etc. (see the figure)

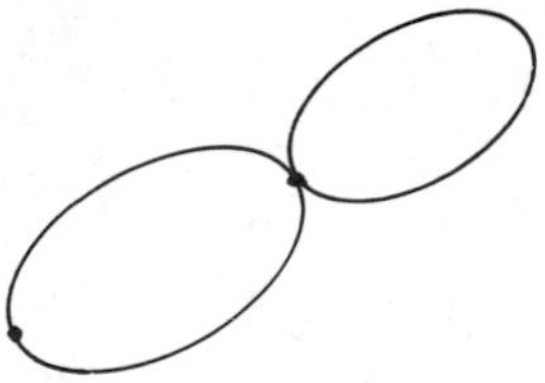

that it suffices to prove the formula when C and D differ in at most one position. If $C = D$ there is nothing to prove. Suppose that

$$C = \begin{pmatrix} +1 & & & & 0 \\ & \pm 1 & & & \\ & & \cdot & & \\ & & & \cdot & \\ & & & & \cdot \\ & 0 & & & \pm 1 \end{pmatrix} \quad \text{and } D = \begin{pmatrix} -1 & & & & 0 \\ & \pm 1 & & & \\ & & \cdot & & \\ & & & \cdot & \\ & & & & \cdot \\ & 0 & & & \pm 1 \end{pmatrix}$$

so that

$$\text{ind } D - \text{ind } C = 1.$$

Thus C is spanned by the vectors

$$e_1 + f_1, g_2, \ldots, g_n \qquad \text{where } g_i = e_i \pm f_i$$

or, if we like, by

$$\cos \frac{\pi}{4} e_1 + \sin \frac{\pi}{4} f_1, g_2, \ldots, g_n.$$

Similarly D is spanned by

$$-e_1 + f_1, g_2, \ldots, g_n$$

or by

$$\cos \frac{3\pi}{4} e_1 + \sin \frac{3\pi}{4} f_1, g_2, \ldots, g_n.$$

Now we define the curve γ_{CD} by

$$\gamma_{CD}(\theta) = (\cos \theta e_1 + \sin \theta f_1, g_2, \ldots, g_n), \qquad \frac{\pi}{4} \leqslant \theta \leqslant \frac{3\pi}{4}.$$

Throughout this range of θ the coefficient of f_1 does not vanish so that γ_{CD} lies in $\mathfrak{L}_A$. Similarly define

$$\gamma_{DC}(\theta) = (\cos \theta e_1 + \sin \theta f_1, g_2, \ldots, g_n), \qquad \frac{3\pi}{4} \leqslant \theta \leqslant \frac{5\pi}{4}.$$

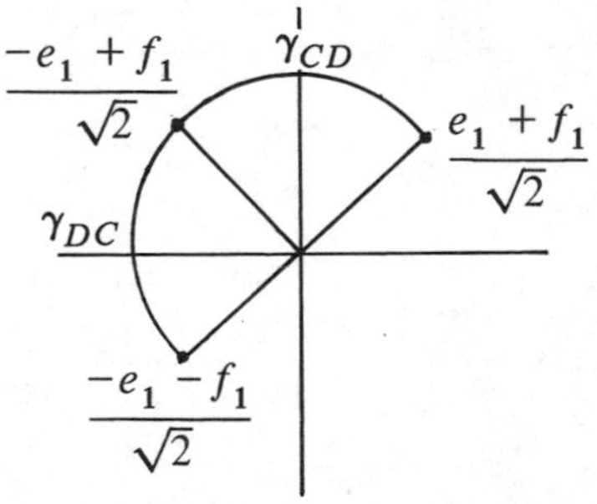

It is now clear that the projection of this curve onto the e_1, f_1 plane describes the projective line once in the proper orientation. In terms of $U(n)/O(n)$ it is clear that the curve γ_{CDC} is given by the equivalence class of the curve of unitary matrices

$$\begin{pmatrix} e^{i\theta} & & & & 0 \\ & c_2 & & & \\ & & \cdot & & \\ & & & \cdot & \\ & & & & \cdot \\ 0 & & & & c_n \end{pmatrix}, \qquad \frac{\pi}{4} \leqslant \theta \leqslant \frac{5\pi}{4},$$

and hence the function $\det^2$ goes around the circle once in the counter clockwise direction. This proves (3.3) and, incidentally, the consistency of the inductive definition of the generator of H^1 with the definition coming from $\det^2$.

PROPOSITION 3.3. *The symbol* (A, B, C, D) *satisfies*

$$(A, B, C, D) = -(C, D, A, B). \tag{3.4}$$

For the case $\dim V = 2$ we can get the result by examining the cases described in the figures given above: If $A = B$ then we can deform C into D remaining transversal to both A and B so that both sides of (3.4) vanish. If C and D lie in the same component of $\mathfrak{L}_A \cap \mathfrak{L}_B$ then we can deform C into D and then A into B so we are back in the preceding case. If C and D lie in different components of $\mathfrak{L}_A \cap \mathfrak{L}_B$ then it is clear from the figure that γ_{CDC} is oriented in the opposite direction from γ_{ABA}.

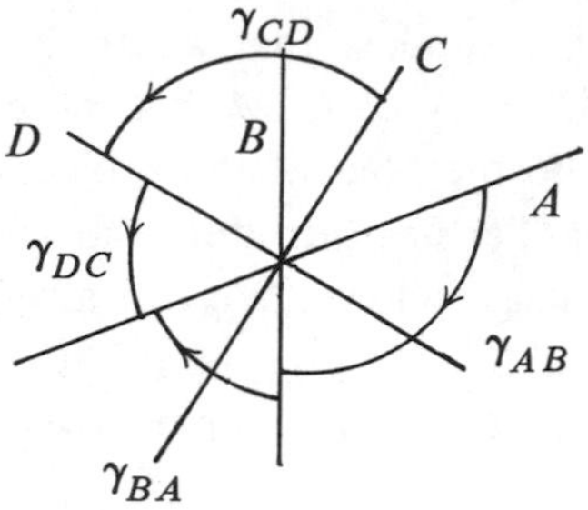

For the general case we shall use the following observation: Let V_1 and V_2 be two symplectic vector spaces. Then $V_1 + V_2$ is again a symplectic vector space and if $X_1 \in L(V_1)$ and $X_2 \in L(V_2)$ then $X_1 + X_2 \in L(V_1 + V_2)$. Thus we have a map of

$$L(V_1) \times L(V_2) \to L(V_1 + V_2).$$

We can clearly choose the Hermitian structure on $V_1 + V_2$ to be consistent with this direct sum decomposition and thus the map corresponds to the block diagonal embedding of $U(n) \times U(m) \to U(m + n)$. Thus we obtain

$$\frac{U(n) \times U(m)}{O(n) \times O(m)} \xrightarrow{f} \frac{U(n+m)}{O(n+m)} \xrightarrow{\det^2_{m+n}} S^1$$

where

$$\det{}^2_{m+n} \circ f = \det{}^2_m \det{}^2_n$$

with the obvious notation. Thus

$$f^* (\det{}^2_{m+n})^* \frac{d \log z}{2\pi i} = df^* \log \det{}^2_{m+n}/2\pi i = (\det{}^2_m)^* \frac{dz}{2\pi i z} + (\det{}^2_n)^* \frac{dz}{2\pi i z}.$$

Therefore in computing (A, B, C, D), if we could arrange that $A = A_1 + A_2$, etc., we would conclude that

$$(A, B, C, D) = (A_1, B_1, C_1, D_1) + (A_2, B_2, C_2, D_2).$$

Now by the deformation argument presented in the proof of the preceding proposition we know that we can arrange that $A = \{e_1, \ldots, e_n\}$, $B = \{f_1, \ldots, f_n\}$, $C = \{g_1, \ldots, g_n\}$, $D = \{h_1, \ldots, h_n\}$ where $g_i = e_i \pm f_i$ and $h_j = e_j \pm f_j$ for suitable choices of $\pm$. Thus we are reduced to the two dimensional case, which has already been established.

Now let $E \to N$ be a symplectic vector bundle. (Thus E is a vector bundle such that each fiber, E_n, has a symplectic structure varying smoothly with n.) The main application we have in mind will be the situation where $N = \Lambda$ is a Lagrangian submanifold of some cotangent bundle T^*M and where $E_n = T_n(T^*M)$. Of course, we then get a fiber bundle $L(E) \to N$ where $L(E)_n$ consists of all Lagrangian subspaces of E_n. Suppose that we are given two sections, A and B, of $L(E)$. (For example, if $N = \Lambda \subset T^*M$ then we could take $A_\lambda = T_\lambda(\Lambda)$ and take B_λ to be the tangent to the fiber of the projection $T^*M \to M$.) With this data, $\{E; A \text{ and } B\}$, Hormander has introduced an element of $H^1(N)$, which is defined, in the Čech theory, as follows: We can always locally find sections, C, D, etc. defined on open sets U_C, U_D, etc. such that C_x is transversal to both A_x and B_x for each $x \in U_C$. The U's form an open cover of N and we define a Čech 1-cocycle

$$\mathfrak{L}(U_C, U_D) = (A, B, C, D),$$

where the right hand side is taken to mean the function $x \to (A_x, B_x, C_x, D_x)$ defined on $U_C \cap U_D$. Since (A, B, C, D) is continuous and integer valued, it defines a Čech cochain $z(A, B)$ on N. Since for each $x \in U_C \cap U_D$ we have

$$(A_x, B_x, C_x, D_x) = i(A_x, B_x, C_x) - i(A_x, B_x, D_x)$$

we see that $\delta z(A, B) = 0$. (This is just (3.2).) Notice that $i(A, B, C)$ need not be continuous so that $z(A, B)$ is not a coboundary, in general. We will denote the corresponding cohomology class by α or $\alpha(E; A, B)$.

Let us compute the cohomology class α for the following situation. Let V be a symplectic vector space, and let $N = L(V)$. We define the symplectic vector bundle $E \to N$ by assigning a copy of V to each point of N. (In other words, E is the pull back to N of the vector bundle $V \to$ pt. under the constant map.) Then $L(E)$ has a canonical (tautologous) section, B, namely $B(n) = n$ where n is considered as a subspace of $E_n = V$. Let A be a constant section of E (i.e., the pull back of a "section" of $L(V) \to$ pt.) We then obtain an element $\alpha(A, B) \in H^1(L(V))$. We claim that

PROPOSITION 3.4 (HÖRMANDER). *The class $\alpha(A, B)$ coincides with the fundamental generating class of $H^1(L(V))$ introduced above.*

As before, it suffices to verify the proposition for the case dim $V = 2$. (Indeed, we need only check that the two classes coincide when evaluated over some nontrivial cycle, since we know that $H^1(L(V)) = \mathbf{Z}$. We can then choose this cycle as a curve $n(t)$ such that $n(t) \cap A = F$ is a fixed space of dimension $n - 1$ where dim $V = 2n$. Then all formulas are obtained by projecting onto a two dimensional space.) Let us choose a fixed vector space A. Then if we pick two lines, C and D, transversal to A, then C is transversal to A and B on $U_C = L(V) - \{C\}$ and D is transversal to A and B on $U_D = L(V) - \{D\}$. (See the figure.)

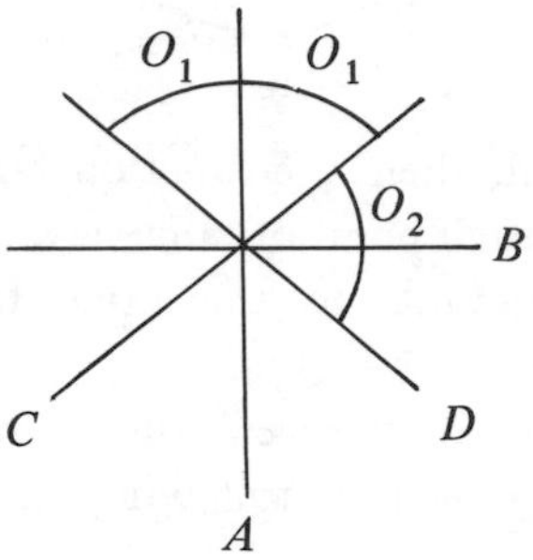

Now $U_C \cap U_D$ has two components, one, O_1, containing A, and the other, O_2, not containing A. We clearly have

$$z(C, D) = (A, B, C, D) = \begin{cases} 0 & \text{on } O_1, \\ -1 & \text{on } O_2. \end{cases}$$

It is easy to check directly, that this defines the generating cohomology class. It is instructive to compute this class via the passage from the Čech to the de Rham theory. Recall that this is done as follows: We choose (smooth) functions f_C defined on U_C (and f_D defined on U_D) such that $f_C - f_D = z(C, D)$ on $U_C \cap U_D$. Then $df_C = df_D$ on $U_C \cap U_D$ and so defines a one form, β, on the whole space. The integral of this one form over the cycle is the value of α on the cycle. In our case, let us take

$$f_C(B) = \tfrac{1}{2}\,\mathrm{sig}\,(A, B, C) \qquad \text{and} \qquad f_D(B) = \tfrac{1}{2}\,\mathrm{sig}\,(A, B, D).$$

Notice that f_C isn't quite smooth—it has a jump, from $-\frac{1}{2}$ to $+\frac{1}{2}$ as B goes through the point A. We could replace f_C by any smooth function which agrees with it in some neighborhood of A. It is simpler to allow differential forms with distribution coefficients, in which case $df_C = \delta_A\, ds$ where δ_A is the δ-function at A and ds is the fundamental form on S^1. Since $\int \delta_A\, ds = +1$ we see that $\alpha(A, B)$ is indeed the fundamental class.

If $N = \Lambda \subset T^*M$ is a Lagrangian manifold and if we take $A(\lambda) = T_\lambda(\Lambda)$ and $B(\lambda) =$ tangent space to the fiber, then the corresponding class is called the *Maslov class* of Λ. For example, the same computation as we just gave shows that if Λ is a simple closed curve in $\mathbf{R}^2 = T^*(\mathbf{R}^1)$ then the Maslov class is exactly twice the fundamental generator: If Λ has only isolated (non-degenerate)

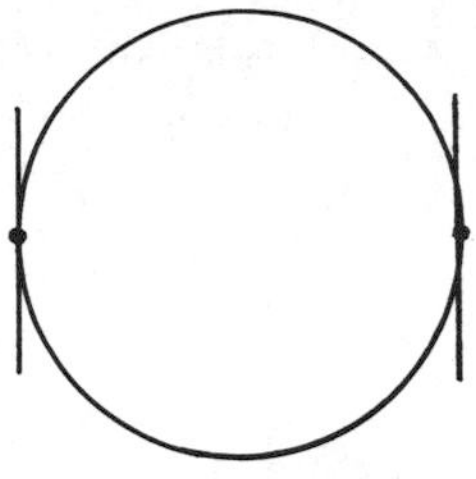

tangencies with the vertical, then a δ-function contribution occurs (with the appropriate orientation) at each point of tangency, i.e., at each point where $d\pi$ is not injective, or, what amounts to the same thing, at each point where $A \cap B \neq \{0\}$.

We can give such a geometric interpretation of the Hormander class $\alpha(A, B)$ in general. For this purpose we need the following fact, which we shall occasion to use quite a bit later on.

PROPOSITION 3.5. *Let Y be a Lagrangian subspace of V. Then the set*

$$L_k(V, Y) = \{W \mid \dim(W \cap Y) = k\}$$

is a submanifold of $L(V)$ of codimension $k(k+1)/2$.

PROOF. It suffices to verify the proposition locally, so that we may assume that Y and W are in $\mathfrak{L}_X$. Then, by Proposition 2.3, all $W \in L(V)$ are parametrized by $S^2(Y)$, and, it is clear from the proof of Proposition 4.3, that $L_k(V, Y) \cap \mathfrak{L}_X$ corresponds to symmetric matrices of corank k. Thus, we are reduced to proving

PROPOSITION 3.6. *The set of symmetric matrices of corank k is a submanifold of codimension $k(k+1)/2$ in the space of all symmetric matrices.*

Let

$$\begin{pmatrix} P_0 & Q_0 \\ R_0 & S_0 \end{pmatrix}$$

be a symmetric matrix, where, with no loss of generality, we may assume that the upper left hand $(n-k) \times (n-k)$ block is non-singular. Then all nearby matrices have the form

$$\begin{pmatrix} P & Q \\ R & S \end{pmatrix}$$

with P non-singular. Now

$$\begin{pmatrix} P^{-1} & 0 \\ -RP^{-1} & I \end{pmatrix} \begin{pmatrix} P & Q \\ R & S \end{pmatrix} = \begin{pmatrix} I & P^{-1}Q \\ 0 & S - RP^{-1}Q \end{pmatrix},$$

and this matrix has rank k if and only if the symmetric $k \times k$ matrix $S - RP^{-1}Q$ vanishes. This imposes $k(k+1)/2$ conditions, proving Proposition 3.6 and hence also Proposition 3.5.

Let E be a symplectic vector bundle over N, with two sections, A and B, of $L(E)$. Then we obtain a subbundle, $L_k(E, A)$ for each integer, k, whose codimension in $L(E)$ is $k(k+1)/2$. We will say that $(E; A, B)$ is *in general position*, if B intersects each of these subbundles transversally. (Notice that by the Thom transversality theorem, we can modify B by an arbitrarily small amount, and hence not change $\alpha(A, B)$, so that $(E; A, B)$ is in general position.) We let $S_k(E, A, B) = B^{-1}(L_k(E, A))$, so that, if $(E; A, B)$ is in general position then $S_k(E, A, B)$ is a submanifold of N of codimension $k(k+1)/2$. Notice that

$$\overline{S}_k = S_k \cup S_{k+1} \cup \cdots \cup S_n.$$

In particular, if $(E; A, B)$ is in general position, then

$$\begin{aligned} \overline{S}_1 &= \{x \mid A(x) \cap B(x) \neq \{0\}\} \\ &= S_1 \cup \overline{S}_2 \end{aligned}$$

where $\overline{S}_2$ is a union of submanifolds of codim ≥ 3. In particular, every (smooth) curve can be deformed into a curve intersecting S_1 transversally, and also every homotopy of curves. We now claim that S_1 is also oriented in N, i.e. has a $+$ and $-$ side in N. In fact, let x be any point of S_1. In some neighborhood, U_C, of x, we can find a C transversal to both A and B. Then at all points of U_C not on S_1 the spaces A and B intersect transversally, and hence C defines a quadratic form, Q_C,

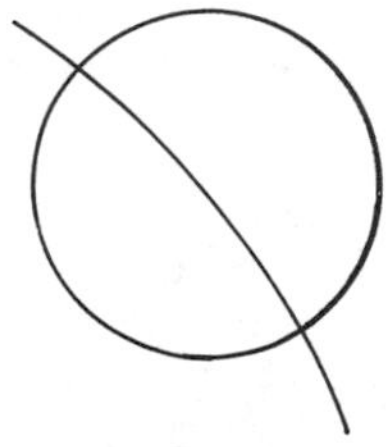

on B. We may assume that U_C is connected and that $U_C - U_C \cap S_1$ has two components.
Then sig Q_C is clearly constant on each component and the difference in sig Q_C between the two components is independent of the choice of C. Indeed, if D were another section transverse to A and B, then

$$\tfrac{1}{2}[\operatorname{sig} Q_C - \operatorname{sig} Q_D] = (A, B, C, D)$$

off S_1, but (A, B, C, D) is well defined and continuous even across S_1. Now we can clearly choose a trivialization of E near x such that $A = \{e_1, \ldots, e_n\}$ and $C = \{f_1, \ldots, f_n\}$ where the e_i and f_j form dual bases. We may also arrange that the basis is chosen so that $B \cap A = \{e_1\}$ on S_1 near x. Then

$$B = \{e_1 + \varphi_1 f_1, e_2 + \varphi_2 f_2, \ldots, e_n + \varphi_n f_n\}$$

where $\varphi_2, \ldots, \varphi_n$ are all non-zero near x and S_1 is given locally by $\varphi_1 = 0$. Transversality requires that $d\varphi_1 \neq 0$, so that φ_1 changes sign across S_1 and sig Q_C clearly changes by exactly 2 as we cross S_1. Now if γ is any smooth oriented closed curve which intersects S_1 transversally we can apply the argument used in the proof of Proposition 3.4 to conclude

PROPOSITION 3.7. *If $(E; A, B)$ is in general position and γ is a smooth closed curve intersecting S_1 transversally, then the class $\alpha(A, B)$, when evaluated on γ, is given by the number of intersections of γ with S_1, each counted with sign ± 1 according to whether the crossing is in the positive or negative direction.*

In the case of $\Lambda \subset T^*M$ the set $\bar{S}_1$ consists of exactly the Maslov cycle introduced in Chapter II, and the Maslov class is the cohomology class discussed there. (We must still establish that by a slight perturbation we can bring every Lagrangian manifold into general position, i.e. that the vertical section and the tangential section be in general position.)

We close this section with an alternative description, also due to Hormander, of the class $\alpha(A, B)$ on a symplectic bundle, E. Consider the pullback, $\tilde{E}$, of E to $L(E) \to N$. We also obtain a bundle $L(\tilde{E}) \to L(E)$ and a natural section, S, where $S(x) = x$ where $x \in L(E)_n$ and we have identified $\tilde{E}_x$ with E_n. A section A of $L(E)$ pulls back to a section $\tilde{A}$ of $L(\tilde{E})$. Notice that if s: $N \to L(E)$ is a section of $L(E)$ then

$$s^* \tilde{A} = A \qquad \text{and} \qquad s^* S = s,$$

Finally, we have

$$\alpha(\tilde{A}, \tilde{B}) = \alpha(\tilde{A}, S) - \alpha(\tilde{B}, S) \in H^1(L(E)).$$

Now if g: $N_1 \to N_2$ is a continuous map and E_1 is the pullback of a symplectic bundle E_2 over N_2 then $\alpha(g^* A, g^* B) = g^* \alpha(A, B)$; in other words the assignment of α to $(E; A, B)$ is functorial. Therefore, taking $B = s$ we get

$$s^* \alpha(A, B) = B^* \alpha(A, S)$$

since $\alpha(B, B) = 0$. Now each section, A, of $L(E)$ determines the class

$$\alpha(\tilde{A}, S) \stackrel{\text{def}}{=} \alpha_A$$

on $L(E)$. This class has the property

(i) that its restriction to each fiber is exactly the generating class of the fiber. This is the content of Proposition 3.4.

It further has the property that

(ii) $A^* \alpha_A = 0$.

Now it is a consequence of a standard theorem in the topology of fiber bundles (the Leray-Hirsch theorem, cf. for example Spanier [19, p. 258]) that any one form, β, on $L(E)$ must be of the form $\beta = \pi^* c + k\alpha_A$. Now if β satisfies (ii) then $A^* \beta = (\pi \circ A)^* c = c = 0$ and if it satisfies (i) clearly $k = 1$. *Thus* (i) *and* (ii) *characterize* α_A.

Thus if β_A is any form satisfying (i) and (ii) then

$$\alpha(A, B) = B^* \beta_A.$$

For example, given A, we can always choose a section everywhere transversal to A and a Riemann metric on A. This determines a Hermitian structure on E, which together with A allows us to identity $L(E)$ with $U(n)/O(n)$ for each N. Thus

$$\sigma = (\det^2)^* \left(\frac{dz}{2\pi i z} \right)$$

is a well defined form on E which clearly satisfies (i) and (ii). Thus the form

$$B^* \sigma$$

will define the class $\alpha(A, B)$ on N.

As above let $E \to N$ be a symplectic vector bundle and A and B sections of $L(E)$. Let $\{\mathcal{U}\}$ be a contractible open cover of N. We recall again how $\alpha(A, B)$ is defined. We choose sections

$$C_{\mathcal{U}} \colon \mathcal{U} \to L(E) \tag{3.5}$$

transversal to A and B. The Cech cocycle defining $\alpha(A, B)$ has the value

$$\mathfrak{L}(\mathcal{U}, V) = (A, B, C_{\mathcal{U}}, C_v)$$

on the pair of open sets $(\mathcal{U}, V)$. Let us set

$$\tau(\mathcal{U}, V) = e^{(\pi i/2)\mathfrak{L}(\mathcal{U}, \mathcal{V})}. \tag{3.6}$$

We can regard (3.6) as defining transition functions for a line bundle on N; i.e., we can define a line bundle on N by requiring that it have local trivializations, $S_{\mathcal{U}}$, on $\mathcal{U}$ and that these be related by $S_{\mathcal{U}} = \tau(\mathcal{U}, v)S_v$ on $\mathcal{U} \cap v$. The line bundle defined this way is called the *Maslov bundle* associated with (E, A, B) and denoted $\mathfrak{M} = \mathfrak{M}_{A,B}$. It is a *locally constant* bundle: i.e., we define a section of $\mathfrak{M}$ over a subset Z of N to be *constant* if for each $\mathcal{U}$ intersecting Z it is a constant multiple of $S_{\mathcal{U}}$ on $\mathcal{U} \cap Z$. Since the transition functions (3.6) are constant this definition is independent of the choice of $\mathcal{U}$. Moreover, locally constant sections exist (e.g. the $S_{\mathcal{U}}$'s).

In [3] Hörmander gives an alternative definition of $\mathfrak{M}$ which avoids the use of transition functions. This definition goes as follows. First given a fixed symplectic vector space V and fixed Lagrangian subspaces A and B, we attach to this data a one dimensional vector space $\mathfrak{M}_{A,B}(V)$; Let $\mathcal{O}$ be the open subset of $L(V)$ consisting of all C such that $C \cap A = C \cap B = \{0\}$. Then $\mathfrak{M}_{A,B}(V)$ is the space of all functions

$$f\colon \mathcal{O} \to \mathbf{C},\ f(C) = e^{i\pi/2(A,B,C,D)}f(D) \tag{3.7}$$

for $C, D \in \mathcal{O}$. Such a function is determined by its value at one point, so the space $\mathfrak{M}_{A,B}(V)$ is one dimensional.

Now let $E \to N$ be a symplectic vector bundle and A and B sections of $L(E) \to N$. We define a line bundle $\mathfrak{M} \to N$ by defining its fiber at $p \in N$ to be the vector space $\mathfrak{M}_{A,B}(E_p)$. Let us show that $\mathfrak{M}$ is identical with the bundle defined by the transition functions (3.6). Over the open set $\mathcal{U}$ a section $S_{\mathcal{U}}$ is defined by choosing $S(p)$ to be the function (3.7) taking the value 1 at the point $C_{\mathcal{U}}(p)$ of $L(E_p)$, $C_{\mathcal{U}}$ being the section (3.5). It is clear from (3.6) and (3.7) that $S_{\mathcal{U}} = \tau(\mathcal{U}, v)S_v$; so $\mathfrak{M}_{A,B}$ is the bundle defined by (3.6) as claimed.

§4. Functorial properties of Lagrangian submanifolds.

Let X and Y be symplectic manifolds with forms ω_X and ω_Y. Then $X \times Y$ becomes a symplectic manifold with two form

$$\omega_{X\times Y} = \rho_X^* \omega_X + \rho_Y^* \omega_Y$$

where $\rho_X\colon X \times Y \to X$ is the projection onto X and ρ_Y the projection onto Y. Let $\Lambda_X \subset X$ be a Lagrangian submanifold, and let us set

$$H = \Lambda_X \times Y.$$

Let $\iota_H\colon H \to X \times Y$ be the immersion of H into $X \times Y$. Notice that since Λ_X is Lagrangian we have

$$\iota_H^* \omega_{X\times Y} = \iota_H^* \rho_Y^* \omega_Y. \tag{4.1}$$

Now suppose that $\Lambda_{X\times Y}$ is a Lagrangian submanifold of $X \times Y$, and suppose that $\Lambda_{X\times Y}$ intersects H transversally. Notice that since the codimension of H is $\frac{1}{2}\dim X$ and $\dim \Lambda_{X\times Y} = \frac{1}{2}(\dim X + \dim Y)$ we see that $\dim(H \cap \Lambda_{X\times Y}) = \frac{1}{2}\dim Y$. We claim that the *projection* ρ_Y *makes this intersection into an immersed Lagrangian submanifold of* Y. Let us first show that the map is an immersion. Suppose that ξ is a tangent vector to $H \cap \Lambda_{X\times Y}$ and that $d\rho_Y\xi = 0$. Then $\xi \lrcorner \rho_Y^* \omega_Y = 0$. By (4.1) this implies that

$$\xi \lrcorner \iota_H^* \omega_{X\times Y} = 0,$$

i.e., that

$$\langle \xi \wedge \eta, \omega_{X\times Y}\rangle = 0$$

for all η tangent to H. On the other hand, since ξ is tangent to $\Lambda_{X\times Y}$ which is Lagrangian, the above equation must hold for all η tangent to $\Lambda_{X\times Y}$. Since $T(\Lambda_{X\times Y})$ and $T(H)$ span all of $T(X \times Y)$, then the above equation holds for all η, which implies that $\xi = 0$. By (4.1) we see that $\rho_Y^*\omega_Y = 0$ on $\Lambda_{X\times Y} \cap H$ since $\omega_{X\times Y}$ vanishes on $\Lambda_{X\times Y}$. We shall now give several examples of this construction.

(i) *Composition of canonical relations*. Let U, W and Z be symplectic manifolds with corresponding forms ω_U, ω_W and ω_Z. Let us consider $W \times U$ as a symplectic manifold with the two form $\omega_W - \omega_U$ (where, for example, ω_W is considered as a form on $W \times U$ via projection). A Lagrangian submanifold, Λ_1, of $W \times U$ is called a canonical relation. For example, if f: $U \to W$ were a canonical transformation, then $f^*\omega_W - \omega_U = 0$, so that $\omega_W - \omega_U$ would vanish on graph f. Since dim graph $f = \dim U = \frac{1}{2}(\dim W \times U)$ we see that graph f is a Lagrangian submanifold. The concept of a canonical relation is thus a generalization of the notion of a canonical map. Let $Z \times W$ have the symplectic form $\omega_Z - \omega_W$ and let $\Lambda_2 \subset Z \times W$ be a canonical relation. We would like, in favorable circumstances, to know that $\Lambda_2 \circ \Lambda_1$ is a canonical relation in $Z \times U$. As a set the composition $\Lambda_2 \circ \Lambda_1$ consists of all pairs (z, u) such that there is some w with $(z, w) \in \Lambda_2$ and $(w, u) \in \Lambda_1$. This can be described as follows, let Δ denote the diagonal in $W \times W$ and let π denote the projection

$$\pi\colon Z \times W \times W \times U \to Z \times U.$$

Then $Z \times \Delta \times U$ and $\Lambda_2 \times \Lambda_1$ are submanifolds of $Z \times W \times W \times U$ and

$$\Lambda_2 \circ \Lambda_1 = \pi(\Lambda_2 \times \Lambda_1 \cap Z \times \Delta \times U).$$

Let us take $X = W \times W$ with the two forms $\omega_{W_2} - \omega_{W_1}$ (where ω_{W_1} is the pullback to X of the ω_W of the first factor and similarly for ω_{W_2}). Then Δ is a Lagrangian submanifold of X. Let us take $Y = Z \times U$ with form $\omega_Z - \omega_U$. We can now apply our construction to conclude that *if $\Lambda_2 \times \Lambda_1$ intersects $Z \times \Delta \times U$ transversally then $\Lambda_2 \circ \Lambda_1$ is a canonical relation in $Z \times U$.*

For example, if Λ_2 is the graph of a map then for any $\eta \in TW$ there is ζ such that $(\zeta, \eta, 0, 0)$ is tangent to $\Lambda_2 \times \Lambda_1$. Thus it is easy to see that in this case the intersection will be transversal.

(ii) *Pullback of a Lagrangian submanifold of the cotangent bundle.* Let M and N be differentiable manifolds and $f\colon M \to N$ a differentiable map. Let $\Lambda \subset T^*N$ be a Lagrangian submanifold. Then

$$df^*\Lambda = \{(m,\xi) \in T^*M \mid \exists(n,\eta) \in \Lambda \text{ with } f(m) = n \text{ and } df_m^*\eta = \xi\}$$

is a subset of T^*M and we would like to know whether it is a Lagrangian submanifold. Here let us take $X = T^*N$ and $Y = T^*M$. Let us write graph $f \subset N \times M$ as $\{(f(x), x) \mid x \in M\}$, and let us take $\Lambda_{X\times Y} = \mathfrak{N}(\text{graph } f)$. Thus a point of $\Lambda_{X\times Y}$ is of the form

$$(f(x), x, \gamma, -df^*\gamma) \qquad \gamma \in T^*N_{f(x)}.$$

We take $\Lambda_X = \Lambda$ so that H consists of all points of the form

$$(y, x, \eta, \xi) \qquad (y, \eta) \in \Lambda.$$

It is clear that

$$\pi_Y(H \cap \mathfrak{N}(\text{graph } f)) = -df^*\Lambda.$$

Of course, $-df^*\Lambda$ is a Lagrangian submanifold if and only if $df^*\Lambda$ is. Rewriting the graphs in the usual way we have proved

Let $f\colon M \to N$ be a smooth map and $\Lambda \subset T^*N$ a Lagrangian manifold. If $\mathfrak{N}(\text{graph } f)$ and $T^*M \times \Lambda$ intersect transversally in $T^*(M \times N)$ then $df^*\Lambda$ is a Lagrangian submanifold of T^*M.

Examining the above form of H and $\mathfrak{N}(\text{graph } f)$ we see that all values of the last three components can be achieved for any f and Λ, and that the intersection will be transversal if and only if the maps $f\colon M \to N$ and $\pi_N\colon \Lambda \to N$ are transversal. Thus

PROPOSITION 4.1. *Let $f\colon M \to N$ be a smooth map and $\Lambda \subset T^*N$ a Lagrangian manifold. Let $\pi\colon \Lambda \to N$ be the restriction to Λ of the projection of $T^*N \to N$. If f and π are transversal then $df^*\Lambda$ is a Lagrangian submanifold of T^*M.*

For example, if $\pi\colon \Lambda \to N$ is locally a diffeomorphism then the hypothesis is fulfilled. Here $\Lambda =$ graph $d\varphi$ (locally) and $df^* \Lambda =$ graph $df^* \varphi$.

As a second example, suppose that $\Lambda = \mathfrak{N}(S)$ where S is a submanifold of N. Then $\pi\Lambda = S$ and the hypothesis becomes that f intersects S transversally. In this case $f^{-1}S$ is a submanifold of M and

$$df^* \Lambda = \mathfrak{N}(f^{-1}S).$$

(iii) *Pushforward of Lagrangian manifolds.* Let $f\colon M \to N$ be a smooth map and let Λ be a Lagrangian submanifold of T^*M. Then

$$df_* \Lambda = \{(y,\eta) \mid y = f(x), (x, df^*\eta) \in \Lambda\}.$$

Take $X = T^*M$ and $Y = T^*N$ and

$$\Lambda_{X\times Y} = \mathfrak{N}(\text{graph } f) = \{(x, f(x), df^*\eta, -\eta) | \eta \in T^*N_{f(x)}\}.$$

Then, with $\Lambda_X = \Lambda$,

$$H = \{(x, y, \xi, \eta) \mid (x, \xi) \in \Lambda\}$$

so that $\pi_Y(H \cap \mathfrak{N}(\text{graph } f)) = -df_* \Lambda$. Thus, *if $\Lambda \times T^*N$ intersects $\mathfrak{N}$(graph f) transversally, then $df_* \Lambda$ is a Lagrangian submanifold of T^*N.*

Notice that if df has constant rank there this condition takes on a somewhat simpler form. In this case the dimension of $df_x^* T^*N_{f(x)}$ does not vary so that $df^* T^*N$ is a sub-bundle of T^*M. The transversality condition is then clearly the condition that this subbundle intersect Λ transversally. Thus

PROPOSITION 4.2. *Let $f\colon M \to N$ be a smooth map with df of constant rank and let Λ be a Lagrangian submanifold of T^*M. If Λ intersects $df^* T^*N$ transversally then $df_* \Lambda$ is a Lagrangian submanifold of T^*N.*

For example, if f is an immersion (so that df^* is surjective everywhere and thus $df^* T^*N = T^*M$) the conditions are verified and $df_* \Lambda$ is a Lagrangian submanifold of T^*N for any Λ.

At the other extreme, suppose $f\colon M \to N$ is a fibration. Then $df^* T^*N = H$ is the bundle of those covectors which vanish on vectors tangent to the fiber. If Λ intersects H transversally then $df_* \Lambda$ is a Lagrangian submanifold. For example, if $\Lambda =$ graph $d\varphi$ then $\Lambda \cap H$ consists of those points $(m, d\varphi(m))$ on Λ where the vertical derivative, $d_V\varphi$, vanishes. At such points $d\varphi$ clearly defines a covector at $n = f(m)$ and thus gives a map from $\Lambda \cap H \to T^*N$. According to the general theory this map is a Lagrangian immersion. If we consider the Lagrangian manifold, Λ_1, of $T^*R^1 = R^1 \times R^1$ where $\Lambda_1 = \{(x, 1)\}$ then $\Lambda = d\varphi^* \Lambda_1$ and thus we can think of the pushforward of Λ as described by the diagram

$$\begin{array}{ccc} M & \xrightarrow{\varphi} & \mathbf{R} \\ {\scriptstyle f}\downarrow & & \\ N & & \end{array}$$

We shall soon see that the most general Lagrangian manifold on T^*N can locally be described as $df_* \, d\varphi^* \Lambda_1$, where M, φ, and f are suitably chosen.

Suppose that instead of Λ_1 on $\mathbf{R}$ we take $\mathfrak{N}(\{0\})$, the normal bundle to the origin. Then, if in the above diagram φ is transversal to $\{0\}$, i.e., if 0 is a regular value of φ, then $d\varphi^*(\mathfrak{N}\{0\}) = \mathfrak{N}(\varphi^{-1}(0))$. If $\mathfrak{N}(\varphi^{-1}(0))$ intersects H transversally then $df_* \mathfrak{N}(\varphi^{-1}(0))$ is a Lagrangian submanifold of N. This construction generalizes the classical notion of an *envelope*: Suppose that $M = N \times S$ where S is some auxiliary parameter space. We have assumed that the map

$$\varphi \colon N \times S \to \mathbf{R}$$

has zero as a regular value, so that $\varphi^{-1}(0) = Z$ is a hypersurface (of codimension one) in $N \times S$. Let $\varphi_s \colon N \to \mathbf{R}$ be defined by $\varphi_s(x) = \varphi(x, s)$. We can make the stronger hypothesis that φ_s has zero as a regular value for each s. If we set $N_s = \varphi_s^{-1}(0)$ then N_s is a hypersurface in N for each s, and $N_s = Z \cap N \times \{s\}$ so that, as a set, $Z = \bigcup_s N_s$. Now the Lagrangian manifold $\mathfrak{N}(Z)$ consists of all points of the form $\{(x, s, t d_N \varphi, t d_S \varphi) \mid \varphi(x, s) = 0, t \in \mathbf{R}\}$ and our transversality condition asserts that the rank of $d(d_S \varphi)$ be equal to dim S on Z. The Lagrangian manifold $\Lambda = df_*(\mathfrak{N}(Z))$ then consists of all covectors $t d_N \varphi$ where

$$\varphi(x, s) = 0 \qquad d_S \varphi(x, s) = 0.$$

These represent $p + 1$ equations in $p + n$ variables where $p = \dim S$ and $n = \dim N$. Our transversality assumption asserts that these equations define a submanifold of $N \times S$, i.e., that $0 \in \mathbf{R}^{p+1}$ is a regular value for $(\varphi, d_S \varphi)$. If we make the stronger assumption that the equations $d_S \varphi(x, s) = 0$ can be solved for s as a function of x the first equation becomes

$$\varphi(x, s(x)) = 0$$

which defines a hypersurface, $\mathcal{E}$, called the envelope of the hypersurfaces N_s. Furthermore

$$\begin{aligned} d\varphi(\cdot, s(\cdot)) &= d_N \varphi(\cdot, s(\cdot)) + d_S \varphi(\cdot, s(\cdot)) \\ &= d_N \varphi(\cdot, s(\cdot)) \end{aligned}$$

since $d_S \varphi = 0$. Thus $\Lambda = \mathfrak{N}(\mathcal{E})$. From our point of view it is more natural to consider the Lagrangian manifolds than the hypersurfaces. But even classically one is obliged to consider the Lagrangian manifold rather than the hypersurfaces if one wants to avoid singularities. For example, let S be a submanifold of

$N = \mathbf{R}^n$ and let N_s be the sphere of radius r centered at $s \in S$. (We are thus in the situation envisaged in our treatment of Huygens' principle in Chapter I.) Then the (classical) envelope of the spheres N_s will develop singularities, if r is larger than the minimum radius of curvature of S. However, the Lagrangian submanifold $\Lambda \subset T^*(\mathbf{R}^n)$ will still be perfectly well defined. It just won't project onto a hypersurface in $\mathbf{R}^n$.

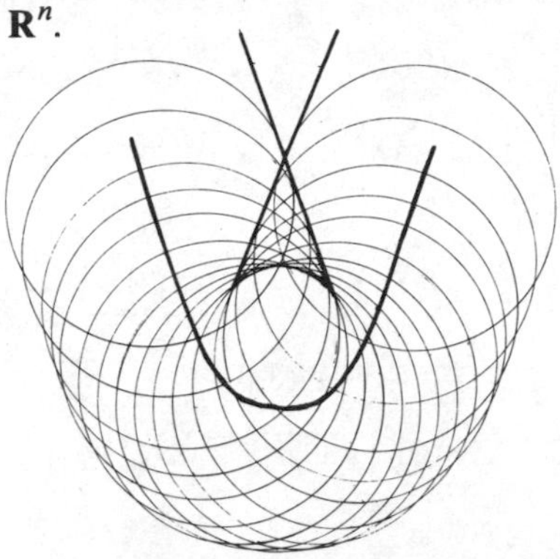

Later on we shall show how to associate to any distribution, μ, on M, a subset of $T^* M$, describing the singularities of μ. A particular class of distributions will have their singularities along (the normal bundle to) hypersurfaces. We will be able to show that when we superimpose distributions concentrated along N_s we obtain a distribution concentrated along the envelope, $\mathcal{E}$. In this way we shall be able to give a purely geometric version of Huygens' principle.

§5. Local parametrizations of Lagrangian submanifolds.

In this section we discuss local presentations of Lagrangian submanifolds of $T^* M$. The first basic observation is that, locally, we can represent any Lagrangian manifold as the push forward of graph $d\varphi$. More precisely

PROPOSITION 5.1. *Given $\Lambda \subset T^* M$ we can find, near each point $\lambda \in \Lambda$, a fibered manifold $N \xrightarrow{\pi} M$ and a function φ: $N \to \mathbf{R}$ such that*

$$\begin{aligned}\Lambda &= d\pi_* \text{ graph } d\varphi \\ &= d\pi_* \, d\varphi^* \{(x, 1)\}.\end{aligned}$$

PROOF. Let $x^1, \ldots, x^n, \xi^1, \ldots, \xi^n$ be dual local coordinates. By a *linear* change of variables, we can arrange that $\xi^1, \ldots, \xi^k, x^{k+1}, \ldots, x^n$ are independent on Λ for some suitable k. (Indeed let X and Y be two complementary Lagrangian subspaces of a symplectic vector space V. For any Lagrangian W we know that $W \cap X$ and $PW \subset Y$ are orthogonal to each other under the symplectic form where P is the projection on Y through X. We can choose $e^1, \ldots, e^k$ as a basis of $W \cap X$ and $f^{k+1}, \ldots, f^n$ as a basis of PW and extend to get a dual basis. Applying this result to $T_\lambda \Lambda$ gives the desired result.)

We can thus find functions $f^1, \ldots, f^k, f^{k+1}, \ldots, f^n$ such that Λ is described by the equations

$$\begin{aligned} x^1 &= -f^1(x^{k+1}, \ldots, x^n; \xi^1, \ldots, \xi^k) \\ &\vdots \\ x^k &= -f^k(x^{k+1}, \ldots, x^n; \xi^1, \ldots, \xi^k) \\ \xi^{k+1} &= f^{k+1}(x^{k+1}, \ldots, x^n; \xi^1, \ldots, \xi^k) \\ &\vdots \\ \xi^n &= f^n(x^{k+1}, \ldots, x^n; \xi^1, \ldots, \xi^k). \end{aligned}$$

Now the fundamental two form, ω, is given by

$$-d(x^1 d\xi^1 + \cdots + x^k d\xi^k) + d(\xi^{k+1} dx^{k+1} + \cdots + \xi^n dx^n).$$

Thus on Λ

$$0 = d(f^1 d\xi^1 + \cdots + fd\xi^k + f^{k+1} dx^{k+1} + \cdots + f^n dx^n).$$

We can thus find a function $F = F(x^{k+1}, \ldots, x^n; \xi^1, \ldots, \xi^k)$ such that

$$f^1 = \frac{\partial F}{\partial \xi^1} \quad \cdots \quad f^n = \frac{\partial F}{\partial x^n}.$$

Consider φ, defined on $T^* M$, by the formula

$$\varphi(x^1, \ldots, x^n, \xi^1, \ldots, \xi^n) = \xi^1 x^1 + \cdots + \xi^k x^k + F + \sum_{k+1}^{n} (\xi^j - f^j)^2.$$

Now (x, ξ) are coordinates on $T^* M$. Let (x, ξ, z, ζ) be the corresponding coordinates on $T^*(T^* M)$ so that, for example, the subbundle $H = d\pi^* T^* M$ is given by $\zeta = 0$. Then graph $d\varphi$ consists of all points of the form (x, ξ, z, ζ) where

$$\begin{aligned} z^1 &= \xi^1 \\ &\vdots \\ z^k &= \xi^k \\ z^{k+1} &= \frac{\partial}{\partial x^{k+1}}[F + \sum (\xi^j - f^j)^2] \\ &\vdots \\ z^n &= \frac{\partial}{\partial x^n}[F + \sum (\xi^j - f^j)^2] \\ \zeta^1 &= x^1 + \frac{\partial F}{\partial \xi^1} + \frac{\partial}{\partial \xi^1} \sum_{k+1}^{n} (\xi^j - f^j)^2 \\ &\vdots \\ \zeta^k &= x^k + \frac{\partial F}{\partial \xi^k} + \frac{\partial}{\partial \xi^k} \sum_{k+1}^{n} (\xi^j - f^j)^2 \end{aligned}$$

$$\zeta^{k+1} = 2(\xi^{k+1} - f^{k+1})$$

$$\vdots$$

$$\zeta^n = 2(\xi^n - f^n).$$

Now graph $d\varphi \pitchfork H$ implies that $\zeta^j = 0$ so that

$$\xi^{k+1} = f^{k+1}, \ldots, \xi^n = f^n, \qquad x^1 = -\frac{\partial F}{\partial \xi_1}, \ldots, x^k = -\frac{\partial F}{\partial \xi_k},$$

and

$$z^i = \xi^i, \qquad i = 1, \ldots, n.$$

This, of course, gives Λ.

Notice that it is a peculiarity of the particular representation that we have chosen that $N = T^*M$ and, setting $\Lambda_N = $ graph $d\varphi$, that in the diagram

$$T^*M = N \xleftarrow{\pi_N} T^*N \leftarrow \Lambda_N \leftarrow \Lambda_N \cap H$$

$$\downarrow \pi_M$$

$$M \leftarrow T^*M$$

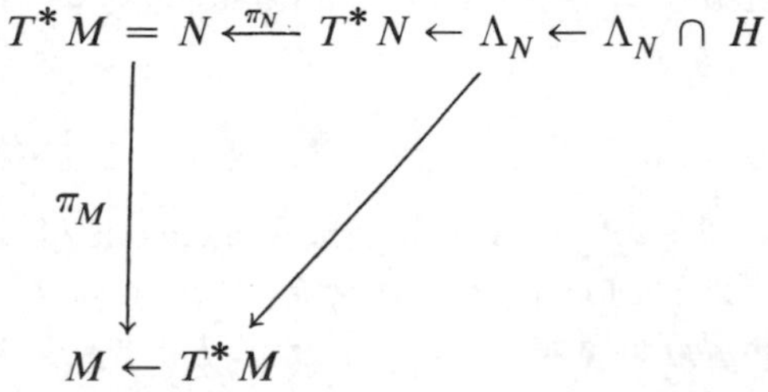

we have $d\pi_{M*}\Lambda_N = \Lambda$. We shall discuss the significance of this particular type of representation in a later section.

It is of interest to know when by a possibly non-linear change of coordinates, we can arrange that $k = n$, and so eliminate the unpleasant quadratic terms that appear in the expression for φ. Thus, we wish to choose coordinates so that $\xi^1, \ldots, \xi^n$ are linearly independent on Λ. Now this is certainly not always going to be possible. For example, if Λ is the zero section, then $\xi^1 = \cdots = \xi^n \equiv 0$ in any coordinate system. Suppose, on the other hand, that $\lambda \in \Lambda$ with $\lambda \neq 0$. Let us choose a Lagrangian subspace of $T_\lambda(T^*X)$ which is transversal both to $T_\lambda\Lambda$ and to the vertical. This is always possible by Proposition 2.4. Let us pass a Lagrangian submanifold, K, tangent to this subspace. Since K is transversal to the vertical, it is of the form graph $d\psi$. If $x = \pi\lambda$ then $d\psi(x) = \lambda \neq 0$. We can thus introduce a coordinate system $x^1, \ldots, x^n$ with $\psi = x^1$. Then, in terms of these coordinates $K = \{(x^1, \ldots, x^n, 1, 0, \ldots, 0)\}$. At λ the tangent space to K is exactly the kernel of the projection onto the $\xi^1, \ldots, \xi^n$. Since Λ is transversal to K we conclude that the ξ_i are independent on Λ near λ. We thus have

PROPOSITION 5.2. *If* $0 \neq \lambda \in \Lambda$ *then near* λ *we can parametrize* Λ *as follows: We can introduce coordinates* $x^1, \ldots, x^n$ *near* $\pi\lambda$, *with corresponding coordinates* $x^1, \ldots, x^n, \xi^1, \ldots, \xi^n$ *on* T^*M *near* λ, *and find a function* $f(\xi^1, \ldots, \xi^n)$ *such that, near* λ,

$$\Lambda = d\pi_* \text{ graph } d\varphi$$

where

$$\varphi = x \cdot \xi - f. \tag{5.1}$$

If Λ *is homogeneous, i.e., invariant under multiplication by* $\mathbf{R}^+$, *then we can choose* φ *to be homogeneous of degree one in* ξ.

The only assertion that remains to be proved is the last one. If Λ is invariant under $\mathbf{R}^+$ then $\xi\partial/\partial\xi$ is tangent to Λ and hence $\alpha = \xi dx = (\xi\partial/\partial\xi) \lrcorner\, \omega$ must vanish when restricted to Λ. Also the f^i are homogeneous of degree zero so that on Λ

$$0 = \alpha = \sum \xi^i dx^i = d\left(\sum f^i \xi^i\right) - \sum f^i d\xi^i$$

so for f we may take $f = \sum f^i \xi^i$ which is homogeneous of degree one.

Let us call $\varphi: N \to \mathbf{R}$ a local *phase function* for $\Lambda \subset T^*M$ if Λ is locally of the form $d\pi_*$ (graph $d\varphi$) as above. Here $N \xrightarrow{\pi} M$ is a submersion. We shall set Λ_φ = graph $d\varphi$.

We shall set $C_\varphi = \pi_N(\Lambda_\varphi \cap H)$ where $\pi_N: T^*N \to N$. Thus we have the commutative diagram

$$\begin{array}{ccc}
N \leftarrow C_\varphi & \xleftarrow{\pi_N} & \Lambda_\varphi \cap H \\
\pi \big\downarrow & \swarrow & \\
M & \xleftarrow[\pi_M]{} & T^*M.
\end{array}$$

Now $\pi_N: \Lambda_\varphi \to N$ is a diffeomorphism so its restriction to $\Lambda_\varphi \cap H$ is an immersion. Thus, for $\lambda \in \Lambda_\varphi \cap H = \Lambda$ let

$$k_\lambda = \ker d\pi_M: T_\lambda \Lambda \to M$$

and let $n = \pi_N \lambda$ and

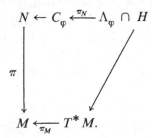

$$l_n = \ker d\pi\colon T_n(C_\varphi) \to T_{\pi n}(M)$$

so that $l_n = T_n C_\varphi \cap T_n F$ where F is the fiber through n. Then

$$d\pi_N k_\lambda = l_n$$

and in particular

$$\dim k_\lambda = \dim l_n .$$

Let $x^1, \ldots, x^n$ be coordinates on M and $x^1, \ldots, x^n, \theta^1, \ldots, \theta^k$ be coordinates on N. Then $C_\varphi = \{(x, \theta) \mid \partial\varphi/\partial\theta^i = 0, i = 1, \ldots, k\}$. The tangent to the fiber is spanned by $\partial/\partial\theta^i$. Notice that $\boldsymbol{\eta} = \sum \alpha^i \partial/\partial\theta^i$ is tangent to C_φ if and only if

$$\sum_i a^i \frac{\partial^2\varphi}{\partial\theta^i\partial\theta^j} = 0 \qquad \text{all } j$$

i.e., if $\boldsymbol{\eta}$ is in the null space of the Hessian

$$H_\theta(\varphi) = d_\theta^2\,\varphi = \left(\frac{\partial^2\varphi}{\partial\theta^i\partial\theta^j}\right).$$

Thus

$$\dim k_\lambda = \text{nullity } H_\theta(\varphi) \qquad \text{at } \pi_N\lambda. \tag{5.2}$$

Notice that this proves that

$$\text{fiber dim } N \geqslant \dim k_\lambda \tag{5.3}$$

(where fiber dim N is the dimension of the fibers of π, i.e., fiber dim N $= \dim N - \dim M$) .

Suppose $H_\theta(\varphi)$ is non-singular. Then $\Lambda_\varphi \cap H$ projects diffeomorphically on M and is of the form graph $d\psi$. Notice that then

$$\begin{array}{ccc} C_\varphi & \leftrightarrow & \Lambda_\varphi \cap H \\ \big\downarrow & \swarrow & \\ M & & \end{array}$$

are all diffeomorphisms. In particular there is a map $s\colon M \to C_\varphi$, i.e., a section of N and $d\varphi \circ ds = d\psi$ so we may take $\psi = s^*\varphi$.

We will say that the phase function φ is *reduced* (at λ) if fiber dim $N = \dim k_\lambda$. Notice that if φ is reduced at λ then $H_\theta\varphi$ is identically zero at $\pi_N\lambda$.

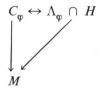

PROPOSITION 5.3. *We can always factor N and φ through a reduced parametrization of Λ. In other words, we can locally find an intermediate fibration*

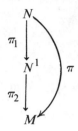

$$\begin{array}{ccc} & N & \\ \pi_1 & \downarrow & \\ & N^1 & \pi \\ \pi_2 & \downarrow & \\ & M & \end{array}$$

and a function φ^1 on N^1 such that $d\pi_{1*}\Lambda_{\varphi^1} = \Lambda_{\varphi^1}$ and $d\pi_{2*}\Lambda_{\varphi^1} = \Lambda$, with φ^1 a reduced phase function at λ.

PROOF. Choose a complement in the tangent space to the fiber at $\pi_N\lambda$ to the null space of $H_\theta(\varphi)$. Extend this to an integrable foliation in the fibers. This makes N fibered over some N^1, locally, where $H_\theta\varphi$ is non-degenerate when restricted to the tangent space to the fibers of π_1.

Let $H_1 = d\pi_1^* T^* N^1$ so that $H \subset H_1$ where $H = d\pi^* T^* M = d\pi_1^*(d\pi_2^* T^* M)$. Since Λ_φ intersects H transversally, it certainly intersects H_1 transversally and we have

$$\begin{array}{ccc} N \leftarrow \Lambda_\varphi \cap H_1 \\ \pi \Big\downarrow \qquad\qquad \\ N^1 \qquad\qquad \end{array}$$

Since $H_\theta\varphi$ is non-degenerate on fibers of N over N_1, $\Lambda_\varphi \cap H_1$ projects diffeomorphically onto N^1, and hence is of the form graph $d\varphi^1 = \Lambda_{\varphi^1}$, with $\varphi^1 = s^*\varphi$ where $s\colon N^1 \to N$ is a section of π_1. The rest follows easily.

PROPOSITION 5.4 (*Hormander-Morse lemma*). *Let (N_1, φ_1) and (N_2, φ_2) be two parametrizations of Λ near λ with* dim N_1 = dim N_2. *Let $z_1 = \pi_{N_1}\lambda$ and $z_2 = \pi_{N_2}\lambda$. Suppose that*

$$\operatorname{sig} H_\theta(\varphi_1)(z_1) = \operatorname{sig} H_\theta(\varphi_2)(z_2).$$

Then there exists a fiber preserving diffeomorphism f of $N_1 \to N_2$ defined near z_1 with $f(z_1) = z_2$ and

$$\varphi_2 \circ f = \varphi_1 + \text{const}.$$

In other words, Λ, dim N *and* sig $H_\theta\varphi$ *determine φ up to a diffeomorphism.*

(If we take M to be a point, this reduces to the standard Morse lemma. Indeed H is the 0 section of $T^* N$ in this case, and hence the transversality condition is that φ have a non-degenerate critical point. Then we are clearly in the situation described by Morse's lemma.)

We first point out that it is sufficient to consider the case where N_1 and N_2 are reduced. Indeed, suppose that the proposition has been established in the case of reduced parametrizations. Then there are fibrations $N_1 \to N_1^1$ and $N_2 \to N_2^1$ with non-degenerate phase functions. By applying a diffeomorphism, we may assume that $N_1^1 = N_2^1$. Then we have sections s_1 and s_2 with $\varphi_1 \circ s_1 = \varphi_2 \circ s_2 + \text{const.}$

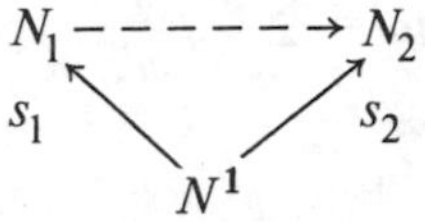

given by $d_\theta \varphi_i = 0, i = 1, 2$. Now $H_\theta(\varphi_i)$ is non-singular, so we may apply the standard Morse's lemma to each fiber. This then gives the desired diffeomorphism from N_1 to N_2.

We thus may assume that (N_1, φ_1) and (N_2, φ_2) are reduced. Our next step will be to show that we can assume that $N_1 = N_2$, $C_{\varphi_1} = C_{\varphi_2}$, and the first derivatives of φ_1 and φ_2 agree on $C_{\varphi_1} = C_{\varphi_2}$ and so that h maps $d\varphi_2 1 C_{\varphi_2}$ onto $d\varphi_1 1 C_{\varphi_1}$. For this, we need only exhibit a fiber diffeomorphism h: $N_1 \to N_2$ with $h(C_{\varphi_1}) = C_{\varphi_2}$. Then replace φ_2 by $\varphi_2 \circ h$. For this purpose, let (x, θ) be local coordinates on N_1 and consider the fiber map $g_1 : N_1 \to T^* M$ given by

$$g_1(x, \theta) = \left(x, \frac{\partial \varphi_1}{\partial x}(x, \theta)\right).$$

On C_{φ_1} this gives the diffeomorphism with Λ. Since $T_{z_1}(\text{fiber}) \subset T_{z_1}(C_{\varphi_1})$ we conclude that g_1 is an immersion at z preserving the fiber over M. The same is true for a similar map g_2. By the implicit function theorem, there is some map ρ: $T^* M \to N_2$ commuting with the fibrations over M such that $\rho \circ g_2 = \text{id}$. Then $\rho \circ g_1 : N_1 \to N_2$ is a diffeomorphism, and since $g_1 C_{\varphi_1} = g_2 C_{\varphi_2} = \Lambda$ we have $(\rho \circ g_1) C_{\varphi_1} = C_{\varphi_2}$. Moreover, from the definition of ρ, $d(\rho \circ g_1)^* \varphi_2 = \varphi_1$ on C_{φ_1}.

We are thus in the situation of Appendix I. For convenience we repeat the relevant proofs:

φ_1 and φ_2 are reduced phase functions for Λ on N with $C_{\varphi_1} = C_{\varphi_2}$, with

$$\frac{\partial \varphi_1}{\partial x} = \frac{\partial \varphi_2}{\partial x} \quad \text{and} \quad \frac{\partial \varphi_1}{\partial \theta} = 0 = \frac{\partial \varphi_2}{\partial \theta}$$

on C_φ. Thus $\varphi_1 = \varphi_2$ on C_φ (up to a constant which we shall ignore) and $\varphi_1 - \varphi_2$ vanishes to second order. We set $\varphi_t = (1 - t)\varphi_1 + t\varphi_2$ and seek a fiber diffeomorphism f_t with $f_t^* \varphi_t = \varphi_1$, $f_t = I$ on C_φ. We look for the corresponding vector field ξ_t. It should satisfy

$$\dot{\varphi}_t + D_{\xi_t}\varphi_t = 0.$$

Now $\varphi_2 - \varphi_1$ vanishes to second order on C_φ and $d(\partial\varphi_1/\partial\theta^i)$ are independent by the transversality of Λ_φ with H. Thus

$$\varphi_2 = \varphi_1 + \sum b_{ij}\frac{\partial\varphi_1}{\partial\theta^i}\frac{\partial\varphi_1}{\partial\theta^j} \qquad b_{ij} = b_{ij}(x,\theta)$$

so

$$\dot{\varphi}_t = \sum b_{ij}\frac{\partial\varphi_1}{\partial\theta^i}\frac{\partial\varphi_1}{\partial\theta^j}.$$

We seek

$$\xi_t = \sum \mu_t^i(x,\theta)\frac{\partial}{\partial\theta^i}$$

with $\mu_t^i = 0$ on C_φ so

$$\mu_t^i = \sum \mu_{ij}\frac{\partial\varphi_1}{\partial\theta^j}$$

and

$$\dot{\varphi}_t + D_{\xi_t}\varphi_t = 0$$

becomes

$$0 = \sum b_{ij}\frac{\partial\varphi_1}{\partial\theta^i}\frac{\partial\varphi_1}{\partial\theta^j} + \sum \mu_{ij}\frac{\partial\varphi_1}{\partial\theta^j}\frac{\partial}{\partial\theta^i}\left[\varphi_1 + t\sum b_{kl}\frac{\partial\varphi_1}{\partial\theta^k}\frac{\partial\varphi_1}{\partial\theta^l}\right].$$

Equating the coefficients of $(\partial\varphi_1/\partial\theta^i)(\partial\varphi_1/\partial\theta^j)$ this becomes

$$B + U(I+S) = 0 \qquad B = (b_{ij}) \qquad U = (\mu_{ij})$$

where $S = 0$ at z_1 since φ_1 is reduced. We can thus solve for U near z_1, find ξ_t and integrate to get f_t, proving the proposition.

In many applications we will be dealing with homogeneous Lagrangian manifolds. For these applications we will require a slight modification of Proposition 5.4.

Let $N^+ = N \times \mathbf{R}^+$ and consider parametrizations of the form

$$\begin{array}{ccc} N^+ & \xrightarrow{\varphi^+} & \mathbf{R} \\ \pi\downarrow & & \\ M & & \end{array}$$

which are equivariant with respect to the action of $\mathbf{R}^+$ in the obvious way, namely $\varphi^+(z, sa) = a\varphi^+(z, s)$ and $\pi(z, sa) = \pi(z, s)$. It is clear that the resulting Lagrangian manifold $\Lambda = \pi_*(\text{graph } d\varphi)$ will be invariant under the action of $\mathbf{R}^+$ on T^*M and hence homogeneous. By Proposition 5.2 every homogeneous Lagrangian manifold admits an $\mathbf{R}^+$ equivalent parametrization, at least locally. Moreover, it is easy to modify the proof of Proposition 5.3 so as to show:

PROPOSITION 5.3 (*homogeneous version*). *Let* (Λ, λ) *be a homogeneous Lagrangian manifold (germ) in* T^*M. *Then every* $\mathbf{R}^+$ *equivariant parametrization of* (Λ, λ) *can be factored through a reduced* $\mathbf{R}^+$ *equivariant parametrization.*

PROOF. Let $\varphi(z) = \varphi^+(z, 1)$. Then $\varphi^+(z, a) = a\varphi(z)$. Now if $z^+ = (z, a)$ is in the critical set of φ^+, then

$$\partial\varphi^+/\partial z = 0 = a(\partial\varphi/\partial z);$$

so $\partial\varphi/\partial z = 0$ since $a \neq 0$. Moreover,

$$d^2\varphi^+\left(\frac{\partial}{\partial t}, w\right) = \left(d\left(\frac{\partial\varphi^+}{\partial t}\right), \eta\right) = \langle d_z\varphi, \eta\rangle = 0.$$

Here $\partial/\partial t$ is the unit vector tangent to $\mathbf{R}^+$ at z^+ and η is any vector tangent to N. This shows that $\partial/\partial t$ is in the null-space of the Hessian of φ^+ at z^+ so in the proof of Proposition 5.3 we can choose the fibration π_1 so that its fibers lie in the sets $N \times \text{const.}$; that is, so that π_1 commutes with the action of $\mathbf{R}^+$. Thus the reduced parametrization is also an $\mathbf{R}^+$ equivariant parametrization. Q.E.D.

We will use this result to prove:

PROPOSITION 5.4 (*homogeneous version*). *Let* (N_1^+, φ_1^+) *and* (N_2^+, φ_2^+) *be two parametrizations of* (Λ, λ) *with* $\dim N_1^+ = \dim N_2^+$. *Let* $z_1^* = \pi_{N_1^+}(\lambda)$ *and* $z_2^+ = \pi_{N_2^+}(\lambda)$. *Suppose that*

$$\text{sig } H_\theta(\varphi_1^+)(z_1^+) = \text{sig } H_\theta(\varphi_2^+)(z_2^+).$$

Then there exists an $\mathbf{R}^+$ *equivariant fiber preserving diffeomorphism* $f\colon N_1^+ \to N_2^+$ *defined near* z_1^+ *with* $f(z_1^+) = z_2^+$ *and* $\varphi_2^+ \circ f = \varphi_1^+$.

PROOF. As before it is sufficient to consider the reduced case. Define $g_1^+\colon N_1^+ \to T^*M$ and $g_2^+\colon N_2^+ \to T^*M$ exactly as we defined g_1 and g_2 above, i.e.,

$$g_1^+(x, \theta, s) = \left(x, \frac{\partial\varphi_1^+}{\partial x}\right).$$

It is clear that g_1^+ and g_2^+ are $\mathbf{R}^+$ equivariant , so we can choose $h\colon C_{\varphi_1^+} \to C_{\varphi_2^+}$ as before, but so that it is equivariant, and replacing φ_2^+ by $\varphi_2^+ \circ h$ reduce to the case: $N_1^+ = N_2^+$, $C_{\varphi_1}^+ = C_{\varphi_2}^+$, and $\varphi_1^+ = \varphi_2^+$ on $C_{\varphi_1}^+ = C_{\varphi_2}^+$. Note that $\varphi_1^+(z,t) = t\varphi_1(z)$; so at a critical point

$$z^+ = (z,t), \quad \frac{\partial \varphi_1^+}{\partial t} = 0 = \varphi_1(z);$$

and hence $\varphi^+(z^+) = 0$. Therefore, in the homogeneous case $\varphi_1^+ = 0$ on $C_{\varphi_1}^+$; so not only do the first derivatives of φ_1^+ and φ_2^+ agree on $C_{\varphi_1}^+ = C_{\varphi_2}^+$ but the functions themselves do as well. (Recall that in the inhomogeneous case we had to adjust by an arbitrary constant at this point in the argument.) The rest of the proof now goes as before. A little care must be exercised in choosing the matrix B. Namely it has to be homogeneous of degree 1 in $\mathbf{R}^+$. To do this just define B on a set where the $\mathbf{R}^+$ variable is constant and extend by linearity.

We can reformulate some of the discussion surrounding equation (5.3) of Chapter II in our current framework, together with a generalization of the exponential map and the computations of Chapter I.

PROPOSITION 5.5. *Let Λ be a homogeneous Lagrangian submanifold of T^*M and let a be a function defined on $T^*M - \{0\}_M$ which is homogeneous of degree one, and such that a is nowhere zero on Λ. Then we can find a homogeneous φ defined on T^*M such that*

$$d\pi_*(\text{graph } d\varphi) = \Lambda$$

and

$$\{a, \varphi\} = a.$$

Notice that if χ is any homogeneous phase function for Λ, we have since $\alpha = d\chi$ on Λ, and by Euler's theorem,

$$\{a,\chi\} = \sum \frac{\partial a}{\partial \xi}\frac{\partial \chi}{\partial x} - \frac{\partial a}{\partial x}\frac{\partial \chi}{\partial \xi}$$

$$= \sum \frac{\partial a}{\partial \xi}\xi = a.$$

Thus $\{a,\chi\} = a$ will always hold on Λ for any homogeneous phase functions. Our problem is to adjust χ off Λ. Notice also that

$$\xi_a = \frac{\partial a}{\partial \xi}\frac{\partial}{\partial x} - \frac{\partial a}{\partial x}\frac{\partial}{\partial \xi}$$

is nowhere tangent to Λ. Indeed if ξ_a were tangent to Λ at z, then $\langle \xi_a, \alpha\rangle_z = 0$ since $\alpha\,|_\Lambda = 0$ for homogeneous Λ. But then $0 = (\partial a/\partial \xi)\cdot \xi = a$ by Euler's theorem again, contradicting the assertion that a does not vanish on Λ. Now choose a homogeneous submanifold C, of codimension one transversal to ξ_a and containing Λ and choose φ to be the solution of the non-singular first order partial differential equation with initial condition

$$\{a, \varphi\} = \xi_a \varphi = a \qquad \varphi = \chi \qquad \text{on } C.$$

This defines φ uniquely in a neighborhood of C. Since a, ξ_a and Λ are all homogeneous we conclude that φ is homogeneous. Finally, $\iota_C^* d\varphi = \iota_C^* d\chi$ since $\varphi = \chi$ on C, and $\xi_a \varphi = \xi_a \chi = a$ at points of Λ. Thus $d\varphi = d\chi$ at points of Λ and hence φ is a phase function for Λ. Q.E.D.

Now consider the flow, f_t, generated by ξ_a, and consider the set of points

$$(f_t(z), t, -a) \subset T^*(M \times \mathbf{R}).$$

This consists exactly of the set swept out by the isotropic submanifold

$$\{\Lambda; 0, -a\}$$

under the Hamiltonian flow generated by the vector field $\xi_a + \partial/\partial t$ on $T^*(M \times \mathbf{R})$ and is thus a (homogeneous) Lagrangian submanifold, which we shall denote by Λ^a. Notice that for small values of t we have

$$\left(\frac{\partial}{\partial t} + \xi_a\right)(\varphi - ta) = 0$$

and, at Λ^a with $t = 0$, letting τ denote the dual variable to t,

$$\begin{aligned} d(\varphi - ta) &= d\varphi - adt \\ &= \xi dx + \tau dt \,|_{\Lambda^a, t=0} \\ &= (\alpha_{M \times \mathbf{R}}) \,|_{\Lambda^a, t=0} \,. \end{aligned}$$

Now $D_{\xi_a} \alpha_M = 0$ and $D_{\partial/\partial t}(\tau dt) = 0$ so

$$D_{(\partial/\partial t + \xi_a)}(\alpha_{M \times \mathbf{R}}) = 0.$$

(In fact if b is any homogeneous function on $T^* N$ then it follows from Euler's equation that $D_{\xi_b} \alpha_N = 0$.) Thus

$$d(\varphi - ta) = \alpha_{M \times \mathbf{R}}$$

along Λ^a for sufficiently small t, i.e., $\varphi - ta$ is a phase function for Λ^a.

Associated with the same picture we can consider an inhomogeneous Lagrangian submanifold Λ_1 of $T^* M$, defined as follows. Let $S_1(\Lambda)$ consist of those points of Λ on which $a = 1$. (Notice that da does not vanish on Λ. Indeed $\langle \xi \partial/\partial \xi, da \rangle = a \neq 0$ and $\xi(\partial/\partial \xi)$ is tangent to Λ. Thus $S_1(\Lambda)$ defines a submanifold of Λ of codimension one.) Then $S_1(\Lambda)$ is an isotropic submanifold which is transversal to ξ_a (since ξ_a is in fact transversal to Λ). Let us map $S_1(\Lambda) \times \mathbf{R} \to T^*M$ by $f(\lambda, t) = f_t(\lambda)$ where f_t is the flow generated by ξ_a. Then $\Lambda_1 = f(S_1(\Lambda) \times \mathbf{R})$ is a Lagrangian manifold of T^*M. Notice that since $\xi_a a = 0$, we can conclude that $a \circ f \equiv 1$ on $S_1(\Lambda) \times \mathbf{R}$. We can now define the exponential map, exp: $S_1(\Lambda) \times \mathbf{R} \to M$ by setting $\exp = \pi \circ f$. For example, let a denote length relative to a Riemann metric and let Λ be the cotangent space at a point y. Then $S_1(\Lambda)$ is the unit sphere at the point in question, and exp is the usual exponential map. Similarly, we can let Λ be the normal bundle to a submanifold.

Let us set $S_1 = \{z \in T^*M \mid a(z) = 1\}$ so that, in the preceding notation $S_1(\Lambda) = \Lambda \cap S_1$. Let us assume that S_1 is a submanifold of T^*M, let $\pi_1 : S_1 \to M$ be the restriction of π to S_1 and let φ_1 be the restriction of φ to S_1, where φ is the previously constructed phase function for Λ. We claim that

$$\Lambda_1 = d\pi_1(\text{graph } d\varphi_1).$$

Indeed the set C_{φ_1} is given (by the method of Lagrange multipliers) by the equations

$$d_\xi \varphi - \mu d_\xi a = 0$$

$$a = 1$$

for some μ. Thus $(x, \xi) \in C_{\varphi_1}$ if and only if $(x, \xi, \mu, -1) \in \Lambda^a$. Now on Λ^a we have $d(\varphi - ta) = \alpha + \tau dt$ so

$$d\varphi - t da = \alpha.$$

On S_1 this becomes

$$d\varphi_1 = \iota^* \alpha \qquad \iota: S_1 \to T^*M$$

so, on C_{φ_1}

$$\frac{\partial \varphi_1}{\partial x^i} = \xi^i$$

which is exactly the required assertion.

Notice that the value of the Lagrange multiplier μ has a very simple geometric interpretation. Indeed, on the one hand, since φ is homogeneous, we have

$$\varphi = \sum \frac{\partial \varphi}{\partial \xi_i} \xi_i = \mu \sum \frac{\partial a}{\partial \xi_i} \xi_i = \mu a = \mu$$

on S_1. If $S_1(x)$ denotes $S_1 \cap T^*M_x$ and $x \in \pi\Lambda_1$ then $x = \pi(z)$ where z is a critical point of φ_1, i.e. μ is a critical value of φ_1. Notice that μ is also the time parameter such that $f_t(\lambda) = z$, where $\lambda \in S_1(\Lambda)$. Thus μ is a critical value of

$\varphi_{1|S_1(x)}$ and represents the length of time for an extremal from $S_1(\Lambda)$ to reach T^*M_x. In the Riemannian case the values of μ are exactly the length of the geodesics joining x to the point y.

[1]We close this section by examining the behavior of the Maslov cochain under functorial operations on Lagrangian submanifolds of cotangent bundles. Let $f\colon M \to N$ be a submersion. Then df^*T^*N is a subbundle of T^*M and we have the diagram

$$\begin{array}{ccccc} M & \xleftarrow{\pi_M} & T^*M & \xleftarrow{\iota} & df^*TN = \mathcal{H} \\ f\downarrow & & \downarrow g & & \\ N & \xleftarrow{\pi_N} & T^*N & & \end{array}$$

Let $p \in \mathcal{H}$ and $x = \pi_M p \in M$, while q is the image of p in T^*N and $y = f(x)$ is the image of q in N. Thus

$$p = df_x^*q \qquad q = gp$$

$$x = \pi_M p \qquad y = \pi_N q.$$

Let α_M and α_N denote the fundamental one forms on T^*M and T^*N. We claim that

$$\iota^*\alpha_M = g^*\alpha_N.$$

Indeed, let ξ be tangent to $\mathcal{H}$ at p. Then

$$\begin{aligned} \langle \xi, g^*\alpha_N\rangle &= \langle dg\xi, \alpha_N\rangle \\ &= \langle d\pi_N \circ dg\xi, q\rangle \\ &= \langle df \circ d\pi_M\xi, q\rangle \\ &= \langle d\pi_M\xi, df^*q\rangle \\ &= \langle d\pi_M\xi, p\rangle = \langle \xi, \alpha_M\rangle. \end{aligned}$$

If $\omega_M = d\alpha_M$ and then, *a fortiori*, we have (as in (4.1))

$$g^*\omega_N = \iota^*\omega_M. \tag{5.4}$$

If Λ_M is a Lagrangian submanifold of T^*M which intersects $\mathcal{H}$ transversally, then $df_*\Lambda_M$ is defined to be $\Lambda_M \cap \mathcal{H}$ (when mapped in T^*N by g). Let $j\colon \Lambda_M \cap \mathcal{H} \to \Lambda_M$, and let $\mathfrak{M}_{\Lambda_M}$ be the Maslov class of Λ_M. We claim that

$$j^*\mathfrak{M}_{\Lambda_M} = \mathfrak{M}_{df_*\Lambda_M} \tag{5.5}$$

where $\mathfrak{M}_{df_*\Lambda_M}$ is the Maslov class of $df_*\Lambda_M$.

Similarly, suppose that Λ_N is a Lagrangian submanifold of T^*N. Then $df^*\Lambda_N = g^{-1}\Lambda_N$ and we claim that

[1]The rest of this section should be omitted on first reading.

$$g^* \mathfrak{M}_{\Lambda_N} = \mathfrak{M}_{df^*\Lambda_N} \tag{5.6}$$

where g: $df^* \Lambda_N \to \Lambda_N$. (We have used the same letter, g, to denote the restriction of g to $g^{-1}\Lambda_N$.)

We first prove (5.5):

Let us set

$$\begin{aligned} h &= T_p \mathcal{H} \\ v_M &= \text{tangent space to fiber (of } \pi_M) \text{ at } p \\ v_N &= \text{tangent space to fiber (of } \pi_N) \text{ at } q \\ T_p &= T_p(T^* M) \\ T_q &= T_q(T^* N) \end{aligned}$$

and k the kernel of dg: $h \to T_q$ so that

$$0 \to k \to h \to T_q \to 0$$

is exact. We claim that k is an isotropic subspace and that, in fact $h = k^{\perp}$ under ω_M so that $T_q = k^{\perp}/k$. Indeed, if $\boldsymbol{\xi} \in k$ and $\boldsymbol{\eta} \in h$ then

$$\langle \boldsymbol{\xi} \wedge \boldsymbol{\eta}, \omega_M \rangle = \langle dg\boldsymbol{\xi} \wedge dg\boldsymbol{\eta}, \omega_N \rangle = 0$$

so that $h \subset k^{\perp}$ on the other hand $\dim k = \dim M - \dim N$ while $\dim h = \dim M + \dim N$ so that we must have $h = k^{\perp}$.

In particular, we have a map, ρ, from $L(T_p) \to L(T_q)$ sending any $u \subset T_p$ into $u \cap k^{\perp}/u \cap k \subset T_q$. Let Λ be a Lagrangian submanifold of $T^* M$ intersecting $\mathcal{H}$ transversally. We have

$$T_p(\Lambda \cap \mathcal{H}) = T_p \Lambda \cap h$$

while

$$\begin{aligned} (T_p \Lambda \cap k)^{\perp} &= T_p \Lambda + k^{\perp} \\ &= T_p \Lambda + h \\ &= T_p(T^* M), \end{aligned}$$

by transversality, so that

$$T_p \Lambda \cap k = 0.$$

Thus

$$T_q(\Lambda \cap \mathcal{H}) = \rho T_p \Lambda.$$

Of course this holds at all of points of $\Lambda \cap \mathcal{H}$ so we can write

$$T(\Lambda \cap \mathcal{H}) = \rho T\Lambda.$$

Since v_M is transversal to h, the same argument shows that

$$v_N = \rho(v_M).$$

Now let C_N and D_N be sections of $L(\Lambda \cap \mathcal{H})$. Then $C_N(p)$ is a subspace of $k^\perp/k$ and hence determines a unique Lagrangian subspace of $k^\perp$ which we denote by $C_M(p)$. We claim that if $C_N(p) \cap T_p(\Lambda \cap \mathcal{H}) = 0$ then $C_M(p) \cap T_p(\Lambda) = 0$. Indeed

$$C_M(p) \cap T_p(\Lambda) \subset k$$

since $C_M(p) \subset k^\perp$ and $C_M(p) \cap T_p(\Lambda)/k = 0$. But then

$$C_M(p) \cap T_p(\Lambda) \subset C_M(p) \cap T_p(\Lambda) \cap k = 0$$

since $T_p(\Lambda) \cap k = 0$. The same argument works for v_N.

Thus, if C_N is transversal to v_N and $T(\Lambda \cap \mathcal{H})$, then C_M is transversal to v_M and $T(\Lambda)$ along $\Lambda \cap \mathcal{H}$, and hence can be extended to a section of $L(\Lambda)$ defined near $\Lambda \cap \mathcal{H}$ which is still transversal to v_M and $T(\Lambda)$. We claim that

$$(T(\Lambda), v_M, C_M, D_M) = (T(\Lambda \cap \mathcal{H}), v_N, C_N, D_N),$$

if D_N is a second such section. Indeed,

$$(T(\Lambda), v_M, C_M, D_M) = -(C_M, D_M, T(\Lambda), v_M)$$

and since $C_M \cap D_M \supset k$ we have

$$\begin{aligned}(C_M, D_M, T(\Lambda), v_M) &= (C_N, D_N, \rho T(\Lambda), \rho v_M)\\ &= (C_N, D_N, T(\Lambda \cap \mathcal{H}), v_N).\end{aligned}$$

This shows explicitly how to relate the Čech cochains on Λ and $\Lambda \cap \mathcal{H}$ so that the Maslov class on Λ pulls back to the Maslov class on $\Lambda \cap \mathcal{H}$, proving (5.5).

We leave (5.6) as an exercise for the reader. The crucial point is to observe that at any point of $g^{-1}\Lambda_M$ all of the intersection $v_M \cap T(g^{-1}\Lambda_N)$ arises from N, in other words comes from $v_N \cap \Lambda_N$.

The functoriality of the Maslov class implies a corresponding functoriality for the Maslov line bundle. This follows from general nonsense (the isomorphism between cohomology classes and equivalence classes of line bundles) but can also be seen directly as follows. Let $f\colon M \to N$ be a submersion, let $y = f(x)$ and $p = df_x^* q$, $p \in \Lambda_M$ and $q \in \Lambda_N = df_* \Lambda_M$, the notations being as in the preceding paragraph. We recall from Section 3 that the fiber of the Maslov bundle at p is just the set of all functions

$$f: \mathcal{O} \to \mathbf{C}, f(C) = e^{(i\pi/2)\alpha(C,D)} f(D),$$

$\alpha(C, D)$ being the cross-ratio $(T_p(\Lambda), v_M(p), C, D)$, and $\mathcal{O}$ being the subset of $L(T_p)$ consisting of those Lagrangian spaces which are transversal both to $T_p(\Lambda)$ and $v_M(p)$. Now the map ρ defined in the preceding paragraph maps $\mathcal{O}$ into the corresponding subset of $L(T_q)$ and preserves cross-ratios. Thus the fiber of the Maslov bundle at q gets mapped bijectively onto the fiber of the Maslov bundle at p by ρ^*.

§6. Periodic Hamiltonian systems

In this section we depart briefly from the study of general properties of symplectic manifolds and examine the special case of a Hamiltonian flow with manifolds of periodic solutions. We have already encountered such a situation in conjunction with the phenomenon of perfect focusing in Chapter III. We shall elaborate on some of the properties of such systems in this section and describe some important examples. The reader who prefers to continue with the general theory can skip to the next section.

Let ξ be a Hamiltonian vector field on a symplectic manifold X so that

$$\xi \rfloor \omega = -dH.$$

Suppose we are given a one parameter family of trajectories of ξ. That is, suppose we have a map F of the region $0 \leqslant t \leqslant T(s)$, $0 \leqslant s \leqslant a$ of the (t, s) plane into X such that

$$dF\left(\frac{\partial}{\partial t}\right)_{(t,s)} = \xi_{F(t,s)}.$$

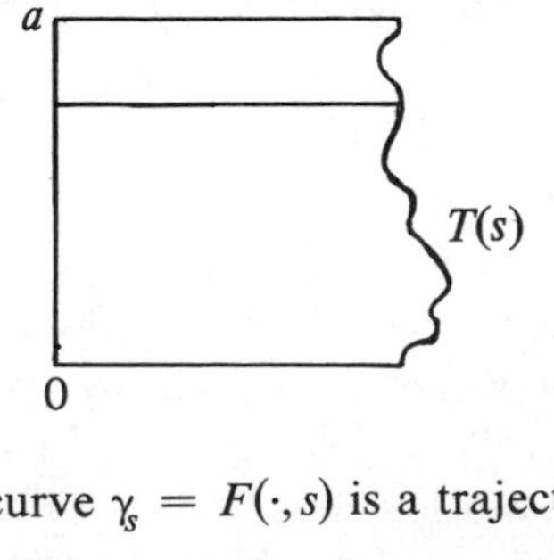

Thus, for each fixed s, the curve $\gamma_s = F(\cdot, s)$ is a trajectory of ξ. Now

$$\left\langle \frac{\partial}{\partial t} \wedge \frac{\partial}{\partial s}, F^* \omega \right\rangle = \left\langle dF\frac{\partial}{\partial t} \wedge dF\frac{\partial}{\partial s}, \omega \right\rangle$$
$$= \left\langle dF\frac{\partial}{\partial s}, \xi \lrcorner \omega \right\rangle$$
$$= -\left\langle dF\frac{\partial}{\partial s}, dH \right\rangle$$
$$= -\left\langle \frac{\partial}{\partial s}, d(H \circ F) \right\rangle$$
$$= \left\langle \frac{\partial}{\partial s}, \frac{\partial}{\partial t} \lrcorner (d(H \circ F) \wedge dt) \right\rangle$$
$$= \left\langle \frac{\partial}{\partial t} \wedge \frac{\partial}{\partial s}, d(H \circ F) \wedge dt \right\rangle$$

or, in short, we obtain Cartan's formula,

$$F^* \omega - d(H \circ F) \wedge dt = 0. \tag{6.1}$$

Now suppose there is a form α on X with $d\alpha = \omega$. Then the above equation becomes

$$dF^*(\alpha + tdH) = 0.$$

Let b_0 denote the curve $t = 0$ and b_1 the curve $s \rightsquigarrow (T(s), s)$. We can now apply Stokes' theorem to conclude, since H is constant along γ_0 and γ_a, that

$$\int_{\gamma_a} \alpha - \int_{\gamma_0} \alpha = \int_{b_1} F^* \alpha - \int_{b_0} F^* \alpha + \int_{b_1} T(s)\, d(H \circ F). \tag{6.2}$$

We have already encountered versions (and applications) of this formula in Chapter II and Chapter III. Let us now apply this formula to the case where each trajectory γ_s is periodic, i.e., that

$$F(0, s) = F(T(s), s).$$

Then

$$\int_{b_1} F^* \alpha = \int_{b_0} F^* \alpha$$

and (6.2) becomes

$$\int_{\gamma_a} \alpha = \int_{\gamma_0} \alpha + \int_0^a T(s)\, d\overline{H} \tag{6.3}$$

where $\overline{H}(s) = H(F(T(s), s))$. In particular, we obtain the following result, due to W. B. Gordon [4],

If γ_s is a one parameter family of periodic orbits lying on fixed energy surface $H = \text{const}$ then $\int_{\gamma_s} \alpha$ is independent of s.

Now suppose that all the trajectories of ξ are periodic and that the surfaces $H =$ const. are connected. Then the integral $\int \alpha$ over any periodic trajectory depends only on H so we can write it as $I(H)$. The equation (6.3) shows that

$$I'(H) = T(s) \tag{6.4}$$

showing that all orbits on the same energy surface have the same period. (The above argument is also due to Gordon, [4]. The actual result in various special cases had been known for some time, cf. Wintner [5] and the references cited in Moser [6] and Gordon [4].) We now present a generalization of this result due to Weinstein [8]. Let $f\colon X \times \mathbf{R} \to X$ be the map giving the flow generated by $\boldsymbol{\xi}$ (so that $f(x, \cdot)$ is the trajectory through x). If the flow is not globally defined then the domain of definition of f is understood to be the appropriate subset of $X \times \mathbf{R}$. We shall consider ω and H to be defined on $X \times \mathbf{R}$ via the projection of $X \times \mathbf{R}$ onto X. Let $\boldsymbol{\eta}$ be any tangent vector to X, considered as a tangent vector to $X \times \mathbf{R}$. Then the computation we did in proving (6.1) shows that

$$\left\langle \frac{\partial}{\partial t} \wedge \boldsymbol{\eta}, f^* \omega \right\rangle = \left\langle \frac{\partial}{\partial t} \wedge \boldsymbol{\eta}, dH \wedge dt \right\rangle.$$

If $\boldsymbol{\eta}$ and $\boldsymbol{\zeta}$ are two tangent vectors to X then

$$\langle \boldsymbol{\eta} \wedge \boldsymbol{\zeta}, f^* \omega \rangle = \langle \boldsymbol{\eta} \wedge \boldsymbol{\zeta}, \omega \rangle$$

since the map $x \to f(x, t)$ is a symplectic diffeomorphism for each fixed t. We can combine these last two equations into the formula

$$f^* \omega = \omega + dH \wedge dt \tag{6.5}$$

Now let $g\colon X \times \mathbf{R} \to X \times X$ be the graph of f so that $g(x, t) = (x, f(x, t))$. (Again g may not be defined on all of $X \times \mathbf{R}$, it has the same domain of definition as f.) On $X \times X$ we put the symplectic structure $\omega_{X \times X} = \pi_2^* \omega - \pi_1^* \omega$ where π_1 and π_2 denote the projections of $X \times X$ onto the first and second factors, so that, for example $\pi_2 \circ g = f$. Then, by (6.5)

$$\begin{aligned} g^* \omega_{X \times X} &= g^* \pi_2^* \omega - g^* \pi_1^* \omega \\ &= f^* \omega - \omega \end{aligned}$$

or

$$g^* \omega_{X \times X} = dH \wedge dt. \tag{6.6}$$

In particular, if $M \subset X \times \mathbf{R}$ is a submanifold such that $g_{|M}$ is an isotropic submanifold of $X \times X$, so that $(g_{|M})^* \omega_{X \times X} = 0$, we conclude that $dH \wedge dt$ vanishes on M. For example, suppose that all the trajectories of $\boldsymbol{\xi}$ are periodic, and that the period of the trajectory through x is $T(x)$. Then take M

$= \{(x, T(x))\}$ and we have $g(x, T(x)) = (x, x)$. Since the diagonal is a Lagrangian submanifold of $X \times X$ we conclude that

$$dH \wedge dT = 0$$

showing once again that the period and the energy are "functionally dependent".

Suppose that $H = c$ defines an energy surface, Z_c, and that the trajectories of ξ make this energy surface into a fiber bundle over Y. Thus Y is the "space of orbits" of ξ. Notice ω, when restricted to Z_c, is of rank $2n - 2$ (where $\dim X = 2n$) and on this surface $\xi \rfloor \omega = dH_{|\{H=c\}} = 0$. We can identify TY_y with the space $TZ_x/\{\xi_x\}$ where x is any point in the orbit given by y. Since ω induces a nondegenerate antisymmetric bilinear form on $TZ_x/\{\xi_x\}$, we obtain an induced form $\bar{\omega}_y$ on TY_y. As the flow generated by ξ preserves ω this induced form does not depend on x. We thus get a form, $\bar{\omega}$, on Y and it is easy to check that this makes Y into a symplectic manifold. If h is any symplectic automorphism preserving ξ it clearly induces a symplectic automorphism of Y.

For example, let us consider the geodesic flow on the sphere S^n (considered as the unit sphere in $\mathbf{R}^{n+1}$) . The energy surfaces correspond to tangent (or cotangent) vectors of constant length and all trajectories are great circles, traversed at different speeds, the period being inversely proportional to the speed, and hence constant on each energy hypersurface. The group $O(n + 1)$ acts as automorphisms of the flow, and hence also as automorphisms of the associated orbit spaces. A most important special case is where $n = 3$. In this case S^3 is itself a group, and the metric invariant under right and left translations. Since for any group G we can identify $T(G)$ with $G \times g$, where g is the Lie algebra of G, we can identify the unit tangent vectors with $S^3 \times S^2$. As the metric is invariant, the corresponding geodesic flow is just the exponential map and its translates. See, Sternberg [2]. The trajectories are hence of the form $\gamma \times v$ where γ is a great circle determined by the unit vector v and v is a constant unit vector in g. For each fixed v this induces the Hopf fibration of S^3 and we thus see that the space of orbits is $S^2 \times S^2$. What is remarkable about this example is the beautiful result, due to Moser [6], (see also Souriau [9]) that, up to compactification and "regularization", the geodesic flow on a fixed energy surface is the same as the flow on a fixed surface of constant negative energy for the Kepler problem, i.e., for a particle moving according to the inverse square law of attraction from a fixed center. In particular the orbit spaces are the same, i.e., $S^2 \times S^2$. As we shall see, this accounts for the role of the group $O(4)$ in the quantization of the hydrogen atom.

We first make a preliminary remark about "regularizing" a Hamiltonian vector field. Suppose ξ is a Hamiltonian vector field with $\xi \rfloor \omega = -dH$, and f is a smooth function. Then on any fixed energy hypersurface $H = c$ the vector field $f\xi$ coincides with the restriction to $H = c$ of a Hamiltonian vector field. Indeed let η be the Hamiltonian vector field determined by

$$\boldsymbol{\eta} \,\lrcorner\, \omega = -d((H-c)f) = -(H-c)df - fdH.$$

On $H = c$ we get $\boldsymbol{\eta} \,\lrcorner\, \omega = -fdH$ so $\boldsymbol{\eta} = f\boldsymbol{\xi}$, and thus $f\boldsymbol{\xi}$ coincides with the Hamiltonian vector field $\boldsymbol{\eta}$ on $H = c$. (Of course $f\boldsymbol{\xi}$ will not be Hamiltonian everywhere if f and H are not functionally dependent.)

Now multiplying a vector field, $\boldsymbol{\xi}$, by a function f can be thought of as making a "change of independent variable" $t = \int f ds$ (i.e., $dt/ds = f(z(t))$ where $z(t)$ is a trajectory of $\boldsymbol{\xi}$). Now if f is nowhere zero, nothing much happens. If f is allowed to become zero, then $\boldsymbol{\eta}$ might have nicer properties than $\boldsymbol{\xi}$. For example, $\boldsymbol{\xi}$ might be defined on an open submanifold, U, of a manifold W, becoming "infinite" at ∂U, while $\boldsymbol{\eta} = f\boldsymbol{\xi}$ is defined on all of W. This is precisely what happens in the "collision orbits" of the Kepler problem, i.e., those orbits on which the particle moves on a straight line course to the origin. The velocity becomes infinite as the particle approaches the origin, and so $\boldsymbol{\xi}$ blows up. On the other hand, after multiplication by an appropriate f, these orbits will turn out to be well behaved, indeed to correspond to the tangent vectors to great circles passing through the north pole, while the energy surface will correspond to those tangent vectors lying over the sphere with the north pole removed. Furthermore by suitable normalization it turns out that the relation between t and s will be given by the celebrated Kepler equation $t = s - e \sin s$ where e is the eccentricity of the ellipse. For $e = 1$ the ellipse degenerates into a straight line collision orbit with collision at $t = 0$. For t near zero $t = s - \sin s = s^3/3! + \cdots$ yielding the result that $t^{1/3}$ is a "regularizing parameter".

Let us now give the details. We follow Moser's elegant discussion almost verbatum, with the exception of minor notational changes. Let $x = (x_0, \ldots, x_n)$ be Euclidean coordinates on $\mathbf{R}^{n+1}$ and $\xi = (\xi_0, \ldots, \xi_n)$ the dual coordinates so that (x, ξ) are coordinates on $\mathbf{R}^{n+1} + (\mathbf{R}^{n+1})^* = T^*\mathbf{R}^{n+1}$. The Hamiltonian flow generated by the function $\Phi = \frac{1}{2}\|\xi\|^2\|x\|^2$ gives the differential equations

$$\frac{dx}{ds} = \frac{\partial \Phi}{\partial \xi} = \|x\|^2 \xi \qquad \frac{d\xi}{ds} = -\frac{\partial \Phi}{\partial x} = -\|\xi\|^2 x.$$

It is clear that

$$\frac{d(x \cdot \xi)}{ds} = 0, \qquad \frac{d\|x\|^2}{ds} = 2\|x\|^2 x \cdot \xi,$$

and

$$d\|\xi\|^2/ds = -2\|\xi\|^2(x \cdot \xi).$$

Thus, the (co)tangent bundle to the unit sphere, T^*S^n, given by

$$\|x\|^2 = 1, \qquad x \cdot \xi = 0,$$

is preserved under the flow and on T^*S^n the trajectories are given by

$$\frac{dx}{ds} = \xi \qquad \frac{d\xi}{ds} = -\|\xi\|^2 x$$

or

$$\frac{d^2x}{ds^2} + \|\xi\|^2 x = 0$$

which is clearly the equation of a great circle. Identifying $T^* S^n$ as a submanifold of $T^* \mathbf{R}^{n+1}$ as indicated above thus yields the fact that the restriction of Φ to $T^* S^n$ generates the geodesic flow on $T^* S^n$. We shall continue to denote the restriction of Φ by Φ. The unit tangent vectors constitute the "energy" hypersurface $\Phi = \frac{1}{2}$.

We now let $y = (y_1, \ldots, y_n)$ denote Euclidean coordinates on $\mathbf{R}^n$, with dual coordinates η, and consider stereographic projection from the north pole

$$y_k = \frac{x_k}{1 - x_0}, \qquad k = 1, \ldots, n$$

so that

$$\|y\|^2 = \frac{1 - x_0^2}{(1 - x_0)^2} = \frac{1 + x_0}{1 - x_0}$$

and hence

$$x_0 = \frac{\|y\|^2 - 1}{\|y\|^2 + 1} \qquad x_k = \frac{2y_k}{\|y\|^2 + 1}$$

gives the inverse map. It is easy to check the well-known fact that this map is conformal, more particularly that

$$dx^2 = \sum_0^n dx_k^2 = \frac{4}{(\|y\|^2 + 1)} dy^2.$$

Let S_0^n denote the sphere with the north pole removed. The stereographic projection is a diffeomorphism of S_0^n onto $\mathbf{R}^n$ and hence induces a diffeomorphism of $T^* S_0^n$ with $T^* \mathbf{R}^n$ (which carries the canonical one form of $T^* S_0^n$ onto the canonical one form of $T^* \mathbf{R}^n$) . Thus we will get $\xi = \xi(y, \eta)$ and $\eta = \eta(x, \xi)$ with $\xi \cdot dx = \eta \cdot dy$, and where, in fact, the $\xi = \xi(y, \eta)$ and the $\eta = \eta(x, \xi)$ are determined by this equation. Using the fact that $\xi \cdot x = 0$ and $\sum x_i^2 = 1$ it is easy to check that $\eta(x, \xi)$ is given by $\eta_k = (1 - x_0)\xi_k + \xi_0 x_k$ and since $\eta \cdot y = \sum \eta_k y_k = \xi_0$ we get

$$\xi_0 = \eta \cdot y \qquad \xi_k = \frac{\|y\|^2 + 1}{2} \eta_k - (\eta \cdot y) y_k$$

so that

$$\|\xi\| = \frac{\|y\|^2 + 1}{2}\|\eta\|.$$

We can now use the diffeomorphism of $T^*S_0^n$ with $T^*\mathbf{R}^n$ to transfer the Hamiltonian which becomes

$$F(y,\eta) = \frac{(\|y\|^2 + 1)^2\|\eta\|^2}{8}.$$

If u is any function of a real variable with $u'(\frac{1}{2}) = 1$, then $u(F)$ and F will define the same trajectories on the energy surface $F = \frac{1}{2}$. Let us take $G = u(F) = \sqrt{2F} - 1$. Then

$$G = \frac{(\|y\|^2 + 1)\|\eta\|}{2} - 1$$

and the corresponding Hamiltonian vector field, $\boldsymbol{\eta}$, is defined by $\boldsymbol{\eta} \rfloor \omega = -dG$. Let us now introduce the vector field $\boldsymbol{\xi}$ given by

$$\boldsymbol{\eta} = \|\eta\|\boldsymbol{\xi}$$

so that $\boldsymbol{\xi}$ is defined only for $\|\eta\| \neq 0$. Now $F = \frac{1}{2}$ corresponds to $G = 0$ and so the vector field $\boldsymbol{\xi}$ is generated by the Hamiltonian

$$H = \|\eta\|^{-1}G - \tfrac{1}{2} = \|\eta\|^{-1}(\sqrt{2F} - 1) - \tfrac{1}{2} = \frac{\|y\|^2}{2} - \frac{1}{\|\eta\|},$$

Let us set $p = -y$ and $q = \eta$ so that $dp \wedge dq = d\eta \wedge dy$. Then

$$H = \tfrac{1}{2}\|p\|^2 - 1/\|q\|$$

which is the Hamiltonian for the Kepler problem. The energy surface $\Phi = \frac{1}{2}$ corresponds to $H = -\frac{1}{2}$. We may get a more general energy hypersurface as follows: Let us consider the covectors of length ρ, i.e. the set of all (x,ξ) with $\|\xi\| = \rho$, and the image of this set under stereographic projection, i.e. the hypersurface $F = \frac{1}{2}\rho^2$. Now set

$$G = (2\rho^2 F)^{\frac{1}{2}} - \rho^2 = \tfrac{1}{2}\rho(\|y\|^2 + 1)\|\eta\| - \rho^2,$$

so that the surface is given by $G = 0$. Then

$$\rho^{-3}\|\eta\|^{-1}G = \frac{\|y\|^2}{2\rho^2} - \frac{1}{\rho\|\eta\|} + \frac{1}{2\rho^2}.$$

We can now make the symplectic change of coordinates $p = -y/\rho$ and $q = \rho\eta$, and we see that the surface $G = 0$ is carried over into the energy surface $H = -1/2\rho^2$ for the Kepler problem. We can summarize the discussion as follows: Each of the energy surfaces $\|\xi\|^2 = \frac{1}{2}\rho^2$ is a manifold of dimension $2n - 2$ which is fibered by circles (the geodesics) over a symplectic manifold of dimension $2n-2$. Under stereographic projection, the $2n-2$ dimensional manifold (less the points sitting over the north pole) goes over into a manifold that can be identified with the space of orbits of the Kepler problem with energy $H = -1/2\rho^2$. In particular, the induced map of the $2n-2$ dimensional manifold onto the $2n-2$ dimensional space of orbits of the Kepler problem is a symplectic diffeomorphism. For $n = 3$, the "completed" space of orbits is topologically $S^2 \times S^2$.

Let us analyze, in terms of more classical terminology, the parameters that enter into the Kepler orbits in terms of the stereographic projection, $(x, \xi) \to (y, \eta) \to (-p, q)$. A given orbit corresponds to a great circle. Since the whole picture is invariant under all rotations about the x_0 axis, we may assume that the orbit lies in the subspace $x_3 = \cdots = x_n = 0$ and is described by the equations

$$x_0 = \sin\alpha \cos s \qquad x_1 = \sin s \qquad x_2 = -\cos\alpha \cos s$$

so

$$\xi_0 = -\sin\alpha \sin s \qquad \xi_1 = \cos s \qquad \xi_2 = \cos\alpha \sin s$$

and thus setting $e = \sin\alpha$

$$p_1 = -\frac{\sin s}{1 - e\cos s} \qquad p_2 = \frac{\sqrt{1-e^2}\cos s}{1 - e\cos s}$$

$$q_1 = \cos s - e \qquad q_2 = \sqrt{1-e^2}\sin s$$

so that the p's describe a circle while the q's describe an ellipse of eccentricity e parametrized by the eccentric anomaly, s; (see Sternberg [2]). Now $\|q\|^2 = (1 - e\cos s)^2$ so

$$t = \int_0^s |q|\,ds = \int_0^s (1 - e\cos s)\,ds = s - e\sin s$$

showing that our regularization procedure does indeed reduce to Kepler's equation.

We refer the reader to Moser [6] and Weinstein [8] for applications of these results to the existence of periodic orbits for the three body problem under conditions which can be regarded as small perturbation of Kepler motion. More generally these papers discuss the problem of establishing periodic orbits for systems which are perturbations of a system having manifolds of periodic orbits. Our main use of these examples will be to the study of a "quantization procedure" to be described in the next chapter and to the study of the asymptotic distribution of eigenvalues of elliptic operators.

The preceding considerations show why the group $O(4)$ enters as a symmetry group of the Kepler problem, on the set of orbits of some fixed (negative) energy.

(This fact is strikingly confirmed in the study of atomic spectra. Each energy level of the hydrogen atom occurs with a multiplicity corresponding to an irreducible representation of $O(4)$. In comparison, the alkali metals exhibit a spectrum corresponding only to $O(3)$ symmetry. However for the higher energy levels of the outer electron, where the potential is approximately of the form k/r, an approximate $O(4)$ degeneracy appears.) If we consider the set of orbits of all negative energies, it is known that a still larger group, the group $SO(2,4)$, of all orthogonal transformations in six dimensions preserving a quadratic form of signature $++----$ acts as a group of symmetries. That is, let T^+S^3 denote the subset of T^*S^3 consisting of the non-zero covectors. We shall show that the group $SO(2,4)$ acts transitively as a group of symplectic diffeomorphisms on T^+S^3. (In all that follows we could replace 3 by n and $SO(2,4)$ by $SO(2, n+1)$ etc.)

Let G be a Lie group and g its Lie algebra. If $\xi \in g$, then $\exp t\xi$ is a one parameter subgroup of G, where exp: $g \subset G$ denotes the exponential map, and the most general one parameter subgroup of G is of this form. If a is some element of G, then $a(\exp t\xi)a^{-1}$ is again a one parameter subgroup, and so is of the form $\exp t\zeta$. The map $\xi \rightsquigarrow \zeta$ is linear–we write $\zeta = \mathrm{Ad}_a \xi$. This assigns to each $a \in G$ a linear transformation; in this way we get a linear representation of G on G known as the adjoint representation. Since G acts on g, it has a contragredient action on the dual space, g^*: for any $l \in g^*$ we define the element $a \cdot l$ by

$$\langle \xi, a \cdot l \rangle = \langle \mathrm{Ad}_{a^{-1}} \xi, l \rangle.$$

This representation is known as the coadjoint representation. For any $\eta \in g$ and any $l \in g^*$ we can consider the curve $(\exp t\eta) \cdot l$. We denote the tangent to this curve at $t = 0$ by η_l. Since

$$\frac{d}{dt} \mathrm{Ad}_{(\exp t\eta)}\, \xi_{|t=0} = [\eta, \xi],$$

where [,] denotes Lie bracket, we have

$$\langle \xi, \eta_l \rangle = \frac{d}{dt} \langle \mathrm{Ad}_{(\exp - t\eta)}\, \xi, l \rangle_{|t=0} = -\langle [\eta, \xi], l \rangle.$$

The vectors η_l span the tangent space to the orbit, $G \cdot l$, at the point l. In the next section we shall prove that for any Lie group, G, and any $l \in g^*$, the orbit $G \cdot l$ is always a symplectic manifold, and the symplectic form ω is given by

$$\langle \xi_l \wedge \eta_l, \omega \rangle = -\langle [\xi, \eta], l \rangle.$$

(In fact, we shall prove that all symplectic manifolds on which G acts transitively as a group of symplectic diffeomorphisms can be obtained by a slight modification of the preceding construction.) We shall now show how T^+S^3 can be regarded as a particular orbit in the dual of the Lie algebra of $SO(2,4)$. For this purpose, it is convenient to have the following description of the Lie algebra: Let V be a real vector space equipped with a non-degenerate scalar product, (,). Let $o(V)$ denote the Lie algebra of the orthogonal group of this scalar product. We can identify $\wedge^2(V)$ with $o(V)$ by letting $u \wedge v$ act as the linear transformation

$$(u \wedge v)w = (v,w)u - (u,w)v.$$

It is easy to check that this does indeed define a map which is equivariant with respect to the action of $O(V)$ on both sides. Similarly, the metric on V induces an isomorphism of V with V^*, and hence of $\wedge^2(V)$ with its dual space. Hence we may identify $o(V)^*$ with $\wedge^2(V)$ as well. In particular, we have the identification of $o(2,4)$ with $\wedge^2(\mathbf{R}^{2,4})$.

Let us consider the set of all elements of $o(2,4) \sim \wedge^2(\mathbf{R}^{2,4})$ of the form

$$0 \neq f \wedge f' \quad \text{where } \|f\|^2 = \|f'\|^2 = 0 \text{ and } (f,f') = 0.$$

(As linear transformations, these elements can be characterized as those nilpotent elements of rank two whose range (which is two dimensional) is totally isotropic.) By Witt's theorem, any two such elements are conjugate under $O(2,4)$. We shall now show that under the connected group $SO(2,4)$ they split into two components. Let us choose an orthonormal basis, e_{-1}, e_0, e_1, e_2, e_3, e_4 of $\mathbf{R}^{2,4}$, with e_{-1} and e_0 positive definite and the rest negative. We claim that for any such element $f \wedge f'$ we must have

$$(e_{-1} \wedge e_0, f \wedge f') \neq 0. \tag{6.7}$$

Indeed, suppose that $(e_{-1} \wedge e_0, f \wedge f') = 0$. Then the space spanned by f and f', i.e. the range of $f \wedge f'$, would have to contain some non-zero vector orthogonal to both e_{-1} and e_0. But such a vector would lie in $\mathbf{R}^4$ and could not be isotropic. Thus, under the connected group, $SO(2,4)$, the set of all $f \wedge f'$ splits into two components according to whether the scalar product in (6.7) is positive or negative. Let $\mathcal{V}$ denote the set of $f \wedge f'$ for which the scalar product is positive. Without changing the element $f \wedge f'$, we can choose f and f' so that $(f, e_0) = 0 = (f', e_{-1}) = 0$, while (f, e_{-1}) and (f', e_0) are both positive. This shows that every element of $\mathcal{V}$ has a unique representation as

$$s(e_{-1} + p) \wedge (e_0 + q) \quad \text{where } s > 0,\ p, q \in \mathbf{R}^4,\ (p,q) = 0 \\ \text{and } \|p\|^2 = \|q\|^2 = -1. \tag{6.8}$$

Now we can think of p as ranging over the unit sphere in $\mathbf{R}^4$, and then for fixed p, the q ranges over the unit tangent vectors to the sphere. Thus (6.8) becomes identified with a non-zero tangent vector to the three sphere. Using the Riemann metric we may identify TS^3 and $T^* S^3$, and so we see that $\mathcal{V}$ is diffeomorphic to

$T + S^3$. We shall soon see that this is a symplectic diffeomorphism, when we put the canonical symplectic form on $T^* S^3$.

Let us first examine the action of $SO(2)$ and $SO(4)$. Any $A \in SO(4)$ preserves e_{-1} and e_0 and sends p into Ap and q into Aq. This is clearly the induced action of $SO(4)$ on TS^3. The element R_θ in $SO(2)$ sends e_{-1} to $\cos\theta e_{-1} + \sin\theta e_0$ and e_0 to $-\sin\theta e_{-1} + \cos\theta e_0$. Thus

$$
\begin{aligned}
R_\theta[(e_{-1} + p) \wedge (e_0 + q)]& \\
&= (\cos\theta e_{-1} + \sin\theta e_0 + p) \wedge (-\sin\theta e_{-1} + \cos\theta e_0 + q) \\
&= (e_{-1} + \cos\theta p - \sin\theta q) \wedge (e_0 + \sin\theta_p + \cos\theta q).
\end{aligned}
$$

In other words, from the point of view of TS^3, the map R_θ moves (p, q) through angle θ along the circle determined by p and q. But this is precisely the image of (p, q) under the geodesic flow. Thus for *unit* vectors, the group $SO(2)$ acts as geodesic flow. Let F_t denote the geodesic flow. The above computation shows that for vectors of length s, we have $R_\theta(p, q) = F_{s^{-1}\theta}(p, q)$. To express this slightly differently, let $H = \frac{1}{2}\|q\|^2 = -\frac{1}{2}s^2$. The geodesic flow has, as its infinitesimal generator, the vector field ξ_H corresponding to $-dH$ under the identification of differential forms and vector fields on $T^* S^3$ given by the symplectic form. Suppose we consider the function $\sqrt{-2H}$. It gives rise to the vector field $-(-2H)^{-1/2}\xi_H = -s^{-1}\xi_H$. We thus see that the action of R_θ is the same as that generated by a vector field corresponding to the function $(-2H)^{1/2}$.

Let $O(1,4)$ be the subgroup of $O(2,4)$ which fixes the vector e_{-1}. It is clear that $SO(1,4)$ acts transitively on $\mathcal{V}$. Indeed, let $\mathbf{R}^{1,4}$ be the subspace of $\mathbf{R}^{2,4}$ spanned by e_0, e_1, e_2, e_3, and e_4. By Witt's theorem, $O(1,4)$ acts transitively on the set of all pairs of the form (p, f') where $\|p\|^2 = -1$, $\|f'\|^2 = 0$ and $\langle p, f'\rangle = 0$, with p and $f' \neq 0$ in $\mathbf{R}^{1,4}$. Now f' cannot be orthogonal to e_0, since it would then lie in $\mathbf{R}^4$ and could not be isotropic. Thus any such f' can be written as $f' = s(e_0 + q)$ where $\|q\|^2 = -1$ and $\langle p, q\rangle = 0$, $s \neq 0$. The group $SO(1,4)$ acts transitively on those pairs for which $s > 0$, and these can clearly be identified with points of $\mathcal{V}$. The map of $\mathcal{V} \to o(1,4) \sim o(1,4)^*$ sending $s(e_{-1} + p) \wedge (e_0 + q)$ to $p \wedge s(e_0 + q)$ is manifestly equivariant with respect to $SO(1,4)$ and sends $\mathcal{V}$ onto one of the two components of the set of (non-zero) nilpotent elements in $o(1,4)$.

For the group $SO(1,4)$ we can give a rather direct interpretation of its action on $T^* S^3$. Indeed, $SO(1,4)$ can be identified as the (connected component of the) group of all conformal transformations of S^3, and hence has an induced action on $T^* S^3$, and we claim that this induced action is exactly the orbit action described above. To make the identification, we regard S^3 as the set of "forward light-like directions" in $\mathbf{R}^{1,4}$. That is, we consider the "forward light cone" in $\mathbf{R}^{1,4}$ consisting of all null vectors f' with $\langle e_0, f'\rangle > 0$. The group $SO(1,4)$ acts transitively on all such vectors, and hence on the set of all rays, i.e. on equivalence classes of such vectors where two vectors are identified if they differ by a positive scalar multiple. The set of such rays is topologically S^3, and we can parametrize them explicitly by S^3 once we choose the unit vector e_0, namely each such ray has a unique representative of the form $e_0 + q$ with $\|q\|^2 = -1$. We thus obtain an action of $SO(1,4)$ on S^3 which is easily checked to be conformal. The cotangent space at f' to S^3 can be identified as the quotient space of the space of all vectors orthogonal to the light cone, modulo the line spanned by f'. If $f' = e_0 + q$, this space can be identified with the set of all vectors of the form sp where $p \in \mathbf{R}^4$, $\langle p, q\rangle = 0$, and $\|p\|^2 = -1$. It now follows readily that the action of $SO(1,4)$ on $\mathcal{V}$ can be identified with the induced action of $SO(1,4)$ on T^+S^3 coming from the conformal action on S^3. Furthermore, if we think of p as a covector, and if ξ is any element of $o(1,4)$, whose corresponding tangent vector at $[(e_0 + q)]$ is denoted by $\xi_{[(e_0+q)]}$, then the value of the covector p on the tangent vector $\xi_{[(e_0+q)]}$ is given by

$$-(p, \xi \cdot (e_0 + q)) = -\langle p \wedge (e_0 + q), \xi\rangle$$

(where the scalar product on the left is between two vectors in $\mathbf{R}^{1,4}$ and the scalar product on the right is between two vectors in $o(1,4)$).

We can now use the above remarks to show that the three different identifications that we have made of $\mathcal{V}$—as nilpotent elements in $o(2,4)$, as nilpotent elements of $o(1,4)$ and as T^+S^3 are symplectic diffeomorphisms. Since the passage from $o(2,4)^*$ to $o(1,4)^*$ is the restriction map, it follows from general considerations, or directly from the formula

$$\langle \xi_l \wedge \eta_l, \omega_{\mathcal{V}}\rangle = \langle [\xi, \eta], l\rangle$$

for the value of the symplectic form, $\omega_{\mathcal{V}}$, of $\mathcal{V}$ evaluated at the image vectors, ξ_l and η_l of ξ and $\eta \in o(2,4)$ at $l \in \mathcal{V}$ that the identification of $\mathcal{V}$ either as a subvariety in $o(2,4)^*$ or in $o(1,4)^*$ is a symplectic diffeomorphism. We now show that the identification of $\mathcal{V}$ as an orbit in $o(1,4)^*$ or as T^+S^3 is symplectic. For this purpose it is convenient to make use of the following lemma:

Let ξ and η be vector fields on the differentiable manifold, M, and let $\hat{\xi}$ and $\hat{\eta}$ be the induced vector fields on T^*M. Let α be the fundamental one form on T^*M, let $\omega = d\alpha$ be the fundamental two form, let z be some point of T^*M, and let $x = \pi z$ be its base point in M. Then

$$\langle \hat{\xi}_z \wedge \hat{\eta}_z, \omega\rangle = -\langle [\hat{\xi}, \hat{\eta}]_z, \alpha\rangle = -\langle [\xi, \eta]_x, z\rangle.$$

PROOF OF LEMMA. Since $\hat{\xi}$ is an infinitesimal automorphism of the cotangent bundle structure, we have $D_{\hat{\xi}}\alpha = 0$, where D denotes Lie derivative. Thus $0 = D_{\hat{\xi}}\alpha = \hat{\xi} \lrcorner d\alpha + d(\hat{\xi} \lrcorner \alpha)$ or

$$\hat{\xi} \lrcorner \omega = -d\langle \xi, \alpha\rangle.$$

Thus

$$\langle \hat{\xi} \wedge \hat{\eta}, \omega \rangle = -\hat{\eta} \rfloor \hat{\xi} \rfloor \omega = -\hat{\eta} \rfloor d(\hat{\xi} \rfloor \alpha) = -D_{\hat{\eta}}(\hat{\xi} \rfloor \alpha) = [\hat{\xi}, \hat{\eta}] \rfloor \alpha$$

since $D_{\hat{\eta}}\alpha = 0$. Evaluating at z and using the definition $\langle \zeta_z, \alpha \rangle = \langle d\pi\zeta_z, z \rangle$ for any tangent vector ζ_z at z proves the lemma. Taking $z = p$ and $x = [e_0 + q]$, and using the formulas derived above and the lemma, proves that the identifications of $\mathcal{V}$ with T^+S^3 is a symplectic diffeomorphism.

Notice that the group $SO(1,4)$ not only preserves the symplectic structure on T^*S^3, it also preserves the cotangent fibration. On the other hand geodesic flow certainly does not preserve the cotangent fibration, and hence $SO(2)$ does not. Since $SO(1,4)$ is a maximal connected subgroup of $SO(2,4)$ we conclude that $SO(1,4)$ is the largest connected subgroup preserving the cotangent fibration.

§7. Homogeneous symplectic spaces.

Let the Lie group G act on the manifold X. Thus we are given a smooth map $G \times X \to X$, $(a,x) \to ax$ which is a group action. The differential of this map induces a map of $T_aG \times T_xX \to T_{ax}X$. In particular, taking $a = e$, the identity of G, then for any $\xi \in g = T_eG$, the Lie algebra of G, we obtain a vector field $\bar{\xi}$ on X where $\bar{\xi}_x$ is the image of $(\xi, 0)$ under the map $T_eG \times T_xX \to T_xX$. Thus $\bar{\xi}$ is the infinitesimal generator of the flow $x \to (\exp t\xi)x$.

As

$$\frac{d}{dt}\left\{\left[a^{-1}(\exp t\xi)a\right]x\right\} = (a^*\bar{\xi})(x)$$

it follows that

$$a^*\bar{\xi} = (Ada^{-1})\xi.$$

If we set $a = \exp s\eta$ and differentiate with respect to s we get

$$[\bar{\eta}, \bar{\xi}] = -[\eta, \xi].$$

(The reason for the confusing change of sign is that we are identifying the Lie algebra of G with the set of left invariant vector fields G, and hence they generate *right* translations. The opposite convention is sometimes used, for example by Souriau [9].)

If we set

$$\hat{\xi} = -\bar{\xi}$$

then the map $\xi \to \hat{\xi}$ gives a homomorphism of g into the Lie algebra of vector fields of X.

Now suppose that X is a symplectic manifold, and that G acts as a group of symplectic automorphisms. Then ξ satisfies

$$D_{\hat{\xi}}\omega = 0 \qquad \text{or} \qquad d(\hat{\xi} \lrcorner \omega) = 0.$$

We say that the action is *strongly symplectic if* $\hat{\xi} \lrcorner \omega = df_{\xi}$ for some function f_{ξ}. Notice that

$$\begin{aligned}[\hat{\xi}, \hat{\eta}] \lrcorner \omega &= D_{\hat{\xi}}(\hat{\eta} \lrcorner \omega) \\ &= \hat{\xi} \lrcorner d(\hat{\eta} \lrcorner \omega) + d(\hat{\xi} \lrcorner (\hat{\eta} \lrcorner \omega)) \\ &= d(\hat{\xi} \lrcorner \hat{\eta} \lrcorner \omega)\end{aligned}$$

so that $[\xi, \eta]^{\wedge} \lrcorner \omega$ is always of the form $-df$. Thus if $[g, g] = g$, then all symplectic actions are strongly symplectic. (This happens, for example, if g is semi-simple or if g is the semi-direct product of a semi-simple algebra k acting on a finite dimensional vector space V with no trivial component in the representation of k on V.) Suppose that the action is strongly symplectic. By choosing a basis of the Lie algebra and an f_{ξ_i} for each ξ_i in the basis, we can write

$$\xi \lrcorner \omega = -d \sum a_i f_{\xi_i} \qquad \text{if } \xi = \sum a_i \xi_i.$$

Put another way, we can define a function $f\colon X \to g^*$ where

$$\langle \xi, f \rangle = \sum a_i f_{\xi_i}$$

and then

$$\hat{\xi} \lrcorner \omega = -d(\langle \xi, f \rangle).$$

Notice that f is determined only up to an additive constant (in g^*). Now for any $a \in G$ we have

$$a^*(\hat{\xi} \lrcorner \omega) = d\langle \xi, a^* f \rangle$$

since ξ is a constant in this equation. But $a^*\omega = \omega$ and $a^*\hat{\xi} = (Ada^{-1}\xi)^{\wedge}$ so setting $\eta = Ada^{-1}\xi$ we see that

$$\begin{aligned}\hat{\eta} \lrcorner \omega &= d\langle Ada\eta, a^* f \rangle \\ &= d\langle \eta, (Ada)^*(a^* f) \rangle.\end{aligned}$$

Here $(Ada)^*\colon g^* \to g^*$ is just the adjoint of Ada. In particular

$$(Ada)^*(a^* f) - f$$

is a constant on X. Replacing a by a^{-1} gives us the fact that

$$\rho_a f - f = z_a$$

is a constant where ρ_a is the representation

$$\rho_a f(x) = (Ad^{\#}\, a) f(a^{-1} x)$$

and $Ad^{\#}$ is the representation of G on g^* given by the contragredient to the adjoint representation. It is clear that $a \to z_a$ is a cocycle of G with values in g^*, i.e., satisfies the identity

$$z_{ab} = (Ad^{\#}\, a) z_b + z_a.$$

Replacing f by $f + c$ where $c \in g^*$ is a constant has the effect of changing z_a by adding the coboundary $(Ad^*\, a)c - c$, and thus the cohomology class is well defined. It represents the obstruction to choosing $f\colon X \to g^*$ to be a G map. If this obstruction vanishes then

$$\langle \eta, a^* f \rangle = \langle Ada^{-1} \eta, f \rangle$$

and setting $a = \operatorname{expt} \xi$ and differentiating with respect to t gives

$$D_{\hat{\xi}} \langle \eta, f \rangle = \langle [\xi, \eta], f \rangle$$

or

$$D_{\hat{\xi}} f_\eta = f_{[\xi,\eta]}.$$

But $D_{\hat{\xi}} f_\eta = \hat{\xi} \rfloor df_\eta$ is just the definition of the Poisson bracket, $\{f_\xi, f_\eta\}$. Thus in this case

$$\{f_\xi, f_\eta\} = f_{[\xi,\eta]}$$

i.e., the map $\xi \to f_\xi$ is a homomorphism of Lie algebras. An action of g on X together with such a lifting of the homomorphism $\xi \to \hat{\xi}$ to $\xi \to f_\xi$ (if it is possible) is called a *Hamiltonian* action of G on X. The symplectic actions are of basic importance in classical mechanics. (See for example Souriau [9] for an elegant discussion of many of the classical laws of physics from this point of view.) In particular, a symplectic action is called *elementary* if the action of G on X is transitive. (This is the classical analogue of the quantum notion of an elementary particle as an irreducible representation of G.) If G acts transitively on X and this action is Hamiltonian, then the equation

$$(Ad^{\#}\, a) f = f \circ a$$

shows that f maps X onto an orbit. Notice also that f is an *immersion*. Indeed since G acts transitively, the vector fields ξ span the tangent space at each point.

If $df_x(\hat{\xi}) = \langle \xi, df \rangle_x = 0$ then $\langle \hat{\xi}, df_\eta \rangle_x = 0$ for all η, implying that $\hat{\xi}_x = 0$ since the $df_\eta = \hat{\eta} \lrcorner \omega$ span the cotangent space. Thus X is a covering space of an orbit in g^*. This suggests that the orbits in g^* are symplectic manifolds which is indeed true, a fact due to Kirillov, Kostant and Souriau. We shall develop these facts from a more general point of view as developed by Chu [10].

Let G be a Lie group and $X = G/H$ a homogeneous space for G where H is a closed subgroup, and let $\pi\colon G \to G/H = X$ be the projection. If Ω is an invariant form on X then it is clear that $\sigma = \pi^* \Omega$ is a left invariant form on G which satisfies

(i) $\xi \lrcorner \sigma = 0$ for all $\xi \in h$ where h is the Lie algebra of H;

(ii) σ is invariant under right multiplication by elements of H, and hence under Ad for elements of H.

Conversely, it is clear that any left invariant form on G satisfying (i) and (ii) arises from G/H. If Ω is a symplectic form then it is clear that a left invariant vector field will satisfy $\xi \lrcorner \sigma = 0$ if and only if $\xi \in h$. Furthermore, since $d\sigma = 0$, the set of all vector fields satisfying $\xi \lrcorner \sigma = 0$ forms an integrable subbundle of TG, and in particular, the left invariant ones form a subalgebra of the Lie algebra of G. Let us call it h_σ. We have thus recovered h. Let H_σ be the group generated by h_σ. Notice that for any $\xi \in h_\sigma$ we have $D_\xi \sigma = \xi \lrcorner d\sigma + d(\xi \lrcorner \sigma) = 0$ so that σ is invariant under H_σ. The only problem is that H_σ need not be closed. Let us say that σ is *regular* if H_σ is closed. Notice that if G/H is a symplectic homogeneous space, so that H is a closed subgroup, and if we construct σ as above then H_σ is just the connected component of the identity in H (and hence a closed subgroup of G). We have thus established:

PROPOSITION 7.1 (*Chu* [10]). *Each 2p-dimensional homogeneous symplectic space determines a left invariant regular closed two form of rank 2p on G. Conversely, a regular closed two form determines a homogeneous symplectic space. Up to covering, the space of all homogeneous symplectic manifolds for G is the same as the space of orbits of G acting on* $Z^2_{\mathrm{reg}}(g)$, *where* $Z^2_{\mathrm{reg}}(g)$ *denotes the set of regular two cocycles of g.*

Notice that if $\sigma = d\beta$ where β is a left invariant one form then σ is automatically regular. Indeed $D_\xi \beta = \xi \lrcorner d\beta + d(\xi \lrcorner \beta) = \xi \lrcorner \sigma$ for any left invariant vector field ξ since $\xi \lrcorner \beta$ is constant. Thus $\xi \in h_\sigma$ if and only if $D_\xi \beta = 0$. Now $H = \{a \mid (Ad^\# a)\beta = \beta\}$ is clearly a closed subgroup, and H_σ is the identity component. Thus, if $H^2(g) = \{0\}$, every cocycle is regular. We can also show that if $H^1(g) = \{0\}$ then every cocycle is regular. Indeed, to say that $H^1(g) = \{0\}$ is the same as saying that $d\colon g^* \to \wedge^2(g^*)$ is injective. But then $\xi \lrcorner \sigma = 0$ is equivalent to $D_\xi \sigma = d(\xi \lrcorner \sigma) = 0$ so h_σ is the Lie algebra of the isotropy group of σ. This argument was pointed out to us by Ofer Gabber. In any event, it is clear from the foregoing discussion that:

PROPOSITION 7.2 (*Kirillov-Kostant-Souriau*). *Each orbit, $Ad^{\#}(G)\beta$ for $\beta \in g^*$ is a symplectic manifold whose symplectic structure is induced from $d\beta$.*

We now assert:

PROPOSITION 7.3 (*Chu* [4]). *If G is a simply connected Lie group then every left invariant closed two form is regular.*

We sketch the proof. Let σ be a closed two form. We can think of σ as a one cocycle, f, from g to g^*, where $f(\xi) = \xi \lrcorner \sigma$. Here f is a cocycle relative to the action, $ad^{\#}$, of g on g^*. Hence f defines an action of g as affine transformations on g^* via

$$\begin{aligned} \xi \cdot \theta &= (ad^{\#}\xi)\theta + f(\xi) \\ &= (ad^{\#}\xi)\theta + \xi \lrcorner \sigma. \end{aligned}$$

Since G is simply connected this defines an affine action of G on g^*. It is clear that $\xi \in h_\sigma$ if and only if $\xi \cdot 0 = 0$. Thus H_σ is the identity component of the isotropy group of the origin and hence closed.

As the orbits in g^* represent (up to coverings) the "universal" elementary symplectic homogeneous spaces, it becomes important to analyze them. In this regard, for complex semi-simple groups see Kostant [11], for real semi-simple groups, see Rothschild [12], while for nilpotent groups see [20].

Let σ be a left invariant closed two form on G and suppose that the subalgebra h has minimum dimension among all subalgebras of the form h_σ. This implies that if σ_t is a curve of closed two forms with $\sigma_0 = \sigma$, then any $\xi \in h_\sigma$ can be extended to a curve ξ_t with $\xi_0 = \xi_0$ and $\xi_t \in h_{\sigma_t}$. (Indeed, choose a subspace m complementary to h_σ in the Lie algebra TG_e. Since dim h_σ is minimal, this implies that dim h_σ = dim $h_{\sigma'}$ for all σ' close to σ. Then projection along m defines an isomorphism of h_σ with $h_{\sigma'}$ for all σ' close to σ.) In particular, let θ be any left invariant one form and consider

$$\sigma_t = \sigma + td\theta.$$

We can write $\xi_t = \xi + t\xi' + O(t^2)$. Examining the coefficient of t in the equation $\xi_t \lrcorner \sigma_t = 0$ gives

$$\xi \lrcorner d\theta + \xi' \lrcorner \sigma = 0.$$

Let η be some other element of h_σ and take the interior product of this last equation with η. The term $\eta \lrcorner \xi' \lrcorner \sigma = -\xi' \lrcorner \eta \lrcorner \sigma = 0$ and we get

$$\eta \lrcorner \xi \lrcorner d\theta = 0.$$

Now, since η and θ are both left invariant, $\eta \lrcorner \theta$ is a constant and therefore $0 = D_\xi(\eta \lrcorner \theta) = D_\xi \eta \lrcorner \theta + \eta \lrcorner D_\xi \theta = [\xi, \eta] \lrcorner \theta + \eta \lrcorner \xi \lrcorner d\theta$, since $\xi \lrcorner \theta$ is also constant. Thus

$$[\xi, \eta] \lrcorner \theta = 0.$$

Since this holds for arbitrary θ we conclude that $[\xi, \eta] = 0$. We have thus proved

PROPOSITION 7.4. *Let σ be a left invariant closed two form such that h_σ has minimal dimension. Then h_σ is commutative. In particular, let X be a homogeneous symplectic manifold of G with maximal dimension. Then the connected component of the isotropy group of any point of X is commutative.*

For the case that $\sigma = d\theta$ is an exact two form this result was obtained by Duflo and Vergne [14]. (It is just a trivial observation to remark that their proof works just as well for the case of closed two forms.) For the case where G is a semi-simple group, the dual of the Lie algebra can be identified with the Lie algebra via the Killing form. In this case, to say that $h_{d\theta}$ has minimal dimension becomes the assertion that the centralizer of the corresponding element, θ, has minimal dimension, and Proposition 7.4 reduces to the classical assertion that for such regular elements the centralizer is abelian. For regular semi-simple elements the subalgebra $h_{d\theta}$ is a Cartan subalgebra.

For semi-simple subalgebras one has a conjugacy theorem for Cartan subalgebras, which, in the real case, can be formulated as asserting that if θ is generic, then $h_{d\theta'}$ is conjugate to $h_{d\theta}$ under the adjoint group if θ' is sufficiently close to θ. One can ask to what extent this remains true in the general case. It is not true for all Lie algebras as is shown by the following example: let $g = \mathbf{R} + V$ where V is the trivial Lie algebra (a vector space with trivial bracket) and $[r, v] = rv$ for $r \in \mathbf{R}$ and $v \in V$. It is easy to see that for any $\theta \in g^*$ which does not vanish on V the subalgebra $h_{d\theta}$ consists of the hyperplane in V defined by the equation $\theta(v) = 0$, corresponding to two dimensional orbits in g^*. It is clear that no two such subalgebras are conjugate to one another if they are distinct. Let us call a θ in g^* *stable* if $h_{d\theta'}$ is conjugate to $h_{d\theta'}$ for all θ' close to θ.

PROPOSITION 7.5. *Suppose that $h_{d\theta}$ has minimal dimension and that $[g, h_{d\theta}] \cap h_{d\theta} = \{0\}$. Then θ is stable and conversely.*

PROOF. It is clear that for any θ' on the orbit of θ the algebras $h_{d\theta}$ and $h_{d\theta'}$ are conjugate. Thus we will be done if we can find a submanifold, W, transversal to the orbit through θ with the property that $h_{d\theta'} = h_{d\theta}$ for all $\theta' \in W$ (near θ). By the implicit function theorem we can reduce the problem to the corresponding infinitesimal problem: to show that every θ' can be written as $\theta_1 + \theta_2$ where $\theta_1 \in g \lrcorner d\theta$ (the tangent space to the orbit) and $h_{d\theta} \lrcorner d\theta_2 = 0$ (which, on account of the minimality of dim $h_{d\theta}$ is the same as saying that $h_{d(\theta+\theta_2)} = h_{d\theta}$ if θ_2 is sufficiently small). It therefore suffices to show that no vector in g can be annihilated by all such θ_1 and θ_2. Now to say that $\langle \xi, g \lrcorner d\theta \rangle = 0$ is the same as saying that $\xi \lrcorner d\theta = 0$, i.e. that $\xi \in h_{d\theta}$. To say that $\langle \xi, \theta_2 \rangle = 0$ for all θ_2 means that $\langle \xi, \theta_2 \rangle = 0$ for all θ_2 with the property that $\langle [g, h_{d\theta}], \theta_2 \rangle = 0$, i.e. that $\xi \in [g, h_{d\theta}]$. By hypothesis this implies that $\xi = 0$.

If θ is stable, then $h_{d\theta}$ must have the generic dimension, which is the minimal dimension. Suppose that there are some η in $h_{d\theta}$ with $0 \neq \sum [\eta, \zeta]$ in $h_{d\theta}$ for some ζ in g. Choose γ with $\langle \sum [\eta, \zeta], \gamma \rangle \neq 0$. If we apply the condition for the existence of a conjugacy of $h_{d(\theta+t\gamma)}$ with $h_{d\theta}$ and compare coefficients of t, it is easy to see that we must be able to solve the equations

$$\langle [[\xi, \eta], \zeta], \theta \rangle = \langle [\eta, \zeta], \gamma \rangle$$

for all η in h and ζ in g. Choosing $\sum [\eta, \zeta] \in h_{d\theta}$ and using Jacobi's identity on the left gives zero while the right side does not vanish, giving a contradiction.

Observe that Proposition 7.5 is not true if we replace the coboundary $d\theta$ by a cocycle, σ. Indeed, consider the trivial three dimensional algebra. Here every two form is a cocycle and, for non-zero σ, the subalgebra h_σ consists of the line $\xi \lrcorner \sigma = 0$, and no distinct lines are conjugate since the adjoint group acts trivially. On the other hand, $[g, g] = 0$, so the condition $[g, h_\sigma] \cap h_\sigma = 0$ is certainly satisfied.

In order to understand this example it is useful to observe that for any Lie algebra g, we can form the central extension of g by $H^2(g)$ as follows: choose a basis $c_1, \ldots, c_k$ for $H^2(g)$ and cycles $z_1, \ldots, z_k$ representing the c's. Then define $[(v, x), (w, y)] = (z_1(x, y)c_1 + \cdots + z_k(x, y)c_k, [x, y])$ where v and w are in $H^2(g)$ and x and y are in g. This gives a Lie algebra structure to $H^2(g) + g$.

If $\theta \in (H^2(g) + g)^*$ is given by $\theta(v, x) = a_i$ where $v = \sum a_i c_i$ then it is clear that $d\theta = z_i$. In this way every cocycle of g can be regarded as a coboundary in the extended algebra. If σ is a cocycle of g corresponding to the coboundary $d\theta$ of the extended algebra, it is clear that $h_{d\theta} = H^2(g) + h_\sigma$. If σ is stable then so is θ and conversely. We must therefore require the stability criterion in the extended algebra.

We can extend the assertion and proof of Proposition 7.5 as follows: Suppose that ℓ is a Lie algebra, and g is an ideal in ℓ. Then any element of ℓ acts on g by Lie bracket and hence on g^* by the contragredient action. For $\xi \in \ell$ and $\theta \in g^*$, we let $\xi \lrcorner d\theta$ denote the action of ξ on θ so that $(\xi \lrcorner d\theta)(\eta) = -[\xi, \eta] \lrcorner \theta$. Let L be a Lie group with Lie algebra ℓ, and suppose that L acts on g so that its infinitesimal action is the given action of bracketing by ℓ. We say that θ is L stable if $h_{d\theta'}$ is conjugate to $h_{d\theta}$ by an element of L, for all θ' close to θ. Then the condition for L stability is that

$$[g, h_{d\theta}] \cap (\ell \lrcorner d\theta)^\perp = 0.$$

The proof is as before.

In particular, we can let L be the group of all automorphisms of g, in which case ℓ is the algebra of all derivations of g, and we get a condition for stability under all automorphisms. One can construct algebras for which no form θ is stable under the group of all automorphisms, cf. [15].

We would now like to classify the homogeneous symplectic manifolds for various interesting Lie groups. We will do this by reducing the problem to studying the behavior of closed two forms with respect to certain subgroups. In particular, we will make the following assumption about the Lie algebra, g, of G. We will assume that there are two subspaces, k and p, of g such that

$$g = k + p \qquad k \cap p = \{0\}$$

$$[k, k] \subset k \qquad \text{and} \qquad [k, p] \subset p.$$

Thus we are assuming that k is a subalgebra of g and that p is a supplementary subspace to k which is stable under the action of k. We do not make any further assumptions at the moment about p. Thus $[p, p]$ will have both a k and a p component which we denote by r and s respectively: for η and η' in p we have $[\eta, \eta'] = r(\eta, \eta') + s(\eta, \eta')$ where $r(\eta, \eta') \in k$ and $s(\eta, \eta') \in p$. Jacobi's identity implies some identities on r and s. It is easy to check that these are

$$\mathfrak{S} r(s(\eta, \eta'), \eta'') = 0$$

$$\mathfrak{S}\{s(s(\eta, \eta'), \eta'') + [r(\eta, \eta'), \eta'']\} = 0$$

where $\mathfrak{S}$ denotes cyclic sum. Also

$$[\xi, r(\eta, \eta')] = r(\xi \cdot \eta, \eta') + r(\eta, \xi \cdot \eta')$$

where $\xi \in k$ and η, $\eta' \in p$ and we have written $\xi \cdot \eta$ for $[\xi, \eta]$, thinking of k acting on p. We also have the equation

$$\xi \cdot s(\eta, \eta') = s(\xi \cdot \eta, \eta') + s(\eta, \xi \cdot \eta').$$

In addition we have the identity asserting that k acts as a Lie algebra of linear transformations on p and Jacobi's identity for k. Conversely, starting from any action of a Lie algebra k on a vector space p together with r and s satisfying the above identities it is clear that $g = k + p$ becomes a Lie algebra. Let us give some illustration of this situation:

A) $r = s = 0$. In this case p is a supplementary abelian ideal, and k acts as linear transformations on p. In other words, g is the semidirect product of k and p where k is a Lie algebra with a given linear representation of k on p. Any such linear representation of k gives rise to a Lie algebra, g, which is called the associated affine algebra.

B) $r = 0$. Here all that is assumed is that p is a supplementary ideal to k. An important illustration of this situation is the case of the Galilean group. Recall that the Galilean group can be regarded as the group of all five by five matrices of the form

$$\begin{bmatrix} A & v & x \\ 0 & 1 & t \\ 0 & 0 & 1 \end{bmatrix}$$

where $A \in O(3)$, $v \in \mathbf{R}^3$, $x \in \mathbf{R}^3$ and $t \in \mathbf{R}$. Such a matrix carries the space time point (x_0, t_0) into the space time point $(Ax_0 + x + t_0 v, t + t_0)$. The corresponding Lie algebra consists of all matrices of the form

$$\begin{bmatrix} a & v & x \\ 0 & 0 & t \\ 0 & 0 & 0 \end{bmatrix}$$

where $a \in o(3)$ and v, x, t as before. Here we can take $k \sim o(3)$ to consist of the subalgebra with $x = v = t = 0$ and p to be the seven dimensional subalgebra with $a = 0$. Denoting an element of p by (v, x, t) we see that $[(v,x,t),(v',x',t')] = s((v,x,t),(v',x',t')) = (0, t'v - tv', 0)$ and $\xi \cdot (v,x,t) = (\xi \cdot v, \xi \cdot x, 0)$ where $\xi \cdot v$ denotes the usual action of $\xi \in o(3)$ on $v \in \mathbf{R}^3$ and similarly for $\xi \cdot x$.

C) The case where g is semi-simple and k, p corresponds to a Cartan decomposition. Here $s = 0$.

D) The case where k is an ideal. Here the action of k on p is trivial. For example, in the case of the Heisenberg algebra we can take k to be the center. For this case p is a symplectic vector space, $k = \mathbf{R}$ acts trivially on p and r is the symplectic two form, while $s = 0$.

Let $f \in \wedge^2 g^*$ be a two form. Identifying $\wedge^2 g^*$ with $\wedge^2 k^* \oplus k^* \otimes p^* \oplus \wedge^2 p^*$ allows us to write $f = a + b + c$ so that

$$f(\xi + \eta, \xi' + \eta') = a(\xi, \xi') + b(\xi, \eta') - b(\xi', \eta) + c(\eta, \eta').$$

Now $df \in \wedge^3 g^*$ is given by $df(\chi, \chi', \chi'') = \mathfrak{S} f([\chi, \chi'], \chi'')$ where $\mathfrak{S}$ denotes cyclic sum. Writing $\chi = \xi + \eta$ etc. the equation $df = 0$ becomes

$$\begin{aligned} \mathfrak{S}\{a([\xi,\xi'] + r(\eta,\eta'),\xi'') + b([\xi,\xi'] + r(\eta,\eta'),\eta'') \\ - b(\xi'', \xi \cdot \eta' - \xi' \cdot \eta + s(\eta,\eta')) \\ + c(\xi \cdot \eta - \xi' \cdot \eta + s(\eta,\eta'),\eta'')\} = 0. \end{aligned}$$

We now derive various identities for a, b, and c by considering special cases of this identity:

i) $\xi = \xi' = \xi'' = 0$. In this case the identity becomes

$$\mathfrak{S}\{b(r(\eta,\eta'),\eta'') + c(s(\eta,\eta'),\eta'')\} = 0. \qquad (*)$$

For the case of the affine algebra this identity is vacuous. If p is a subalgebra so that $r = 0$, only the identity involving c remains. For example, a direct computation in the case of the Galilean group shows that $(*)$ reduces to the condition $c((0,x,0),(0,x',0)) = 0$. For the case of the Cartan decomposition only the identity involving b remains. Similarly for the case of the Heisenberg algebra.

ii) $\xi = \xi' = 0$, $\eta'' = 0$. In this case the identity becomes

$$a(r(\eta,\eta'),\xi'') - b(\xi'',s(\eta,\eta')) + c(\xi''\cdot\eta,\eta') + c(\eta,\xi''\cdot\eta') = 0. \qquad (**)$$

For the case of the affine algebra both r and s vanish and this identity becomes

$$c(\xi\cdot\eta,\eta') + c(\eta,\xi\cdot\eta') = 0 \qquad (**)_{\mathrm{A}}$$

which asserts that the antisymmetric form c is invariant under the action of k. For example, in the case of the Poincaré algebra where $k = o(3,1)$ and $p = \mathbf{R}^4$ there is no invariant antisymmetric form so we conclude that $c = 0$.

In the case that we only assume that p is a subalgebra so that r vanishes the identity becomes

$$c(\xi\cdot\eta,\eta') + c(\eta,\xi\cdot\eta') = b(\xi,s(\eta,\eta')) \qquad (**)_{\mathrm{B}}$$

For example, in the case of the Galilean algebra, if we apply this identity to $\eta = (v,x,0)$ and $\eta' = (v',x',0)$ the right hand side vanishes and we conclude that c, when restricted to $(\mathbf{R}^3 + \mathbf{R}^3)\wedge(\mathbf{R}^3 + \mathbf{R}^3)$ is invariant under the action of $o(3)$, which acts diagonally on $\mathbf{R}^3 + \mathbf{R}^3$. There is obviously only one such invariant (up to scalar multiples) and it is given by

$$c((v,x,0),(v',x',0)) = m(\langle v,x'\rangle - \langle v',x\rangle)$$

where $\langle\ ,\ \rangle$ denotes the Euclidean scalar product. If we take $\eta = (0,x,0)$ and $\eta' = (0,0,t)$ the right side of $(**)_{\mathrm{B}}$ still vanishes. On the left the term $\xi\cdot\eta'$ vanishes and $\xi\cdot x$ is arbitrary. We conclude that

$$c((0,x,0),(0,0,t)) = 0.$$

Thus

$$c((v,x,t),(v',x',t')) = m(\langle v,x'\rangle - \langle v',x\rangle) + \langle \ell, t'v - tv'\rangle$$

for some $\ell \in \mathbf{R}^3$ where $(**)_{\mathrm{B}}$ implies that

$$\langle \ell, \xi\cdot v\rangle = b(\xi,(0,v,0)).$$

In the case of a Cartan decomposition, or, more generally when $s = 0$ the identity $(**)$ becomes

$$c(\xi\cdot\eta,\eta') + c(\eta,\xi\cdot\eta') = a(\xi,r(\eta,\eta')). \qquad (**)_{\mathrm{C}}$$

For the case where k is an ideal $(**)$ becomes

$$a(\xi, r(\eta, \eta')) + b(\xi, s(\eta, \eta')) = 0. \qquad (**)_D$$

iii) $\xi = 0$, $\eta' = \eta'' = 0$. In this case neither a nor c contributes and we obtain the identity

$$b([\xi', \xi''], \eta) + b(\xi'', \xi' \cdot \eta) - b(\xi', \xi'' \cdot \eta) = 0. \qquad (***)$$

This identity says that the map from k to p^* sending $\xi \rightsquigarrow b(\xi, \cdot)$ is a cocycle. If k is semisimple, then Whitehead's lemma asserts that b must be a coboundary, i.e. that there exists a $\theta \in p^*$ such that

$$b(\xi, \eta) = \theta(\xi \cdot \eta). \qquad (***)_S$$

Suppose that instead of assuming that k is semi-simple we assume that k contains an element in its center which acts as the identity transformation on p. Taking ξ' to be this element and ξ'' to be an arbitrary ξ in $(***)$ we see that $(***)_S$ holds with $\theta(\eta) = b(\xi', \eta)$. Thus

if either k is semi-simple or k contains an element in its center acting as the identity transformation on p then $(***)_S$ holds.

For example, in the case of the Galilean algebra, we see that the bilinear form b is given by

$$b(\xi, (v, x, t)) = \langle \ell', \xi \cdot v \rangle + \langle \ell, \xi \cdot x \rangle$$

where ℓ' and ℓ are elements of $\mathbf{R}^3$.

iv) $\eta = \eta' = \eta'' = 0$. In this case we simply obtain the identity which asserts that a is a cocycle in $\wedge^2 k^*$. Again, if k is semi-simple we can conclude that a must be a coboundary. In the case of the Galilean algebra we have thus established that the most general cocycle can be written as

$$f((\xi, v, x, t), (\xi', v', x', t')) = \tau([\xi, \xi']) + \langle \ell', \xi v' - \xi' v \rangle$$
$$+ \langle \ell, \xi x' - \xi' x + t' v - t v' \rangle + m(\langle v, x' \rangle - \langle v', x \rangle),$$

where $\tau \in o(3)^*$. Now the sum of the first three terms can be written as $\theta([(\xi, v, x, t), (\xi', v', x', t')])$ where $\theta = (\tau, \ell', \ell, 0) \in g^*$, i.e., as a coboundary. On the other hand it is clear that the last term is definitely not a coboundary. We have thus recovered a result first proved by Bargmann.

If G is the Galilean group then $H^2(g)$ is one dimensional and, up to coboundaries, a cocycle can be written as

$$f((\xi, v, x, t), (\xi', v', x', t')) = m(\langle v, x' \rangle - \langle v', x \rangle).$$

We now turn to the problem of describing the action of G on the space of two-cycles in order to determine when two such cycles define equivalent symplectic structures. We begin with the case of the semi-direct product, i.e. the affine algebra. Every element of the simply connected group corresponding to g can be written as $m \exp \eta$ where $m \in K = \exp k$, and $\eta \in p$. Now K leaves both k and p invariant so that the action of K on $f = a + b + c$ does not mix the summands and the action on each summand is the appropriate exterior or tensor product of the contragredient representation. In case $f = d\theta$ for $\theta \in g^* = k^* + p^*$, $mf = dm\theta$ where K acts on g^* via the contragredient representation. We must therefore examine the action of $\exp \eta$. Now

$$[\eta, \eta'] = 0 \qquad \text{and} \qquad [\eta, \xi] = -\xi \cdot \eta.$$

Thus

$$Ad\,(\exp -\eta)(\xi' + \eta') = \exp\,(ad\, -\eta)(\xi' + \eta') = (\xi' + \xi' \cdot \eta + \eta').$$

Therefore

$$\begin{aligned}(\exp \eta)f(\xi' + \eta', \xi'' + \eta'') &= f(\exp ad\, -\eta(\xi' + \eta'), \exp ad\, -\eta(\xi'' + \eta''))\\ &= a(\xi', \xi'') + b(\xi', \xi'' \cdot \eta) - b(\xi'', \xi' \cdot \eta) + c(\xi' \cdot \eta, \xi'' \cdot \eta)\\ &= f(\xi' + \xi' \cdot \eta + \eta', \xi'' + \xi'' \cdot \eta + \eta'')\\ &\quad + b(\xi', \eta'') - b(\xi'', \eta') + c(\xi' \cdot \eta, \eta'') + c(\eta', \xi'' \cdot \eta) + c(\eta', \eta'').\end{aligned}$$

Now by $(***)$, $b(\xi', \xi'' \cdot \eta) - b(\xi'', \xi' \cdot \eta) = b([\xi', \xi''], \eta)$ and, by $(**)_A$

$$c(\xi' \cdot \eta, \xi'' \cdot \eta) = -c(\eta, \xi' \cdot \xi'' \cdot \eta) = c(\xi' \cdot \xi'' \cdot \eta, \eta) = \tfrac{1}{2}c([\xi', \xi''] \cdot \eta, \eta).$$

We can thus write

$$(\exp \eta)(a + b + c) = (a + d(b_\eta + \tfrac{1}{2}c_{\eta\eta})) + (b + dc_\eta) + c$$

where b_η and $c_{\eta\eta} \in k^*$ are defined by

$$b_\eta(\xi) = b(\xi, \eta) \qquad \text{and} \qquad c_{\eta\eta}(\xi) = c(\xi \cdot \eta, \eta)$$

while $c_\eta \in p^*$ is defined by

$$c_\eta(\eta') = -c(\eta, \eta').$$

In the important special case where $(a + b) = d(\tau + \theta)$ is exact, where $\tau \in k^*$ and $\theta \in p^*$ we can write

$$(\exp \eta)(d(\tau + \theta) + c) = d((\tau + b_\eta + \tfrac{1}{2}c_{\eta\eta}) + (\theta + c_\eta)) + c.$$

We can therefore describe the situation as follows. The c component is invariant under the action of G. It is invariant under $\exp p$ by the above computation and it is invariant under the action of K by $(**)_A$. For a given choice of c we can move θ into $(h^{*-1})\theta + c_\eta$ where $h \in K$ and $\eta \in p$. This determines an action of $K \times p$ on p^*. Suppose that we have parametrized the orbits of this action and have, in fact, chosen a cross-section for these orbits. For a given orbit we have thus picked a fixed θ. This determines a subgroup of G, the isotropy group of θ. The corresponding algebra consists of those (ξ, η) for which $\xi\theta + c_\eta = 0$. The set of ξ' which occur form a subalgebra of k which we denote by k_θ. Thus $\xi \in k_\theta$ if and only if there exists an $\eta_\xi \in p$ such that $\theta(\xi \cdot \eta) = c(\eta_\xi, \eta)$ for all $\eta \in p$. It is easy to check that the identity $(**)_{\text{A}}$ implies that the assignment $\xi \rightsquigarrow \eta_\xi$ is a cocycle of k_θ with values in p. If this cocycle is a coboundary (for instance if k_θ is semi-simple or contains the identity operator) then we can find an $\bar{\eta}$ such that $\xi\theta - c_{\xi\bar{\eta}} = \xi(\theta - c_{\bar{\eta}}) = 0$. Thus by changing our choice of θ within the orbit we can arrange that k_θ consist exactly of those ξ for which $\xi\theta = 0$. Notice that this equation is equivalent to the equation $\theta_\eta(\xi) = 0$ for all η. If we consider the action of $\exp \eta$ on the k^* component, it adds exactly $\theta_\eta + \frac{1}{2}c_{\eta\eta}$. If $c_{\eta\eta} = 0$, we see that the orbit of τ is just the complete inverse image of the orbit of $\rho_\theta(\tau)$ under K_θ where $\rho_\theta\colon k^* \to k_\theta^*$ is the projection dual to the injection of $k_\theta \to k$. In this case the cocycles are thus parametrized by c, θ ranging over a cross-section of the action of G on k^* determined by c, and χ ranging over a cross-section for the action of K_θ on k_θ^*.

For example, for the case of the Lie algebra of the Poincaré group we have already seen that $c = 0$. The orbits of G on p^* are thus the same as the orbits of $K = SO(3,1)$ on p^* and consist of single sheeted hyperboloids $\theta^2 = m^2 > 0$, $\theta_0 > 0$ and $\theta^2 = m^2 > \theta$, $\theta_0 < 0$; the forward light cone $\theta^2 = 0$, $\theta_0 > 0$, the backward light cone $\theta^2 = 0$, $\theta_0 < 0$, the single sheeted hyperboloids $\theta^2 = -m^2 < 0$; and the origin. We thus choose cross-sections for these orbits as follows:

$$(m,0,0,0) \qquad (-m,0,0,0) \qquad (1,1,0,0)$$

$$(-1,1,0,0) \qquad (0,m,0,0) \qquad \text{and} \qquad (0,0,0,0).$$

It is easy to see that the group $K_{(m,0,0,0)}$ is exactly $SO(3)$. Its orbits acting on the dual of its Lie algebra are spheres. If we call s the radius of these spheres, we see that a family of orbits is parametrized by the two real parameters $m > 0$ and $s \geq 0$. Here m is the "mass" and s is the "spin". For "mass zero" i.e. for (1, 1, 0, 0) or (−1, 1, 0, 0) it is easy to see that the corresponding isotropy group is the Euclidean group, $E(2)$. The orbits of $E(2)$ in the dual of its Lie algebra are easily seen to be cylinders (of radius r, say) and points on the axis $r = 0$. If we let the real parameter s describe the points on this axis we see that the symplectic structures corresponding to (1, 1, 0, 0) are parametrized by $r > 0$ and, if $r = 0$ by an arbitrary real parameter, s. The case $r > 0$ does not arise in known physical systems; for $r = 0$ the parameter s is also called the "spin". The isotropy group of $(0,m,0,0)$ is $SL(2)$. Its orbits are again hyperboloids, forward and backward "light cone" and the origin. No particles with negative mass^2 ("tachyons") seem to occur.

Let us now do a slightly more complicated computation–determining the symplectic homogeneous spaces for the Galilean group. Here p is not abelian. However, it is easy to check that

$$Ad\,(\exp -(0, v, x, t))(\xi, w, y, s) = (\xi, w + \xi v, y + \xi(x + \tfrac{1}{2}tv) + tw - sv, s).$$

We have already written the form of the most general cocycle, f, of the Galilean algebra. Under the action of $\exp(\eta)$ it is easy to check that ℓ' is moved into $\ell' + mx$ and ℓ is changed into $\ell - mv$. Thus by suitable choice of x and v we can arrange that both ℓ and ℓ' vanish, provided that $m \neq 0$. Now $\vec{p} = -\ell$, being dual to the translation vector, x, has the character of linear momentum. By applying a pure "velocity transformation" $\exp(0, v, 0, 0)$, p is moved into $\vec{p} + mv$. Thus m is just the ratio between momentum and velocity, and hence corresponds to the usual notion of mass. Our choice of v amounts to making a change to a new frame of reference in which the center of gravity is at rest. The physical interpretation of $-\ell'/m$ is that it is the position of the center of mass in the frame in which it is at rest. By shifting the origin of the coordinate system we can arrange that this is the origin.) Once we have arranged that $\ell = \ell' = 0$, the only possibility left for $\boldsymbol{\eta}$ (in the case of non-zero m) is $\boldsymbol{\eta} = (0, 0, t)$ and it is clear that $\exp(0, 0, -t)$ acts trivially. Thus we are left with the action of $SO(3)$ on $k^* = o(3)^*$. Again, the orbits are spheres, parametrized by, their radius, a non-negative parameter, s. Thus for $m \neq 0$ the homogeneous symplectic manifolds for the Galilean group are parametrized by m and $s \geq 0$, the "mass" and the spin. For $m = 0$, we cannot change ℓ while ℓ' is moved into $\ell' + t\ell$. On the other hand τ is moved into

$$\tau + \langle \ell', \cdot\, v\rangle + \langle \ell, \cdot\, (x + \tfrac{1}{2}tv)\rangle.$$

If we identify τ as a vector in three space this last expression can be written as

$$\tau + \ell' \times v + \ell \times (x + \tfrac{1}{2}tv)$$

where $\times$ denotes vector product. In this case it is more convenient to let G act by letting $SO(3)$ act first and then $\exp p$. By applying a suitable element of $SO(3)$ we can arrange that $\ell = (f, 0, 0)$ and then, if $f \neq 0$ that $\ell' = (0, b, c)$. If ℓ and ℓ' are independent by suitable choice of v and x we can arrange that $\tau = 0$. If $f \neq 0$ and $b = c = 0$, then we can arrange that $\tau = (\pm sf, 0, 0)$ where $s \geq 0$ and $f > 0$. (This corresponds to the case of a particle of zero mass, travelling with infinite velocity. Here the condition that $b = c = 0$ amounts to the condition that the "disturbance is transverse" and the parameter f is the "inverse of the wave length", i.e. the "color". The parameter s is called the spin and the + or − is called the helicity. For details, see Souriau [9, p. 195].)

Let us give a procedure for interpreting, in terms of "particles" moving in space time, the meaning of the symplectic manifolds that we have descibed above for the Poincaré and Galilean groups. (We are indebted to Thomas Ungar for help in the ensuing discussion.) Let G be any Lie group and $M = G/L$ a homogeneous space for G, where L is some closed subgroup. (In the case at hand, G is either the Poincaré or Galilean group and M consists of space time.) Let S be some homogeneous symplectic space for G. We would like to find a homogeneous space, N, for G which is fibered over S (and so carries a presymplectic structure coming from S) and is also fibered over M (so that it makes sense to talk of the "position in M" of a point of N). Thus we wish to have the following double fibration:

$$\begin{array}{ccc} & N & \\ \swarrow & & \searrow \\ S & & M \end{array}$$

If $S = G/H$ and $M = G/L$, then the "smallest" N that will do will be of the form $G/H \cap aLa^{-1}$ for some $a \in G$, where a is chosen so that $H \cap aLa^{-1}$ has maximal dimension. The image in M of a typical fiber of N over S will look like the orbit through the point aL of N under the action of the group H. In particular the dimension of the image in M of a typical fiber will be equal to the dimension of $H/H \cap aLa^{-1}$.

For example, let us consider the case where G is the Poincaré group and $L = O(1,3)$ is the Lorentz group so that $M = G/L$ is just space time. Suppose we first consider a positive mass orbit of the Poincaré group as described above. A typical point on this orbit is (p, τ) where $p = (m, 0, 0, 0)$ and $\tau \in o(1,3)^*$. The isotropy algebra of p is a subalgebra $o(3)$ and we can consider τ as an element of $o(3)^*$. The isotropy group of (p, τ) then consists of all translations through vectors tp and all elements of $O(3)$ which preserve τ. Thus $H \sim \mathbf{R} \times O(2)$ if $\tau \neq 0$ and $H \sim \mathbf{R} \times O(3)$ if $\tau = 0$. If a_v is translation through the vector v then $a_v L a_v^{-1}$ consists of all transformations of Minkowski space which send the vector w into $Aw + v - Av$ where $A \in O(1,3) = L$. Thus $\dim(H/H \cap aLa^{-1}) = 1$ for any element of the Poincaré group, so we might as well take $aLa^{-1} = L$. Thus the image in space time of a typical fiber will look like the set $x + tp$. In the case of $s = 0$, we see that N is seven dimensional and the six dimensional manifold S consists of the space of "world lines" of particles of a given rest mass moving in space time. If $s \neq 0$, we can think of the particle as "spinning" (in the space like three space orthogonal to its world line) with total angular momentum ms. The "spin axis" can vary over a two sphere. Thus N is nine dimensional in this case and S is eight dimensional. The analysis for the non-zero mass symplectic homogeneous manifolds for the Galilean group is entirely analogous.

Let us examine the situation for the six dimensional mass zero orbits of the Poincaré group with non-zero spin. Here a typical point is of the form (u, κ), where u is the null vector $(1,1,0,0)$ and $\kappa \in o(1,3)^*$ induces a non-zero point orbit in $e(2)^*$ where $e(2)$ is the Lie algebra of the subgroup $E(2)$ fixing the point

u (and this subgroup is isomorphic to the two dimensional Euclidean group). The isotropy algebra h, in this case is four dimensional, and can be described as follows: Let us write (B, b) for the element of the Poincaré algebra whose linear component is B and whose translation component is b. Let $e_2 = (0, 0, 1, 0)$ and $e_3 = (0, 0, 0, 1)$; let $A_2 \in o(1, 3)$ be defined by $A_2 e_2 = u$, $A_2(1, -1, 0, 0) = -2e_2$ and similarly $A_3 e_3 = u$ and $A_3(1, -1, 0, 0) = 2e_3$. Finally let B denote infinitesimal rotation in the e_2, e_3 plane, so that $Bu = B(1, -1, 0, 0) = 0$ and $Be_2 = e_3$, $Be_3 = -e_2$. Then h is spanned by

$$(0, u),\ (A_3, se_2),\ (-A_2, se_3),\ (B, 0)$$

where s is the "spin" of the mass zero particle and $h \cap o(1, 3)$ is one dimensional. It is easy to check that this is the maximal dimension of intersection. So again, we take $L = O(1, 3)$. This time the fibers are three dimensional. The image of a typical fiber is now a set of the form $x + u^\perp$ where $u^\perp$ is the three dimensional space of all vectors orthogonal to u. We can think of this as a plane in space moving with the speed of light in the direction determined by u. We leave the corresponding computation for the Galilean group to the reader.

Let us now compare the symplectic homogeneous spaces of the Galilean group with those of the Poincaré group. To do so, we wish to regard the Poincaré group as a "deformation" of the Galilean group, or as is more commonly stated in the physical literature, we wish to regard the Galilean group as a "contraction" of the Poincaré group as the speed of light goes to infinity. We first describe what this means. Suppose that we choose a definite splitting of space time into space and time and write a point in space time as a column vector with entries (t, x_1, x_2, x_3) which we shall write as $\binom{t}{x}$. We can write the most general element of the Poincaré algebra as the five by five matrix of the form

$$\begin{bmatrix} 0 & v_1 & v_2 & v_3 & t \\ v_1 & 0 & a_{12} & a_{13} & x_1 \\ v_2 & -a_{12} & 0 & a_{23} & x_2 \\ v_3 & -a_{13} & -a_{23} & 0 & x_3 \\ 0 & 0 & 0 & 0 & 0 \end{bmatrix}$$

which we shall denote more simply by

$$\begin{bmatrix} 0 & v^t & t \\ v & A & x \\ 0 & 0 & 0, \end{bmatrix}, \qquad A \in o(3).$$

This is the form of the Poincaré algebra relative to a space time splitting and a coordinate system in which the speed of light is one. To find the form of the Poincaré algebra relative to coordinates in which the speed of light is c, we must expand the space coordinates by a factor of c. This has the effect of conjugating the above matrices by the matrix

$$\begin{bmatrix} 1 & 0 & 0 & 0 & 0 \\ 0 & c & 0 & 0 & 0 \\ 0 & 0 & c & 0 & 0 \\ 0 & 0 & 0 & c & 0 \\ 0 & 0 & 0 & 0 & 1 \end{bmatrix}$$

so as to obtain matrices of the form

$$\begin{bmatrix} 0 & c^{-1}v^t & t \\ cv & A & cx \\ 0 & 0 & 0 \end{bmatrix}.$$

For each positive value of c we obtain a different subalgebra of five by five matrices. They are all isomorphic (and indeed conjugate) to the Poincaré algebra. Within the framework of our distended coordinates, the vector cv gives a "velocity" (or "boost") transformation and the vector cx gives a spatial translation. If we are accustomed to dealing with velocities and displacements which are "small relative to the velocity of light", it makes sense to introduce $\bar{v} = cv$ and $\bar{x} = cx$ and so parametrize the elements of one such algebra (corresponding to a fixed c) as

$$\begin{bmatrix} 0 & c^{-2}\bar{v}^t & t \\ \bar{v} & A & \bar{x} \\ 0 & 0 & 0 \end{bmatrix}.$$

Let us now set $\epsilon = c^{-2}$ and drop the bars over the v's and x's. We thus obtain, for each $\epsilon > 0$ a map of the ten dimensional space spanned by the A's, v's, x's and t's into the space of five by five matrices. Explicitly, this map is

$$(A, v, x, t) \rightsquigarrow \begin{bmatrix} 0 & \epsilon v^t & t \\ v & A & x \\ 0 & 0 & 0 \end{bmatrix}.$$

Notice that for $\epsilon = 0$ the image is exactly the Galilean algebra. The limit $\epsilon \to 0$ is, of course, the same as $c \to \infty$. Let us denote the underlying ten dimensional vector space spanned by the A's, v's, x's and t's by g. Then for any $\epsilon \geqslant 0$ we have defined a linear map of g into a subalgebra of the Lie algebra of five by five matrices. This induces a Lie bracket, depending on ϵ on the space g, which we shall denote by $[\ ,\ ]_\epsilon$. Explicitly, if

$$\xi_1 = (A_1, v_1, x_1, t_1) \quad \text{and} \quad \xi_2 = (A_2, v_2, x_2, t_2)$$

then

$$[\xi_1, \xi_2]_\epsilon = ([A_1, A_2] + \epsilon(v_1 \otimes v_2^t - v_2 \otimes v_1^t), A_1 v_2 - A_2 v_1,$$
$$A_1 x_2 + t_2 v_1 - A_2 x_1 - t_1 v_2, \epsilon(\langle v_1, x_2 \rangle - \langle v_2, x_1 \rangle)).$$

Similarly, for each ϵ we get a map, $d_\epsilon\colon \wedge^k(g^*) \longrightarrow \wedge^{k+1}(g^*)$. In particular we get a varying space of 2-cocycles, $Z_\epsilon^2 \subset \wedge^2(g^*)$. Notice that in the case at hand, the dimension does not vary for all $\epsilon \geqslant 0$. Indeed, for $\epsilon > 0$ the algebras are all isomorphic to the Poincaré algebra, so we know that $\dim Z_\epsilon^2 = \dim g = 10$. For $\epsilon = 0$, we have already seen that $\dim H^2(g) = 1$ since the algebra is the Galilean algebra. On the other hand $[g,g]_0$ consists of all elements with t component equal to zero. Thus d_0 has a one dimensional kernel and hence a nine dimensional image. Let us introduce "coordinates" $(\tau, L, \vec{p}, E)$ dual to (A, v, x, t). (Here $\vec{p}$ is dual to x, and hence as we have already remarked is to be interpreted as linear momentum and E is dual to time translation and hence should be regarded as "energy".) Let $\nu \in \wedge^2(g^*)$ be the bilinear form given by

$$\nu(\xi_1, \xi_2) = \langle v_1, x_2 \rangle - \langle v_2, x_1 \rangle.$$

We have already seen that for $\epsilon = 0$ the bilinear form ν is a cocycle, and in fact, the cohomology class of ν generates $H_0^2(g)$, i.e., that the most general cohomology class is of the form $m[\nu]$ where m is the mass parameter for the Galilean group. We claim that ν is a cocycle for $\epsilon > 0$ as well (and hence of course a coboundary for positive values of ϵ). Indeed, let $\theta \in g^*$ be the element

$$\theta = (0, 0, 0, 1)$$

in terms of the coordinates that we are using. Thus $\theta(\xi) = t$ is just the t component of ξ. Then, using the above formula for $[\ ,\ ]_\epsilon$ we see that

$$d_\epsilon \theta(\xi_1, \xi_2) = \theta([\xi_1, \xi_2]_\epsilon) = \epsilon(\langle v_1, x_2 \rangle - \langle v_2, x_1 \rangle)$$

or

$$d_\epsilon \theta = \epsilon \nu.$$

Notice that for $\epsilon > 0$, θ is uniquely determined by this equation. Thus, for $\epsilon > 0$ the "mass cocycle" $m\nu$ is a coboundary, and indeed the coboundary of the unique element $(0, 0, 0, E)$ where $E = \epsilon^{-1} m$. If we substitute our definition of $\epsilon = c^{-2}$ we obtain Einstein's famous formula $E = mc^2$ relating mass and energy. The orbit through the point $(\tau, 0, 0, E)$, $\tau \in o(3)^*$ of the Poincaré group (corresponding to speed of light c) then goes over, as $c \to \infty$ to the symplectic homogenous space of the Galilean group (with no trivial cohomology class) with mass m (and spin $\|\tau\|$). The mass² of the orbit (in the Poincaré group) is $m^2 c^4$. If $(\tau_1, L_1, \vec{p}_1, E_1)$ is some other point on this orbit, then

$$E_1^2 - c^2 \vec{p}_1^{\,2} = m^2 c^4$$

gives the relation between mass, energy and momentum.

We can compute the space of cocycles, and the corresponding symplectic manifolds for the Galilean group from a slightly different point of view. Let $SO(3) \times \mathbf{R}^3$ act on $\mathbf{R}^4$ by $(A, v) \cdot (x, t) = (Ax + tv, t)$. Here (x, t) is a vector in $\mathbf{R}^4$, with x a vector in $\mathbf{R}^3$ and t in $\mathbf{R}$. We can regard the Galilean group as the semi-direct product of $SO(3) \times \mathbf{R}^3$ with $\mathbf{R}^4$. Again we have a (k, p) decomposition but this time with $k = o(3) \oplus \mathbf{R}^3$ (semi-direct) and $p = \mathbf{R}^4$. It is easy to check that there are no invariant antisymmetric two forms on $\mathbf{R}^4$, so that $c = 0$. We can write b as

$$b = b_1(\xi, x) + b_2(\xi, t) + b_3(v, x) + b_4(v, t)$$

where ξ is in $o(3)$ and v in $\mathbf{R}^3$. Condition (***) implies that

$$b_3(\xi \cdot v, x) + b_4(\xi \cdot v, t) = b_1(\xi, tv) - b_3(v, \xi \cdot x) + b_2(\xi, t)$$

and

$$b_1([\xi, \xi'], x) + b_2([\xi, \xi'], t) = b_1(\xi, \xi' \cdot x) - b_1(\xi', \xi \cdot x).$$

The second equation implies that b_1 is a cocycle of $o(3)$ with values in $\mathbf{R}^{3*}$, and hence a coboundary, and that $b_2 = 0$. The first of these equations, with $t = 0$, implies that $b_3(v, x) = m\langle v, x\rangle$ while for $t \neq 0$ it implies that $b_4(\xi \cdot v, t) = b_1(\xi, tv)$. Thus

$$b = \langle \ell, \xi \cdot x + tv\rangle + m\langle v, x\rangle.$$

The fact that a is a cocycle in $\wedge^2 k^*$ implies that it is a coboundary, $a = d(\tau + \ell')$ where $\tau \in o(3)^*$ and $\ell' \in \mathbf{R}^{3*}$. We thus recover the expression that we derived above for the most general cocycle for the Galilean algebra. The analysis of the operation of the Galilean group on the space of cocycles proceeds as before.

Let us now do a computation at the opposite extreme—the Heisenberg algebra. In this case k is the one dimensional center that we may identify with $\mathbf{R}$, and p is a symplectic vector space with r identified as its symplectic two form. The action of k on p is trivial and $s = 0$. Since k is one dimensional, $a = 0$ and conditions (**) and (***) are vacuous. Condition (*) can be interpreted as follows: let ω denote the symplectic form on p and let $\kappa \in p^*$ be defined by $\kappa = b(1, \cdot)$. Then (*) becomes $\omega \wedge \kappa = 0$. If $\dim p = 2$, this imposes no condition. If $\dim p > 2$, then this implies $\kappa = 0$. Indeed, for $\dim p > 2$ we can write $\omega = \theta \wedge \kappa + \omega'$ for some suitable $\theta \in p^*$ and where $\omega' \wedge \kappa \neq 0$. Any $c \in \wedge^2 p^*$ gives a cocycle. The element $\omega \in \wedge^2 p^*$ is a coboundary, $\omega = d\ell$ where ℓ is any element of g^* satisfying $\ell(1) = 1$, with 1 denoting the basis element of the center which we have identified with **R**. It is clear that the only coboundaries are the multiples of ω. Now

$$[(s,v),(t,w)] = (\omega(v,w),0) \qquad \text{where } s,t \in k = \mathbf{R} \quad \text{and} \quad v,w \in p.$$

Therefore

$$Ad\,(\exp(s,v))(t,w) = (t + \omega(v,w), w).$$

Thus the group acts trivially on $\wedge^2 p^*$. For $\dim p = 2$ it maps a non-zero $b \in k^* \otimes p^*$ onto $b + c$ where c ranges all over $\wedge^2 p^*$. For $\dim p > 2$ the action of the group on the space of cocycles is completely trivial. Thus $H^2(g) \sim k^* \otimes p^*$ for $\dim p = 2$ while $H^2(g) \sim \wedge^2 p^*/\{\omega\}$ for $\dim p > 2$. The orbits of G acting on g^* via the contragredient representation are the hyperplanes $\ell(1,0) = \text{const}$ for $\ell(1,0) \neq 0$, and the points in the subspace $\ell(1,0) = 0$. Thus these orbits either have dimension equal to $\dim p$ or are zero dimensional. The symplectic manifolds corresponding to the non-vanishing cohomology classes for $\dim p = 2$ are all two dimensional while for $\dim p > 2$ the dimension corresponding to an element of $\wedge^2 p^*$ is equal to its rank. We now list the homogeneous symplectic manifolds for the low dimensional Lie algebras.

$n = 1$.

There is only one Lie algebra of dimension one, the trivial Lie algebra, the corresponding simply connected group is just the additive group of real numbers, which obviously does not act transitively on any symplectic manifold of positive dimension. Hence the only homogeneous symplectic manifold is a point. Nevertheless, even in this most trivial example there are a number of interesting lessons to be learned. The action of the adjoint group is trivial, and hence so is the co-adjoint action. Thus the orbits of G in g^* consist entirely of points. This is obviously the case for any commutative Lie algebra. Although the orbits are all distinct, they all correspond to the same symplectic manifold, because the operator d is trivial, and hence all orbits give the zero cocycle. From the point of view of "classical mechanics" these orbits are all the same. But from the "quantum" viewpoint, i.e. from the point of view of representation theory, they all correspond to different infinitesimal characters, to different representations. This is already true at the level of "prequantization", cf. Chapter V, where one is interested in classifying homogeneous Hermitian line bundles with connection. Another comment is in order, even at the level of "classical mechanics". While it is obvious that the real line cannot act transitively on any manifold whose dimension is greater than one, the action can be "ergodic" in any of the various senses, e.g. topologically transitive, or metrically transitive, or mixing, etc. Each of these concepts corresponds to a different mathematical formulation of the intuitive notion of "irreducibility" for a mechanical system. The notion of transitivity, the one we are dealing with, is just the most simple of these concepts.

$n = 2$.

There are two Lie algebras of dimension two, the trivial Lie algebra and the Lie algebra with basis $\{\xi, \eta\}$ and bracket relations $[\xi, \eta] = \eta$. We shall call this second algebra the scale algebra. It corresponds to the group of symmetries in which one can change the origin of measurement (time translation) and the choice of units (scaling). For both algebras all two forms are cocycles since the algebra is two dimensional. However the operator $d\colon g^* \to \wedge^2 g^*$ behaves differently in each of the two cases.

The trivial algebra.

Here the operator d is trivial. Thus each element of $\wedge^2 g^*$ represents a different cohomology class and the action of G on $\wedge^2 g^*$ is trivial. Each element of $\wedge^2 g^*$ gives a distinct symplectic space. It is easy to see that the explicit realization of these spaces are given by $\omega = c\,dx \wedge dy$, $c \neq 0$, with $\xi \to \partial/\partial x$ and $\eta \to \partial/\partial y$. In addition, of course, there is the zero dimensional symplectic manifold corresponding to the orbits of g^* to which the remarks made above in the one dimensional case apply.

The scale algebra.

We have a $k + p$ decomposition with both k and p one dimensional. Hence $a = c = 0$, and, since d is non-trivial (or, by applying the general argument using the fact that k contains the identity), we know that every cocycle is a coboundary. Thus every b has the form $b(\xi, \cdot) = \ell([\xi, \cdot])$ for some $\ell \in p^*$. The group K acts on p and hence on p^* by multiplication by positive numbers, while $\exp p$ does not change $d\ell$. Hence there are three symplectic homogeneous spaces corresponding to the alternatives $\ell(\eta) > 0$, $\ell(\eta) = 0$, and $\ell(\eta) < 0$. Let (a, b) be the coordinates on g^* given by ξ and η, so that $\theta(\xi) = a$ and $\theta(\eta) = b$ if $\theta \in g^*$ has coordinates (a, b). Then a direct computation shows that

$$Ad^*_{\exp t\xi}(a, b) = (a, e^{-t}b) \qquad \text{and} \qquad Ad^*_{\exp t\eta}(a, b) = (a + tb, b).$$

For $b > 0$, say, we get a symplectic manifold, and

$$\hat{\xi} = b\frac{\partial}{\partial b}, \qquad \hat{\eta} = -b\frac{\partial}{\partial a}.$$

It is clear that the invariant two form ω must be given by

$$\omega = \kappa b^{-1}\, db \wedge da$$

for $\kappa \neq 0$. On the other hand, replacing (a, b) by (sa, sb), where s is an arbitrary constant, does not change $\hat{\xi}$ or $\hat{\eta}$, but replaces κ by $s\kappa$. This shows that the symplectic manifolds corresponding to $\ell(\eta) > 0$ and $\ell(\eta) < 0$ are equivalent, so that there is exactly one two dimensional homogeneous symplectic manifold for the scale algebra. There is, of course, also the zero dimensional manifold as well.

If we introduce the variables $x = a/b$, $y = b$, we can rewrite the two dimensional action as

$$(\exp t\xi)(x,y) = (e^t x, e^{-t} y) \qquad (\exp t\eta)(x,y) = (x + t, y).$$

In terms of these coordinates we have $\omega = dx \wedge dy$.

If we set $u = a$ and $v = -\log b$ we can describe the action as $\omega = du \wedge dv$, $\exp t\xi(u,v) = (u, v + t)$ and $\exp t\eta(u,v) = (u + te^{-v}, v)$.

$n = 3$.

The three dimensional Lie algebras over **R** are classified as follows (cf. Jacobson, *Lie algebras*, pp. 11–13):

(i) the trivial algebra,

(ii) the Heisenberg algebra,

(iii) the direct sum $\mathbf{R} + h$ where h is the two dimensional scale algebra.

$(\text{iv})_A$ the affine algebra $k + p$ where k is one dimensional and p is two dimensional. Here, a basis element of k acts on p via the linear transformation A, which is non-singular. Here A is determined only up to conjugacy on p and multiplication by any non-zero real number (since the basis of k was chosen arbitrarily). We will distinguish several possibilities, according to whether the trace of A is or is not zero. If $\operatorname{tr} A = 0$, and if A has real eigenvalues then we may arrange that A is diagonal with eigenvalues ± 1. Thus A is the matrix which infinitesimally preserves the indefinite metric xy on the (x,y) plane, and g is the algebra of (infinitesimal) motions for this metric. We list this algebra as:

(iv) $e(1,1)$. If $\operatorname{tr} A = 0$ and A has comlex eigenvalues, then the eigenvalues must be purely imaginary, and we can arrange that they are $\pm i$. Thus A is an infinitesimal rotation for the Euclidean metric in the plane and we are in the case,

(v) $e(2)$, the Lie algebra of the group of Euclidean motions in the plane,

(vi) the case of the affine algebra $\text{iv})_A$ where $trA \neq 0$,

(vii) the orthogonal algebra $o(3)$, and

(viii) $sl(2,\mathbf{R})$.

For the case of the trivial algebra, every element of $\wedge^2 g^*$ is a cocycle and no element is a coboundary; each non-zero element has a one dimensional kernel which will act trivially on the corresponding symplectic manifold, which is a homogeneous symplectic manifold for the quotient two dimensional trivial algebra. We have already discussed the Heisenberg algebra. For the case (iii) we can apply the k, p decomposition with $k = \mathbf{R}$ and with $p = h$, both ideals. Then $a = 0$ and c is a cocycle for the scale algebra, while condition $(**)_D$ implies that $b(k, [h,h]) = 0$. The space of possible b's is thus one dimensional and none of them is a coboundary, and thus $H^2(g)$ is one dimensional. Let x, y, and z be a basis of g with the bracket relations

$$[x,y] = y, \qquad [x,z] = [y,z] = 0,$$

and let

$$A_t = \exp tx, \qquad B_t = \exp ty, \qquad \text{and} \qquad C_t = \exp tz$$

be the corresponding one parameter subgroups. Then it follows from the above computation that the space of all two dimensional symplectic manifolds is parametrized by the constant $m = b(z, x)$. The actual manifolds are all $\mathbf{R}^2$ with coordinates (u, v) and $\omega = du\,dv$ and actions given by

$$A_t(u, v) = (u, v + t) \qquad tB_t(u, v) = (u + te^v, v) \quad \text{and} \quad C_t(u, v) = (u + mt, v).$$

We now turn to the remaining cases:

(iv) the algebra $e(1, 1)$. Here we can choose a basis with

$$[z, x] = x, \qquad [z, y] = -y \qquad \text{and} \qquad [x, y] = 0.$$

Here k is spanned by z and p is spanned by x and y. Any c in $\wedge^2 p^*$ is a cocycle and is not a coboundary if $c \neq 0$. The operator d: $g^* \to \wedge^2 g^*$ has a one dimensional kernel and hence the space of coboundaries is two dimensional. Thus $dg^* = k^* \otimes p^*$, i.e. all b's are coboundaries. If $c \neq 0$, the map of $p \to k^* \otimes p^*$ sending $v \rightsquigarrow c(\cdot\, v, \cdot)$ is non-singular and hence surjective. Thus if $c \neq 0$, we can eliminate b by the action of G. There are thus two families of two dimensional symplectic manifolds, one parametrized by non-zero $c \in \wedge^2 p^*$ and the other parametrized by a cross-section for the orbits of C_t acting on p^*. The first family all consist of $\mathbf{R}^2$ with coordinates (u, v) and with varying forms $\omega = m\,du\,dv$ (where $m = c(x, y)$) and action

$$A_t(u, v) = (u + t, v) \qquad B_t(u, v) = (u, v + t) \qquad \text{and} \qquad C_t(u, v) = (e^t u, e^{-t} v).$$

To describe the orbits in p^* observe that x and y can be thought of as functions on g^* and hence on p^*, and the orbits of C_t are the various components of the hyperbolas $xy = \kappa$ for different values of the constant, κ. The actual orbits of G in g^* are given by the same equations, and are, in fact, just cylinders over these curves, with generators in the k^* direction. Again the orbits can all be identified with $\mathbf{R}^2$, with coordinates (u, v) and $\omega = du\,dv$. For the orbits on which $x \neq 0$ we can use the vectors $(1, \kappa)$ as cross-sections to the orbits; the corresponding actions are given by

$$A_t(u, v) = (u + te^{-v}, v), \qquad B_t(u, v) = (u - \kappa e^{+v}, v), \qquad C_t(u, v) = (u, v + t).$$

The case $\kappa = \infty$ is obtained in the limit as A_t acting as the identity and $B_t(u, v) = (u \pm te^v, v)$ and $C_t(u, v) = (u, v + t)$. It is interesting to give some interpretation to the parameters m and κ. Notice that the algebra $e(1, 1)$ contains two copies of the scale algebra, namely z, x and z, y, with the group C_t multiplying x by e^t and multiplying y by e^{-t}. Now there are two situations where making a change in scale of one variable induces the inverse change of scale of a second variable, if the variables are dual to one another (i.e. represent coordinates in dual one dimensional vector spaces) or if the variables are inverse to one another. The first family of orbits corresponds to the duality situation, with the parameter m giving the duality between u and b. The second family of orbits corresponds to the situation where the scale algebra is acting on variables r and s related by $rs = \kappa$.

(v) The situation for the Euclidean algebra $e(2)$ is quite similar to that for $e(1, 1)$. The cohomology is one dimensional, each non-zero element of $\bigwedge^2 p^*$ corresponding to a non-zero cohomology class and giving rise to a symplectic manifold with x and y acting as constant vector fields. The remaining symplectic manifolds are given by orbits in g^* which are the cylinders $x^2 + y^2 = r^2$, for r positive, together with the zero dimensional orbits on the z-axis.

(vi) For the affine algebra with tr $A \neq 0$, there is no non-zero invariant c in $\bigwedge^2 p^*$, and thus the space of cocycles is two dimensional, the cohomology vanishes. The orbits in g^* are seen to be cylinders over the orbits of exp tz acting on p^*, and these provide all the two dimensional symplectic manifolds. The z-axis again splits into zero dimensional orbits.

(vii) The orthogonal algebra is semi-simple, so its cohomology vanishes. The orbits in g^* are given by the spheres $x^2 + y^2 + z^2 = r^2$, where x, y, and z are the usual basis of $o(3)$, with the bracket relations

$$[x,y] = z, \qquad [z,x] = y, \qquad [z,y] = -x.$$

(viii) The algebra $sl(2)$ is also semi-simple, so its cohomology also vanishes. We may choose a basis of, and z with the bracket relations

$$[x,y] = z, \qquad [z,x] = x, \qquad \text{and} \qquad [z,y] = -y.$$

Then $xy + z^2/2$ is invariant under the action of g, and thus, when considered as a function on g^* defines two dimensional surfaces which are invariant under G. The connected components of these level surfaces are clearly the orbits of G; they are the single sheeted hyperboloids, the double sheeted hyperboloids, and the two components of the light cone.

We now study the behavior of homogeneous symplectic manifolds under deformation of the Lie algebra structure. As an illustration of what can happen let us consider the deformation of $sl(2)$ into $e(1, 1)$. We consider a three dimensional vector space with basis x, y, and z, and with bracket relations

$$[z,x] = x, \qquad [z,y] = -y, \qquad [x,y] = \epsilon z.$$

For $\epsilon \neq 0$, this algebra is isomorphic to $sl(2)$, while for $\epsilon = 0$ the algebra is $e(1, 1)$. For all values of ϵ the function $xy + \epsilon z^2/2$ is invariant. The double sheeted hyperboloids, corresponding to positive values of this function for $\epsilon < 0$ (and to negative values for $\epsilon > 0$) clearly deform into the cylinders $xy = c$ for $\epsilon = 0$. It is interesting to examine the behavior of the single sheeted hyperboloids. They provide both the other cylinders and also the symplectic manifolds of

$e(1,1)$ corresponding to non-vanishing cohomology classes of $e(1,1)$: (Recall that $H^2(e(1,1)) = \mathbf{R}$ while $H^2(sl(2)) = 0$). Indeed, for a *fixed* value of $xy + \epsilon z^2/2$ the points near $x = 0$ of the hyperboloid (or near $y = 0$) clearly move off to infinity at the rate $\epsilon^{-1/2}$ and the hyperboloid splits into two cylinders. As to the orbits with non-vanishing cohomology, observe that the cocycles are of the form $hx^* \wedge y^*$, and, for all ϵ, we have

$$dz^* = \epsilon x^* \wedge y^* .$$

For $\epsilon = 0$ we know that $hx^* \wedge y^*$ is not a coboundary (for non-vanishing h) while for $\epsilon \neq 0$ the above equation shows that $hx^* \wedge y^* = d(h\epsilon^{-1}z^*)$. This suggests looking at the orbit through the point with $x = 0, y = 0$ and $z = h\epsilon^{-1}$, i.e. the orbit $xy + \varepsilon z^2/2 = m_\epsilon$ where $m_\epsilon = (2\epsilon)^{-1}h^2$. A direct computation shows that if we consider a bounded region of x and y, the action on this portion of the orbit tends to the desired limiting action for $e(1,1)$.

§8. Multisymplectic structures and the calculus of variations.

The principal motivation that we have discussed for the study of symplectic manifolds has been via asymptotics. There is another route which historically led to symplectic manifolds, and that is classical mechanics, and in particular, its relation to variational principles, cf. Sternberg [2, Chapter III, §7 and Chapter IV], Loomis-Sternberg [13, Chapter XIII] and Souriau [9]. For systems with "finitely many degrees of freedom", i.e. for variational problems with one independent parameter—curves on a finite dimensional manifold—the "space of extremals" forms a finite dimensional symplectic manifold, cf., in this connection, Hermann [20] and [21], Garcia [22] and Dedecker [23].

In this section we sketch the geometry involved for variational problems in one or several independent variables. For the case of several independent variables our results will be rather formal. They would acquire more content in special instances if appropriate existence theorems in the theory of elliptic or of hyperbolic partial differential equations are introduced. However, we shall not go into these kinds of questions. The treatment here follows [17], and the "symplectic structure" is based also on [18]. We will restrict attention to the case of Lagrangians involving at most the first order derivatives. Let $Y \to X$ be a fibered manifold, i.e. we are given a differentiable map, π, from Y to X, such that near every point of Y we can introduce coordinates in which the map π is just projection onto a factor. Thus we can introduce local coordinates on Y of the form $(x,y) = (x^1, \ldots, x^n, y^1, \ldots, y^f)$ where n is the dimension of X and $n + f$ is the dimension of Y. A section of Y over X is a map $s: X \to Y$ satisfying $\pi \circ s = \mathrm{id}$. Thus s assigns, to each $x \in X$, a point, $s(x) \in \pi^{-1}(x)$. (The set $\pi^{-1}(x)$ is called the fiber over x. It is automatically a differentiable submanifold of Y.) Locally, a section is given by f functions $y^I = y^I(x^1, \ldots, x^n)$ where

$\ell = 1, \ldots, f$. Two sections, s_1 and s_2, are said to agree to first order at some point x_0 if $s_1(x_0) = s_2(x_0)$ and the functions $y_1^\ell(x)$ and $y_2^\ell(x)$ have the same first derivatives at x_0. (It is easy to check that this condition does not depend on the choice of coordinates of the form (x,y).) To agree to first order at a point x_0 is an equivalence relation. The equivalence class of a section s at x_0 is called the one jet of s at x_0 and is denoted by $j_1(s)(x_0)$. Since we will not be concerned in the main, with jets of higher order, we shall drop the subscript 1, and simply write $js(x_0)$. The jet, $js(x_0)$ is determined, locally, by the coordinates $(x,y,(y)) = (x^i, y^\ell, y_i^\ell)$, $i = 1, \ldots, n$, $\ell = 1, \ldots, f$, where the x^i are the coordinates of x_0, the y^ℓ are the coordinates of $s(x_0)$ and the y_i^ℓ the coordinates of the first partial derivatives of s evaluated at x_0. Thus the set of all jets at all possible points of X forms a manifold, which we denote by JY. We have the projections:

$$\pi_Y : JY \to Y \qquad \pi_Y(x,y,(y)) = (x,y)$$

and

$$\pi_X : JY \to X \qquad \pi_X(x,y,(y)) = x$$

so that

$$\pi_X = \pi \circ \pi_Y.$$

Thus JY is a fibered manifold over Y and is also a fibered manifold over X. If s is a section of Y, then s determines a section, js, of JY over X, where js assigns, to each point of X, the jet of s at that point. In terms of local coordinates, if s is given by the functions $y^\ell(x) = s^\ell(x)$, then js is given by $y^\ell = s^\ell(x)$, $y_i^\ell = (\partial s^\ell/\partial x^i)(x)$. Not every section of JY will be of the form js. Indeed, if u is a section of JY then u, in local coordinates gives y^ℓ and y_i^ℓ as functions of x, and they must be related by the equations

$$dy^\ell(x) - y_i^\ell(x)dx^i = 0 \qquad \text{(summation convention).}$$

Put another way, let ω be the linear differential form which assigns tangent vectors to Y to tangent vectors to JY according to the formula

$$\omega = (\partial/\partial y^\ell) \otimes [dy^\ell - y_i^\ell\, dx^i]. \tag{8.1}$$

Then a section u, of JY over X is of the form $u = js$ if and only if

$$u^* \omega = 0. \tag{8.2}$$

Notice that we have given a definition of ω in terms of local coordinates. Actually, ω has an invariant definition: let ξ be a tangent vector to JY at a point z, where $z = js(x_0)$, with $x_0 \in X$. Then $d\pi_Y \xi$ is a tangent vector to Y at $y_0 = s(x_0)$ and $d\pi_X \xi$ is a tangent vector to X at x_0. If η is any tangent to X at x_0

then $ds_{x_0}\eta$ is a tangent vector to Y at y_0 which depends only on $z = js(x_0)$. Then we claim that

$$\langle \xi, \omega \rangle = d\pi_Y \xi - ds(d\pi_X \xi). \tag{8.3}$$

Indeed, in terms of the local coordinates, we can write

$$\xi = a^i(\partial/\partial x^i) + b^l(\partial/\partial y^l) + c_i^l(\partial/\partial y_i^l),$$

so that

$$d\pi_X \xi = a^i(\partial/\partial x^i),$$
$$d\pi_Y \xi = a^i(\partial/\partial x^i) + b^l(\partial/\partial y^l),$$

and

$$ds(d\pi_X \xi) = a^i(\partial/\partial x^i) + y_i^l a^i(\partial/\partial y^l),$$

which establishes the formula.

Notice that if ξ is a *vertical* tangent vector, i.e.

$$\text{if } d\pi_X \xi = 0 \text{ then } \langle \xi, \omega \rangle = d\pi_Y \xi. \tag{8.4}$$

Let L be a real valued function on JY, and let (vol) be an n-form on X. The basic problem of the calculus of variations is to find the extremals for integrals of the form

$$I_A[s] = \int_A L(js)(\text{vol}), \tag{8.5}$$

where A is some bounded region of X. Here s is allowed to vary over some class of sections of X. The usual problem (the fixed boundary problem) is the situation where A is a region with smooth boundary, ∂A, and s is restricted to take on assigned values on the boundary.

The main point of the *Hamilton-Cartan formalism* is to replace the integral (8.5) by an integral of the form $\int_A u^* \Theta$ where Θ is a suitable n-form defined on JY and u is a section of JY over X. This integral satisfies

$$\int_A u^* \Theta = \int_A L(js)(\text{vol})$$

if $u = js$. The n-form Θ has the property that an extremal for $\int_A u^* \Theta$, where u is allowed to vary over *all* sections of JY, is automatically of the form $u = js$, if L is "regular".

We now describe the construction of the form Θ. We begin by defining a form, θ, which maps $TJY \to TX$, i.e. which assigns, to each tangent vector, ξ, to JY at z a tangent vector, $\langle \xi, \theta \rangle$, to X at πz. We first give the formula in terms of local coordinates: Define the functions p^i_ℓ on JY by

$$p^i_\ell = \partial L/\partial y^\ell_i$$

and set

$$\theta = (\partial/\partial x^i) \otimes \left[\frac{1}{n} L dx^i + p^i_\ell(dy^\ell - y^\ell_j\, dx^j)\right]. \tag{8.6}$$

To give an invariant definition of θ we observe that L defines a bundle map, σ, from $JY \to Y$ to the vector bundle Hom $(VY, TX) \to Y$, where VY denotes the bundle of vertical tangent vectors to Y. The map σ (which is the Legendre transformation) is defined as $\sigma(z) = d_v L$ where d_v means computing the differential of L with respect to the fiber of JY over Y. (This makes invariant sense because JY is an "affine bundle" over Y whose associated vector bundle is Hom (TX, VY): Given two sections with $s_1(x_0) = s_2(x_0) = y_0$ the map $ds_{1*} - ds_{2*} \in$ Hom (TX_{x_0}, VY_{y_0}) depends only on $js_1(x_0)$ and $js_2(x_0)$.) Then, for $\xi \in TJY_z$,

$$\langle \xi, \theta \rangle = \left(\frac{1}{n}\right) L(z) d\pi_X \xi + \sigma(z)\langle \xi, \omega \rangle \tag{8.7}$$

where $\langle \xi, \omega \rangle \in VY$ so $\sigma(z)\langle \xi, \omega\rangle \in TX$. It follows from the definitions that if s is any section of Y then

$$\begin{aligned} (js)^* \theta &= \left(\frac{1}{n}\right) L(js) \text{ id} \\ &= \left(\frac{1}{n}\right) L(js) \partial/\partial x^i \otimes dx^i \end{aligned} \tag{8.8}$$

in local coordinates.

The form Θ is defined, in local coordinates as

$$\begin{aligned} \Theta &= (L - p^i_\ell y^\ell_i) dx^1 \wedge \cdots \wedge dx^n \\ &\quad + \sum_{\ell, i} (-1)^{i+1} p^i_\ell\, dy^\ell \wedge dx^1 \wedge \cdots \wedge \widehat{dx^i} \wedge \cdots \wedge dx^n \end{aligned} \tag{8.9}$$

where we have chosen our coordinates so that (vol) $= dx^1 \wedge \cdots \wedge dx^n$.

Invariantly, the definition of Θ is

$$\Theta = \theta \,\overline{\wedge}\, \pi^*(\text{vol}) \tag{8.10}$$

where, the operation, $\overline{\wedge}$, pairs a TX valued p-form on JY with a q-form on X to get a $p + q - 1$ form on JY according to the rule

$$(\alpha \otimes \eta) \,\overline{\wedge}\, \tau = \alpha \wedge (\eta \lrcorner \tau)$$

if η is a vector field on X and α a p-form on JY. (For the details we refer the reader to [14].)

It follows from (8.10) that for any section, u, of JY over X we have

$$u^* \Theta = u^* \theta \overline{\wedge} \text{ vol}$$

and therefore

$$u^* \Theta = L(u)(\text{vol}) + \sigma(u) \cdot u^* \omega. \tag{8.11}$$

In particular, if $u = js$, so that $u^* \omega = 0$, we get

$$(js)^* \Theta = L(js)(\text{vol}). \tag{8.12}$$

Suppose that u_t is some one parameter family of sections of JY, $u_0 = u$. Let ξ denote the vector field along u giving the tangent to the deformation, so that $\xi(x) \in TJY_{u(x)}$ is the tangent vector to the curve $t \rightsquigarrow u_t(x)$. Then the basic formula of the differential calculus asserts that

$$\frac{d}{dt} u_t^* \Theta \,|_{t=0} = u^* (\xi \lrcorner d\Theta) + du^* (\xi \lrcorner \Theta).$$

If u is to be an extremal for the integral $\int_A u_t^* \Theta$ among all u_t which satisfy the condition $\pi_Y u_t = \pi_Y u$ on ∂A then we obtain the "Euler equations"

$$u^* (\xi \lrcorner \Omega) = 0 \tag{8.13}$$

where $\Omega = d\Theta$ and ξ is allowed to be any vector field on JY. A computation (which can either be done in local coordinates or invariantly, as in [17, pp. 219–220]) shows that if η is any vector field on JY satisfying $d\pi_Y \eta = 0$ then

$$u^* (\eta \lrcorner \Omega) = \text{tr}\, [(u^* D_\eta \sigma(L)) \circ u^* \omega] \text{ vol} \tag{8.14}$$

where $u^* \omega \in \text{Hom}\,(TX, VY)$ and $u^* D_\eta \sigma \in \text{Hom}\,(VY, TX)$, so that the trace makes sense. If $u = js$ then $u^* \omega = 0$ so that $u^* (\eta \lrcorner \Omega) = 0$ for all η satisfying $d\pi_Y \eta = 0$. If s_t is a one parameter family of sections of Y with $s_0 = s$ and $s_{t/\partial A} = s_{/\partial A}$, then js_t is a one parameter family of sections of JY where the tangent vector, η, satisfies $d\pi_Y \eta = \xi$ where ξ is the tangent field along s. Then if

$$\frac{d}{dt} I_A(s_t) = 0$$

we conclude that

$$\int u^* (\eta \lrcorner \Omega) = 0$$

for all such η which implies that

$$u^* (\eta \lrcorner \Omega) = 0$$

for all such η. This implies that $u^*(\eta \lrcorner \Omega) = 0$ for all η, satisfying $d\pi_X \eta = 0$. But since $u^* \Omega = 0$ (as Ω is an $n+1$ form and X is only n dimensional) we conclude that $u^*(\eta \lrcorner \Omega) = 0$ for all η. We thus see that *if s is an extremal for I_A then $u = js$ satisfies* (8.13).

The same argument shows that *if $u = js$* and u satisfies (8.13) then u is an extremal. But we can say a lot more. Suppose we merely assume that u satisfies (8.13). By (8.14) this implies that $u^* \omega$, a section of Hom (TX, VY) is perpendicular to the sub-bundle of Hom (VY, TX) spanned by all $D_\eta \sigma$. If the $D_\eta \sigma$ span all of Hom $(VY, TX)_z$ then condition (8.13) automatically implies that $u^* \omega = 0$, i.e. that $u = js$. The condition that $D_\eta \sigma$ span all of Hom $(VY, TX)_z$ is known as the "regularity condition" on the Lagrangian L. In terms of local coordinates it says that the Hessian

$$\partial^2 L / \partial y_i^l \, \partial y_j^k$$

be a non-degenerate fn by fn matrix. Thus, for regular Lagrangians, the *equation*

$$u^*(\eta \lrcorner \Omega) = 0$$

on all η is equivalent to the pair of conditions

$$u = js \qquad s \text{ is an extremal of } I.$$

Let ξ be a vector field on JY. The condition that ξ (infinitesimally) preserve Ω is

$$0 = D_\xi \Omega = d(\xi \lrcorner \Omega).$$

Locally, this is equivalent to the stronger condition

$$\xi \lrcorner \Omega = -d\tau. \tag{8.15}$$

Let us call a vector field satisfying (8.15) *Hamiltonian*. Notice that if u satisfies (8.13) then

$$du^*(\tau) = 0,$$

i.e. the form $u^*(\tau)$ is closed. We will want to think of τ as if it defines a functional on extremals by the "formula"

$$\tau(u) = \int_C u^* \tau,$$

where C is a suitable $n-1$ dimensional submanifold of X. If, for instance, $X = M \times \mathbf{R}$ and τ had "compact support" in the M-variables, then we would choose C to be a "space like surface" $\{(m, t(m))\}$ and the value of the integral would not depend on the particular choice of the space like surface, C, i.e. on the choice of the function, t. Moreover, modifying τ by adding an exact form, $d\nu$, (of compact support in M) would not change the value of the integral. With this in mind, we define the "algebra of currents" to consist of all $n-1$ forms τ for which there exists a ξ such that (8.15) holds. We define the "algebra of charges" to consist of equivalence classes $[\tau]$ where τ satisfies (8.15) and where

$$[\tau] = [\tau'] \qquad \text{if } \tau' = \tau + d\nu, \qquad \text{for some } n-2\text{-form } \nu.$$

We define the Poisson bracket

$$\{\tau_1, \tau_2\} = \xi_1 \lrcorner d\tau_2$$

where $\xi_i \lrcorner \Omega = -d\tau_i$. Observe that this does not depend on the particular choice of ξ_i. Indeed

$$\xi_1 \lrcorner d\tau_2 = -\xi_1 \lrcorner \xi_2 \lrcorner \Omega = \xi_1 \lrcorner \xi_2 \lrcorner \Omega = -\xi_2 \lrcorner d\tau_1$$

which shows that if $\xi_1 \lrcorner \Omega = 0$ then $\xi_1 \lrcorner d\tau_2 = 0$ and so $\{\tau_1, \tau_2\}$ is independent of which ξ_1 we choose to satisfy $\xi_1 \lrcorner \Omega = -d\tau_1$. Also we see that $\{\tau_1, \tau_2\}$ is antisymmetric in τ_1 and τ_2. Also observe that

$$\xi_1 \lrcorner d(d\nu) = 0$$

so the Poisson bracket induces an operation:

$$\{[\tau_1], [\tau_2]\} = [\{\tau_1, \tau_2\}]$$

which we also call the Poisson bracket. Notice that

$$D_1(\xi_2 \lrcorner \Omega) = [\xi_1, \xi_2] \lrcorner \Omega$$

and

$$\begin{aligned} D_{\xi_1}(\xi_2 \lrcorner \Omega) &= D_{\xi_1} d\tau_2 = d(D_{\xi_1} \tau_2) \\ &= d(\xi_1 \lrcorner d\tau_2) + d(d(\xi_1 \lrcorner \tau_2)) = d(\xi_1 \lrcorner d\tau_2) \end{aligned}$$

which shows that

$$[\xi_1, \xi_2] \lrcorner \Omega = -d\{\tau_1, \tau_2\}.$$

Also, we see that

$$\begin{aligned} &\{\tau_1, \{\tau_2, \tau_3\}\} = [\xi_2, \xi_3] \lrcorner d\tau_1 \\ &= -D_{\xi_2}(\xi_3 \lrcorner d\tau_1) + \xi_3 \lrcorner D_{\xi_2} d\tau_1 \\ &= -D_{\xi_2}(\xi_1 \lrcorner d\tau_3) + \{\{\tau_1, \tau_2\}, \tau_3\} \\ &= \xi_2 \lrcorner d(\xi_1 \lrcorner d\tau_3) + d(\xi_2 \lrcorner \xi_1 \lrcorner \tau_3) + \{\{\tau_1, \tau_2\}, \tau_3\} \\ &= \{\{\tau_1, \tau_2\}, \tau_3\} + \{\tau_2, \{\tau_1, \tau_3\}\} + d(\xi_1 \lrcorner \xi_2 \lrcorner \xi_3 \lrcorner \Omega). \end{aligned}$$

Thus the algebra of currents need not satisfy Jacobi's identity but the algebra of charges does. Of course, if dim $X = 1$, the last term vanishes.

Let φ_t be a one parameter family of automorphisms of JY which satisfies

$$\varphi_t^* \Theta = \Theta + d\alpha_t .$$

(For example φ_t might arise from a one parameter family of automorphisms of Y which preserves L.) If η is the corresponding vector field, then

$$D_\eta \Theta = d\dot{\alpha}_0 \qquad \eta \lrcorner \Omega = d(\dot{\alpha}_0 - \eta \lrcorner \Theta)$$

so that η is Hamiltonian with $\tau = \dot{\alpha}_0 - \eta \lrcorner \Theta$. Thus φ_t gives rise to the "conserved current" $u^* \tau$ for any extremal, u. This is the content of "Noether's theorem".

For the case where $n = \dim X = 1$, the condition on τ (which is now a form of degree zero, i.e. a function) that is imposed by equation (8.15) is that τ be constant along extremals. Indeed, in this case, the extremals are the integral curves to the line element field spanned by η satisfying $\eta \lrcorner \Omega = 0$, and the condition $\langle \eta, d\tau \rangle = 0$ is equivalent to (8.15) since the two form Ω has rank $2f$ and JY has dimension $2f + 1$. Thus, for the case $n = 1$, the "algebra of currents" is an honest Lie algebra, can be identified with the "algebra of charges", and consists of *all* smooth functions on extremals.

For $n > 1$, the condition (8.15) is much more restrictive. The "functions on extremals" corresponding to $[\tau]$ where τ satisfies (8.15) will, in general, constitute only a small subspace of the space of functions on extremals. In fact, for many interesting Lagrangians, the space of such $[\tau]$ will be finite dimensional. If the Lagrangian is "quadratic", that is can be expressed as a quadratic function of the coordinates of JY when suitable coordinates are chosen on X and Y, then there is an infinite dimensional space of $[\tau]$ which are "sufficient" in a sense that we will not specify here.

The problem then arises as to how to introduce a reasonable class of functions on extremals. One possible method, suggested by [18], is to consider an "infinitesimal version" of the construction of $[\tau]$. This involves considering the so called "second variation" and Jacobi fields. Roughly speaking, the situation is as follows: let u_t be a one parameter family of extremals, and let ξ be the vector field along u tangent to u_t at $t = 0$. Then ξ satisfies the Jacobi equation

$$u^* (\eta \lrcorner d(\xi \lrcorner \Omega)) = 0, \qquad \text{for all vector fields } \eta \text{ along } u.$$

The set of ξ satisfying the above equation can be described as the extremals of a quadratic Lagrangian defined on the vector bundle of all vector fields along u, and this quadratic Lagrangian is the second variation. We refer to [17] for details, especially pp. 255–262. We can think of the set of ξ satisfying the above equation as the "tangent space" to the set of extremals at u. We can then define a bilinear antisymmetric form Ξ from this "tangent space" to $\bigwedge^{n-1} X$, sending ξ_1, ξ_2 into

$u^*(\xi_1 \lrcorner \xi_2 \lrcorner \Omega)$. One checks that this form actually is closed for any ξ_1, ξ_2 in the "tangent space" to u, and hence if ξ_1 or ξ_2 had suitable "compact support in the space like direction" would define an antisymmetric two form on the "tangent space" by integration over some space like surface. This two form would then be the candidate for the symplectic form on the "manifold of all extremals". With this symplectic form one then considers those "functions on extremals" f, which satisfy $df_u = \int \xi \lrcorner \Xi$, the integral being taken over the space like surface. We refer the reader to [18] for details.

References, Chapter IV

1. A. Weinstein, Advances in Math. **6** (1971), 329–346.
2. S. Sternberg, *Lectures on differential geometry*, Prentice-Hall, Englewood Cliffs, N. J., 1964. MR **33** #1797.
3. L. Hormander, Acta Math. **127** (1971), 79–183.
4. W. B. Gordon, J. Math. Mech. **19** (1969/70), 111–114. MR **39** #7236.
5. A. Wintner, *The analytical foundations of celestial mechanics*, Princeton Math. Ser., vol. 5, Princeton Univ. Press, Princeton, N. J., 1941. MR **3**, 215.
6. J. K. Moser, Comm. Pure Appl. Math. **23** (1970), 609–636. MR **42** #4824.
7. S. Sternberg, *Celestial mechanics*. I, Benjamin, New York, 1969.
8. A. Weinstein, Ann. of Math. **98** (1973), 377–410.
9. J.-M. Souriau, *Structure des systèmes dynamiques*, Maîtrises de mathématiques, Dunod, Paris, 1970. MR **41** #4866.
10. B. Y. Chu, Trans. Amer. Math. Soc. **197** (1974), 145–159.
11. B. Kostant, Amer. J. Math. **85** (1963), 327–404. MR **28** #1252.
12. L. F. Rothschild, Trans. Amer. Math. Soc. **168** (1972), 403–421.
13. A. A. Kirillov, Uspehi Mat. Nauk **17** (1962), no. 4 (106), 57–110 = Russian Math. Surveys **17** (1962), no. 4, 53–104. MR **25** #5396.
14. M. Duflo and M. Vergne, C. R. Acad. Sci. Paris Sér. A-B **268** (1969), A583–A585. MR **39** #6935.
15. Y. Kosmann and S. Sternberg, C. R. Acad. Sci. Paris Sér. A-B **279** (1974), A777–A779.
16. S. Sternberg, Trans. Amer. Math. Soc. **212** (1975), 113–130.
17. H. Goldschmidt and S. Sternberg, Ann. Inst. Fourier (Grenoble) **23** (1973), fasc. 1, 203–267. MR **49** #6279.
18. J. Kijowski and W. Szczyrba, in Geometrie symplectique et physique mathematique, Ed., CNRS Paris, 1975, pp. 347–378.
19. E. Spanier, *Algebraic topology*, McGraw-Hill, 1966.
20. L. Pukanszky, *Leçons sur les representations des groupes*, Dunod, Paris, 1967.
21. R. Hermann, *Differential geometry and the calculus of variations*, Academic Press, New York, 1968.
22. P. L. Garcia, Symposia Mathematica XIV (1974), 219–246.
23. P. Dedecker, Colloque Inter. de Geom. Diff. Strasbourg (1953).
24. J. Leray, Symposia Math. XIV (1974).
25. J.-M. Souriau, *Construction explicite de l'indice de Maslov applications* (to appear).

Chapter V. Geometric Quantization

In this chapter we describe some of the recent methods developed to provide a geometrical construction relating classical to quantum mechanics. The first step (called prequantization by Kostant) consists of realizing the symplectic form, ω, on a symplectic manifold, X, as the curvature form of a line bundle, L, over X. (This cannot be done for all X, but only when the form ω defines an integral cohomology class. This condition already imposes interesting "quantum" conditions on X.) The functions on X then operate on sections of L. However the space of all sections of L is too large. One wants to consider sections which are "constant in certain directions" and for this one needs to introduce the concept of a "polarization". To get a Hilbert space structure which behaves properly under transformations, one needs to tensor in objects known as "half forms". (These objects were hinted at in Chapter II and will play an important role in the symbol calculus of Chapters VI and VII). One also has a sesquilinear pairing between objects associated to different polarizations (provided suitable transversality conditions are satisfied). We will attempt to explain these ideas in this chapter, and illustrate them with some quantum mechanical examples in the last section. The idea of considering the line bundle and polarizations was introduced independently by Kostant [1] and Souriau [2] and exploited (especially using complex polarizations) by Kostant and Auslander [3] in the representation theory of solvable Lie groups. The idea of the half form and the pairing is joint work of Blattner, Kostant, and Sternberg, [4], [5]; the quantum mechanical examples are based on work of Simms [6], [7].

§1. Curvature forms and vector bundles.

In §7 of Chapter IV, we discussed a Lie algebra acting as infinitesimal automorphisms of a symplectic manifold. In this context we saw that there was a difference between the notion of infinitesimal automorphism, a vector field satisfying

$$D_\xi \omega = 0$$

or, what is the same thing,

$$d(\xi \rfloor \omega) = 0,$$

and a Hamiltonian vector field which satisfies the stronger condition

$$\xi \rfloor \omega = -df_\xi.$$

Among the Hamiltonian vector fields, the function f_ξ is the only determined up to an additive constant. Put another way, let P denote the algebra of C^∞ functions on X under Poisson bracket, and let A denote the algebra of Hamiltonian vector fields. Then we clearly have the exact sequence

$$0 \to \mathbf{R} \to P \to A \to 0$$

where $\mathbf{R}$ is regarded as the constants. In classical mechanics, it is the algebra A that plays the key role. In quantum mechanics, however, the algebra P is fundamental. (Indeed, one way of phrasing the Heisenberg uncertainty principle is to assert that the constants have a non-trivial representation in the algebra of observables.) It is therefore of interest to us to find a group theoretical analogue of the above exact sequence. This was done by Kostant in his fundamental paper [1]. See also Souriau. The main new ingredient is that in order to construct the group theoretical version, we must be able to regard the symplectic form ω as the curvature form of a line bundle with connection. For this reason we review, in this section, some of the basic ideas concerning vector bundles, connections and curvature. We assume the reader is familiar with the definition of a smooth vector bundle E over a manifold X.

We require that all our vector bundles be locally trivial. That is, each $x \in \mathbf{X}$ must have a nighborhood U such that E_U is trivial, and in fact, that $E_U \sim U \times V$ for some fixed V (independent of U).

Our vector spaces are either over the real or the complex numbers. For each different type of vector space V we get a different type of vector bundle. It is therfore convenient to choose a collection of model vector spaces. For finite dimensional vector bundles we can choose $V = \mathbf{R}^n$ $(n = 0, 1, 2, \ldots)$ or $V = \mathbf{C}^n$ $(n = 0, 1, 2, \ldots)$.

Each of these spaces comes equipped with a standard basis. We let $GL(V)$ denote the group of all non-singular linear transformation of V. Thus, for each of our model spaces, $GL(V)$ can be thought of as the group of $n \times n$ real or complex matrices.

Let $E \to X$ be a vector bundle. A differentiable map $s\colon X \to E$ is called a *section* if $\pi \circ s = \mathrm{id}$; in other words $s(x) \in E_x$ for all x. We can add two sections

$((s_1 + s_2)(x) = s_1(x) + s_2(x))$ and multiply a section by a function $(fs)(x) = f(x)s(x)$. Thus the space of all sections forms a vector space over the real (or complex) numbers, and in fact is a module over the ring of differentiable functions on X.

If E_1 and E_2 are vector bundles over the same space X then we can form the vector bundle $E_1 \oplus E_2$ (direct sum) in the obvious way—thus $(E_1 \oplus E_2)_x = E_{1x} \oplus E_{2x}$. Similarly, we can form $E_1 \otimes E_2$. We get corresponding operations on sections. For example, if s_1 is a section of E_1 and s_2 is a section of E_2 then $s_1 \otimes s_2$ is a section of $E_1 \otimes E_2$.

If ξ is a vector field on X and s is a section of $E \to X$, we would like to be able to define the derivative of s in the direction ξ. We cannot do this without imposing some additional structure since there is no way of comparing vectors which belong to different spaces E_x. Such additional structure is called a linear connection on E. Thus we want a rule which assigns a linear operator ∇_ξ, on sections, to each vector field ξ so that

$$\nabla_\xi(s_1 + s_2) = \nabla_\xi s_1 + \nabla_\xi s_2,$$

$$\nabla_\xi(fs) = (D_\xi f)s + f\nabla_\xi s,$$

$$\nabla_{\xi_1+\xi_2} s = \nabla_{\xi_1} s + \nabla_{\xi_2} s,$$

and

$$\nabla_{f\xi} s = f\nabla_\xi s$$

hold. The last two equations say that for each fixed s, the map $\xi \to \nabla_\xi s$ is linear over functions. Thus we can think of this map as a section of $E \otimes T^*$ where $T^* = T^*(X)$ is the cotangent bundle. Thus the last two equations say that we have assigned a section, Ds, of $E \otimes T^*$ to each section s of E. The first equation now becomes

$$D(s_1 + s_2) = Ds_1 + Ds_2$$

and the second becomes

$$D(fs) = s \otimes df + fDs. \tag{1.1}$$

Conversely any map $s \to Ds$ which satisfies (1.1) determines a linear connection by setting

$$\nabla_\xi s = (Ds \circ \xi)$$

(where $\circ$ denotes the obvious contraction). Suppose that $E_{|U}$ is isomorphic to the trivial bundle $U \times V$, thus $E_{|U} \xrightarrow{\varphi} U \times V$. If $\delta_1, \ldots, \delta_n$ are the (preferred) basis vectors of V $(= \mathbf{R}^n$ or $\mathbf{C}^n)$ then setting

$$s_i(x) = \varphi^{-1}(\delta_i), \qquad x \in U,$$

gives us n sections such that $s_1(x), \ldots, s_n(x)$ forms a basis of E_x at each x. (Conversely, given any n sections $s_1, \ldots, s_n$ which are linearly independent at each $x \in U$ we get an isomorphism of $E_{|U}$ with $U \times V$ by sending $e = a_1 s_1(x) + \cdots + a_n s_n(x) \in E_x$ to $(x, a_1 \delta_1 + \cdots + a_n \delta_n) \in U \times V$.) We call $\mathbf{s} = (s_1, s_2, \ldots, s_n)$ a *frame* of $E_{|U}$. Thus $\mathbf{s}(x)$ is a basis of E_x. If $\mathbf{s}'$ is a frame of $E_{|W}$ then on $U \cap W$ we can write $s_j' = \sum A_{ij} s_i$ or, for short

$$\mathbf{s}' = \mathbf{s}A$$

where $A = (A_{ij})$ is an $n \times n$ matrix of functions, defined on $U \cap W$. Now, on U,

$$Ds_j = \sum s_i \times \theta_{ij}$$

where the θ_{ij} are linear differential forms, which depend on the choice of frame. We let

$$\theta_{\mathbf{s}} = (\theta_{ij})$$

denote the matrix of forms and we can write the previous equation more compactly as

$$D\mathbf{s} = \mathbf{s} \times \theta_{\mathbf{s}}.$$

The most general section in U can be written as $\mathbf{s} \cdot \mathbf{a}$ where $\mathbf{a}$ is a column vector of functions and we think of $\mathbf{s}$ as a row:

$$\mathbf{a} = \begin{bmatrix} a_1 \\ \vdots \\ a_n \end{bmatrix}, \qquad \mathbf{s} = (s_1, \ldots, s_n)$$

and

$$\mathbf{s} \cdot \mathbf{a} = a_1 s_1 + \cdots + a_n s_n.$$

Then equation (1.1) becomes

$$D(\mathbf{s} \cdot \mathbf{a}) = s_1 \times da_1 + \cdots + s_n \times da_n + a_1 Ds_1 + \cdots + a_n Ds_n$$

or, with the obvious notations,

$$D(\mathbf{s} \cdot \mathbf{a}) = \mathbf{s} \times d\mathbf{a} + D\mathbf{s} \cdot \mathbf{a} = \mathbf{s} \times d\mathbf{a} + \mathbf{s} \times \theta_{\mathbf{s}} \mathbf{a}. \tag{1.2}$$

Applying this to the equation $\mathbf{s}' = \mathbf{s}A$ we see that

$$\begin{aligned} D(\mathbf{s}') = D(\mathbf{s}A) &= \mathbf{s} \times dA + \mathbf{s} \times \theta_{\mathbf{s}} A = \mathbf{s}'A^{-1} \times dA + \mathbf{s}'A^{-1} \times \theta_{\mathbf{s}} A \\ &= \mathbf{s}' \times A^{-1} dA + \mathbf{s}' \times A^{-1} \theta_{\mathbf{s}} A = \mathbf{s}' \times [A^{-1} dA + A^{-1} \theta_{\mathbf{s}} A]. \end{aligned}$$

Thus the transition law for $\theta_{\mathbf{s}}$ is given by

$$\theta_{\mathbf{s}'} = A^{-1}dA + A^{-1}\theta_{\mathbf{s}}A. \tag{1.3}$$

Conversely, suppose we are given $\theta_{\mathbf{s}}$ for every frame **s**, each **s** being defined on an open set U where the $\{U\}$ cover X. If (1.3) holds for every pair **s** and **s**$'$ then it is easy to check that (1.2) defines a connection.

A connection is called *flat* if (about each point) it is possible to choose a frame **s** such that $\theta_{\mathbf{s}} = 0$. According to (1.3), if a connection is flat then, for *any* frame, the connection form θ has the form $A^{-1}dA$. Let us examine necessary conditions for θ to have this form. Notice that

$$0 = dI = d(A^{-1}A) = dA^{-1} \cdot A + A^{-1}dA$$

so that

$$dA^{-1} = -A^{-1} \cdot dA \cdot A^{-1}. \tag{1.4}$$

Now

$$d(A^{-1}dA) = dA^{-1} \wedge dA = -(A^{-1}dA) \wedge (A^{-1}dA).$$

Thus, if $\theta_{\mathbf{s}} = A^{-1}dA$ the expression

$$K(\theta_{\mathbf{s}}) = d\theta_{\mathbf{s}} + \theta_{\mathbf{s}} \wedge \theta_{\mathbf{s}},$$

called the curvature form of $\theta_{\mathbf{s}}$ must vanish. (It turns out that this is also sufficient.) Let us compute $K(\theta_{\mathbf{s}'})$ where $\mathbf{s}' = \mathbf{s}A$. Then, by (1.3),

$$D(\theta_{\mathbf{s}'}) = d[A^{-1}dA] + A^{-1}dA \wedge A^{-1}dA + d(A^{-1}\theta_{\mathbf{s}}A) + A^{-1}\theta_{\mathbf{s}}A \wedge A^{-1}\theta_{\mathbf{s}}A$$
$$+A^{-1}dA \wedge A^{-1}\theta_{\mathbf{s}}A + A^{-1}\theta_{\mathbf{s}}A \wedge A^{-1}dA.$$

The first two terms cancel, as we have already seen. Also

$$d(A^{-1}\theta_{\mathbf{s}}A) = dA^{-1} \wedge \theta_{\mathbf{s}}A + A^{-1}d\theta_{\mathbf{s}}A - A^{-1}\theta_{\mathbf{s}} \wedge dA$$
$$= -A^{-1}dA \wedge A^{-1}\theta_{\mathbf{s}}A + A^{-1}d\theta_{\mathbf{s}}A - A^{-1}\theta_{\mathbf{s}} \wedge dA.$$

The two negative terms in this expression cancel the last two terms in the expression for $K(\theta_{\mathbf{s}'})$. Thus

$$K(\theta_{\mathbf{s}'}) = A^{-1}K(\theta_{\mathbf{s}})A. \tag{1.5}$$

This last equation shows that $K(\theta_{\mathbf{s}})$ defines a section of $\mathrm{Hom}(E, E) \otimes \wedge^2 T^*$ called the curvature of D.

If $F \to X$ is a vector bundle and $f\colon Y \to X$ is a differentiable map then we obtain a bundle $f^{\#}E \to Y$ where $(f^{\#}E)_y = E_{f(y)}$. (In case Y is a submanifold of X we have $f^{\#} E = E_Y$) . If s is a section of E defined over a set U then $f^* s$ is the section over $f^{-1}(U)$ defined by $(f^* s)(y) = s(f(y))$. If $\mathbf{s}$ is a frame over U then $f^* \mathbf{s}$ is the corresponding frame over $f^{-1}(U)$. If D is a connection then $f^{\#} D$ is defined to be the connection on $f^{\#} E$ whose connection forms with respect to $f^* \mathbf{s}$ are given by

$$(f^{\#} \theta)_{f^* \mathbf{s}} = f^* \theta_{\mathbf{s}}.$$

In particular, let us consider the case where $Y = (-1, 1)$ is an interval so that f defines a curve on X. Let t be the standard coordinate on Y. Then $f^{\#} E$ is a vector bundle over $(-1, 1)$ and we can write $\mathbf{r} = f^* \mathbf{s}$ (on some open interval) and

$$f^* \theta_{\mathbf{s}} = B_{\mathbf{r}} \, dt$$

where B is some matrix valued function of t. Then for any section $v = f^* \mathbf{s} \cdot \mathbf{a}$ over Y,

$$Dv = \mathbf{r} \times d\mathbf{a} + \mathbf{a} D\mathbf{r} = \mathbf{r} \times \mathbf{a}' dt + \mathbf{r} \times B \cdot \mathbf{a} dt = \mathbf{r} \times [\mathbf{a}' + B\mathbf{a}] dt.$$

Thus

$$\nabla_{\partial/\partial t} v = (\mathbf{a}' + B \cdot \mathbf{a}) \mathbf{r}.$$

We say that the section v is parallel if $\nabla_{\partial/\partial t} v = 0$.

This reduces to the ordinary differential equation

$$\mathbf{a}' = -B\mathbf{a}$$

where B is a given function of t.

We can solve this linear differential equation to obtain "parallel translation" along the curve. Here

$$\mathbf{a}(t) = A(t) \mathbf{a}(0)$$

where $A(t)$ is the unique solution of the matrix differential equation

$$A'(t) = -B(t) A,$$

$$A(0) = \text{id}.$$

This matrix valued function is sometimes written as

$$A(t) = T\Big(\exp - \int_0^t B(u) du\Big)$$

where the symbol on the right is the "time ordered exponential integral"

$$T\Big(\exp - \int_0^t B(u) du\Big) = \lim_{n \to \infty} \exp - \frac{1}{n} B(t) \exp - \frac{1}{n} B\Big(\frac{n-1}{n} t\Big) \cdots \exp - \frac{1}{n} B\Big(\frac{t}{n}\Big).$$

With this notation we can write the operation of parallel translation along any curve $C\colon [0,t] \to X$ lying in the domain of the frame $\mathbf{s}$ as sending the vector $\mathbf{s}(0)\cdot \mathbf{a}(0) \in E_{c(0)}$ into the vector

$$\mathbf{s}(t)\cdot \left[T \exp - \int_C \theta_{\mathbf{s}}\right]\mathbf{a}(0) \in E_{c(t)}.$$

In the particular case where E is a *line* bundle (i.e., the fibers of E are *one* dimensional) we do not need to worry about the order of multiplication since one by one matrices commute. In this case parallel translation can be written as multiplication by the *number* $\exp - \int_C \theta_{\mathbf{s}}$. If C is a closed curve contained in the domain of definition of s and $C = \partial D$ is the boundary of some two dimensional surface D then, by Stokes' theorem, the total effect of parallel translating around a closed curve consists of multiplication by

$$\exp\left(-\int_D d\theta_s\right) = \exp\left(-\int_D K(\theta_s)\right).$$

We return to the case of a general vector bundle. For any frame $\mathbf{s}$, the matrix $K(\theta_{\mathbf{s}})$ is a matrix of two forms. In exterior multiplication, a two form commutes with any other form. With this in mind we claim that $K(\theta_{\mathbf{s}})$ satisfies the identity (known as Bianchi's identity)

$$dK(\theta_{\mathbf{s}}) - [K(\theta_{\mathbf{s}}), \theta_{\mathbf{s}}] = 0$$

where the bracket denotes the commutator of the matrices, with exterior multiplication of the entries: If $A = (A_{ij})$ and $B = (B_{ki})$ are matrices of two forms then $[A, B] = C$ where

$$C_{i,k} = \sum A_{ij} \wedge B_{jk} - B_{ij} \wedge A_{jk}.$$

Notice that since $K(\theta_{\mathbf{s}})$ is a matrix of *two* forms, the exterior multiplication becomes commutative. To prove Bianchi's identity notice that

$$dK(\theta_{\mathbf{s}}) = dd\theta_{\mathbf{s}} + d(\theta_{\mathbf{s}} \wedge \theta_{\mathbf{s}}) = d\theta_{\mathbf{s}} \wedge \theta_{\mathbf{s}} - \theta_{\mathbf{s}} \wedge d\theta_{\mathbf{s}}$$

while

$$[K(\theta_{\mathbf{s}}), \theta_{\mathbf{s}}] = d\theta_{\mathbf{s}} \wedge \theta_{\mathbf{s}} + \theta_{\mathbf{s}} \wedge \theta_{\mathbf{s}} \wedge \theta_{\mathbf{s}} - \theta_{\mathbf{s}} \wedge d\theta_{\mathbf{s}} - \theta_{\mathbf{s}} \wedge \theta_{\mathbf{s}} \wedge \theta_{\mathbf{s}}.$$

If E_1 and E_2 are vector bundles with connections then we get an obvious connection on $E_1 \oplus E_2$:

$$\nabla_\xi(s_1 \oplus s_2) = \nabla_\xi s_1 + \nabla_\xi s_2$$

and a unique connection on $E_1 \otimes E_2$ satisfying

$$\nabla_\xi(s_1 \otimes s_2) = \nabla_\xi s_1 \otimes s_2 + s_1 \otimes \nabla_\xi s_2.$$

It is easy to check that we do indeed get a connection in this way.

Let $E \to X$ and $H \to Y$ be vector bundles. A morphism (f, r) from E to H is a pair consisting of a differentiable map, $f\colon X \to Y$ and a linear map $r(x)\colon H_{f(x)} \to E_x$ for each $x \in X$, such that $r(x)$ depends smoothly on X. If s is a section of H then f^*s is the section of E defined by

$$f^*s(x) = r(x)s(f(x)).$$

If $f\colon X \to Y$ is a differentiable map then we get a bundle morphism $T^*(f)$: $T^*(X) \to T^*(Y)$ defined by $T^*(f) = (f, df^*)$, where $df_x^*\colon T^*Y_{f(x)} \to T^*X_x$ is the transpose of the linear map $df_x\colon TX_x \to TY_{f(x)}$. If $E \to X$ and $H \to Y$ are vector bundles with connections then we say that f is a morphism of vector bundles with connection if the diagram

$$\begin{array}{ccc} C^\infty(E) & \xrightarrow{(f,r)} & C^\infty(H) \\ D\downarrow & & D\downarrow \\ C^\infty(E \otimes T^*X) & \longrightarrow & C^\infty(H \otimes T^*Y) \end{array}$$

commutes. Put another way, this says that for any $\xi \in TX_x$ and any section s of H,

$$D_\xi f^*s = r(x)(D_{df_x\xi}s)(x).$$

Let $E \to X$ be a vector bundle and $f\colon E \to E$ be a diffeomorphism such that f maps fibers into fibers and $f\colon E_x \to E_{\bar{f}(x)}$ is a linear isomorphism. Then f determines a vector bundle morphism $\mathbf{f} = (\bar{f}, r)$ where $r(x) = f(x)^{-1}$ and $\bar{f}(x) = \pi f(e)$ where $e \in E_x$. In this case f is called a vector bundle *automorphism* of E. If E has a connection and $\mathbf{f}$ is a morphism of vector bundles with connection then we say that f is an automorphism of E, considered as a vector bundle with connection. Notice that if f is an automorphism of E then f induces an automorphism of $\mathrm{Hom}(E, E)$ and also of $\bigwedge^2 T^*X \otimes \mathrm{Hom}(E, E)$, which we shall also denote by f. If f preserves a connection on E then clearly

$$f^*K = K,$$

where K is the curvature, regarded as a section of $\mathrm{Hom}(E, E) \otimes \bigwedge^2 T^*X$. In particular, if E is a *line* bundle, then K can be regarded as a closed two form, which is then preserved by f.

If E is a real or complex vector bundle such that E_x has a scalar product varying smoothly with x, then we can demand that our connection preserve the scalar product in the sense that

$$\nabla_\xi(s_1, s_2) = (\nabla_\xi s_1, s_2) + (s_1, \nabla_\xi s_2).$$

This is easily seen to be the same as demanding that for any orthonormal frame $\mathbf{s}$ the matrix $\theta_{\mathbf{s}}$ be orthogonal (resp. unitary) in the sense that $\langle \xi\theta_{\mathbf{s}} \rangle \in O(n)$ or

$U(n)$, for any vector field ξ. This, in turn, is the same as requiring that parallel translation preserve the scalar product. We can then talk, for example, of automorphisms of E which preserve both the scalar product and the connection. Thus, for example, let $L \to X$ be a Hermitian line bundle with connection. If $\mathbf{s}$ is an orthonormal frame, i.e., a section with $(\mathbf{s}, \mathbf{s}) \equiv 1$, then the connection form $\theta_{\mathbf{s}}$ is purely imaginary. We can write $\theta_{\mathbf{s}} = i\alpha_{\mathbf{s}}$ where $\alpha_{\mathbf{s}}$ is a real form and $K(\theta_{\mathbf{s}}) = i\omega$ where $\omega = d\alpha_{\mathbf{s}}$ is independent of $\mathbf{s}$. Here ω is a real closed two form and any automorphism of L induces a diffeomorphism of X preserving ω.

The principal case of interest to us will be the situation where this closed form ω is non-degenerate, and thus determines the structure of a symplectic manifold on X. Here every automorphism of L as a line bundle with connection gives rise to a canonical transformation of X. We shall study this point in detail in the next section.

Let E be a vector bundle. For each $x \in X$ we can consider the set of all bases, i.e., frames, of E_x. If $\mathbf{e} = (e_1, \ldots, e_n)$ is such a frame then so is

$$\mathbf{e}A = (\textstyle\sum A_{i1} e_i, \ldots, \sum A_{in} e_i),$$

for any non-singular matrix $A = (A_{ij})$. The group $GL(n)$ acts transitively and freely on the space of all frames at x. If $B(E_x)$ denotes the space of all frames at x then we let $B(E)$ denote the set of all bases at all x. It forms a manifold in the obvious way and $GL(n)$ acts to the right on it in a free manner. It is called the bundle of frames of E. We shall need to use this bundle in §4 for the definition of half forms and similar objects.

§2. The group of automorphisms of an Hermitian line bundle.

Let $E \to X$ and $H \to Y$ be vector bundles with connection and let $\mathbf{f}$: $E \to H$ be a morphism of vector bundles with connection. Suppose that $\mathbf{f} = (f, r)$ where $r(x)$ is an isomorphism for all x. Let C_1: $[0, 1] \to X$ be a curve on X and $C_2 = f \circ C_1$ be the image curve on Y. Then $\mathbf{f}$ takes parallel translation along C_1 onto parallel translation along C_2. More precisely, if $h \in H_{C_2(0)}$ and $h(t) \in H_{C_2(t)}$ is the image of h by parallel translation along C_2, then

$$e(t) = r(C_1(t))h(t) \in E_{C_1(t)} \tag{2.1}$$

is parallel along $C_1(t)$. This shows that the map $r(C_1(t))$ is determined by the map $r(C_1(0))$. If X is connected, this means that $r(x)$ is determined at all x by its value at any single point, x_0. Thus $\mathbf{f}$ is determined by f up to an element of $GL(E_{x_0})$. Suppose we start with f and $r(x_0) \in \mathrm{Hom}(H_{f(x_0)}, E_{x_0})$ and try to construct $\mathbf{f}$. For any $x \in X$ we would join x_0 to x by a curve C_1 and then use (2.1) to define $r(x)$. For this definition to be consistent, we need to know that the $r(x)$ so defined is independent of the choice of curve joining x_0 to x. What amounts to the same thing, we need to know that this procedure, when applied to a closed curve at x_0 yields the identity. Let us formulate this condition a little more clearly. Let γ be a closed curve at $x_0 \in X$. Then parallel translation around γ gives an element ϕ_γ of Hom(V, V) where $V = E_{x_0}$. The map f carries γ into the closed curve $f \circ \gamma$ at $y_0 = f(x_0)$. Parallel translation around any closed curve β at y_0 gives a $\phi_\beta \in$ Hom(W, W) where $W = H_{y_0}$. We wish to know that there exists a linear map r_0: $W \to V$ such that

$$\phi_\gamma \circ r_0 = r_0 \circ \phi_{f \circ \gamma} \tag{2.2}$$

for all closed curves γ. If such an r_0 exists then (2.1) clearly defines an $r(x)$ for each $x \in X$ with $r(x_0) = r_0$. It is not hard to verify that this indeed defines an $\mathbf{f} = (f, r)$, i.e., that the r so defined is smooth.

If E and H are line bundles, then the situation is somewhat simpler, in that Hom(V, V) and Hom(W, W) are canonically isomorphic to the scalars, and (2.2) becomes

$$\phi_\gamma = \phi_{f \circ \gamma}. \tag{2.3}$$

If γ is completely contained in the domain of a non-zero section $\mathbf{s}$, then ϕ_γ is given as scalar multiplication by

$$\exp \int_\gamma \theta_{\mathbf{s}}.$$

If, in addition, γ is the boundary of a surface σ then, by Stokes' theorem, ϕ_γ is multiplication by

$$\exp \int_\sigma K$$

where K is the curvature. Notice that if we drop the restriction that γ be contained in a domain of definition of a single $\mathbf{s}$, but retain the requirement that $\gamma = \partial\sigma$, this last formula is still valid. This can be easily seen by subdividing σ into small pieces, the interior line integrals cancelling each other out. In particular, if X is simply connected, condition (2.3) can be written as

$$\exp \int_\sigma K_E = \exp \int_{f\sigma} K_H$$

where K_E and K_H are the curvature forms on E and H. This is to hold for all surfaces σ and so (2.3) becomes

$$f^* K_H = K_E. \tag{2.4}$$

Let us apply these considerations to the case where $X = Y$ and $E = H = L$ is a Hermitian line bundle with curvature $K = i\omega$. We suppose that ω is a non-singular exterior two form on X. Thus X is a symplectic manifold, and we let H denote the group of symplectic diffeomorphisms of X, i.e., the group of all diffeomorphisms $f: X \to X$ such that $f^* \omega = \omega$. We let P denote the group of all automorphisms of L as a Hermitian line bundle with connection. If $\mathbf{f}_1 \in P$ and $\mathbf{f}_2 \in P$ determine the same map $f \in H$ then $f_1 = cf_2$ where c is multiplication by a complex scalar of absolute value one. We thus have an exact sequence of groups

$$0 \to T^1 \to P \to H$$

where T^1 is the group of complex numbers of absolute value one. The above considerations show:

PROPOSITION 2.1. *If X is connected and simply connected, the sequence of groups*

$$0 \to T^1 \to P \to H \to 1 \tag{2.5}$$

is exact.

If X is not simply connected then the group of transformations satisfying (2.3) *may be a proper subgroup, call it H_L of H. All that we can say in this case is that*

$$0 \to T^1 \to P \to H_L \to 1$$

is exact.

Suppose we start with a symplectic manifold X. When does its two form, ω, arise from a Hermitian line bundle L with connection? If $\omega = d\alpha$ (as in the case of a cotangent bundle) then we can use α to define a connection globally: Simply take $L = X \times \mathbf{C}$ as the trivial bundle, where $\mathbf{C}$ has its usual Hermitian structure, and let $\theta = i\alpha$ be the connection form corresponding to the section 1. To say that $\omega = d\alpha$ is the same as saying that the cohomology class of ω, which we denote by $[\omega]$, vanishes. More generally, we would like to choose a bundle with connection such that parallel transport around any closed curve bounding a surface σ should be given by multiplication by $\exp i \int_\sigma \omega$. If σ_1 and σ_2 are surfaces with the same boundary then we must have

$$\int_{\sigma_1} \omega = \int_{\sigma_2} \omega + 2k\pi.$$

More precisely it turns out that the proper requirement is that

$$\frac{1}{2\pi}[\omega]$$

be an integral cohomology class. For details, we refer the reader to [1].

Let $L \to X$ be a Hermitian line bundle with connection giving a symplectic structure to X. The Lie algebra of the group H is the algebra h_L of all locally Hamiltonian vector fields, i.e., all vector fields satisfying

$$D_\xi \omega = 0, \qquad \text{i.e.,} \qquad d(\xi \rfloor \omega) = 0.$$

The set of ξ satisfying

$$\xi \rfloor \omega = -dg, \qquad g \text{ a smooth function on } X, \tag{2.6}$$

is an ideal, h, in h_L which we call the algebra of Hamiltonian vector fields, or the Hamiltonian algebra. In case X is simply connected clearly $h = h_L$.

In any event, to each smooth function g on X we get a unique vector field, ξ_g, defined by (2.6). The set of all functions becomes a Lie algebra under the Poisson bracket defined by

$$\{g, h\} = \xi_g h.$$

We recall from Chapter IV, §8 that

$$\{g, h\} = \xi_g h = \langle \xi_g, \xi_h \rfloor \omega \rangle$$

is antisymmetric in g and h. Also

$$d(\xi_g h) = dD_{\xi_g} h = D_{\xi_g} dh = -D_{\xi_g}(\xi_h \rfloor \omega) = -[\xi_g, \xi_h] \rfloor \omega$$

shows that

$$[\xi_g, \xi_h] = \xi_{\{g,h\}}. \tag{2.7}$$

Then

$$\begin{aligned} \{f, \{g, h\}\} = \xi_f \{g, h\} &= D_{\xi_f} \langle \xi_g \wedge \xi_h, \omega \rangle \\ &= \langle [\xi_f, \xi_g] \wedge \xi_h, \omega \rangle + \langle \xi_g \wedge [\xi_f, \xi_h], \omega \rangle \\ &= \{\{f, g\}, h\} + \{g, \{f, h\}\} \end{aligned}$$

so that Jacobi's identity holds. We let p denote the algebra of functions under Poisson bracket and we get an exact sequence of Lie algebras

$$0 \to \mathrm{R} \to p \to h \to 0. \tag{2.8}$$

We claim that p can be identified with the Lie algebra of P, so that (2.8) is the infinitesimal version of (2.5). Indeed, let f_t be a one parameter family of automorphisms of L which preserves the connection and the Hermitian structure of L. Then the $f_t^* \mathbf{s} = (e^{i\psi_t})\mathbf{s}$ for any section $\mathbf{s}$ of unit length and

$$f_t^*(D\mathbf{s}) = Df_t^* \mathbf{s}$$

so

$$f_t^*\theta_{\mathbf{s}} \cdot f_t^*\mathbf{s} = \theta_{f_t^*\mathbf{s}} f_t^*\mathbf{s}, \qquad \text{i.e.,} \qquad f_t^*\theta_{\mathbf{s}} = \theta_{f_t^*\mathbf{s}} = i d\psi_t + \theta_{\mathbf{s}}.$$

If ξ denotes the infinitesimal generator of f_t this last equation implies that

$$D_\xi \theta_{\mathbf{s}} = i d\dot{\psi}$$

where $\dot{\psi} = (d\psi/dt)_{t=0}$. Now $D_\xi\theta_{\mathbf{s}} = \xi \lrcorner d\theta_x + d(\xi\lrcorner\theta_{\mathbf{s}})$ and $d\theta_{\mathbf{s}} = i\omega$ so

$$\xi\lrcorner\omega = d(\dot{\psi} + i\langle \xi, \theta_{\mathbf{s}}\rangle) = d(\dot{\psi} - \langle \xi, \alpha_{\mathbf{s}}\rangle).$$

Notice that the function

$$\phi_\xi = \dot{\psi} - \langle \xi, \alpha_{\mathbf{s}}\rangle$$

does not depend on the choice of section $\mathbf{s}$. Indeed, replacing $\mathbf{s}$ by $\mathbf{s}' = e^{iu}\mathbf{s}$ has the effect of replacing $\alpha_{\mathbf{s}}$ by

$$\alpha_{\mathbf{s}'} = du + \alpha_{\mathbf{s}}$$

while

$$f_t^*(e^{iu}\mathbf{s}) = f_t^*(e^{iu}) f_t^*\mathbf{s} = e^{i(f_t^* u - u)} e^{i\psi_t}(e^{iu}\mathbf{s})$$

so that ψ is replaced by $\psi + \langle \xi, du\rangle$.

Thus we get a linear map, ρ, sending ξ into ϕ_ξ, from the space of infinitesimal automorphisms of L into the space of all functions on X. The map ρ is clearly injective. Let g be an automorphism of L and f_t a one parameter family of automorphisms. If

$$g^* s = e^{iv} s$$

then

$$(g^{-1} f_t g)^* \mathbf{s} = e^{i[v + g^*\psi_t - g^* f_t^* g^{-1*} v]}.$$

Computing the derivative at $t = 0$ shows that the ψ component becomes

$$g^*\dot{\psi} - g^*\langle \xi, d(g^{-1*}v)\rangle = g^*\dot{\psi} - \langle g^*\xi, dv\rangle.$$

On the other hand ξ is replaced by $g^*\xi$ so that the function corresponding to the infinitesimal generator of $g^{-1} f_t g$ is

$$\begin{aligned} g^*\dot{\psi} - \langle g^*\xi, dv\rangle - \langle g^*\xi, \alpha_{\mathbf{s}}\rangle &= g^*\dot{\psi} - \langle g^*\xi, \alpha_{\mathbf{s}} + dv\rangle \\ &= g^*\dot{\psi} - \langle g^*\xi, \alpha_{g^*\mathbf{s}}\rangle \\ &= g^*\dot{\psi} - \langle g^*\xi, g^*\alpha_{\mathbf{s}}\rangle \\ &= g^*(\dot{\psi} - \langle \xi, \alpha_{\mathbf{s}}\rangle). \end{aligned}$$

Thus, if $\text{Ad}\, g$ denotes the induced action on the space of infinitesimal automorphisms of L we see that

$$\rho(\text{Ad}\, g^{-1}) = g^* \rho. \tag{2.9}$$

This implies that ρ is a Lie algebra isomorphism. Notice that the constant function c corresponds to the infinitesimal generator of multiplication by constant e^{itc}. The proof that ρ is surjective is also straightforward;we refer the reader to [1] for the proof in the general case. If X is simply connected, then any one parameter group of symplectic automorphisms of X can be lifted to P. Since we can always adjust the additive constant this proves that ρ is surjective. Thus the sequence (2.8) is indeed the infinitesimal version of (2.5).

Notice that we have also produced a distinguished representation of the Poisson algebra, P. It acts as infinitesimal automorphisms of the Hermitian line bundle L with connection, and hence on the vector space of all sections of this bundle.

Suppose that G is a Lie group and that X is an orbit of G in the dual of the Lie algebra, g, of G. Then each $\xi \in g$ determines a function on all of g^* and hence, by restriction, to X, and this gives a representation of g by Hamiltonian vector fields on X. Thus we have a Hamiltonian action of G on X. The condition that the class $(1/2\pi)[\omega]$ be integral then imposes conditions on X. For example, we suppose we take the group $SO(3)$. The dual of the Lie algebra is then a Euclidean three space and the orbits are three spheres. The symplectic form is just $\omega = (1/r)\sigma_r$ where σ_r is the area form on the sphere of radius r. The condition is thus that the $(1/r) \times$ (total area of the sphere) be of the form $2\pi n$, or, in other words, that r be a half integer. Similarly, if we examine the symplectic homogeneous spaces for the Lorentz group, and we look at the orbits with positive mass, we find that the integrality constraint imposes no condition on m, but restricts s to half integer values.

More generally, there is a useful criterion due to Kostant [1, Theorem 5.7.1] which tells when an orbit $X \subset g^*$ satisfies the integrality condition, provided that G is simply connected. It is the following: Let θ be an element of g^*, and let $H_{d\theta}$ be its isotropy group and $h_{d\theta}$ its isotropy algebra so that $h_{d\theta}$ is the Lie algebra of $H_{d\theta}$. Notice that θ defines (by restriction) a linear function on $h = h_{d\theta}$, and θ vanishes on $[h, h]$. Then the orbit through θ satisfies the integrality condition if and only if the linear function $2\pi i\theta$ exponentiates to a well defined function on $H_{d\theta}$.

If we apply this to the group $SU(2)$ (the simply connected covering group of $SO(3)$) we get precisely the condition $r = n/2$ since the group $H_{d\theta}$ is, in this case, a double covering of the circle. Similarly, applied to $SL(2, \mathbf{C}) \times \mathbf{R}^{1,3}$, which is the covering group of the Poincaré group, we obtain the restriction $s = n/2$ on the spin.

We can apply similar considerations to the study of the hydrogen atom, for as we shall see, the integrality condition actually gives the quantization of the energy levels. Before going into the details, we describe the method. Let Z_E be an energy surface of some Hamiltonian, H, on a symplectic manifold, X, and assume that the trajectories fiber Z_E over some manifold Y_E. Then, as we have seen in Chapter IV, §6, Y_E is a symplectic manifold. Let $\omega_{|Z_E}$ denote the restriction of the form ω to Z_E and let ω_E denote the symplectic form on Y_E. These two forms are related by

$$\omega_{|Z_E} = \pi^* \omega_E$$

where π denotes the projection of Z_E onto Y_E.

The integrality condition on ω_E is thus a condition on the "energy" E. In case the trajectories of H on Z_E are all periodic, and $\omega = d\alpha$, we have already seen that the integral $\int \alpha$ around a trajectory is independent of the trajectory. From the fact that $d\alpha_{Z_E} = \pi^* \omega_E$ it follows from topological considerations[1] that the class ω_E satisfies the integrality condition if and only if $(1/2\pi) \int \alpha$ is an integer; cf. [15].

For the case of the hydrogen atom (the Kepler problem in three dimensions) we now explicitly work out these conditions. We follow the presentation of Simms [6] (and, in particular, rederive some of the results of Chapter IV, §6, using notation prevalent in the physical literature). Let $p = (p_1, p_2, p_3)$ and $q = (q_1, q_2, q_3)$ with $q \neq 0$. The symplectic manifold, X, consists of all such pairs (p, q) with $\omega = \sum dp_i \wedge dq_i$. The Hamiltonian, H, is given by

$$H = \frac{1}{2m}\|p\|^2 - \frac{K}{\|q\|},$$

where m and K are constants. We introduce the angular momentum vector

$$(\ell_1, \ell_2, \ell_3) = \ell = q \times p$$

where $\times$ denotes the vector product in $\mathbf{R}^3$ and the Runge-Lenz vector

$$(a_1, a_2, a_3) = a = \ell \times p + mK\frac{q}{\|q\|}.$$

Notice that

$$a \cdot \ell = 0 \qquad \text{and} \qquad \{H, \ell\} = \{H, a\} = 0$$

while we have bracket relations like

$$\{\ell_1, \ell_2\} = -\ell_3, \qquad \{\ell_1, a_2\} = -a_3, \qquad \{a_1, a_2\} = 2mH\ell_3$$

[1] The equation $d\alpha_{Z_E} = \pi^* \omega_E$ says that the cohomology class $[\alpha]$ in the fiber is transgressive and that $\delta_{2,-1}[\alpha] = [\omega_E]$ in the Serre spectral sequence. If this is true in real cohomology and $[\alpha]$ is integral so will $[\omega_E]$ be.

among the components of a and ℓ.

For some $E < 0$ let Z_E be the energy hypersurface and let Y_E be the corresponding manifold of orbits. Let $\rho = \sqrt{-2mE}$ and set $x = \rho\ell + a$ and $y = \rho\ell - a$. A direct computation shows that $\|x\|^2 = \|y\|^2 = m^2K^2$ so that the map (x, y) sends Y_E onto the product, $S^2 \times S^2$, of two spheres of radius mK.

We can compute the Poisson brackets of the components of x and y on Y_E by computing the corresponding Poisson brackets for $\rho\ell + a$ and $\rho\ell - a$ on X (with ρ constant). Thus we get

$$\{x_1, x_2\} = -2\rho x_3, \qquad \{x_i, y_j\} = 0, \qquad \{y_1, y_2\} = -2\rho y_3,$$

etc. From this it follows that

$$\xi_{x_1} = -2\rho x_3 \frac{\partial}{\partial x_2}, \qquad \xi_{x_2} = 2\rho x_3 \frac{\partial}{\partial x_1},$$

$$\xi_{y_1} = -2\rho y_3 \frac{\partial}{\partial y_2}, \qquad \xi_{y_2} = 2\rho y_3 \frac{\partial}{\partial y_1}.$$

This means that if we use x_1, x_2, y_1, y_2 as local coordinates, the form ω_E is given by

$$\omega_E = \frac{dx_1 \wedge dx_2}{2\rho x_3} + \frac{dy_1 \wedge dy_2}{2\rho y_3}.$$

Now the area element of a sphere of radius mK in x space is

$$mK \frac{dx_1 \wedge dx_2}{x_3}$$

in terms of x_1, x_2 coordinates. Thus the integral of ω_E over each sphere is

$$\frac{1}{2\rho mK} \cdot 4\pi(mK)^2 = 2\pi mK/\rho$$

and this must be an integer. Since $\rho = \sqrt{-2mE}$ we obtain the quantum restrictions

$$E = \frac{-2\pi^2 mK}{n^2}, \qquad n \text{ an integer.}$$

§3. Polarizations.

We return to the study of general symplectic manifolds. Let X be a symplectic manifold with two form, ω. A (real) *polarization* of X consists of smooth assignment of a Lagrangian subspace of TX_x to each $x \in X$ in such a manner that the corresponding assignment of subspace is integrable. We shall also admit complex polarizations: Let $T_{\mathbf{C}}X$ denote the complexified tangent bundle. Then $T_{\mathbf{C}}X_x = TX_x \otimes \mathbf{C}$ is a complex symplectic vector space and it makes sense to talk of a complex Lagrangian subspace. Then a polarization, $\mathcal{F}$, of X is a smooth assignment of a (complex) Lagrangian subspace $\mathcal{F}_x$, to each $x \in X$ which is integrable in the sense that if $\boldsymbol{\xi}$ and $\boldsymbol{\eta}$ are complex vector fields with $\boldsymbol{\xi}_x \in \mathcal{F}_x$ and $\boldsymbol{\eta}_x \in \mathcal{F}_x$ for all x then $[\boldsymbol{\xi}_x, \boldsymbol{\eta}_x] \in \mathcal{F}_x$.

If $\boldsymbol{\xi} \to \bar{\boldsymbol{\xi}}$ denotes the operation of complex conjugation then $\mathcal{F}$ is called real if $\mathcal{F}_x = \bar{\mathcal{F}}_x$. At the opposite extreme we can have the situation where $\mathcal{F}_x \cap \bar{\mathcal{F}}_x = \{0\}$. For the time being we would like to investigate real polarizations. In this case $\mathcal{F}_x = \mathcal{G}_x \otimes \mathbf{C}$ where $\mathcal{G}_x$ is a real Lagrangian subspace of TX_x. The integrability means that $\mathcal{G}$ determines a foliation of X by Lagrangian submanifolds. The most obvious example of such a polarization is given by taking $X = T^* M$ and the foliation induced by the projection π: $X \to M$. In this case, the zero section is a Lagrangian submanifold of X which is transverse to the foliation. Locally, this is the only such example. Indeed we have

PROPOSITION 3.1 (KOSTANT-WEINSTEIN). *Suppose that $\mathcal{G}$ is a real polarization of X which is transverse to a Lagrangian submanifold M. Then there is a symplectic diffeomorphism, f, of some neighborhood of M in X onto some neighborhood of the zero section of $T^* M$ carrying the leaves of $\mathcal{G}$ onto the fibers of $T^* M$.*

In fact, if we look at the proof of Proposition 1.1 of Chapter IV, we see that the polarization $\mathcal{G}$ picks out the transverse Lagrangian sub-bundle. Thus we can apply Proposition 1.1 to conclude that there is a symplectic diffeomorphism near M onto a neighborhood of the zero section in $T^* M$ and which takes $\mathcal{G}$ into a foliation which is is tangent to the cotangent foliation. We are thus reduced to the situation where $X = T^* M$ where M is the zero section of $T^* M$ and where $\mathcal{G}$ is tangent to the cotangent foliation. Let h be a diffeomorphism carrying the leaves of $\mathcal{G}$ onto those of the cotangent foliation and such that $dh = \text{id}$ along M. Such an h certainly exists, but we have not specified that h preserves ω. Let $\omega_1 = h^* \omega$. We will now carefully choose the g in Theorem 1.1 of Chapter IV so that $g^* \omega_1 = \omega$ and so that g preserves the cotangent foliation. If we then set $f = h \circ g$ we will be done. Now choose the φ_t in the proof of Theorem 1.1 of Chapter IV to be simply multiplication by t. Then to show that $f = f_1$ preserves the fibers we need only show that $\boldsymbol{\eta}_t$ is tangent to the fibers. Now for any $\boldsymbol{\xi}$ tangent to the fiber we have

$$\langle \boldsymbol{\eta}_t \wedge \zeta, \omega_t \rangle = \langle \zeta, \beta_t \rangle = \int_0^1 \langle \xi_t \wedge \zeta, \varphi_t^* (\omega_1 - \omega) \rangle \, dt = 0$$

since the fibers are Lagrangian for ω_1 and ω. Since the fibers are then Lagrangian for ω_t we conclude that $\boldsymbol{\eta}_t$ must be tangent to the fibers, proving the proposition.

Now the symplectic manifold $T^* M$ carries a one form, α, and not merely the two form $\omega = d\alpha$. Notice that the form α can be characterized as follows:

(i) $d\alpha = \omega$,

(ii) $\langle \eta, \alpha \rangle = 0$ if η is tangent to the cotangent foliation, i.e., $d\pi\eta = 0$ and

(iii) $\alpha_{|z_M} = 0$ where z_M is the zero section of $T^* M$.

Indeed if α' were a second such form, then (i) implies that, locally, $\alpha - \alpha' = d\psi$ for some function ψ. From (ii) we conclude that ψ is constant along the fibers and from (iii) that ψ is constant along the base so that $d\psi = 0$.

If now $\mathcal{G}$ is any real polarization of a symplectic manifold X which is transversal to a Lagrangian submanifold M then locally, near M, there is uniquely defined one form β satisfying

(i) $d\beta = \omega$,

(ii) $\langle \eta, \beta \rangle = 0$ for η lying in $\mathcal{G}$ and

(iii) $\beta_{|M} = 0$.

(The form β need not by globally defined. For example, consider the (two dimensional) cylinder with $\mathcal{G}$ given by the circles. Then if $\omega = d\theta \wedge dh$ where h is the parameter along the axis and θ is the "parameter" along the circle, and where M is given by $\theta = 0$, then the only candidate for β is θdh which is not globally defined on the cylinder.)

Suppose that we start with a one form β such that $d\beta = \omega$. Now we would expect, generally, that the set of x where $\beta_x = 0$ should consist of isolated points. After all, β is a section of $T^* X$ and we are looking for the intersection of this section with the zero section, i.e., the intersection of two $2n$-dimensional submanifolds of the $4n$-dimensional manifold $T^* X$. What is peculiar about the β arising from a polarization is the fact that its set of zeroes consists of an entire n dimensional submanifold instead of isolated points. We shall now show that such a β completely determines the polarization. The following result was obtained jointly with B. Kostant.

PROPOSITION 3.2. *Let β be a one form on X such that $d\beta = \omega$ and such that the set $\{x \mid \beta_x = 0\}$ is an n dimensional submanifold, M. Then M is Lagrangian and there is a unique polarization defined on X near M which is transversal to M and whose associated one form is β. We say that M is the base associated to β.*

Let ξ be the vector field determined by β so that

$$\xi \lrcorner \omega = \beta$$

and therefore

$$D_\xi \omega = d(\xi \lrcorner \omega) = d\beta = \omega. \tag{3.1}$$

We see that if f_t is the flow generated by ξ then

$$f_t^* \omega = e^t \omega. \tag{3.2}$$

Let x be a zero of β, i.e., of ξ. Then ξ determines a linear transformation A_x on TX_x given by

$$A_x \eta_x = [\xi, \eta]_x$$

where η is any vector field. (The right hand side depends only on the value η_x since ξ vanishes at x. In local coordinates A_x is given by the Jacobian matrix of ξ at x.) It follows that $A_x \in csp$: For any $\eta, \zeta \in TX_x$ we have

$$(A_x \eta, \zeta)_x + (\eta, A_x \zeta)_x = (\eta, \zeta)_x \tag{3.3}$$

where we have set

$$(\eta, \zeta)_x = \langle \eta \wedge \zeta, \omega_x \rangle.$$

Now if η is a vector field tangent to M we clearly have $[\xi, \eta] = 0$. Thus $A_x \eta = 0$ for all η tangent to M. From (3.3) it now follows that $(\eta, \zeta)_x = 0$ for any pair of vectors tangent to M. Since M is n dimensional we conclude that M is Lagrangian. Furthermore, by Proposition 2.1 of Chapter IV, we know that A_x has an n dimensional space of eigenvectors corresponding to the eigenvalue 1. Let $\mathcal{G}_x$ be this subspace of TX_x. We now make use of a theorem on the behavior of ordinary differential equations near a singular point (see, for example, Sternberg [8] or Hartman [9]) which asserts that there passes through x a unique manifold tangent to $\mathcal{G}_x$ and invariant under f_t. For a vector, η, tangent to this invariant manifold we have $|df_{-t}\eta| = O(e^{-t})$. In particular for η and ζ both tangent to the invariant manifold we have

$$\langle df_{-t}\eta \wedge df_{-t}\zeta, \omega \rangle = O(e^{-2t}).$$

On the other hand, by (3.2),

$$\langle df_{-t}\eta \wedge df_{-t}\zeta, \omega \rangle = e^{-t} \langle \eta \wedge \zeta, \omega \rangle$$

from which we conclude that the invariant manifold is Lagrangian. Since we get an invariant manifold passing through every point of M, we conclude that we have a polarization of X near M transversal to M. Now each leaf is Lagrangian and is tangent to ξ. Therefore, if ζ is any other tangent to the leaf, we have

$$\begin{aligned} 0 &= \langle \xi \wedge \zeta, \omega \rangle \\ &= \langle \zeta, \xi \lrcorner \omega \rangle \\ &= \langle \zeta, \beta \rangle. \end{aligned}$$

Thus β vanishes on tangents to the polarization, i.e., satisfies (ii). By assumption it satisfies (i) and (iii) and hence is the one form given by the polarization. This completes the proof of the proposition.

We are now in a position to appreciate the role of the special type of parametrization of Lagrangian manifolds coming from the cotangent fibration as described in the beginning of §5 of Chapter IV. Indeed, let α be the standard one form on T^*M, and let Λ be a Lagrangian submanifold. Let φ be a function on T^*M. Suppose that $\Lambda = \pi_*(\text{graph } d\varphi)$. In terms of local coordinates x, ξ on T^*M this means that

$$\frac{\partial \varphi}{\partial \xi} = 0 \qquad \text{and} \qquad \frac{\partial \varphi}{\partial x} = \xi_0$$

at any point $(x_0, \xi_0) \in \Lambda$. We can reformulate this as

$$\alpha - d\varphi = 0 \qquad \text{on } \Lambda.$$

Now $d(\alpha - d\varphi) = d\alpha = \omega$. Thus giving such a φ picks out a polarization transversal to Λ of T^*M. Thus the parametrizations coming from T^*M correspond (near Λ) to polarizations of T^*M transversal to Λ. On the other hand, not every real polarization corresponds to a parametrization by a φ as described in Chapter IV, §4 in that the transversality requirement, that graph $d\varphi$ intersect $\mathcal{H}$ transversally need not be satisfied. Indeed, we claim that the intersection is transversal precisely when the polarization induced by φ is transversal to the vertical polarization. Indeed, let (x, ξ) be local coordinates about the point in question. Then

$$\alpha - d\varphi = \left(\xi - \frac{\partial \varphi}{\partial x}\right) dx - \frac{\partial \varphi}{\partial \xi} d\xi$$

and so the associated vector field is

$$\frac{\partial \varphi}{\partial \xi} \frac{\partial}{\partial x} + \left(\xi - \frac{\partial \varphi}{\partial x}\right) \frac{\partial}{\partial \xi}$$

and the associated transformation, A, has the matrix

$$\begin{bmatrix} \dfrac{\partial^2 \varphi}{\partial \xi^i \partial x^j} & \dfrac{\partial^2 \varphi}{\partial \xi^i \partial \xi^j} \\ \dfrac{-\partial^2 \varphi}{\partial x^j \partial x^i} & \delta_{ij} - \dfrac{\partial^2 \varphi}{\partial x^i \partial \xi^j} \end{bmatrix}.$$

Now to say that the polarization intersects the vertical (at the point in question) means that there is some vector $v = \sum b_j \partial/\partial \xi_j$ such that $Av = v$. This means that

$$\sum b_j \frac{\partial^2 \varphi}{\partial \xi^i \partial \xi^j} = 0 \quad \text{and} \quad \sum b_j \frac{\partial^2 \varphi}{\partial x^i \partial \xi^j} = 0$$

or that

$$\sum b_j d\left(\frac{\partial \varphi}{\partial \xi^j}\right) = 0,$$

i.e., that the differentials $d(\partial\varphi/\partial\xi^j)$ are not linearly independent, in other words that the intersection of graph $d\varphi$ with $\mathcal{H}$ is not transversal.

Given any $\Lambda \subset T^*M$ we can always find a real polarization of T^*M defined near Λ and a form β with Λ the associated base. However, we will not, in general, be able to find a polarization which is also transversal to the cotangent foliation. Indeed, the Maslov class provides a first obstruction to finding such a polarization.

As an example of an important real polarization consider the case where M is a vector space with linear coordinates x and where ξ are the dual coordinates on M^*. Then $T^*M \sim M + M^*$ and consider the function $\varphi(x, \xi) = \xi \cdot x$. Then $\beta = \xi \cdot dx - d(\xi \cdot x) = -x \cdot d\xi$ so that the polarization consists of the "horizontal" planes $\xi =$ constant and the base is the cotangent space at the origin, $\{(0, \xi)\}$. We shall see later that with each polarization there is an associated Hilbert space and that associated with certain pairs of polarizations there is a transform between the Hilbert spaces. In the case where one polarization is the vertical and the second is the horizontal polarization just introduced this transform can be identified with the Fourier transform.

Let us now look at the notion of a complex polarization. The most important example of a complex polarization is given by a Kahler manifold. Let us begin by examining this example. A Kahler manifold is a complex manifold X together with a positive definite Hermitian form $\langle\, ,\, \rangle$ such that the associated antisymmetric form $\omega = \operatorname{Im}\langle\, ,\, \rangle$ is symplectic. To say that X is a complex manifold implies that the *real* tangent space TX_x at any $x \in X$ has a complex structure, i.e., admits an automorphism, J, such that $J^2 = -1$. (The existence of such a J is known as an almost complex structure). Then the complexified tangent space splits up into the two eigenspaces, $\mathcal{F}_x$ where $J = i$ and $\overline{\mathcal{F}}_x$ where $J = -i$. (The almost complex structure is called complex if we can introduce local coordinates $x_1, \ldots, x_n, y_1, \ldots, y_n$ such that the $+i$ eigenspace, $\mathcal{F}_x$, is spanned by the vectors

$$\frac{\partial}{\partial x_j} - i\frac{\partial}{\partial y_j}$$

and the $-i$ eigenspace, $\overline{\mathcal{F}}_x$, is spanned by $\partial/\partial x_j + i\partial/\partial y_j$.) If $\langle\, ,\, \rangle$ is a Hermitian form on TX_x this means that

$$\langle J\xi, \eta\rangle = i\langle \xi, \eta\rangle \quad \text{and} \quad \langle \xi, J\eta\rangle = -i\langle \xi, \eta\rangle.$$

If we extend this to $TX_x \otimes \mathbf{C}$ we see that for $\xi, \eta \in \mathfrak{F}_x$ we have

$$\begin{aligned}\langle \xi, \eta\rangle &= \langle J\xi, J\eta\rangle \\ &= \langle i\xi, i\eta\rangle \\ &= -\langle \xi, \eta\rangle\end{aligned}$$

so that $\langle \xi, \eta\rangle = 0$ and, in particular, $\langle \xi \wedge \eta, \omega\rangle = \mathrm{Im}\langle \xi, \eta\rangle = 0$. In other words we obtain a complex polarization such that $\mathfrak{F}_x \oplus \bar{\mathfrak{F}}_x = TX_x \otimes \mathbf{C}$.

Conversely, suppose we start with a complex polarization $\mathfrak{F}$ relative to ω such that $\mathfrak{F} \oplus \bar{\mathfrak{F}} = TX \otimes \mathbf{C}$. This defines a J on $TX \otimes \mathbf{C}$ by $J = i$ on $\mathfrak{F}$ and $J = -i$ on $\bar{\mathfrak{F}}$ and if $\xi = \zeta + \bar{\zeta}$ is any real vector this then determines $J\xi$ as $J\xi = i\zeta - i\bar{\zeta}$, giving an almost complex structure. The fact that $\mathfrak{F}$ is closed under Lie bracket then implies that this is a complex structure by a theorem of Newlander and Nirenberg [10]. The fact that $\mathfrak{F}$ and $\bar{\mathfrak{F}}$ are isotropic for ω means that if

$$\begin{aligned}\xi_1 &= \zeta_1 + \bar{\zeta}_1, \\ \xi_2 &= \zeta_2 + \bar{\zeta}_2,\end{aligned}$$

then

$$\begin{aligned}\langle \xi_1 \wedge \xi_2, \omega\rangle &= \langle (\zeta_1 + \bar{\zeta}_1) \wedge (\zeta_2 + \bar{\zeta}_2), \omega\rangle \\ &= \langle \zeta_1 \wedge \bar{\zeta}_2, \omega\rangle + \langle \bar{\zeta}_1 \wedge \zeta_2, \omega\rangle \\ &= \langle i\zeta_1 \wedge (-i)\bar{\zeta}_2, \omega\rangle + \langle (-i)i\xi_1 \wedge i\xi_2, \omega\rangle \\ &= \langle J\xi_1 \wedge J\xi_2, \omega\rangle.\end{aligned}$$

As we have seen in Chapter IV, §3, the fact that J preserves ω implies that ω can be written as the imaginary part of a non-degenerate Hermitian form. We thus almost have a Kahler structure, the only thing missing is the fact that the Hermitian form need not be positive definite. Rather than introduce a new term such as pseudo-Kahler, we shall call any polarization with $\mathfrak{F} + \bar{\mathfrak{F}} = TX \otimes \mathbf{C}$ a polarization of Kahler type. Thus the two extreme cases are the real polarizations, $\mathfrak{F} = \bar{\mathfrak{F}}$ and the polarizations of Kahler type, $\mathfrak{F} + \bar{\mathfrak{F}} = TX \otimes \mathbf{C}$.

In general we can consider $\mathfrak{F} \cap \bar{\mathfrak{F}}$ and $\mathfrak{F} + \bar{\mathfrak{F}}$. We shall assume that both of these give sub-bundles of $TX \otimes \mathbf{C}$, i.e., that their dimension does not vary from point to point. We let $\dim \mathfrak{F} \cap \bar{\mathfrak{F}} = k$ so that $\dim \mathfrak{F} + \bar{\mathfrak{F}} = 2n - k$. We shall assume that $\mathfrak{F} + \bar{\mathfrak{F}}$ is integrable. Then $\mathfrak{F} \cap \bar{\mathfrak{F}} = \mathfrak{D} \otimes \mathbf{C}$ and $\mathfrak{F} + \bar{\mathfrak{F}} = \mathfrak{E} \otimes \mathbf{C}$ where $\mathfrak{D} \subset \mathfrak{E}$ are integrable sub-bundles of TX. In the real case $\mathfrak{D} = \mathfrak{E}$ and $k = n$. In the Kahler case $\mathfrak{D} = 0$, $\mathfrak{E} = TX \otimes \mathbf{C}$ and $k = 0$.

Let X be a symplectic manifold on which the group G acts as a group of symplectic diffeomorphisms. Let $\mathcal{F}$ be a polarization of X. For any $a \in G$ we define the polarization $a\mathcal{F}$ by the formula

$$(a\mathcal{F})_x = da_{a^{-1}x}\mathcal{F}_{a^{-1}x}.$$

We say that a polarization, $\mathcal{F}$, is G invariant if $a\mathcal{F} = \mathcal{F}$ for all $a \in G$. Suppose that G acts transitively on X. Then TX_x is spanned by vectors of the form $\hat{\xi}(x)$ as ξ ranges over the Lie algebra, g, of G, and $\hat{\xi}$ is the corresponding vector field on X. If we fix some point $x \in X$ then a subspace of $TX_x \otimes \mathbf{C}$ determines a subspace of $g \otimes \mathbf{C}$. The map $G \xrightarrow{\rho x} X$ sending a into $a \cdot x$ commutes with the (left) action of G. If ξ is a left invariant vector field and $\mathcal{F}$ is an invariant family of subspaces of $TX \otimes \mathbf{C}$ then $\hat{\xi} \in \mathcal{F}_x$ implies that $d\rho_x \xi(a) \in \mathcal{F}_{ax}$. Since $\rho\colon G \to X$ is a foliation, we conclude that $\mathcal{F}$ is integrable if and only if the ξ's span a subalgebra of $g \otimes \mathbf{C}$. In this way, polarizations of X correspond to certain subalgebras of $g \otimes \mathbf{C}$.

Let us describe the situation more explicitly in terms of an orbit in g^*. Here $X = (\mathrm{Ad}^{\#}\, G) \cdot \beta$ for some $\beta \in g^*$. The invariant two form, $d\beta$, is given by

$$\langle \xi \wedge \eta, d\beta \rangle = \langle \eta, \xi \rfloor d\beta \rangle = \langle \eta, D_\xi \beta \rangle$$

since $D_\xi \beta = \xi \rfloor d\beta + d\langle \xi, \beta \rangle$ and $\langle \xi, \beta \rangle$ is a constant. But, since $\langle \eta, \beta \rangle$ is a constant

$$0 = D_\xi \langle \eta, \beta \rangle = \langle [\xi, \eta], \beta \rangle + \langle \eta, D_\xi \beta \rangle$$

and thus

$$\langle \xi \wedge \eta, d\beta \rangle = -\langle [\xi, \eta], \beta \rangle.$$

We thus see that an invariant polarization on the orbit through β corresponds to a subalgebra, m, of $g \otimes \mathbf{C}$ such that

(i) $m \supset g_\beta$ where $g_\beta = \{\xi \mid \xi \rfloor d\beta = 0\}$,
(ii) $\dim_{\mathbf{C}} m/g_\beta = \dim_{\mathbf{C}} g \otimes \mathbf{C}/m$,
(iii) $\langle [m, m], \beta \rangle = 0$,
(iv) m is invariant under Ad G_β where G_β is the stabilizer of β.

See, in this connection, Auslander-Kostant [3]. Let us give some examples of orbits in g^* and corresponding polarizations. If G is commutative then the adjoint representation is trivial and hence so is the coadjoint representation. In this case all the orbits reduce to points.

Let us consider the group $SO(3)$. Its Lie algebra is just the algebra of anti-symmetric (three by three) matrices. The quadratic form $Q(A) = \frac{1}{2}\operatorname{tr} A^2$ is negative definite on the space of anti-symmetric matrices, indeed $\operatorname{tr} AA = -\operatorname{tr} AA^* = \sum_{i,j} A_{ij}^2$ if $A = (A_{ij}) = -(A_{ji})$. Since $Q(A)$ is non-singular, we may use it to identify g with g^*. (More generally, if g is any real semi-simple Lie algebra, its Killing form $Q(A) = \operatorname{tr}(\operatorname{ad} A)^2$ is non-singular and we may use it to identify g with g^*.) In the case of $SO(3)$ we may identify anti-symmetric three by three matrices with $\mathbf{R}^3$ in the usual way with $SO(3)$ acting as rotations. The orbits are thus spheres (two-dimensional) and the origin (zero-dimensional). The sphere does not admit any real polarization (invariant or not) and so we expect to deal with complex polarizations. Let us take $\beta = (0, 0, r)$ so that G_β consists of rotations about the z-axis, and g_β is spanned by the matrix

$$A_z = \begin{bmatrix} 0 & -1 & 0 \\ 1 & 0 & 0 \\ 0 & 0 & 0 \end{bmatrix}.$$

In $g \otimes \mathbf{C}$ the vectors

$$A_x + iA_y \qquad \text{and} \qquad A_x - iA_y$$

are eigenvectors of ad A_z with eigenvalues i and $-i$. There are thus two invariant complex polarizations (corresponding to the holomorphic and anti-holomorphic structures on the Riemann sphere.)

Next let us consider the group $SL(2, \mathbf{R})$ of all two by two matrices of determinant one. The Lie algebra, g, consists of all matrices of the form

$$A = \begin{bmatrix} a & b \\ c & -a \end{bmatrix}$$

and the quadratic form $Q(A) = \frac{1}{2} \operatorname{tr} A^2$ is given by

$$Q(A) = a^2 + bc.$$

This form is again non-singular, so we identify g^* with g, but no longer positive definite. There are three kinds of two dimensional orbits corresponding to $Q(A) > 0$ (a single sheeted hyperboloid), $Q(A) = 0$ (which is a cone, removing the origin from this cone gives two orbits), and $Q(A) < 0$ (a two sheeted hyperboloid, each sheet being an orbit).

Let us first examine the single sheeted hyperboloid, $Q(A) = \lambda^2 > 0$. We can take

$$\beta = \begin{bmatrix} \lambda & 0 \\ 0 & -\lambda \end{bmatrix}$$

so that

$$G_\beta = \begin{bmatrix} s & 0 \\ 0 & s^{-1} \end{bmatrix}$$

and g_β is the line spanned by β. If we set

$$A = \begin{bmatrix} 0 & 1 \\ 0 & 0 \end{bmatrix} \quad \text{and} \quad C = \begin{bmatrix} 0 & 0 \\ 1 & 0 \end{bmatrix}$$

then the planes spanned by β and A and by β and C are invariant under G_β. Each plane intersects the hyperboloid in a straight line. There are two real polarizations each corresponding to the ruling of the hyperboloid, these are the only polarizations. (This construction can be generalized: If G is a real semi-simple Lie group and β is a regular element which lies on a totally real Cartan subalgebra then any system of positive root vectors will determine a real polarization. In general there will be various different conjugacy classes of Cartan subalgebras, corresponding to how many simple roots are real; cf. Kostant [5]. This will determine the nature of the polarization. For $sl(2)$ there are two types, corresponding to $Q(A) > 0$ and $Q(A) < 0$.)

For a hyperboloid $Q(A) = -\lambda^2 < 0$ we may take

$$\beta = \begin{bmatrix} 0 & \lambda \\ -\lambda & 0 \end{bmatrix}$$

for one sheet and $-\beta$ for the other. Let us examine the sheet corresponding to β. Then G_β is just the rotation group and g_β is spanned by β. There are no real polarizations. A straightforward computation shows (cf., for example, Renouard [11]) that we may identify the orbit with the upper half plane (or the interior of the unit disk) with the form ω becoming λ times the volume form of the standard hyperbolic metric. The two polarizations becomes the ones spanned by $\partial/\partial z$ and $\partial/\partial \bar{z}$.

For the cone $Q(A) = 0$ the orbit through the matrix

$$\beta = \begin{bmatrix} 0 & 1 \\ 0 & 0 \end{bmatrix}$$

gives G_β as the set of all matrices of the form

$$\begin{bmatrix} 1 & a \\ 0 & 1 \end{bmatrix}.$$

It is easy to check that there is only one line in $g \otimes \mathbf{C}/g_\beta$ invariant under G_β and it is real. There is thus one invariant polarization and it is given by the generators of the cone.

Let X be a symplectic manifold and let $\mathfrak{M}$ be a polarization of X. Suppose that g is a symplectic diffeomorphism of X. We then get a new polarization, $g\mathfrak{M}$, by setting

$$(g\mathfrak{M})_x = (dg)_{g^{-1}x}\mathfrak{M}_{g^{-1}x}.$$

It is easy to check that

$$g_1(g_2\mathfrak{M}) = (g_1 g_2)\mathfrak{M}.$$

In this section, we shall be concerned with real polarizations which satisfy a condition, which makes the $\mathfrak{M}_x$ appear as the tangent spaces to the leaves of a fibration. More precisely, let $\mathfrak{M}$ be a real polarization of X. We say that $\mathfrak{M}$ is a *fibrating polarization* if there is an n dimensional smooth manifold, N, and a smooth map $f: X \to N$, such that

$$\mathfrak{M}_x = \ker df_x \qquad \textit{for each } x \in X \tag{i}$$

and

$$\textit{for each } y \in N \textit{ the inverse image, } f^{-1}(y), \textit{ is connected and simply connected}. \tag{ii}$$

Notice that if $\mathfrak{M}$ is a fibrating polarization, and if g is a symplectic automorphism, then $g\mathfrak{M}$ is again a fibrating polarization, where we simply take $f \circ g^{-1}$ as the map associated with $g\mathfrak{M}$. Indeed

$$\ker d(f \circ g^{-1})_x = (dg^{-1})^{-1}(\ker df)_{g^{-1}x} = (dg)(\ker df)_{g^{-1}x} = (g\mathfrak{M})_x.$$

Notice that $f^{-1}(y)$ is a Lagrangian manifold.

Suppose the symplectic structure on X arises from a Hermitian line bundle L with connection. If Λ is a Lagrangian manifold then the restriction of L to Λ, the line bundle $L_{|\Lambda}$, has curvature zero. If in addition Λ is connected and simply connected then it makes sense to talk of a *parallel section* of $L_{|\Lambda}$, independent of the path. This is because parallel translation around any closed path is the identity. We have thus associated a complex line, the space of parallel sections L_Λ, to Λ. We can do this for each $f^{-1}(y)$. We obtain in this way a complex line associated to each $y \in N$. This gives rise to a line bundle $L^{\mathfrak{M}}$ over N. A section of $L^{\mathfrak{M}}$ can be thought of as a section of L which is parallel on each $f^{-1}(y)$. Notice that $L^{\mathfrak{M}}$ inherits a Hermitian structure from L.

Let $g \in P$ and let s be a smooth section of $L^{\mathfrak{M}}$, so that s is a section of L which is parallel, i.e., covariant constant, along the leaves of $\mathfrak{M}$. Thus

$$\langle \xi, Ds \rangle = 0 \qquad \text{all } \xi \in \mathfrak{M}_x \qquad \text{for all } x.$$

We will use the same letter, g, to denote the image of g in H. Then for $\eta \in (g\mathfrak{M})_x$ we can write $\eta = (dg)_{g^{-1}x}\xi$ with $\xi \in \mathfrak{M}_{g^{-1}x}$. Since g^{-1} preserves the connection

$$\langle \eta, D(g^{-1*}s)\rangle = \langle dg^{-1}\eta, Ds\rangle = \langle \xi, Ds\rangle = 0$$

so that

$$g^{-1*}s \text{ is locally constant along } g\mathfrak{M}.$$

We would now like to associate a Hilbert space to $\mathfrak{M}$. Our current definition will be provisional; we will make the correct definition in §5. Let $|\wedge^n|^{1/2}(N)$ denote the space of half densities on N, so that an element of $|\wedge^n|^{1/2}(N)$ is a smooth section of the line bundle $|\wedge^n|^{1/2}T^*(N)$. Let $L^{\mathfrak{M}\frac{1}{2}}$ denote the tensor product

$$L^{\mathfrak{M}\frac{1}{2}} = L^{\mathfrak{M}} \times |\wedge^n|^{1/2}T^*(N).$$

Let $r_1 = s_1 \times \rho_1$ and $r_2 = s_2 \times \rho_2$ be two sections of $L^{\mathfrak{M}\frac{1}{2}}$. Then $\langle s_1, s_2\rangle\rho_1\bar{\rho}_2$ is a density depending only on r_1 and r_2. If r_1 and r_2 have compact support then we define

$$(r_1, r_2) = \int \langle s_1, s_2\rangle \rho_1\bar{\rho}_2.$$

These sections form a pre-Hilbert space which we denote by $|H_0(\mathfrak{M})|$. Its completion we denote by $|H(\mathfrak{M})|$. It is clear from the previous discussion that each $g \in P$ defines an isometry of $|H_0(\mathfrak{M})|$ with $|H_0(g\mathfrak{M})|$.

Let $\mathfrak{M}_1$ and $\mathfrak{M}_2$ be two fibrating polarizations associated with maps $f_1: X \to N_1$ and $f_2: X \to N_2$. We say that these polarizations are *transverse*, if

$$f_1 \times f_2: X \to N_1 \times N_2$$

is an immersion. This means that, for each $x \in X$,

$$\mathfrak{M}_{1x} \cap \mathfrak{M}_{2x} = \{0\}.$$

We say that $\mathfrak{M}_1$ and $\mathfrak{M}_2$ are *strongly transverse* if, in addition,

$$f_1 \times f_2: X \to N_1 \times N_2$$

is proper. We propose to define a pairing between $H_0(\mathfrak{M}_1)$ and $H_0(\mathfrak{M}_2)$ if $\mathfrak{M}_1$ and $\mathfrak{M}_2$ are strongly transverse. We shall see that this pairing is a generalization of the Fourier transform. To explain this pairing let us first point out how, at each x, a half density at $y_1 = f_1(x)$ pairs with a half density at $y_2 = f_2(x)$ to give a number. Notice that $(TN_1)_{y_1}$ is isomorphic to $TX_x/\mathfrak{M}_{1x}$. Since $\mathfrak{M}_{1x}$ is Lagrangian the bilinear form ω gives a non-singular pairing between $\mathfrak{M}_{1x}$ and $TX_x/\mathfrak{M}_{1x}$.

Thus $\mathfrak{M}_{1x}$ is identified with $T^*N_{1y_1}$, and $|\wedge^n|^{1/2}TN^*_{y_1}$ is to be identified with $|\wedge^n|^{1/2}\mathfrak{M}_{1x}$. Similarly, $|\wedge^n|^{1/2}TN^*_{2y_2}$ is to be identified with $|\wedge^n|^{1/2}\mathfrak{M}_{2x}$. Now $\mathfrak{M}_{1x}$ and $\mathfrak{M}_{2x}$ are dually paired by ω. If we choose a basis $e = e_1, \ldots, e_n$ of $\mathfrak{M}$, this determines a basis $f = f_1, \ldots, f_n$, of $\mathfrak{M}_{2x}$. Replacing e by $e' = Ae$ has the effect of replacing f by $f' = (A^t)^{-1}f$. If $\rho_1 \in |\wedge^n|^{1/2}\mathfrak{M}$ then we can write

$$\rho_1 = a|e_1 \wedge \cdots \wedge e_n|^{1/2}$$

where $|e_1 \wedge \cdots \wedge e_n|^{1/2}$ is the basis of $|\wedge^n|^{1/2}\mathfrak{M}_{1x}$ determined by e. If we replace e by $e' = Ae$ then

$$|e_1' \wedge \cdots \wedge e_n'|^{1/2} = |\det A|^{1/2}|e_1 \wedge \cdots \wedge e_n|^{1/2}$$

so

$$\rho_1 = \frac{a}{|\det A|^{1/2}}|e_1' \wedge \cdots \wedge e_n'|^{1/2}.$$

If $\rho_2 \in |\wedge^n|^{1/2}\mathfrak{M}_{2x}$ with

$$\rho_2 = b|f_1 \wedge \cdots \wedge f_n|^{1/2}$$

then we set

$$\rho_1 \cdot \bar{\rho}_2 = a\bar{b}.$$

It is clear from the above that this does not depend on the choice of basis.

Now consider $h_1 = s_1 \times \rho_1 \in |H_0(\mathfrak{M}_1)|$ and $h_2 = s_2 \times \rho_2 \in |H_0(\mathfrak{M}_2)|$ where $\mathfrak{M}_1$ and $\mathfrak{M}_2$ are strongly transverse foliating polarizations. Then

$$\langle s_1, s_2\rangle \rho_1 \cdot \bar{\rho}_2$$

is a smooth function of compact support on X. It is clear that this function does not depend on the specific representation of s_1 and s_2. We can integrate this function with respect to ω^n to obtain a sesquilinear pairing

$$\langle h_1, h_2\rangle = \frac{1}{(2\pi)^{n/2}}\int_X \langle s_1, s_2\rangle \rho_1 \cdot \bar{\rho}_2 \,\omega^n.$$

Let us show that this pairing does indeed generalize the classical Fourier transform. For this purpose, let X be the phase space of a linear space, i.e., $X = T^*(V)$ where V is a vector space. Then on X we have the global coordinates $(q,p) = (q^1, \ldots, q^n, p^1, \ldots, p^n)$.

The polarization $\mathfrak{M}_1$, associated to the map $(q,p) \to q$, is clearly strongly transversal to the polarization $\mathfrak{M}_2$ associated with the map $(q,p) \to p$. Here N_1 can be identified with V, regarded as the zero section of $T^*(X)$ and N_2 can be identified with V^*, regarded as $T^*(X)_0$, the cotangent spaces at the origin. Notice that here N_1 and N_2 have been identified as Lagrangian submanifolds of X, which are in fact themselves connected and simply connected. A section of $L^{\mathfrak{M}_1}$, considered as a section of L is determined by its values on N_1. On N_1 there is (up to constant factor) a preferred section, namely the section of unit length which is covariant constant along N_1. Let us call this section r_1. Then sections of $L^{\mathfrak{M}_1}$ are of the form ar_1 where a is a *function* constant along the fibers of $\mathfrak{M}_1$, i.e., a function on N_1. Thus sections on $L^{\mathfrak{M}_1}$ can be identified with functions on N_1. Similarly for N_2.

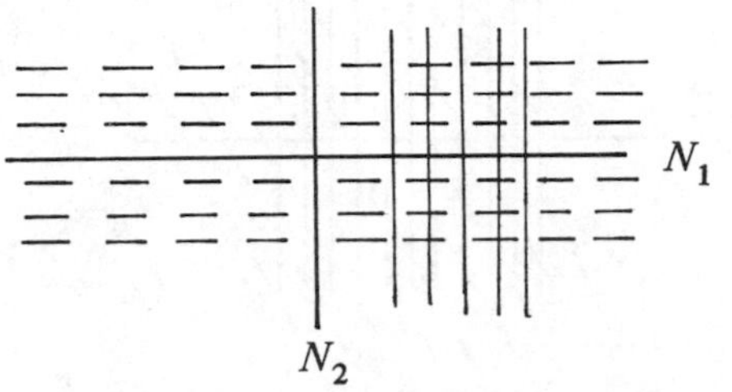

In this way we get a pairing between half densities (of compact support) on N_1 with those on N_2: Let us write the most general section of $L^{\mathfrak{M}_1 \frac{1}{2}}$ as $f(q)r_1\, dq^{1/2}$ and the most general section of $L^{\mathfrak{M}_2 \frac{1}{2}}$ as $h(p)r_2\, dp^{1/2}$. The pairing between f and h is given by

$$(2\pi)^{-n/2} \int f(q)\bar{h}(p)\langle r_1, r_2\rangle dq^{1/2} \cdot dp^{1/2}\, \omega^n.$$

Now $dq^{1/2} \cdot dp^{1/2} = 1$. We must compute $\langle r_1, r_2\rangle$. Since r_1 and r_2 are of unit length we can write $r_2 = e^{i\varphi} r_1$ for some real function φ, and therefore $\langle r_1, r_2\rangle = e^{-i\varphi}$. We must determine φ. Let $\alpha_1 = \alpha_{r_1}$ denote the imaginary part of the connection form: $\theta_{r_1} = i\alpha_1$ and similarly for r_2. Then

$$\alpha_2 = \alpha_1 + d\varphi.$$

We must therefore determine α_1 and α_2. Now r_1 is covariant constant along the leaves $q =$ const. Therefore $\langle \partial/\partial p, Dr_1\rangle = 0$ thus, since $Dr_1 = r_1 \otimes \alpha_1$ we see that $\alpha_1 = \sum f_i\, dq^i$. As $d\alpha_1 = \sum dp^i \wedge dq^i$, it follows that

$$\alpha_1 = \sum p^i\, dq^i + dF$$

where $F = F(q)$. Since r_1 is also parallel along N_1, we see that $\alpha_{1|N_1} = 0$ so that $dF = 0$. Thus $\alpha_1 = \sum p^i\, dq^i$. The same argument shows that $\alpha_2 = -\sum q^i\, dp^i$. Therefore φ is determined (up to constant) as

$$\varphi = -\sum p^j q^j$$

and thus the pairing is

$$(2\pi)^{-n/2} \int e^{-ip\cdot q} f(q)\bar{h}(p)\, dp^1 \wedge \cdots \wedge dp^n dq^1 \wedge \cdots \wedge dq^n.$$

If we think of this pairing as assigning to f a distribution in p, this is exactly the Fourier transform.

Let us now do the following computation: Let $N_1 = N_2 = (p = 0) \subset X = T^*(V)$. Let $\mathfrak{M}_1$ and $\mathfrak{M}_2$ be defined by the maps f_1 and f_2 where

$$f_1(q,p) = q,$$

$$f_2(q,p) = q - tp, \qquad t \neq 0.$$

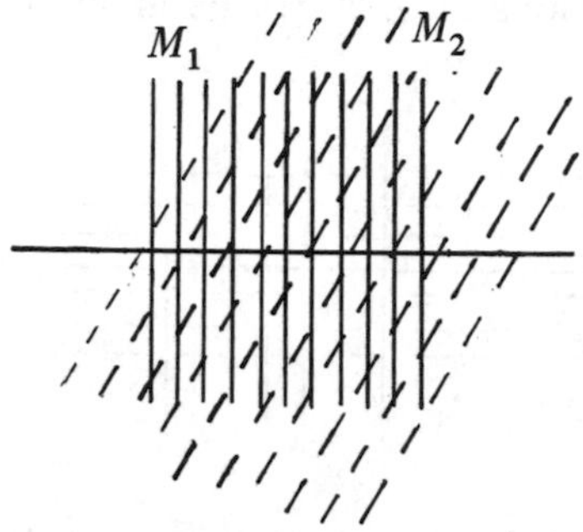

We can identify both $L^{\mathfrak{M}_1}$ and $L^{\mathfrak{M}_2}$ with functions on N_1. Here r_1 is as before, and r_2 is constant along the leaves $q - tp = \text{const}$. Thus $\alpha_2 = \sum f_i(dq^i - tdp^i)$ and since $d\alpha_2 = \omega$ we get $\alpha_2 = \sum p_i(dq^i - tdp^i) + dF$ where $F = F(q - tp)$. Since $\alpha_{2|N_1} = 0$, we conclude that $dF = 0$. Thus

$$\alpha_2 = \sum p_i dq^i + td\varphi$$

where, up to a constant, φ is given by

$$\varphi(q,p) = -\tfrac{1}{2}t \sum p^{j2}.$$

Let us now compute the pairing between the half-densities. At any $x \in X$, the subspace $\mathfrak{M}_{1x}$ is spanned by $\partial/\partial p^1, \ldots, \partial/\partial p^n$ and the subspace $\mathfrak{M}_{2x}$ is spanned by

$$\frac{\partial}{\partial p^1} + t\frac{\partial}{\partial q^1}, \ldots, \frac{\partial}{\partial p^n} + t\frac{\partial}{\partial q^n}.$$

If we choose $\partial/\partial p^1, \ldots, \partial/\partial p^n$ as our basis of $\mathfrak{M}_{1x}$ then the corresponding basis of $\mathfrak{M}_{2x}$ is

$$t^{-1}\frac{\partial}{\partial p^1} + \frac{\partial}{\partial q^1}, \ldots, t^{-1}\frac{\partial}{\partial p^n} + \frac{\partial}{\partial q^n}.$$

The covariant half density $|dq_1 \wedge \cdots \wedge dq_n|^{1/2}$ on N_x corresponds to the covariant half density

$$\left|\frac{\partial}{\partial p^1} \wedge \cdots \wedge \frac{\partial}{\partial p^n}\right|^{1/2}$$

on $\mathfrak{M}_{1x}$ via f_1 and to the half density

$$\left| \frac{\partial}{\partial p^1} + t\frac{\partial}{\partial q^1} \wedge \cdots \wedge \frac{\partial}{\partial p^n} + t\frac{\partial}{\partial q^n} \right|^{1/2}$$

on $\mathfrak{M}_2$ via f_2. This last expression equals

$$t^{n/2}\left| t^{-1}\frac{\partial}{\partial p^1} + \frac{\partial}{\partial q^1} \wedge \cdots \wedge t^{-1}\frac{\partial}{\partial p^n} + \frac{\partial}{\partial q^n} \right|^{1/2}.$$

Therefore

$$|dq|^{1/2} \cdot |dq|^{1/2} = |t|^{n/2}.$$

We thus obtain a pairing on $H(N_1)^{1/2}$ given by

$$\langle f|dq|^{1/2}, g|dq|^{1/2}\rangle_t = \frac{t^{n/2}}{(2\pi)^{n/2}}\int e^{-itp^2/2} f(q)\overline{g(q-tp)}\,dp\,dq.$$

Notice that as t approaches zero the possible domain of the integral, i.e., $f_1^{-1}(\operatorname{supp} f) \cap f_2^{-1}(\operatorname{supp} g)$ gets larger and larger. Let us compute this limit by the method of stationary phase. Set $p' = tp$ so that the right hand side becomes

$$\int \frac{1}{(2\pi t)^{n/2}} \int e^{-ip'^2/2t}\overline{g(q-p')}\,dp' f(q)\,dq.$$

According to the method of stationary phase, the inner integral approaches

$$e^{\pi in/4}\overline{g(q)} \qquad \text{as } t \to 0^+$$

and approaches

$$e^{-\pi in/4}\overline{g(q)} \qquad \text{as } t \to 0^-.$$

If we ignore the phase factors, we would obtain, in the limit, the pairing in the Hilbert space $H_0(N_1)$, i.e.,

$$\lim_{t\to 0} |\langle f|dq|^{1/2}, g|dq|^{1/2}\rangle|_t = |\langle f|dq|^{1/2}, g|dq|^{1/2}\rangle|,$$

the scalar product on the right being in the Hilbert space $H_0(N_1)$. To make the phases come out right we proceed as follows: Notice that on the level of forms, rather than 1/2 densities, the n-form $dq^1 \wedge \cdots \wedge dq^n$ determines the element

$$\frac{\partial}{\partial p^1} \wedge \cdots \wedge \frac{\partial}{\partial p^n} \qquad \text{of } \Lambda^n(\mathfrak{M}_{1x})$$

and the element

$$\frac{\partial}{\partial q^1} + t\frac{\partial}{\partial p^1} \wedge \cdots \wedge \frac{\partial}{\partial q^1} + t\frac{\partial}{\partial p^n} \qquad \text{of } \Lambda^n(\mathfrak{M}_{2x}).$$

Therefore, if we were to define a pairing between forms, we would obtain

$$dq \cdot dq = t^n = \begin{cases} |t|^n & \text{for } t > 0, \\ e^{n\pi i}|t|^n & \text{for } t < 0. \end{cases}$$

If we could define and deal with "half forms", instead of half densities, we would obtain

$$dq^{1/2} \cdot dq^{1/2} = t^{n/2} \qquad \text{for } t > 0$$

and

$$dq^{1/2} \cdot dq^{1/2} = e^{n\pi i/2}|t|^{n/2} \qquad \text{for } t < 0$$

(where we have made a choice of $\sqrt{-1}$) . If, in the pairing, we replace $(2\pi)^{-n/2}$ by $(e^{\pi i/2}2\pi)^{-n/2}$ and $|dq|^{1/2}$ by $dq^{1/2}$ we would obtain, as our pairing,

$$\left(\frac{t}{2\pi i}\right)^{n/2} \int e^{-itp^2/2} f(q)\overline{g(q - tp)}\, dp\, dq,$$

and the method of stationary phase would now give

$$\lim \langle f dq^{1/2}, g dq^{1/2} \rangle_t = \langle f dq^{1/2}, g dq^{1/2} \rangle.$$

We must therefore study the notion of half forms, and how they are paired in a symplectic manifold.

Notice that the function

$$h(q,t) = \left(\frac{1}{2\pi i t}\right)^{n/2} \int e^{i(q-x)^2/2t} h(x)\, dx$$

is exactly the solution of the Schrödinger equation

$$\frac{1}{i}\frac{\partial h}{\partial t} = \frac{\partial^2 h}{\partial q^2}$$

for the free particle, with the initial conditions

$$h(q,0) = h(q).$$

Our interpretation of Schrödinger's equation is thus as follows. The function $\frac{1}{2}p^2$ is considered as an element of the Lie algebra h. It therefore generates a one parameter subgroup g_t. We shall also denote the image of this one parameter group in H by g_t. Thus g_t is also a one parameter group of diffeomorphisms of X whose infinitesimal generator is $p\partial/\partial q = \sum p^i \partial/\partial q^i$. Then

$$g_t(q,p) = (q + tp, p)$$

and thus carries the foliation $\mathfrak{M} = \mathfrak{M}_0$ given by $q =$ const. into the foliation $\mathfrak{M}_t = g_t \mathfrak{M}$ given by $q + tp =$ const. It therefore maps $H_0(\mathfrak{M})$ onto $H_0(\mathfrak{M}_t)$ isometrically, and thus extends to a unitary isomorphism of $H(\mathfrak{M})$ onto $H(\mathfrak{M}_t)$. We will denote this map also by g_t. Then, for $t \neq 0$ we define the map U_t: $H(\mathfrak{M}) \to H(\mathfrak{M})$ by

$$\langle h_1, U_t h_2 \rangle_0 = \langle h_1, g_t h_2 \rangle_t$$

for all $h_1, h_2 \in H(\mathfrak{M})$ and set $U_0 =$ id. Then U_t is a unitary one parameter group on $H(\mathfrak{M})$ whose infinitesimal generator is given by the Schrödinger operator $\partial^2/\partial q^2$.

In the preceding computation, we could replace the scalar t by a diagonal matrix $T_2 = \operatorname{diag}(t_1, \ldots, t_n)$ so that f_2 is now $f_2(q,p) = q - T_2 p$. As long as none of the $t_i = 0$, the computation goes through essentially unchanged:

We let $N = N_1 = N_2$ be the subspace determined by the equation $p = 0$ and we let U_{1N}: $C_0^\infty(N) \to H_0(\mathfrak{M}_1)$ be the identification corresponding to the choice of half density $dq^{1/2}$ and the map f_1 and similarly define U_{2N}: $C_0^\infty(N) \to H_0(\mathfrak{M}_2)$. Then

$$\langle U_{1N} f, U_{2N} g \rangle = \frac{\prod t_j^{\frac{1}{2}}}{(2\pi)^{n/2}} \iint e^{-\frac{1}{2} i \sum t_j p_j^2} f(q) \overline{g(q - Tp)}\, dq\, dp.$$

We can simplify this formula by use of the Fourier transform: If $\tilde{f}$ denotes the Fourier transform of f and if g_T denotes the function $g_T(q) = g(q - Tp)$, then $\tilde{g}_T(\xi) = e^{-i\sum t_j p_j \xi_j} \tilde{g}(\xi)$. Applying Plancherel's formula to the preceding expression yields

$$\langle U_{1N} f, U_{2N} g \rangle = (\tilde{f}, \tilde{\varphi}_T \tilde{g})$$

where

$$\tilde{\varphi}_T(\xi) = \frac{\prod t_j^{1/2}}{(2\pi)^{n/2}} \int e^{\frac{1}{2} i \sum t_j p_j^2 + i \sum t_j p_j \xi_j}\, dp$$

$$= e^{(\operatorname{sig} T)\pi i/4} e^{-i\sum |t_j| \xi_j^2/2}$$

by the basic stationary phase computation on page 4. We see once again the "continuity up to phase" in the pairing. We can, however, deduce another important fact from this formula, by choosing two diagonal matrices, T_2 and T_3

with no common eigenvalues, and hence the corresponding foliations, $\mathfrak{M}_1$, $\mathfrak{M}_2$ and $\mathfrak{M}_3$ are pairwise transverse. We can use the sesquilinear pairing between $H(\mathfrak{M}_1)$ and $H(\mathfrak{M}_2)$ to define a map, $F_{21}\colon H(\mathfrak{M}_1) \to H(\mathfrak{M}_2)$, which we know to be unitary, by the above formula. Similarly we can define the unitary maps F_{23} and F_{13}. Let X_1, X_2 and X_3 be the Lagrangian subspace corresponding to these three polarizations. Then it follows from the above computation that

$$F_{32} \circ F_{21} = e^{(\pi i/4)i(X_1, X_2, X_3)} F_{3,1}$$

where $i(X_1, X_2, X_3)$ is the index of the three transverse Lagrangian subspaces as defined in §2 of Chapter IV. Let X and Y be any two transverse Lagrangian subspaces and $\mathfrak{M}_X$ and $\mathfrak{M}_Y$ the corresponding real polarizations. We then have defined a unitary map

$$F_{YX}\colon H(\mathfrak{M}_X) \to H(\mathfrak{M}_Y).$$

For any element, a, of the symplectic group we have the induced map

$$a_*\colon H(\mathfrak{M}_X) \to H(\mathfrak{M}_{aX})$$

and it follows from the invariance of our definitions that

$$F_{aY,aX}\, a_* = a_* F_{Y,X}. \tag{3.1}$$

We thus know that for *any* three transverse Lagrangian subspaces, X, Y, and Z we have

$$F_{Z,Y} F_{Y,X} = e^{(\pi i/4)i(X,Y,Z)} F_{Z,X}. \tag{3.2}$$

Equation (3.1) suggests that we might try to define a representation of the symplectic group on any one fixed $H(\mathfrak{M}_X)$ by sending a into the unitary map $a_* F_{a^{-1}X,X}$. Equation (3.2) shows that this is not quite a representation, it is only a representation up to a phase factor, i.e. a ray representation. However, we can pass to the universal cover of the symplectic group, and the universal cover of the space of Lagrangian subspaces in order to eliminate the problems with the phase: As in §2 of Chapter IV, let $\tilde{L}(V)$ denote the universal covering space of the space, $L(V)$ of real Lagrangian subspaces of V. If $u = (X, \theta)$ is a point of $\tilde{L}(V)$ we define $\mathfrak{M}_u = \mathfrak{M}_X$. We now define the unitary maps

$$F_{v,u}\colon H(\mathfrak{M}_u) \to H(\mathfrak{M}_v)$$

by

$$F_{v,u} = e^{-\pi i m(u,v)/2} F_{Y,X} \text{ where } u = (X, \theta) \text{ and } v = (Y, \varphi), \tag{3.3}$$

where $m(u, v)$ is the Maslov index and where u and v are assumed to be transverse. It follows from the above considerations that if we define

$$\tilde{a}_*: H(\mathfrak{M}_u) \to H(\mathfrak{M}_{\tilde{a}u}) \quad \text{by} \quad \tilde{a}_* = a_*$$

for any $\tilde{a} \in \tilde{S}p(V)$ covering $a \in Sp(V)$ then

$$F_{\tilde{a}v, \tilde{a}u}\, \tilde{a}_* = \tilde{a}_* F_{v,u} \tag{3.4}$$

$$F_{u,v} = F_{v,u}^{-1} \tag{3.5}$$

and

$$F_{w,v} F_{v,u} = F_{w,u} \tag{3.6}$$

for three transverse elements u, v, and w, where, in (3.6) we have made use of (3.2) and the Leray formula, equation (2.26) of Chapter IV. We now show (following Souriau [16]) how to define the map $F_{v,u}$ for any pair of elements of $\tilde{L}(V)$, transversal, or not, in such a way that $F_{v,u}$ is still unitary, where

$$F_{u,u} = \text{id} \tag{3.7}$$

and (3.4), (3.5) and (3.6) continue to hold. For this purpose, we use the notation of §2 of Chapter IV (in particular, choose a complex structure and identifying V with $\mathbf{C}^n$). For each θ satisfying $0 \leqslant \theta < 2\pi$ we set

$$z_\theta = (e^{i\theta}\mathbf{R}^n, \theta)$$

and let z, z' etc. range over the set of possible z_θ. Given any finite set of elements, u, v, w etc., it is possible to choose some z transverse to all of them, simply by arranging that θ does not coincide with any of the eigenvalues of (X) etc. Furthermore, if $z \neq z'$, then z is transverse to z'. Now simply define the map

$$G_{v,u} = F_{v,z} \circ F_{z,u}$$

where z is chosen transversal to u and v. It is a formal consequence of (3.5) and (3.6) that G satisfies the same equations, and is independent of the choice of z. Similarly, it follows that G coincides with F when u and v are transverse, and that (3.7) holds. It follows from (3.7) and equation (2.21) of Chapter IV

$$F_{\mathrm{Exp}(4k\pi i)u,u} = \text{id}\,. \tag{3.8}$$

We thus see that the representation $\tilde{a} \to \tilde{a}_* F_{\tilde{a}^{-1}u,u}$ of $\tilde{S}p(V)$ is already a representation of the double covering, $Mp(V)$ of the symplectic group. This double covering is called the metaplectic group, and the corresponding representation is called the metaplectic representation. We shall study it in detail in §6.

Let us do a similar computation for the case of the harmonic oscillator. Here, for simplicity, we take $n = 1$. Then the Hamiltonian is given by

$$\frac{1}{2m}p^2 + \frac{m^2}{2}q^2$$

so that the corresponding vector field is

$$\frac{p}{m}\frac{\partial}{\partial q} - m^2 q\frac{\partial}{\partial p},$$

and the flow is

$$g_t(q,p) = \left((\cos ft)q + \frac{1}{mf}(\sin ft)p, -mf(\sin ft)q + (\cos ft)p\right).$$

Let us introduce coordinates p, q, and t so that $m = f = 1$ in order to simplify the calculations. We shall revert to the original coordinates in the final result. Then the Hamiltonian becomes

$$\tfrac{1}{2}(p^2 + q^2)$$

the corresponding vector field

$$p\frac{\partial}{\partial q} - q\frac{\partial}{\partial p}$$

and the corresponding flow, g_t, given by

$$g_t(q,p) = (q\cos t + p\sin t, -q\sin t + p\cos t).$$

The polarization $g_t\mathfrak{M}$ is given by $q\cos t - p\sin t = \text{const}$. The scalar product between $dq^{1/2}$ and $g_t dq^{1/2}$ is of the form $e^{i\varphi_t}$. The function φ_t is specified by the equation

$$(g_t)^{-1*}\alpha - \alpha = d\varphi_t.$$

Let us do the corresponding computation for any

$$g = \begin{pmatrix} A & B \\ C & D \end{pmatrix} \in SL(2, \mathbf{R}),$$

where A, B, C and D are real numbers. We wish to find a φ such that

$$g^{-1*}\alpha - \alpha = d\varphi.$$

Now

$$g^{-1} = \begin{pmatrix} D & -B \\ -C & A \end{pmatrix}$$

and

$$(g^{-1})^*(pdq) = (-Cq + Ap)(Ddq - Bdp).$$

Thus

$$g^{-1*}\alpha - \alpha = -DCqdq - BApdp + BCqdp + CBpdq$$

since $AD - BC = 1$. We thus see that we may take

$$\varphi(p,q) = -\tfrac{1}{2}DCq^2 + (Cq)(Bp) - \tfrac{1}{2}BAp^2.$$

(We will do the corresponding n dimensional computation, which is practically identical with the one dimensional case, in §7.) In our present case we see that

$$\varphi_t = +\tfrac{1}{2}(\sin t \cos t)q^2 - \sin^2 tpq - \tfrac{1}{2}(\sin t \cos t)p^2.$$

Thus if $v_1 = h_1 s_1$ and $v_2 = h_2 s_2$ are elements of $H_0(\mathfrak{M})$ we get

$$\langle v_1, g_t v_2\rangle_t = \frac{1}{\sqrt{2\pi i}} \int e^{-i[-(\sin^2 t)pq + (1/2)\sin t \cos t(q^2 - p^2)]}$$

$$\cdot\, h_1(q)\bar{h}_2(q \cos t - p \sin t)\sin^{1/2} t\, dp\, dq.$$

We can understand this as

$$\langle v_1, g_t v_2\rangle_t = \langle v_1, U(t)v_2\rangle_0$$

where

$$U(t)v(q) = \int K(q,t;z,0)v(z)\,dz,$$

for a suitable kernel K. To find the explicit expression for K we make the substitution

$$z = q \cos t - p \sin t$$

in the above integral. This yields

$$dp = -(\sin t)^{-1} dz$$

and

$$-(\sin^2 t)pq + \tfrac{1}{2}(\sin t \cos t)(q^2 - p^2) = \frac{1}{\sin t}[qz - \tfrac{1}{2}\cos t(q^2 + z^2)].$$

These two equations show that

$$K(q,t;z,0) = \left(\frac{1}{2\pi i \sin t}\right)^{1/2} \exp\left(\frac{i}{\sin t}\left[\frac{1}{2}(q^2 + z^2)\cos t - qz\right]\right).$$

A direct computation shows that for $t \neq 0$ the kernel K satisfies the Schrödinger equation

$$\frac{1}{i}\frac{\partial K}{\partial t} = \frac{1}{2}\left(\frac{\partial^2 K}{\partial q^2} + q^2 K\right)$$

and the method of stationary phase shows that $\lim_{t\to 0} K(q,t;z,0) \to \delta(q-z)$. Reverting to the original coordinates one can write

$$K(q,t;z,0) = \left(\frac{mf}{2\pi i \sin ft}\right)^{1/2} \exp\left\{\frac{imf}{\sin ft}\left[\tfrac{1}{2}(q^2+z^2) - qz\right]\right\}.$$

We see that the function

$$h(t,q) = (U(t)h)(q) = \int K(q,t;z,0)h(z)\,dz$$

is the solution of the Schrödinger equation

$$\frac{1}{i}\frac{\partial h}{\partial t} = \frac{1}{2m}\frac{\partial^2 h}{\partial q^2} + \frac{mf^2}{2}q^2 h$$

and initial conditions

$$h(0,q) = h(q).$$

The operators $U(t)$ are unitary and satisfy $U(t+s) = U(t)U(s)$. Furthermore, the operator $U(t)$ is defined at the values $t = k\pi$ by the method of stationary phase. Notice, however, that $U(t)$ is *not* periodic of period π. Indeed, on account of the square root in $\sin^{1/2} t$ we see that $U(t+2\pi) = -U(t)$ so that U is periodic of period 4π. Thus U provides a representation of the double covering of the circle of rotations in the plane, rather than of the circle itself.

We can generalize the preceding discussion. Let f be a function on T^*M which generates the one parameter group, g_t. Let $\mathfrak{M}_0$ denote the cotangent polarization and let us suppose that, for t sufficiently close to zero, $t \neq 0$, the polarizations $\mathfrak{M}_0$ and $\mathfrak{M}_t = g_t \mathfrak{M}_0$ are completely transverse, so that the map

$$\psi\colon T^*M \to M \times M, \qquad \psi(z) = (\pi z, \pi g_{-t} z)$$

is a diffeomorphism. To compute the pairing, we must find a function, φ_t, such that

$$d\varphi_t = g_{-t}^*\alpha - \alpha.$$

We claim that we may choose

$$\varphi_t \psi^{-1}(x,y) = \int_\gamma \alpha - f\,dt = \int_C \mathfrak{L}\,dt$$

where $\mathfrak{L}$ is the "Lagrangian" corresponding to the "Hamiltonian" f, and C is the classical path joining x to y, with γ the lifting of this path to T^*M (determined by $\mathfrak{L}$). Indeed, we can consider the function $F = f - E$ on $T^*(M \times \mathbf{R})$ and the Hamiltonian flow it generates, where E is the coordinate on $\mathbf{R}^*$. Then the above equation is precisely the content of Hamilton's method of characteristics, described in Chapter III, §3. (The two dimensional case there is replaced by the n dimensional case here, with t now playing the role of the auxiliary variable. The above equation is then a consequence of equation (3.3) of Chapter III.) By applying the same type of calculations done earlier in this section we see that the associated operator, U_t, is given by

$$(U_t h)(x_{-t}) = (2\pi i)^{-n/2} \int e^{iS_t(x_{-t},x)}(\det(\partial^2 S_t/\partial x_{-t}\partial x))^{-1/2} h(x)\,dx$$

where $S_t = \varphi_t \circ \psi$. It is well known that this is not the correct formula for the quantization procedure in general, but only an approximation to the formal expression given by the Feynman path integral method; cf. the end of Chapter II. Thus a more intricate procedure will be required in general. Nevertheless, the current procedure seems to be correct for many of the problems of group representations. At the end of this chapter we present a preliminary computation which gives some indication of the kinds of geometric conditions which are involved in justifying the procedure used here.

§4. Metalinear manifolds and half forms.

In this section we introduce the concept of half form. Let M be a manifold of dimension n. We can think of an n form at a point, in terms of its transformation properties: an n form ω at x assigns a number, $\omega(\mathbf{v})$, to each basis $\mathbf{v} = (v_1, \ldots, v_n)$ of $T(X)_x$. If $A = (A_{ij}) \in GL(n, \mathbf{R})$ and $\mathbf{w} = \mathbf{v}A$, i.e., $\mathbf{w} = (w_1, \ldots, w_n)$ where $w_j = \sum A_{ij} v_i$, then $\omega(\mathbf{w}) = \omega(\mathbf{v}) \det A$. We have defined a density of order s as an object which changes according to $\omega(\mathbf{w}) = \omega(\mathbf{v}) |\det A|^s$. We would like to be able to define a half form as an object which changes according to $\omega(\mathbf{w}) = \omega(\mathbf{v}) \cdot (\det A)^{1/2}$. The trouble is that $(\det A)^{1/2}$ is not a well defined function on $GL(n, \mathbf{R})$. We must, in fact, pass to a double covering of $GL(n, \mathbf{R})$, and a corresponding double covering of the bundle of bases of $T(X)$.

Since $GL(n, \mathbf{R})$ has two components (the matrices of positive and of negative determinants) we expect that the double covering we are looking for should have four components and that $(\det A)^{1/2}$ should take on values in $\mathbf{R}^+$, $i\mathbf{R}^+$, $-\mathbf{R}^+$, $-i\mathbf{R}^+$ in each of the four components. For various purposes, it will be convenient to regard $GL(n, \mathbf{R})$ as the subgroup of real matrices lying in $GL(n, \mathbf{C})$.

Recall that every non-singular complex matrix can be written as PU where P is a positive definite Hermitian matrix and U is unitary. We can write every unitary matrix as $U = e^{i\theta} U_1$ where $U_1 \in SU(n)$. Now the group $SU(n)$ is simply connected. Indeed every closed curve starting at the identity can be written as

$$C(t) = A(t)\begin{bmatrix} e^{i\alpha_1(t)} & & 0 \\ & \ddots & \\ 0 & & e^{i\alpha_n(t)} \end{bmatrix} A(t)^{-1},$$

and this can be deformed into the constant curve by considering the deformation $C(s,t)$ given by

$$C(s,t) = A(t)\begin{bmatrix} e^{i\beta_1(s,t)} & & 0 \\ & \ddots & \\ 0 & & e^{i\beta_n(s,t)} \end{bmatrix} A^{-1}(t)$$

where

$$\beta_k(s,t) = (1-s)\alpha_k(t) + s\sum_1^n \alpha_j(t)/n.$$

Thus $\pi_1\, GL(n,\mathbf{R}) = \pi_1\, U(n) = \pi_1\, S^1 = \mathbf{Z}$.

We now examine this fact from a slightly different point of view. Let $\mathbf{C}^* = GL(1,\mathbf{C})$ be the punctured plane, regarded as the group of invertible one by one matrices. The exponential map sending $u \in \mathbf{C}$ to e^u gives $\mathbf{C}$ as the universal covering group of $\mathbf{C}^*$. The group $\mathbf{Z}$ acts as the group of covering transformations, the map of $\mathbf{Z} \times \mathbf{C} \to \mathbf{C}$ being given by

$$(k,u) \rightsquigarrow 2\pi ik + u$$

so that $\mathbf{C}^* = \mathbf{C}/\mathbf{Z}$.

We now do a corresponding computation for $GL(n,\mathbf{C})$: Let $\mathbf{Z}$ act on $\mathbf{C} \times SL(n,\mathbf{C})$, the map

$$\mathbf{Z} \times (\mathbf{C} \times SL(n,\mathbf{C})) \to \mathbf{C} \times SL(n,\mathbf{C})$$

being given by

$$(k,(u,A)) \rightsquigarrow \left(u + \frac{2\pi ik}{n}, e^{-2\pi ik/n}A\right).$$

We claim that relative to this action we have an identification

$$(\mathbf{C} \times SL(n,\mathbf{C}))/\mathbf{Z} = GL(n,\mathbf{C}).$$

Indeed, we have the map

$$\mathbf{C} \times SL(n,\mathbf{C}) \xrightarrow{\pi} GL(n,\mathbf{C})$$

given by

$$(u,A) \rightsquigarrow e^u A$$

which is invariant under the action of $\mathbf{Z}$ and is easily seen to be an isomorphism of $(\mathbf{C} \times SL(n, \mathbf{C}))/\mathbf{Z}$ with $GL(n, \mathbf{C})$. The function det, defined on $GL(n, \mathbf{C})$ by sending $B \in GL(n, \mathbf{C}) \rightsquigarrow \det B \in \mathbf{C}^*$, pulls back to the function $\det \circ \pi$ on $\mathbf{C} \times SL(n, \mathbf{C})$ where

$$(\det \circ \pi)(u, A) = e^{nu}.$$

This function has a well defined holomorphic square root

$$u \rightsquigarrow e^{nu/2}$$

which is already defined on the group

$$(\mathbf{C} \times SL(n, \mathbf{C}))/2\mathbf{Z}.$$

We set

$$(\mathbf{C} \times SL(n, \mathbf{C}))/2\mathbf{Z} = ML(n, \mathbf{C})$$

and call it the metalinear group. It is a double covering of $GL(n, \mathbf{C})$. We let r denote the projection of $ML(n, \mathbf{C})$ onto $GL(n, \mathbf{C})$ and we let χ denote the holomorphic square root of $\det \circ r$ so that

$$\chi^2(C) = \det(rC), \qquad C \in ML(n, \mathbf{C}). \tag{4.1}$$

We now regard $GL(n, \mathbf{R})$ as the subgroup of $GL(n, \mathbf{C})$ consisting of real matrices, and define

$$ML(n, \mathbf{R}) = r^{-1} GL(n, \mathbf{R})$$

so that $ML(n, \mathbf{R})$ is a subgroup of $ML(n, \mathbf{C})$ and is a double cover of $GL(n, \mathbf{R})$. We shall continue to denote by r the restriction of r to $ML(n, \mathbf{R})$ and by χ, the restriction of χ to $ML(n, \mathbf{R})$. It follows from (4.1) that χ can take on values in the four half lines $\mathbf{R}^+$, $i\mathbf{R}^+$, $-\mathbf{R}^+$ and $-i\mathbf{R}^+$ and thus that the group $ML(n, \mathbf{R})$ has four components.

Now let V be any real n dimensional vector space, and let $B(V)$ be the set of bases of V. The group $GL(n, \mathbf{R})$ acts (on the right) on $B(V)$, the map $B(V) \times GL(n, \mathbf{R}) \to B(V)$ sending $(v_1, \ldots, v_n) \times (A_{ij}) \to (\sum v_i A_{i1}, \ldots, \sum v_i A_{in})$. By a *metalinear structure* on V we mean a covering $MB(V) \xrightarrow{\rho} B(V)$ together with an action $MB(V) \times ML(n, \mathbf{R}) \to MB(V)$ consistent with the covering $ML(n, \mathbf{R}) \to GL(n, \mathbf{R})$, i.e. such that the diagram

$$\begin{array}{ccccc} MB(V) & \times & ML(n, \mathbf{R}) & \longrightarrow & MB(V) \\ \rho \downarrow & & \downarrow r & & \downarrow \rho \\ B(V) & \times & GL(n, \mathbf{R}) & \longrightarrow & B(V) \end{array}$$

commutes.

PROPOSITION 4.1. *There exist exactly two metalinear structures on V.*

The space $B(V)$ splits up into two components, B_+ and B_-, with elements of $GL(n,\mathbf{R})_0$ preserving the components and elements of $GL(n,\mathbf{R})^-$ interchanging the components, where $GL(n,\mathbf{R})^-$ denotes the matrices with negative determinant. As a set, it is clear that $MB(V)$ must consist of four components, which we denote by B_{1+}, B_{2+}, B_{1-}, B_{2-}, where B_{1+} and B_{2+} project onto B_+ and where B_{1-} and B_{2-} project onto B_-. It is then clear that the action of $ML(n,\mathbf{R})$ on $MB(V)$ is determined by the action of any element θ with $\chi(\theta) = i$. We must have either $B_{1+}\theta = B_{1-}$ or $B_{1+}\theta = B_{2-}$. In the first case, we have

$$B_{1+} \xrightarrow{\theta} B_{1-} \xrightarrow{\theta} B_{2+} \xrightarrow{\theta} B_{2-} \xrightarrow{\theta} B_{1+}$$

and in the second case we have

$$B_{1+} \xrightarrow{\theta} B_{2-} \xrightarrow{\theta} B_{2+} \xrightarrow{\theta} B_{1-} \xrightarrow{\theta} B_{1+}.$$

Now while the labelling $MB = B_1 \cup B_2$ is conventional, there is no preferred choice of which bases are to be labelled as positive and which are to be labelled as negative. This depends on a choice of orientation. Notice that if we interchange + and − then the second sequence above becomes the first. We can require the first sequence to be standard, and then the two actions of $ML(n,\mathbf{R})$ on $MB(V)$ are specified by a choice of orientation on V:

PROPOSITION 4.2. *A choice of metalinear structure on a vector space V is equivalent to a choice of orientation of V.*

If we complexify the vector space V, so as to obtain $V^{\mathbf{C}} = V \otimes \mathbf{C}$, we can consider the set of complex bases

$$B(V^{\mathbf{C}}) = B(V) \underset{GL(n,\mathbf{R})}{\times} GL(n,\mathbf{C}).$$

If we are given a real metalinear structure $MB(V)$ on V we can form a complex metalinear structure on $V^{\mathbf{C}}$ by setting

$$MB(V^{\mathbf{C}}) = MB(V) \underset{ML(n,\mathbf{R})}{\times} ML(n,\mathbf{C}).$$

Now the group $ML(n,\mathbf{C})$ has an involutive antiholomorphic automorphism covering the automorphism of $GL(n,\mathbf{C})$ sending a matrix into its complex conjugate. (Namely the map $(z,A) \rightsquigarrow (-z,\bar{A})$ of $\mathbf{C} \times SL(n,\mathbf{C})$ quotients to a well defined automorphism of $ML(n, \mathbf{C})$. We denote the image of any $B \in ML(n, \mathbf{C})$ under this automorphism by B. Notice that on $ML(n, \mathbf{R}) \subset ML(n, \mathbf{C})$ this is just the outer automorphism mentioned above. Given the complex metalinear structure $MB(V^{\mathbf{C}})$ we can define a new complex metalinear structure (and a new complex linear structure) by leaving the set $MB(V^{\mathbf{C}})$ unchanged but defining a new action, by first applying the automorphism $B \rightsquigarrow \overline{B}$. (Thus if $f \in MB(V^{\mathbf{C}})$, the element B, in the old action, sends f into fB and in the new action sends f into $f\overline{B}$.) This is the conjugate complex metalinear structure on $V^{\mathbf{C}}$. If we regard $MB(V)$ as a subset of $MB(V^{\mathbf{C}})$ then the conjugate metalinear structure induces the other real metalinear structure on V. We thus say that the two (real) metalinear structures on V are conjugates of one another.

We recall that an n form on an n dimensional vector space V can be defined as a function, Ω, from $B(V) \to \mathbf{C}$ which satisfies

$$\omega(bA) = (\det A)\omega(b)$$

and a density is a function σ: $B(V) \to \mathbf{C}$ satisfying

$$\sigma(bA) = |\det A|\sigma(b).$$

Given a metalinear structure $MB(V)$, we now define a *half form* to be a map ρ: $MB(V) \to \mathbf{C}$ which satisfies

$$\rho(fB) = \chi(B)\rho(f).$$

Let ρ_1 and ρ_2 be half forms, then, by (4.1),

$$\begin{aligned}\rho_1(fB)\rho_2(fB) &= \rho_1(f)\rho_2(f)\chi^2(B)\\ &= \rho_1(f)\rho_2(f)\det(rB).\end{aligned}$$

In particular, if f and f' project onto the same element of $B(V)$ then $f = f'C$ where $rC = I$ and the above equation shows that $\rho_1(f)\rho_2(f) = \rho_1(f')\rho_2(f')$ and so $\rho_1\rho_2$ gives a well defined function from $B(V) \to \mathbf{C}$ and the above equation shows that $\rho_1\rho_2$ is an n form on V. Taking $\rho_1 = \rho_2 = \rho$ shows that a half form can be regarded as a "square root" of an n form. Let us denote the line of half forms on V by $\wedge^{1/2} V$, let us denote the line of n forms on V by $\wedge^n V$, and let us denote the line of densities on V by $|\wedge|(V)$. Then we have seen that there is a bilinear pairing

$$\wedge^{1/2}(V) \times \wedge^{1/2}(V) \to \wedge^n(V), \qquad \rho_1 \times \rho_2 \to \rho_1\rho_2,$$

and a similar argument shows that there is a sesquilinear pairing

$$\wedge^{1/2}(V) \times \wedge^{1/2}(V) \to |\wedge|(V), \qquad \rho_1 \times \rho_2 \to \rho_1\bar{\rho}_2.$$

We can also introduce the space, $\overline{\wedge}^{1/2}(V)$, of conjugate half forms, τ, which satisfy

$$\tau(fB) = \overline{\chi}(B)\tau(f)$$

(which are half forms relative to the conjugate metalinear structure) and the space of negative half forms, $\wedge^{-1/2}(V)$, which satisfy

$$\tau(fB) = \chi^{-1}(B)\tau(f).$$

PROPOSITION 4.3. *A metalinear structure on V induces a metalinear structure on its dual space, V^*, so that $\wedge^{-1/2}(V) \sim \wedge^{1/2}(V^*)$.*

PROOF. There is a natural isomorphism, as sets, from $B(V)$ to $B(V^*)$, as each basis, b, of V induces a basis, b^*, of V^*. For any $A \in GL(n, \mathbf{R})$ we have $(bA)^* = b^* A^{t-1}$, where A^t denotes the transpose of A. Thus we can identify $B(V^*)$ with $B(V)$ provided that we define a new action of $GL(n, \mathbf{R})$ on $B(V)$ via the automorphism $A \rightsquigarrow A^{t-1}$. Now it is clear that we can define an automorphism of $ML(n, \mathbf{R})$ covering this automorphism of $GL(n, \mathbf{R})$, which we shall continue to denote by $B \rightsquigarrow B^{t-1}$, such that $\chi(B^{t-1}) = \chi(B)^{-1}$. (It is the restriction to $ML(n, \mathbf{R})$ of the unique holomorphic automorphism of $ML(n, \mathbf{C})$ covering the automorphism $A \rightsquigarrow A^{t-1}$ of $GL(n, \mathbf{C})$.) We can now choose $MB(V^*)$ to be identical with $MB(V)$ as a set, together with the new action of $ML(n, \mathbf{R})$, and it is clear that $\wedge^{1/2}(V^*)$ can be identified with $\wedge^{-1/2}(V)$. We will express the construction of Proposition 4.3 by saying that a metalinear frame, f, of V determines a metalinear frame, f^*, of V^* so that $(fA)^* = fA^{t-1}$.

PROPOSITION 4.4. *Let $0 \to U \to V \to W \to 0$ be an exact sequence of vector spaces. Then a choice of metalinear structures on any two of these spaces determines a metalinear structure on the third. When metalinear structures are so chosen consistently, we say that we have an exact sequence of metalinear spaces. Given such an exact sequence of metalinear spaces we have the natural isomorphism*

$$\wedge^{1/2}(V) \cong \wedge^{1/2}(U) \otimes \wedge^{1/2}(W).$$

PROOF. By Proposition 4.2, a choice of metalinear structure is the same as a choice of orientation. It is well known that a choice of orientation on any two of the three spaces, U, V, W in the exact sequence determines an orientation on the third. We review how this goes: Let $B_T(V)$ denote the set of bases of V for which the first m vectors belong to U, where $\dim U = m$ and $\dim W = n$. The set $B_T(V)$ is stable under the group $T(m, n)$ of triangular matrices of the form:

$$\begin{matrix} & m & n \\ \begin{matrix} m \\ n \end{matrix} & \multicolumn{2}{c}{\begin{pmatrix} A & B \\ 0 & D \end{pmatrix}} \end{matrix}.$$

Any element of $B_T(V)$ determines a basis of U (the first m elements) and a basis of W (the image of the last n elements). Two elements of $B_T(V)$ determine the same basis of U and the same basis of W if and only if they differ by the action of a matrix of the form

$$\begin{pmatrix} I_m & B \\ 0 & I_n \end{pmatrix}.$$

We let N be the group of such matrices, so we have the identification

$$B(U) \times B(W) = B_T(V)/N.$$

We also have the group isomorphism

$$GL(m) \times GL(n) \cong T(m,n)/N$$

consistent with the above identification. Then, making a choice of positive frames in U and in W gives a family of frames in V which differ from one another by matrices with positive determinant and hence give an orientation on V. Let us start with an orientation on V and on U. If restriction of the given frames of V to U are positive, the projection of these frames to W are declared positive, otherwise they are negative. Similarly if we start with W and V.

Now suppose we are given an exact sequence of metalinear vector spaces $0 \to U \to V \to W \to 0$. The group N is contractible, and hence lifts to a subgroup of $ML(m + n)$, which we continue to denote by N. We let $MT(m,n)$ denote the inverse image of $T(m,n)$ in $ML(m + n)$, and let $MB_T(V)$ denote the inverse image of $B_T(V)$ in $MB(V)$. Thus $MT(m,n)$ acts on $MB(V)$. We can think of $ML(m)$ as a subgroup of $ML(m + n)$, being the inverse image in $ML(m + n)$ of all matrices of the form

$$\begin{pmatrix} A & 0 \\ 0 & I_n \end{pmatrix}$$

and similarly we can think of $ML(n)$ as a subgroup of $ML(m + n)$. We thus have an isomorphism

$$ML(m) \times ML(n) \cong MT(m,n)/N.$$

We thus get a map

$$MB(U) \times MB(W) \xrightarrow{\cong} MB_T(V)/N$$

consistent with the above action, and specified by the requirements that it project onto the identification $B(U) \times B(W) \to B_T(V)/N$ and that $B_{1+}(U) \times B_{1+}(W)$ be mapped onto $B_{1+}(V)$. Since χ takes on the value one for any element of N (because N is contractible), we see that we do indeed get the identification

$$\wedge^{1/2}(U) \otimes \wedge^{1/2}(W) \cong \wedge^{1/2}(V).$$

If E is a real vector bundle over some manifold, X, we let $B(E)$ denote the bundle of bases of E. Thus $B(E)$ is a principal bundle for the group $ML(n)$, where n is the fiber dimension of E. By a metalinear structure on E we mean a lifting of the $GL(n)$ bundle to a $ML(n)$ bundle, $MB(E)$. Thus a metalinear structure on E is a principal $ML(n)$ bundle, $MB(E)$, with a projection onto $B(E)$ so that the diagram

$$\begin{array}{ccccc} MB(E) & \times & ML(n) & \longrightarrow & MB(E) \\ \downarrow & & \downarrow & & \downarrow \\ B(E) & \times & GL(n) & \longrightarrow & B(E) \end{array}$$

commutes. Not every vector bundle need have a metalinear structure—there may be some topological obstruction to the construction of a lifting of the bundle $B(E)$. (We will briefly touch upon this point at the end of this section.) If E is orientable, then it clearly possesses a metalinear structure. However E might admit a metalinear structure even if it is not orientable. Associated to every metalinear bundle, E, we have a bundle of half forms, $\wedge^{1/2} E$, and the analogues of Propositions 4.3 and 4.4 hold.

In particular, if M is a differentiable manifold, and TM carries a metalinear structure, we say that M is a metalinear manifold. We denote the space of smooth sections of $\wedge^{1/2} TM$ by $\wedge^{1/2} M$, and call such a section a smooth half form on M. If ρ_1 and ρ_2 are half forms, then $\rho_1 \bar{\rho}_2$ is a density (of order one) on M. Thus, if $\wedge_0^{1/2} M$ denotes the space of smooth half forms of compact support, we can make $\wedge_0^{1/2} M$ into a pre-Hilbert space under the scalar product

$$(\rho_1, \rho_2) = \int_M \rho_1 \bar{\rho}_2 .$$

The completion of this Hilbert space will be denoted by $H(M)$.

Let M and M' be two metalinear manifolds. By a morphism, f, from M to M' we shall mean a morphism of the corresponding line bundles, $\wedge^{1/2} TM$ and $\wedge^{1/2} TM'$—thus a map $f\colon M \to M'$ together with maps $\wedge^{1/2} TM'_{f(x)} \to \wedge^{1/2} TM_x$ for each $x \in M$, ranging smoothly with x. This then gives a map, f^*, from $\wedge^{1/2} M' \to \wedge^{1/2} M$, which we call the pull back map on half forms.

For the sake of completeness we shall finish this section by determining when $E \to X$ supports a metalinear structure and exactly how many such structures it supports up to topological equivalence. (Note that a metalinear structure and its conjugate structure are topologically equivalent.)

If V is a vector space then, as we have seen above, the choice of a metalinear structure on V is the choice of an action of the group $Z_4 = \{1, -1, i, -i\}$ on the components of $MB(V)$ compatible with the action of $Z_2 = \{1, -1\}$ on the components of $B(V)$. Thus the choice of a metalinear structure on $E \to X$ is the choice of a Z_4-bundle over X covering the Z_2-bundle $\wedge^k(E)/\mathbf{R}^+$, k being the fiber dimension of E. Consider the short exact sequence of groups

$$0 \to Z_2 \xrightarrow{i} Z_4 \xrightarrow{j} Z_2 \to 0,$$

i being the inclusion map and j being multiplication by 2. This induces a long exact sequence in cohomology:

$$0 \to H^1(X, Z_2) \xrightarrow{i_*} H^1(X, Z_4) \xrightarrow{j_*} H^1(X, Z_2) \xrightarrow{\delta} H^2(X, Z_2) \to \cdots.$$

The homomorphism j_* has a nice geometric interpretation. A Z_n-bundle is classified up to equivalence by its Stieffel-Whitney class which lies in $H^1(X, Z_n)$. The map $j_*: H^1(X, Z_4) \to H^1(X, Z_2)$ corresponds to the covering (of Z_2-bundles by Z_4-bundles) described above. Therefore, if $w = w_2(E)$ is the Stieffel-Whitney class of $\wedge^k(E)/\mathbf{R}^+$, we see that a necessary and sufficient condition for E to admit a metalinear structure is that w be in the image of j_* or, equivalently, in the kernel of δ. However, δ is just the Bockstein homomorphism. In dimension one it is equal to the homomorphism $a \to a^2$. (See Spanier, *Algebraic Topology*, Chapter 5, §9. Also see exercise G5 on p. 280.) Therefore, *E admits a metalinear structure if and only if the square of its first Stieffel-Whitney class is zero.* Assuming E does admit a metalinear structure the set of all such up to topological equivalence is equal to the set of all elements, a, of $H^1(X, Z_4)$ such that $j_* a = w$. Given one such a, every one is of the form $a + i_* b$ where b is in $H^1(X, Z_2)$. Since i_* is injective this shows that *the set of all metalinear structures has the same cardinality as $H^1(X, Z_2)$. In particular if $H^1(X, Z_2) = 0$, X admits a unique metalinear structure.* For the circle, $w_1^2 = 0$ automatically since $H^2(S^1, Z_2) = 0$. Also, $H^1(S^1, Z_2) = Z_2$, so *the circle admits just two inequivalent metalinear structures.* An application of this result will be found in §8 of this Chapter and in Chapter VII, §3.

§5. Metaplectic manifolds.

The correct formulation of the pairing described in §3 requires a choice of metalinear structure on each Lagrangian subspace of $T(X)$, where X is a symplectic manifold. In order to do this in a consistent manner, we have to impose additional structure on X. This is done by requiring that the bundle of symplectic frames on X can be lifted to a double covering much as was the situation in defining a metalinear structure. Here a symplectic frame, $e_1, \ldots, e_n, f_1, \ldots, f_n$, is one such that $\omega(e_i, e_j) = \omega(f_i, f_j) = 0$ and $\omega(e_i, f_j) = \delta_{ij}$. In order to discuss the possible double coverings of the bundle of symplectic frames, we must first go back and discuss the double covering of the symplectic group. This requires a review of some of the facts about the symplectic group itself.

Let $Sp(n, \mathbf{R})$ be the real symplectic group acting on $\mathbf{R}^{2n}$. Its elements are matrices $T \in GL(2n, \mathbf{R})$ with the block form

$$T = \begin{bmatrix} T_1 & T_2 \\ T_3 & T_4 \end{bmatrix}$$

satisfying

$$T_4^* T_1 - T_2^* T_3 = I$$

and

$$T_3^* T_1 \text{ and } T_4^* T_2 \text{ are symmetric.}$$

The matrices of the form

$$\begin{bmatrix} A & -B \\ B & A \end{bmatrix} \quad \text{with } A^* A + B^* B = I$$

form a compact subgroup, which is clearly isomorphic to $U(n)$ under the isomorphism sending the above matrix into the unitary matrix $A + iB$. We already know that this a maximal compact subgroup, and that $Sp(n)$ is diffeomorphic to a space $V \times U(n)$ where V is topologically a cell. (This is the consequence of polar decomposition, about which we shall have a great deal to say further on in this section.) In particular, as we have already seen, the fundamental group of $Sp(n)$ is $\mathbf{Z}$ so that $Sp(n)$ has a unique double covering, which we denote by $Mp(n)$ and call, following Weil [12], the metaplectic group. Notice that $GL(n, \mathbf{R})$ is naturally embedded in $Sp(n, \mathbf{R})$ via the map sending the matrix $A \in GL(n, \mathbf{R})$ into

$$\begin{bmatrix} A & 0 \\ 0 & A^{*-1} \end{bmatrix}$$

in $Sp(n, \mathbf{R})$. Geometrically, the subgroup is the one which leaves invariant each of the pair of Lagrangian subspaces spanned by the first n and by the last n basis vectors. Let G denote the inverse image of this subgroup in $Mp(n, \mathbf{R})$. It follows from the same argument as that presented in the beginning of §4 that G_0 is isomorphic to $GL(n, \mathbf{R})_0$. We claim that G is in fact just $ML(n, \mathbf{R})$; that is, we claim that there exists a unique isomorphism of $ML(n, \mathbf{R})$ to $Mp(n, \mathbf{R})$ which makes the diagram

$$\begin{array}{ccc} ML(n, \mathbf{R}) & \longrightarrow & Mp(n, \mathbf{R}) \\ \downarrow & & \downarrow \\ GL(n, \mathbf{R}) & \longrightarrow & Sp(n, \mathbf{R}) \end{array}$$

commute. Indeed the covering $ML(n, \mathbf{R}) \to GL(n, \mathbf{R})$ is completely determined by the corresponding covering of the orthogonal group. For orthogonal matrices, A, the composite map

$$A \to \begin{bmatrix} A & 0 \\ 0 & A^{*-1} \end{bmatrix} = \begin{bmatrix} A & 0 \\ 0 & A \end{bmatrix} \to A + i0$$

is just the standard embedding of $O(n)$ into $U(n)$. The coverings of $U(n)$ (both in $Mp(n)$ and $ML(n\mathbf{C})$ coincide, and the covering of $O(n)$ is, by definition, the one induced from $ML(n, \mathbf{C})$. From now on we shall regard $ML(n, \mathbf{C})$ as a subgroup of $Mp(n)$ via this identification.

Now let X be a symplectic manifold, and let $Bp(X)$ denote the bundle of symplectic frames of TX. Thus $Bp(X)$ is a (right) principal $Sp(n)$ bundle. A lifting of $Bp(X)$ to a (right) principal $Mp(n)$ bundle is called a *metaplectic structure* on X, and the corresponding bundle is denoted by $Mp(X)$, and called the bundle of metaplectic frames. Of course, such a lifting need not exist. There are some simple topological conditions which will guarantee the existence of such a structure. Furthermore, a given symplectic manifold might admit several metaplectic structures. We will discuss this point at the end of the present section. We claim that the choice of a metaplectic structure on X has the effect of putting a metalinear structure on each Lagrangian subspace $E_x \subset TX_x$ at each $x \in X$. More precisely, let $L(X)$ denote the bundle of all Lagrangian subspaces of $T(X)$, and let E be the tautologous vector bundle assigning to $\ell \in L$ the subspace of TX that it determines. Thus

$$\begin{array}{ccc} E & \longrightarrow & L \\ \downarrow & & \downarrow p \\ TX & \longrightarrow & X \end{array}$$

is a diagram representing the situation. Let $B(E)$ denote the bundle of bases of E, a (right) $GL(n, \mathbf{R})$ bundle. Each basis of E_x extends to a symplectic basis of TX_x. Conversely, any symplectic basis, $(e_1, \ldots, e_n, f_1, \ldots, f_n)$ determines a Lagrangian subspace, namely the one spanned by $e_1, \ldots, e_n$, and the basis, $e_1, \ldots, e_n$ of that subspace. Thus the map, $\lambda\colon Bp(X) \to B(E)$

$$\lambda(e_1, \ldots, e_n, f_1, \ldots, f_n) = (e_1, \ldots, e_n)$$

is a surjective smooth bundle map. Furthermore it is clear any two frames in $\lambda^{-1}(e_1, \ldots, e_n)$ differ by the action of a matrix of the form

$$\begin{bmatrix} I & S \\ 0 & I \end{bmatrix}$$

where S is a symmetric matrix. In other words, if we let N denote the group of all such matrices, then $Bp(X) \xrightarrow{\lambda} B(E)$ is a principal N bundle. Notice also that $GL(n, \mathbf{R}) \subset Sp(n, \mathbf{R})$ normalizes N, i.e.,

$$\begin{pmatrix} A & 0 \\ 0 & A^{*-1} \end{pmatrix}\begin{pmatrix} I & S \\ 0 & I \end{pmatrix}\begin{pmatrix} A^{-1} & 0 \\ 0 & A^* \end{pmatrix} = \begin{pmatrix} I & ASA^* \\ 0 & I \end{pmatrix}.$$

So if we consider $GL(n, \mathbf{R})$ acting on $Bp(X)$ and also on E we see that the map λ is equivariant with respect to the action of $GL(n, \mathbf{R})$. This is of course obvious from the explicit expression of the action. Now the group N is clearly simply connected, and hence lifts isomorphically to a subgroup of $Mp(n)$, which we shall continue to denote by N cf. Chapter IV, Proposition 2.5. Now consider the bundle $Mp(X)/N$. Thus we have the diagram

$$\begin{array}{ccc} Mp(X) & \xrightarrow{\hat{\lambda}} & Mp(X)/N = MB(E) \\ \downarrow s & & \downarrow r \\ Bp(X) & \xrightarrow{\lambda} & B(E) \end{array} \tag{5.1}$$

Since $Mp(X)$ is a double covering of $Bp(X)$ consistent with the action of $Mp(n)$, and since $ML(n, \mathbf{R})$ normalizes N in $Mp(n)$, we conclude that r provides a double covering of $B(E)$ giving E a metalinear structure.

In particular, if $Y \hookrightarrow X$ is a submanifold, and E_1 is a bundle of Lagrangian subspaces over Y then E_1 carries a metalinear structure. If E_1 and E_2 are two transverse bundles of Lagrangian subspaces then each carries a metalinear structure. Now the *symplectic* structure on X allows us to identify the *linear* structure on E_1 with the dual of the linear structure on E_2–the spaces E_{1x} and E_{2x} are non-singularly paired under ω. We shall show that the *metaplectic* structure on X determines an isomorphism between the metalinear structure on E_1 and the complex conjugate dual metalinear structure on E_2. In particular, we shall show that there is a natural sesquilinear pairing between $\wedge^{1/2}(E_1)$ and $\wedge^{1/2}(E_2)$, induced by the metaplectic structure on X.

To motivate our construction, let us redo the linear pairing between E_1 and E_2 in a more group theoretical setting. Let $(e_1, \ldots, e_n)$ be a basis of E_{1x}. Then there is a unique basis $(f_1, \ldots, f_n)$ of E_{2x} such that $\mathbf{u} = (e_1, \ldots, e_n, f_1, \ldots, f_n)$ is a symplectic basis of TX_x. Let J be the matrix

$$J = \begin{bmatrix} 0 & -I \\ I & 0 \end{bmatrix}.$$

Then $J \in Sp(n)$ and

$$\mathbf{u}J = (f_1, \ldots, f_n, -e_1, \ldots, -e_n)$$

so that

$$\lambda(\mathbf{u}) = (e_1, \ldots, e_n) \quad \text{and} \quad \lambda(\mathbf{u}J) = (f_1, \ldots, f_n)$$

where λ is the projection introduced above. The identification of E_1 with E_2^* identifies $(e_1, \ldots, e_n)$ with the dual basis to $(f_1, \ldots, f_n)$. We will perform a similar construction for the bundle of metaplectic frames. We first must find the analogue to J. For this purpose we notice that $J \in Sp(n)$ has the peculiar property that it also belongs to the Lie algebra $sp(n)$ and satisfies

$$J = \exp \frac{\pi}{2} J$$

where $\exp: sp(n) \to Sp(n)$ is the exponential map. We now let $\operatorname{Exp}: sp(n) \to Mp(n)$ be the exponential map in the metaplectic group, and let us set

$$\hat{J} = \operatorname{Exp} \frac{\pi}{2} J$$

so that $\hat{J}$ is one of two elements of $Mp(n)$ which project onto J. Let us be given $\hat{\mathbf{e}} \in MB(E_1)_x$. We now claim that *there is a unique* $\hat{\mathbf{u}} \in Mp(X)_x$ *such that*

$$\hat{\lambda}(\hat{\mathbf{u}}) = \hat{\mathbf{e}} \qquad \text{and} \qquad \hat{\lambda}(\hat{\mathbf{u}}J) = \mathbf{f} \in MB(E_2)_x.$$

The elements $\hat{\mathbf{e}}$ determine a frame $\mathbf{e} = r\hat{\mathbf{e}} \in B(E_1)$, where the map r and the other maps we currently use are given in diagram (5.1). We know that there exists a unique $\mathbf{w} \in Bp(X)$ such that

$$\lambda(\mathbf{w}) = r\mathbf{e} \qquad \text{and} \qquad \lambda(\mathbf{w}J) \in Bp(E_2)_x$$

Now choose $\hat{\mathbf{w}} \in Mp(X)$ with

$$s\hat{\mathbf{w}} = \mathbf{w}, \qquad \hat{\lambda}\hat{\mathbf{w}} = \hat{\mathbf{e}}.$$

This uniquely determines $\hat{\mathbf{w}}$ and thus picks out $\hat{\mathbf{f}} = \hat{\lambda}(\hat{\mathbf{w}}\hat{J})$. Notice that if we allow $\hat{\mathbf{e}}$ to be an arbitrary metalinear basis of E_1 then $\mathbf{w}$ is determined up to an element of $GL(n, \mathbf{R})$, and $\hat{\mathbf{w}}$ is thus determined up to an element of $ML(n, \mathbf{R})$.

We now introduce a pairing between half forms on E_{1x} and half forms on E_{2x}. Let $\sigma_1 \in \wedge^{1/2} E_{1x}$ and $\sigma_2 \in \wedge^{1/2} E_{2x}$. We wish to define a number $\langle \sigma_1, \sigma_2 \rangle$, depending linearly on σ_1, and antilinearly on σ_2. For this purpose, choose a metaplectic frame, $\hat{\mathbf{w}}$ as above, so that $\hat{\mathbf{e}} = \hat{\lambda}(\hat{\mathbf{w}})$ is a metaplectic basis of E and $\hat{\mathbf{f}} = \hat{\lambda}(\hat{\mathbf{w}}\hat{J})$ is a metaplectic basis of E_{2x}. Then we can write

$$\sigma_1(\hat{\mathbf{e}}) = a_1, \sigma_2(\hat{\mathbf{f}}) = a_2 \qquad \text{where } a_1, a_2 \in \mathbf{C}.$$

We then set

$$\langle \sigma_1, \sigma_2 \rangle = (2\pi)^{-n/2} e^{-i\pi n/4} a_1 \bar{a}_2. \tag{5.2}$$

Here the term $(2\pi)^{-n/2} e^{-i\pi n/4}$ is just a normalizing factor introduced to cancel out terms arising from the method of stationary phase. We must show that the right hand side of (5.2) does not depend on the choice of $\hat{\mathbf{w}}$. Suppose we make some other choice, say $\hat{\mathbf{v}}$. Then $\hat{\mathbf{v}} = \hat{\mathbf{w}}\hat{g}$ for some $\hat{g} \in ML(n, \mathbf{R}) \subset Mp(n)$. This has the effect of replacing $\hat{\mathbf{e}}$ by $\hat{\mathbf{e}}\hat{g}$ and $\hat{\mathbf{f}}$ by

$$\hat{\lambda}(\hat{\mathbf{w}}\hat{g}\hat{J}) = \hat{\lambda}(\hat{\mathbf{w}}\hat{J}\hat{J}^{-1}\hat{g}J) = \hat{\mathbf{f}}(\hat{J}^{-1}\hat{g}\hat{J}).$$

This means that

$$a_1 \text{ is replaced by } \chi(\hat{g})a_1$$

and

$$a_2 \text{ is replaced by } \chi(\hat{J}^{-1}\hat{g}\hat{J})a_2.$$

We must show that

$$\chi(\hat{g})\bar{\chi}(\hat{J}^{-1}\hat{g}\hat{J}) = 1. \tag{5.3}$$

Our procedure for establishing (5.3) will be to use the so called "polar decomposition" on $Mp(n)$. We shall show that every element of $Mp(n)$ can be written as

$$\hat{g} = (\exp p)\hat{u}$$

where p is a symmetric element of $sp(n)$ and $\hat{u} \in MO(n)$. We shall also see that

$$\hat{J}^{-1}(\text{Exp } p)\hat{J} = \text{Exp}(-p)$$

and

$$\hat{J}^{-1}\hat{u}J = \hat{u}.$$

Finally, since $(\text{Exp } p) \mathrel{\hat{\in}} ML(n)_0$ we conclude that

$$\chi(\text{Exp } p) > 0 \text{ is real}$$

while

$$|\chi(u)| = 1$$

since $u \in MO(n)$. Thus

$$\chi(\hat{J}^{-1}\hat{g}J) = \chi(\text{Exp } -p)\chi(u) = \bar{\chi}^{-1}(\text{Exp } p)\bar{\chi}^{-1}(u) = \bar{\chi}(\hat{g})^{-1}.$$

In order to carry out this analysis of $Mp(n)$ and the character χ we need to make use of the notion of polar decomposition. We discuss this matter in some detail for the convenience of the reader, even though it is standard, cf. for example, [14]. The knowledgeable reader can skip directly to the end of the discussion. For the group $GL(n, \mathbf{R})$ we recall that every non-singular matrix T can be written as $T = PO$ where P is positive definite and O is orthogonal. (We sketch the proof: TT^* is positive definite hence has a positive definite square root P. We write $T = PP^{-1}T$ and observe that $(P^{-1}T)(P^{-1}T)^* = P^{-1}TT^*P^{-1} = P^{-1}P^2P^{-1} = I$ so $O = P^{-1}T$ is orthogonal.) On the infinitesimal level the analogous decomposition is even simpler. The Lie algebra of the orthogonal group consists of antisymmetric matrices while every positive matrix $P = e^S$ where S is symmetric. Then any matrix B can be written as $B = A + S$ where S is symmetric and A is antisymmetric. Of course this decomposition is not an invariant of $GL(n)$. It turns out that one can describe an analogous polar decomposition for an arbitrary connected real semi-simple Lie group, provided that the corresponding infinitesimal decomposition, known as a *Cartan decomposition*, is given in its Lie algebra.

Let g be a real semi-simple Lie algebra. Let

$$g = k + p$$

be a vector space direct sum decomposition of g satisfying

$$[k,k] \subset k \qquad \text{(so that } k \text{ is a subalgebra)}, [k,p] \subset p \text{ and } [p,p] \subset k.$$

(Notice that for $g = g\ell(n)$ we may take k to consist of the antisymmetric matrices and p to consist of the symmetric ones. This yields such a decomposition.) Let θ: $g \to g$ be defined by

$$\theta(\xi + \eta) = \xi - \eta \qquad \text{for } \xi \in k \text{ and } \eta \in p.$$

Then θ is an automorphism of g, i.e.,

$$[\theta\zeta_1, \theta\zeta_2] = \theta[\zeta_1, \zeta_2] \qquad \text{and} \qquad \theta^2 = \text{id}.$$

Conversely, given an involutary automorphism θ, every $\zeta \in g$ can be written as $\zeta = \xi + \eta$ where $\xi = \frac{1}{2}(\zeta + \theta\zeta)$ and $\eta = \frac{1}{2}(\zeta - \theta\zeta)$, and thus, $\theta\xi = \xi$ and $\theta\eta = -\eta$. If we let

$$k = \{\xi \mid \theta\xi = \xi\} \qquad \text{and} \qquad p = \{\eta \mid \theta\eta = -\eta\},$$

then this provides a decomposition of g as above. Let B denote the Killing form of g. We can then introduce the bilinear form $(\,,)$ by

$$(\zeta_1, \zeta_2) = -B(\zeta_1, \theta\zeta_2).$$

We say that θ is a *Cartan involution* and that $g = k + p$ is a *Cartan decomposition* if $(\,,)$ is positive definite. What amounts to the same thing, $g = k + p$ is a Cartan decomposition if B is negative definite on k and is positive definite on p. For example, in the case of $s\ell(n, \mathbf{R})$ the Killing form is (proportional to)

$$B(Z, W) = \text{tr } ZW, \qquad \text{where } Z \text{ and } W \text{ are regarded as } n \times n \text{ matrices}.$$

Notice that if W is antisymmetric

$$(Z, W) = -\text{tr } ZW = \text{tr } ZW^*$$

while if W is symmetric

$$(Z, W) = \text{tr } ZW = \text{tr } ZW^*$$

so in all cases

$$(Z, W) = \operatorname{tr} ZW^*$$

is positive definite.

Let us now compute the Killing form for $sp(n)$ in terms of the Poisson bracket. Here the elements of $sp(n)$ can be regarded as quadratic functions of the p's and q's, acting on the linear functions via Poisson bracket. Again, up to a scalar,

$$\begin{aligned} B(f, g) &= \operatorname{tr}\{f, \{g, \cdot\}\} = \sum \frac{\partial}{\partial p_i}\{f, \{g, p_i\}\} + \sum \frac{\partial}{\partial q_i}\{f, \{g, q_i\}\} \\ &= \sum \frac{\partial^2 f}{\partial p_i \partial q_j} \frac{\partial^2 g}{\partial p_j \partial q_i} - \frac{\partial^2 f}{\partial p_i \partial p_j} \frac{\partial^2 g}{\partial q_j \partial q_i} - \frac{\partial^2 f}{\partial q_i \partial q_j} \frac{\partial^2 g}{\partial p_j \partial p_i} + \frac{\partial^2 f}{\partial q_i \partial p_j} \frac{\partial^2 g}{\partial q_j \partial p_i}. \end{aligned}$$

Notice that B is negative definite on the set of all f of the form

$$f(p, q) = \sum_{i \leqq j} A_{ij}(p_i p_j + q_i q_j) + \sum_{i<j} B_{ij}(p_i q_j - p_j q_i)$$

which corresponds exactly to the Lie subalgebra of all antisymmetric matrices in $sp(n)$. Thus $\theta T = -T^*$ is a Cartan involution on $sp(n)$. Notice that θ leaves $g\ell(n) \subset sp(n)$ invariant, where $g\ell(n)$ is regarded as the subalgebra of the form

$$\begin{pmatrix} A & 0 \\ 0 & -A^* \end{pmatrix}.$$

Notice, also, that if $T \in sp(n)$ is symmetric then

$$0 = TJ + JT^* = TJ + JT$$

so

$$J^{-1}TJ = -T$$

while if T is antisymmetric we have

$$J^{-1}TJ = T.$$

Thus the automorphism θ can be described by conjugation with J.

Let G be any group having $sp(n)$ as its Lie algebra, for example $G = Mp(n)$. Then

$$\operatorname{Ad}\left(\exp \frac{\pi}{2} J\right) = \exp \frac{\pi}{2} \operatorname{ad} J$$

and so

$$\theta = \operatorname{Ad} \hat{J} \qquad \text{where } \hat{J} = \exp \frac{\pi}{2} J.$$

We now state and prove the basic facts concerning polar decomposition. The result, due to Cartan, is standard; see [14], for instance, whose exposition we follow. *Let G be a real connected semisimple Lie group with Lie algebra g. Let θ be a Cartan involution of g with Cartan decomposition $g = k + p$. Let K be a Lie subgroup of G having Lie algebra k. Then*

(i) *K is closed and connected.*

(ii) *K is the fixed point set of an involutary automorphism, Θ, of G where $d\Theta_e = \theta$.*

(iii) *The center of G, $Z(G)$, is contained in K, and $K/Z(G)$ is compact.*

(iv) *K is its own normalizer, the centralizer of K in G is just $Z(K)$.*

(v) *The map of $p \times K \to G$ given by*

$$(Z,k) \to (\exp Z)\cdot k$$

is a surjective diffeomorphism.

Consider the scalar product $(,)$ in g. Let A be any element of $\mathrm{Aut}(g)_0$ (so that $A = \mathrm{Ad}\, a$ for some $a \in G$) . Our first step is to show that if we form the polar decomposition $A = PO$ relative to $(,)$ then P and O lie in $\mathrm{Aut}(g)_0$. To this effect let us begin by showing that $A^* \in (\mathrm{Aut}\, g)_0$.

We have, for $A \in (\mathrm{Aut}\, g)_0$ and $Z_1, Z_2 \in g$,

$$\begin{aligned}(Z_1, AZ_2) &= -B(Z_1, \theta AZ_2) = -B(\theta Z_1, AZ_2) = -B(A^{-1}\theta Z_1, Z_2)\\ &= -B(\theta A^{-1}\theta Z_1, \theta Z_2) = (\theta A^{-1}\theta Z_1, Z_2)\end{aligned}$$

so

$$A^* = \theta A^{-1}\theta \in (\mathrm{Aut}\, g)_0.$$

In particular $AA^* \in (\mathrm{Aut}\, g)_0$. We now wish to show that $(AA^*)^{1/2} \in (\mathrm{Aut}\, g)_0$. Let $X_1, \ldots, X_n$ be a basis of g consisting of eigenvectors for AA^* with corresponding eigenvalues λ_i. Then

$$[X_i, X_j] = \sum C_{ij}^k X_k$$

and

$$[(AA^*)X_i, (AA^*)X_j] = \sum C_{ij}^k (AA^*)X_k$$

or

$$C_{ij}^k \lambda_i \lambda_j = C_{ij}^k \lambda_k.$$

But this implies that $C_{ij}^k \lambda_i^t \lambda_j^t = C_{ij}^k \lambda_k^t$ for any real number t. Taking $t = \frac{1}{2}$ shows that $P = (AA^*)^{1/2} \in (\mathrm{Aut}\, g)_0$ and hence $O \in (\mathrm{Aut}\, g)_0$. Notice that for W

$\in k$(or p) we have

$$\begin{aligned}([W,Y],Z) &= -B([W,Y],\theta Z) = B(Y,[W,\theta Z]) \\ &= B(Y,\theta[\theta W,Z]) = \pm B(Y,\theta[W,Z]) = \mp(Y,\theta[W,Z])\end{aligned}$$

so that the elements of k are the antisymmetric elements and the elements of p are the symmetric elements of g relative to $(\,,)$. From the above discussion we saw that if $P = e^Y$ is a positive definite element of $(\mathrm{Aut}\, g)_0$, with Y symmetric then $e^{tY} \in \mathrm{Aut}\, g$ for all t so that $Y \in p$. Thus

$$(\mathrm{Aut}\, g)_0 \cap (\text{positive definite elements}) = \{\exp \operatorname{ad} p\}.$$

Let Ad: $G \to \mathrm{Aut}\, g$ be the adjoint homomorphism so that $\operatorname{Ad} G = (\mathrm{Aut}\, g)_0$. Let $\overline{K} = \operatorname{Ad} G \cap \mathbb{O}$ where $\mathbb{O}$ is the orthogonal group relative to $(\,,)$. We have just shown that

$$\operatorname{Ad} G = \overline{K} \times \exp p.$$

Let $K' = \operatorname{Ad}^{-1} \overline{K}$. Then

$$G/K' = \operatorname{Ad} G/\overline{K} \sim \exp p$$

is simply connected. Since G is connected, this implies that K' and $\overline{K}$ are connected, because any path going through two points of K' in G can be deformed to a path in K'. Now let $K \subset G$ be a subgroup with Lie algebra k. Then $\operatorname{Ad} K$ leaves k invariant and preserves B, hence preserves the decomposition. Thus $\operatorname{Ad} K \subset \overline{K}$ so $K \subset K'$. On the other hand K' is connected and has the same Lie algebra as K. Thus $K' = K$. Furthermore $Z(G) \subset \operatorname{Ad}^{-1}(e)$ so $Z(G) \subset K$ and $K/Z(G)$ is compact. This establishes (i) and (ii).

Let $\hat{G}$ be the universal covering group of G, and $N \subset Z(\hat{G})$ the kernel of the covering map. We can find an involutory automorphism Θ of $\hat{G}$ extending θ, and such that $\operatorname{Ad}(\Theta\tilde{a}) = (\operatorname{Ad} \tilde{a})^{*-1}$. Since $N \subset Z(G) \subset K(\operatorname{Ad})^{-1} aK$ we see that Θ is the identity on N and hence induces an involution Θ of G extending θ. This proves (iii).

For any $a \in G$,

$$(\operatorname{Ad} a)(\operatorname{Ad} a)^* = \operatorname{Ad}(a(\Theta a)^{-1})$$

is positive definite. Thus

$$\operatorname{Ad}(a(\Theta a)^{-1}) = \exp 2Y, \qquad Y \in p.$$

Let $O = (\exp -Y)a$. Then

$$(\operatorname{Ad} O)(\operatorname{Ad} O)^* = (\exp -Y)\operatorname{Ad} a(\operatorname{Ad} a)^* \exp -Y = \mathrm{id}$$

so Ad $O \in \bar{K}$ and hence $O \in K$. Thus

$$a = (\exp Y)(\exp -Y)a = PO$$

yields the desired polar decomposition and proves (v).

Now if $a = PO$ and $aKa^{-1} = K$ then $PKP^{-1} = K$. Writing $P = \exp Y$, we see that exp Y leaves k and hence P invariant. Since exp ad Y acts as a positive definite transformation it can be diagonalized. Therefore exp tY also leaves k and p invariant. Thus $[Y,k] \subset k$ and $[Y,k] \subset p$ so $[Y,k] = 0$. But for all $W \in p$ we have

$$B([Y,W],[Y,W]) = -B([Y,[Y,W]],W) = 0$$

since $[Y,W] \subset p$. Therefore $[Y,W] = 0$ and hence $Y = 0$. Thus $N(K) = K$. If $POR = RPO$ then

$$ORO^{-1} = P^{-1}RP.$$

If this holds for all $R \in K$ then P normalizes K and hence $P = e$, and $O \in Z(k)$. This completes the proof.

In the case of $Mp(n)$ we have seen that Θ is given by $\Theta a = \hat{J}^{-1}a\hat{J}$. For any $A \in M\ell(n) \subset Mp(n)$ we can write $A = PO$ where $P = \exp Y$ where $Y \in g\ell(n)$ is symmetric and $O \in K \cap ML(n)$. Thus

$$\hat{J}^{-1}A\hat{J} = (\exp -Y)O.$$

Now $|\chi(O)| = 1$ and $\chi(\exp Y)$ is real. Therefore

$$\overline{\chi(O)}^{-1} = \chi(O) \qquad \text{and} \qquad \overline{\chi(\exp Y)}^{-1} = \chi(\exp -Y).$$

Thus

$$\chi(\hat{J}^{-1}A\hat{J}) = \overline{\chi(A)}^{-1}.$$

From this we see that the pairing $\langle\sigma_1,\sigma_2\rangle$ given by (5.2) is well defined.

We also claim that

$$\langle\sigma_2,\sigma_1\rangle = \overline{\langle\sigma_1,\sigma_2\rangle}. \tag{5.3}$$

To see this, let us first observe that

$$\chi(J^2) = e^{in\pi/2}.$$

Indeed Exp tJ lies in $MU(n)$ and χ is actually defined as a function on $MU(n)$ with

$$\chi^2(\text{Exp } tJ) = \det(\text{Exp } tJ) = \det(e^{it}I) = e^{int}$$

so that

$$\chi(\hat{J}^2) = \chi^2(J) = e^{in\pi/2}.$$

We use the frame $\hat{\mathbf{w}}$ so that $\hat{\lambda}\hat{\mathbf{w}} = \hat{\mathbf{e}}$, and $\hat{\lambda}(\hat{\mathbf{w}}J) = \hat{\mathbf{f}}$ as above with $\sigma_1(\hat{\mathbf{e}}) = a_1$ and $\sigma_2(\hat{\mathbf{e}}) = a_2$, so that $\langle\sigma_1, \sigma_2\rangle = (2\pi)^{-n/2} e^{in\pi/4} a_1 \bar{a}_2$. Then we have

$$\hat{\lambda}(\hat{\mathbf{w}}\hat{J}) = \mathbf{f} \qquad \text{and} \qquad \hat{\lambda}(\hat{\mathbf{w}}\hat{J}^2) = \hat{\mathbf{e}}\hat{J}^2.$$

Then $\sigma_1(\mathbf{e}\hat{J}^2) = e^{in\pi/2}\sigma_1\mathbf{e}$ so

$$\langle\sigma_2, \sigma_1\rangle = (2\pi)^{-h/2} e^{in\pi/4} a_2 \overline{e^{in\pi/2} a_1} = (2\pi)^{-h/2} e^{-in\pi/4} a_2 \bar{a}_1 = \overline{\langle\sigma_1, \sigma_2\rangle}$$

as was to be proved.

Needless to say, we equally well get a sesquilinear pairing between $\overline{\wedge}^{1/2}(E_{1x})$ and $\overline{\wedge}^{1/2}(E_{2x})$, between $\wedge^{-1/2}(E_{1x})$ and $\wedge^{-1/2}(E_{2x})$ and between $\overline{\wedge}^{-1/2}(E_{1x})$ and $\overline{\wedge}^{-1/2}(E_{2x})$ satisfying (5.3) if E_{1x} and E_{2x} are transverse Lagrangian subspaces of TX_x.

Let us now turn to the question of describing the possible metaplectic structures on a symplectic manifold. We will regard two metaplectic structures, $Mp(X)_1$ and $Mp(X)_2$ as equivalent if there is a map φ: $Mp(X)_1 \to Mp(X)_2$ where φ is a morphism of $Mp(n)$ bundles over X, and where

$$\begin{array}{ccc} Mp(X)_1 & \xrightarrow{\ \varphi\ } & Mp(X)_2 \\ \downarrow & & \downarrow \\ Bp(X) & \xrightarrow{\ id\ } & Bp(X) \end{array}$$

commutes so that φ is a morphism of $\mathbf{Z}_2$ bundles over $Bp(X)$, where $\mathbf{Z}_2$ is regarded as the kernel of the homomorphism $Mp(n) \to Sp(n)$. We can describe any principal G bundle by the transition functions c_{ij}: $U_i \cap U_j \to G$, where $\{U_i\}$ covers X and where the bundle has sections defined on U_i. The c_{ij} form a Čech cocycle with values in G, and the corresponding cohomology class determines the bundle. Suppose that $\{c_{ij}\}$ is a cocycle corresponding to a metaplectic structure $MB(X)$, and that $\{z_{ij}\}$ is a cocycle with values in $\mathbf{Z}_2$. Then $\{c_{ij}z_{ij}\}$ is again a cocycle with values in $Mp(n)$, and it is straightforward to check that this defines an action of $H^1(X, \mathbf{Z}_2)$ on the class of metaplectic structures on X so that we obtain the result, due to Kostant:

If the symplectic manifold X admits a metaplectic structure, then the set of equivalence classes of metaplectic structures is a principle homogeneous space for $H^1(X, \mathbf{Z}_2)$. In particular, if $H^1(X, \mathbf{Z}^2) = 0$ then all metaplectic structures on X are equivalent.

There is a simple criterion, also due to Kostant [5, Proposition 3.3.1], for deciding whether a symplectic manifold admits a metaplectic structure, which we briefly describe: Since $U(n)$ is the maximal compact subgroup of $Sp(n)$, we can reduce the group of $Bp(X)$ to $U(n)$, that is, we can find sections whose transition functions, γ_{ij} take values in $U(n)$. The character det: $U(n) \to \mathbf{C}$ then gives rise to a line bundle, $\mathfrak{L}'$, and the line bundle $\mathfrak{L}'$ corresponds to some element, $c' \in H^2(X, \mathbf{Z})$. For a metaplectic structure to exist, we must be able to find a $MU(n)$ bundle which would allow the construction of a bundle associated to χ, where $\chi^2 = \det \circ r$. This is easily seen to be equivalent to the existence of a line bundle $\mathfrak{L}$ such that $\mathfrak{L}' = \mathfrak{L} \otimes \mathfrak{L}$. This is equivalent to the condition that $c' = 2c$ for some $c \in H^2(X, \mathbf{Z})$, i.e. that c' be divisible by two in $H^2(X, \mathbf{Z})$.

§6. The pairing of half form sections.

Let $\mathfrak{M}$ be a fibrating polarization of a metaplectic manifold X. We will assume that the symplectic structure on X arises from a Hermitian line bundle with connection. Let $\rho: X \to Y$ be a fibrating map for the polarization, $\mathfrak{M}$. At each $x \in X$, the subspace $\mathfrak{M}_x$ is the annihilator space of $\rho^* T^* Y_{\rho(x)}$. Since $\mathfrak{M}_x$ is Lagrangian, it is also the annihilator space of its image under the isomorphism of TX with T^*X given by ω. This gives an isomorphism of $B(\mathfrak{M}_x)$ with $B(T^*Y_y)$ where $y = \rho(x)$. Furthermore, $\mathfrak{M}_x$ has a metalinear structure, derived from the metaplectic structure on X. We then can obtain a metalinear structure on $T^* Y_y$ by covering the frame bundle $B(T^*Y_y)$ by the bundle $MB(\mathfrak{M}_x)$, via the isomorphism of $\mathfrak{M}_x$ with $T^* Y_y$. We claim that we obtain, in this manner, a metalinear structure on Y. We must check that the metalinear structure on $T^* Y_y$ is independent of the choice of $x \in \rho^{-1}(y)$, and does give a bundle covering the frame bundle of Y. To this effect, observe that it makes sense to talk of a constant vector field along $\rho^{-1}(y)$, namely one that at each $x \in \rho^{-1}(y)$ corresponds to the same fixed covector in $T^* T_y$. Thus it makes sense to talk of a constant linear frame, i.e., a constant section, s, of $B(\mathfrak{M}_{|\rho^{-1}(y)})$. Such constant sections clearly correspond to a linear basis of $T^* Y_y$. Now we can cover s by a section, $\tilde{s}$, of $MB(\mathfrak{M}_{|\rho^{-1}(y)})$ because $\rho^{-1}(y)$ is *simply connected*. We can thus talk of a constant section of $MB(\mathfrak{M}_{|\rho^{-1}(y)})$, and such a section is clearly determined by its value at any $x \in \rho^{-1}(y)$ since $\rho^{-1}(y)$ is connected. Thus $ML(n, \mathbf{R})$ acts simply and transitively on the constant sections of $MB(\mathfrak{M}_{|\rho^{-1}(y)})$, and this determines the metalinear structure on $T^* Y_y$. Let U be any open subset of X. We can say that a smooth section of $MB(\mathfrak{M}_{|U})$ is fiber constant if it is constant along each $\rho^{-1}(y) \cap U$. It is easy to see that such a smooth section gives a section of $MB(T^* Y)$ over $\rho(U)$ and thus defines a bundle structure on $MB(T^*Y)$. In this way we get a metalinear structure on Y. In particular, we can talk of a half form, σ, on Y, and can identify such a half form σ with a section of $\wedge^{-1/2}(\mathfrak{M})$ which is "constant" along each fiber. We shall denote the space of such half forms either by $\wedge^{1/2}(Y)$, when we think of them as half forms on Y, or by $\wedge^{-1/2}_{\mathfrak{M}}$ when thought of as sections of $\wedge^{-1/2}(\mathfrak{M})$.

We are now in a position to define the pre-Hilbert space, $H_0(\mathfrak{M})$ associated to $\mathfrak{M}$, namely the space of sections of compact support of $L^Y \otimes \wedge^{1/2}(Y)$. Given two such sections $e_1 \otimes \sigma_1$ and $e_2 \otimes \sigma_2$ we define

$$(e_1 \otimes \sigma_1, e_2 \otimes \sigma_2) = \int_Y \langle e_1, e_2 \rangle \sigma_1 \cdot \bar{\sigma}_2.$$

Here $\langle e_1, e_2 \rangle$ is a function on Y and $\sigma_1 \cdot \bar{\sigma}_2$ is a density on Y so that the integrand is a density of compact support on Y.

The completion of this pre-Hilbert space will be denoted by $H(\mathfrak{M})$.

Let $\mathfrak{M}_1$ and $\mathfrak{M}_2$ be fibrating polarizations associated to maps ρ_1 and ρ_2 of X onto Y_1 and Y_2. Let us assume that $\mathfrak{M}_1$ and $\mathfrak{M}_2$ are completely transverse, so that $\rho_1 \times \rho_2 \colon X \to Y_1 \times Y_2$ is a diffeomorphism. We then obtain a sesquilinear pairing of $H_0(\mathfrak{M}_1)$ with $H_0(\mathfrak{M}_2)$ just as in §3. Namely, let $e_1 \otimes \sigma_1$ be an element of $H_0(\mathfrak{M}_1)$ and $e_2 \otimes \sigma_2$ an element of $H_0(\mathfrak{M}_2)$. Then at any x we can form the scalar product $\langle e_1(x), e_2(x) \rangle$ in L_x and the product $\langle \sigma_1(x), \sigma_2(x) \rangle$ given by the metaplectic structure. We can thus define

$$((e_1 \otimes \sigma_1, e_2 \otimes \sigma_2)) = \int_X \langle e_1, e_2 \rangle \langle \sigma_1, \sigma_2 \rangle \omega^n.$$

This gives the desired pairing, and, as we shall see, behaves correctly as far as phase factors are concerned.

Let X be a metaplectic manifold, and MpX the bundle of metaplectic frames. Thus MpX is a $Mp(n)$ bundle over X which covers the $Sp(n)$ bundle, BpX, of symplectic frames. Now a symplectic automorphism, f of X, induces a bundle automorphism, $\tilde{f}$, of BpX. Given a bundle automorphism, g, of BpX, it induces a transformation, h, of X onto itself, and $\tilde{h}$ need not coincide with g. (The condition for $\tilde{h}$ to equal g can be expressed in terms of certain differential equations on BpX which we shall not go into here.) We can thus describe the symplectic automorphisms of X as certain kinds of bundle automorphisms of BpX, namely, those for which $\tilde{h} = g$. Similarly, we can define a *metaplectic automorphism* of X as a bundle automorphism, $\hat{g}$, of MpX whose induced automorphism, g, of BpX is a symplectic automorphism.

Finally, let us suppose that the metaplectic manifold X is such that its symplectic structure comes from a Hermitian line bundle L. By an *automorphism* of X we shall mean a pair $(\hat{g}, a)$ where $\hat{g}$ is a metaplectic automorphism of X and a is an automorphism of L, such that $\hat{g}$ and a induce the *same* symplectic automorphism, g. We shall denote the group of all automorphisms of X by $G(X)$.

If $f \in G(X)$ and $u = e_1 \otimes \sigma_1 \in H_0(\mathfrak{M})$, then it is clear that we obtain an element $f_* u \in H_0(f_* \mathfrak{M})$ in view of all the data given by f, and that the map $f_* \colon H_0(\mathfrak{M}) \to H_0(f_* \mathfrak{M})$ is an isometry and thus extends to a unitary map, which we continue to denote by f_*, of $H(\mathfrak{M}) \to H(f_* \mathfrak{M})$. It is clear that $(fg)_* \colon H(\mathfrak{M}) \to H((fg)_* \mathfrak{M})$ is the composite of $f_* \colon H(g_* \mathfrak{M}) \to H((fg)_* \mathfrak{M})$ with $g_* \colon H(\mathfrak{M}) \to H(g_* \mathfrak{M})$, in other words $G(X)$ acts as morphisms in the category of Hilbert spaces of the form $H(\mathfrak{M})$.

Similarly, let $\mathfrak{M}_1$ and $\mathfrak{M}_2$ be two completely transverse fibrating polarizations. Then if $u_1 \in H_0(\mathfrak{M}_1)$ and $u_2 \in H_0(\mathfrak{M}_2)$ then it is clear that

$$((fu_1, fu_2)) = ((u_1, u_2)).$$

As in §3, we can consider the pairing between $H_0(\mathfrak{M}_1)$ and $H_0(\mathfrak{M}_2)$ as defining a linear map, U, from $H_0(\mathfrak{M}_1)$ to the space of antilinear functions on $H_0(\mathfrak{M}_2)$. Frequently, U can be extended to be a unitary map from $H(\mathfrak{M}_1)$ to $H(\mathfrak{M}_2)$. We will then say that $\mathfrak{M}_1$ and $\mathfrak{M}_2$ are *unitarily related.*

For example, we can consider the situation of the classical Fourier transform, as in §3. There we identified $H_0(\mathfrak{M}_1)$ with $C_0^\infty(dq)$ and $H_0(\mathfrak{M}_2)$ with $C_0^\infty(dp)$. The induced transformation now becomes

$$U(f)(p) = e^{-\pi i n/4}\left(\frac{1}{2\pi}\right)^{n/2} \int e^{-ip\cdot q} f(q)\,dq.$$

Notice the new phase factor that enters into this version of the Fourier transform. Notice also that the theory of the standard Fourier transform asserts that U extends to a unitary map of $L^2(dq)$ into $L^2(dp)$. Thus the polarizations $q =$ const. and $p =$ const. are unitarily related. Thus the effect of using half forms instead of half densities, and making use of the metaplectic structure, is precisely the same as what we achieved in §3 by use of the Maslov index.

If $\mathfrak{M}_1$ and $\mathfrak{M}_2$ are unitarily related, and if f is an automorphism, then it is clear that $f_*\mathfrak{M}_1$ and $f_*\mathfrak{M}_2$ are unitarily related. For example, suppose that X is a symplectic vector space and that $\mathfrak{M}_1$ and $\mathfrak{M}_2$ are polarizations all of whose leaves are affine subspaces. If $\mathfrak{M}_1$ and $\mathfrak{M}_2$ are transverse, then we can find a symplectic transformation carrying $\mathfrak{M}_1$ into $q =$ const. and $\mathfrak{M}_2$ into $p =$ const. and thus $\mathfrak{M}_1$ and $\mathfrak{M}_2$ are unitarily related. (Let us say that two polarizations on a symplectic manifold are *Heisenberg related* if we can find a symplectic diffeomorphism of X with a symplectic vector space carrying the two polarizations onto transverse affine polarizations. It is clear that if two polarizations are Heisenberg related then they are unitarily related.)

Let V be a metaplectic vector space. Let X, Y, and Z be real Lagrangian subspaces, and $\mathfrak{M}_X$, $\mathfrak{M}_Y$ and $\mathfrak{M}_Z$ the corresponding real polarizations. If X, Y, and Z are pairwise transverse, we get the unitary maps

$$U_{Y,X}\colon H(\mathfrak{M}_X) \to H(\mathfrak{M}_Y) \qquad U_{Z,Y}\colon H(\mathfrak{M}_Y) \to H(\mathfrak{M}_Z)$$

and

$$U_{Z,X}\colon H(\mathfrak{M}_X) \to H(\mathfrak{M}_Z)$$

determined by the pairing, and we know from §3 that the transitivity relation

$$U_{Z,X} = U_{Z,Y} \circ U_{Y,Z}$$

holds. As in §3, we extend these unitary maps so as to be defined between non-transverse real polarizations, and then use the invariance of the pairing under the metaplectic group to obtain a representation of $Mp(V)$, which we discuss in some detail in the next section.

§7. The metaplectic representation.

The metaplectic group has a distinguished (infinite dimensional) unitary representation which we have sketched in the preceding section. We shall describe it in detail in the present section. It will play an important role in what follows and will also provide an alternative approach, due to Kostant, to the construction, given in the previous sections, of half forms on Lagrangian subspaces of symplectic manifolds. Our description of this representation will be geometric and based upon the constructs of the preceding section. However the computations will be rather lengthy and involved. For this reason we begin this section with an abstract description of the representation. Let V be a symplectic vector space, with symplectic form $(\,,)$. Then $\mathbf{R} \oplus V$ has the structure of a Lie algebra where $[\mathbf{R}, V] = 0$ and $[v, w] = (v, w) \in \mathbf{R}$. Thus $[V, V] = \mathbf{R}$ so that the Lie algebra is nilpotent. This Lie algebra is known as the *Heisenberg algebra* since (up to constants) the bracket relations in this algebra are the same as the famous Heisenberg commutation relations. The symplectic group, $Sp(V)$, clearly acts as a group of automorphisms of the Heisenberg algebra, where $Sp(V)$ acts trivially on $\mathbf{R}$. Let N denote the simply connected group whose Lie algebra is $\mathbf{R} \oplus V$. It is called the Heisenberg group.

Now a theorem of Stone and von Neumann asserts that, up to unitary equivalence, the Heisenberg group has a unique irreducible unitary representation, such that $1 \in \mathbf{R}$ is represented as scalar multiplication by i. Let σ denote such a representation. Then for any $g \in Sp(V)$ we can define a new representation, σ^g, by setting

$$\sigma^g(a) = \sigma(ga), \qquad a \in \mathbf{R} \oplus V.$$

Then the Stone-von Neumann theorem asserts that there exists a unitary operator, U_g, such that

$$\sigma^g(a) = U_g \sigma(a) U_g^{-1}$$

and, since σ is irreducible, U_g is determined up to a scalar multiple of absolute value one. In particular

$$U_{g_1} U_{g_2} = c(g_1, g_2) U_{g_1 g_2}$$

where $c\colon Sp(V) \times Sp(V) \to S^1$ is a cocycle. Thus $g \rightsquigarrow U_g$ gives a *projective* representation of $Sp(V)$, and, according to general principles, a unitary representation of the universal covering group of $Sp(V)$. It turns out that the representation in question is already a representation of the double covering, i.e., of the metaplectic group, $Mp(V)$. This representation is the metaplectic representation. We shall describe this representation more explicitly, and without depending on the Stone-von Neumann theorem.

Let us begin by describing the group N more explicitly. As a manifold, we can again identify $N = \mathbf{R} \oplus V$. We claim that the group multiplication on N is given by

$$(z_1 \oplus v_1) \cdot (z_2 \oplus v_2) = (z_1 + z_2 + \tfrac{1}{2}(v_1, v_2)) \oplus (v_1 + v_2). \tag{7.1}$$

It is easy to check that this does indeed give a group law, and that

$$t \rightsquigarrow tz \oplus tv$$

defines a one parameter subgroup. Thus the Lie algebra of $\mathbf{R} \oplus V$ is again $\mathbf{R} \oplus V$ and the Lie bracket is obtained by antisymmetrizing the bilinear terms in the group law, which clearly gives the bracket relations in the Heisenberg algebra. The metaplectic group acts as a group of automorphisms of N. We can therefore form the semi-direct product $N \times Mp(V)$ where, as usual, the group multiplication is given by

$$(n_1 \times A_1) \cdot (n_2 \times A_2) = (n_1 \cdot A_1 n_2) \times A_1 A_2$$

where $n_i \in N$ and $A_i \in Mp(V)$. The metaplectic representation will be an irreducible representation of this larger group, whose Lie algebra is clearly $\mathbf{R} \oplus V \oplus sp(V)$. Actually, the condition that $1 \in \mathbf{R}$ be represented by multiplication by i suggests that we consider a slightly different group, namely that we replace $N = \mathbf{R} \oplus V$ by $N' = S^1 \times V$ where S^1 is the circle $\mathbf{R}/2\pi\mathbf{Z}$ via the homomorphism $(t, v) \rightsquigarrow (e^{it}, v)$ with the obvious multiplication. The group N' has the same Lie algebra as N, and the basic group for our purposes will be the semi-direct product $\mathcal{G} = N' \times Mp(V)$.

We claim that $\mathcal{G}$ has a very simple geometric interpretation in terms of the constructs of the preceding section. Indeed, the dual vector space, V^*, also carries a symplectic structure. (If we like, we may identify V with V^* using the symplectic structure on V.) We can choose a metaplectic structure on V^*. We can also construct the trivial line bundle $L = V^* \otimes \mathbf{C}$ with a Hermitian structure and connection whose curvature form is the symplectic form. Thus V^* carries all the required structure and we can consider its group of automorphisms, $G(V^*)$. We claim that $\mathcal{G}$ can be identified with the subgroup of $G(V^*)$ consisting of those elements which act as *affine* transformations on V^*.

Indeed, let $q_1, \ldots, q_n, p_1, \ldots, p_n$ be a symplectic basis of V, so that $\alpha = pdq = \sum p_i dq_i$ where $d\alpha$ is the symplectic form on V^*. We put the standard Hermitian form on $\mathbf{C}$ and the connection form $i\alpha + dz/z$ on the corresponding principal bundle. The choice of basis identifies V^* with $\mathbf{R}^{2n}$ and $B_p(V)$ with $\mathbf{R}^{2n} \times Sp(n, \mathbf{R})$ where (x, T) stands for the symplectic frame at x given by the $2n$ columns of the matrix T. Then $\mathbf{R}^{2n} \times Mp(n)$ gives the metaplectic structure in the obvious way.

The element $e^{it} \in N' \subset \mathcal{G}$ acts on $L = V^* \times \mathbf{C}$ by $e^{it}(v^*, z) = (v^*, e^{it}z)$.

In order to describe the action of $V \subset N'$ on L we will introduce the following notation: We will identify with V with V^* and describe vectors in V as (x,y) or (q,p) where x, y, q, and p are n vectors. Then the vector $v = (q,p)$ acting on the $v^* = (x,y)$ is given by

$$v(v^*) = (x + p, y - q). \tag{7.2}$$

(Indeed, the vector field corresponds to v, considered as a function on V^* is the constant vector field $p(\partial/\partial x) - q(\partial/\partial y)$. Exponentiating gives the action of v on V^* as a translation.) We let $v = (q,p)$ act on $L = V^* \times \mathbf{C}$ by

$$v(x,y;z) = (x + p, y - q; e^{i[q\cdot x + p\cdot q/2]}z). \tag{7.3}$$

It is easy to check that

$$\begin{aligned} v'[v(x,y;z)] &= e^{i[p\cdot q' - p'\cdot q]/2}(v + v')(x,y;z) \\ &= (v' \cdot v)(x,y;z) \end{aligned}$$

where $v' \cdot v$ is the product of v' and v in the group N'.

Let us now describe the action of $Mp(n)$ on L. It acts via the action of $Sp(n)$. We shall specify the action of $Sp(n)$ on $L = \mathbf{R}^{2n} \times \mathbf{C}$ by requiring it to act trivially on the fiber over the origin, i.e., each $T \in Sp(n)$ leaves $\{0\} \times \mathbf{C}$ pointwise fixed. Then

$$T(v, z) = (Tv, e^{i\varphi_T(Tv)}z)$$

where the function φ_T is specified by

$$d\varphi_T = (T^{-1})^*\alpha - \alpha, \qquad \varphi_T(0) = 0.$$

We can now proceed to compute φ as we did for the case $n = 1$ in §3. Let

$$T = \begin{bmatrix} A & B \\ C & D \end{bmatrix} \in Sp(n).$$

Then

$$T^{-1} = \begin{bmatrix} D^* & -B^* \\ -C^* & A^* \end{bmatrix}.$$

Furthermore

$$\begin{aligned}(T^{-1})^* \alpha &= (T^{-1})^* y \cdot dx \\ &= (-C^* x + A^* y) \cdot (D^* dx - B^* dy).\end{aligned}$$

An expression like $C^* x \cdot D^* dx$ can be rewritten as $DC^* x \cdot dx$, where $DC^* = CD^*$, $BA^* = AB^*$ and $DA^* - CB^* = I$ since both T and T^{-1} are symplectic. We can therefore write

$$\begin{aligned}&(-C^* x + A^* y) \cdot (D^* dx - B^* dy) - y \cdot dx \\ &\qquad = -DC^* x \cdot dx - BA^* y \cdot dy + CB^* y \cdot dx + BC^* x \cdot dy \\ &\qquad = d[-\tfrac{1}{2} DC^* x \cdot x + C^* x \cdot B^* y - \tfrac{1}{2} BA^* y \cdot y].\end{aligned}$$

We thus get

$$\varphi_T(x, y) = -\tfrac{1}{2} DC^* x \cdot x + C^* x \cdot B^* y - \tfrac{1}{2} BA^* y \cdot y. \tag{7.4$'$}$$

Recall that for two symplectic transformations T and S we have

$$\begin{aligned}d\varphi_{T \circ S} &= T^{-1*} S^{-1*} \alpha - \alpha \\ &= T^{-1*} S^{-1*} \alpha - T^{-1*} \alpha + T^{-1*} \alpha - \alpha \\ &= T^{-1*} d\varphi_S + d\varphi_T.\end{aligned}$$

Since $\varphi_{T \circ S}(0) = T^{-1*} \varphi_S(0) = \varphi_T(0) = 0$ we conclude that

$$\varphi_{T \circ S}(w) = \varphi_S(T^{-1} w) + \varphi_T(w).$$

In particular, taking $S = T^{-1}$, we obtain

$$\varphi_{T^{-1}}(T^{-1} w) = -\varphi_T(w)$$

or

$$\begin{aligned}\varphi_T(Tw) &= -\varphi_{T^{-1}}(w) \\ &= \tfrac{1}{2} D^* Cx \cdot x + Cx \cdot By + \tfrac{1}{2} B^* Ay \cdot y.\end{aligned} \tag{7.4}$$

Notice that if f and g are any symplectic transformations (so that φ_f and φ_g are defined up to arbitrary constant) we have $\varphi_{f \circ g} = f^{-1*} \varphi_g + \varphi_f + \text{const.}$ In particular, let $v = (q, p)$ and $v' = (p, -q)$, so that v acts on $\mathbf{R}^{2n}$ via translation by v'. Let T be a symplectic transformation. Then $T \circ v \circ T^{-1}$ acts on $\mathbf{R}^{2n}$ via translation by Tv'. Now

$$T(v'(T^{-1}(w; z))) = ((Tv')w; e^{i\psi(w)} z)$$

where

$$\psi(w) = \varphi_{T^{-1}}(T^{-1}w) + q \cdot x(T^{-1}w) + p \cdot q + \varphi_T(w + T(p, -q)).$$

We already know, from the general considerations mentioned above, that

$$\psi(w) = \varphi_u(w) + \text{const.},$$

where $u' = Tv$. We must show that this constant vanishes. Now the constant term in ψ is

$$\begin{aligned} p \cdot q + \varphi_T(T(p, -q)) &= p \cdot q - \varphi_{T^{-1}}((p, -q)) \\ &= -\tfrac{1}{2}A^* Cp \cdot p + B^* Cp \cdot q - \tfrac{1}{2}B^* Dq \cdot q + \tfrac{1}{2}p \cdot q. \end{aligned}$$

On the other hand

$$\begin{bmatrix} A & B \\ C & D \end{bmatrix} \begin{bmatrix} p \\ -q \end{bmatrix} = u' = \begin{bmatrix} \bar{p} \\ -\bar{q} \end{bmatrix}$$

so

$$\tfrac{1}{2}\bar{p} \cdot \bar{q} = -\tfrac{1}{2}(Ap - Bq) \cdot (Cp - Dq).$$

Expansion, using the fact that T is symplectic, shows that this expression coincides with the one given above. This exhibits G as acting as a group of automorphisms of all the structure on V^*. It is clear that G acts as the subgroup of affine transformations of $G(V^*)$.

We can now proceed as indicated in the last section. We choose some fixed affine polarization, $\mathfrak{M}$, say, that given by $x =$ const., where x, y are symplectic coordinates. Then any $a \in G$ induces a unitary map $a_*\colon H(\mathfrak{M}) \to H(a\mathfrak{M})$, and the pairing induces a unitary map $U\colon H(a\mathfrak{M}) \to H(\mathfrak{M})$. The composition of these maps is $Q_a\colon H(\mathfrak{M}) \to H(\mathfrak{M})$. The map $a \rightsquigarrow Q_a$ is the desired unitary representation of G.

A number of remarks are in order. First of all, since $N' \subset G$, we have, in particular, a unitary representation of the Heisenberg group. Furthermore any $t \in \mathbf{R} \subset N$ is represented on the space of sections of L as multiplication by e^{it}, and so, is also represented on $H(\mathfrak{M})$ as multiplication by e^{it}. (We shall see in a moment that this representation is irreducible, and so is the unique irreducible representation of the Heisenberg group as specified by the Stone-von Neumann theorem.) Notice that any element of N acts as a translation on V^*, and therefore $n\mathfrak{M}' = \mathfrak{M}'$ for any polarization $\mathfrak{M}'$ and any $n \in N$. Thus $Q_n = n_*$ for $n \in N$. Furthermore, we obtain a representation $Q_n^{\mathfrak{M}'} = n_*\colon H(\mathfrak{M}') \to H(\mathfrak{M}')$ on each of the Hilbert spaces $H(\mathfrak{M}')$.

Let $a \in \mathcal{G}$. Then $ana^{-1} \in N$ and $Q_{ana^{-1}} = Q_a Q_n Q_a^{-1}$ so that for $a \in Mp(n)$ the operators Q_a provide a unitary equivalence between the representations $n \to Q_n$ and $n \to Q_{ana^{-1}}$ as indicated at the beginning of this section.

Let us write out some explicit expressions for this representation, using (7.3) and (7.4) and the polarization $x =$ const. We can identify $H(\mathfrak{M})$ with $L^2(\mathbf{R}^n)$, an element of $H_0(\mathfrak{M})$ corresponding to a C^∞ function, $f(x)$, of compact support (as we choose, once and for all, the half form $dx^{1/2}$ on $\mathbf{R}^n$). Taking $v = (q,0)$ in (7.3) shows that

$$(Q_q f)^*(x) = e^{-iq\cdot x} f(x). \tag{7.5}$$

Taking $v = (0,p)$ in (7.3) shows that

$$Q_p f(x) = f(x-p). \tag{7.6}$$

It is clear from (7.5) or (7.6) that the representation of N on $H(\mathfrak{M})$ is irreducible. Let A_q and A_p denote the corresponding infinitesimal operators:

$$A_q = \lim_{t\to 0} \frac{1}{t}[Q_{-tq} - I], \qquad A_p = \lim_{t\to 0} \frac{1}{t}[Q_{-tp} - I].$$

Then

$$A_q f(x) = i(q\cdot x) f(x) \qquad \text{and} \qquad A_p f(x) = \sum p_j \frac{\partial f}{\partial x_j}(x).$$

Recall that for any representation of a Lie group on a Hilbert space H the common domain of all the powers of the infinitesimal operators (i.e., the common domain for the universal enveloping algebra) is called the space of C^∞ vectors for the representation. It is clear that in our present circumstance, this common domain consists of the space of $f \in L^2$ such that

$$x^\alpha \frac{\partial^{|\beta|} f}{\partial x^\beta} \in L^2$$

for all α and β. Thus the space of C^∞ vectors is precisely the Schwartz space, $\mathcal{S}$. We can consider N acting on $\mathcal{S}$ and also on its dual space, $\mathcal{S}'$, the space of tempered distributions. We thus have

$$\mathcal{S} \subset H(\mathfrak{M}) \subset \mathcal{S}'.$$

Notice that $\mathcal{S}'$ has a unique one dimensional subspace consisting of eigenvectors with eigenvalue one for all Q_q, namely the space spanned by δ_0. Similarly, the constants form a subspace of eigenvectors with eigenvalue one for the Q_p.

Let us now examine the action of various elements of $Mp(n)$. For $A \in ML(n) \subset Mp(n)$ we clearly have

$$Q_A f(x) = f(A^{-1}x)(\det A)^{-1/2}. \tag{7.7}$$

For an element of the form

$$n' = \begin{bmatrix} I & 0 \\ S & I \end{bmatrix}$$

the corresponding transformation is

$$Q_{n'} f(x) = e^{-ix\cdot x/2} f(x). \tag{7.8}$$

The corresponding expression for Q_n with

$$n = \begin{pmatrix} I & S \\ 0 & I \end{pmatrix}.$$

can be obtained by Fourier transform as indicated in the preceding section. It is constructive to work out the representations of the infinitesimal generators. For this purpose let us write the Lie algebra of the group $\mathcal{G}$ as the algebra, under Poisson bracket, of all (inhomogeneous) quadratic polynomials in the p's and q's. Thus a basis of the Lie algebra consists of

$$1, q_1, \ldots, q_n, p_1, \ldots, p_n, \qquad i \leq j = 1, \ldots, n,$$

$$q_i q_j, q_i p_j, p_i p_j, \qquad i, j = 1, \ldots, n.$$

The bracket relations are given by

$$\{f, g\} = \sum \frac{\partial f}{\partial p_i}\frac{\partial g}{\partial q_i} - \frac{\partial f}{\partial q_i}\frac{\partial g}{\partial p_i}.$$

In particular $\{1, g\} = 0$ for all g and

$$\{p_i, q_j\} = \delta_{ij},$$

$$\{p_i q_j, q_k\} = \delta_{ik} q_j, \qquad \{p_i q_j, p_k\} = -\delta_{jk} p_i,$$

$$\{p_i p_j, q_k\} = \delta_{ik} p_j + \delta_{jk} p_i, \qquad \{p_i p_j, p_k\} = 0,$$

and

$$\{q_i q_j, q_k\} = 0, \qquad \{q_i q_j, p_k\} = -\delta_{ik} q_j - \delta_{jk} q_i.$$

Thus $p_i q_j$ acts as the matrix

$$\begin{pmatrix} E_{ij} & 0 \\ 0 & -E_{ji} \end{pmatrix}$$

where E_{ij} is the matrix which is one in the i,jth position and zero elsewhere. Now for any element ξ in the Lie algebra, the corresponding skew adjoint operator is the infinitesimal generator of $Q_{\exp - t\xi}$. Thus since

$$\exp tE_{ij} = I + tE_{ij} + O(t^2) \quad \text{and} \quad \det \exp tE_{ij} = I + t\delta_{ij} + O(t^2)$$

we see that the infinitesimal generator $A_{p_i q_j}$ is given by

$$x_i \frac{\partial}{\partial x_j} + \frac{1}{2}\delta_{ij}.$$

Continuing in this way we can construct the following table giving the representations of the basic elements:

f	A_f
1	i
q_k	ix_k
p_ℓ	$\frac{\partial}{\partial x_\ell}$
$p_i q_j$	$x_i \frac{\partial}{\partial x_j} + \frac{1}{2}\delta_{ij}$
$p_k p_\ell$	$\frac{1}{i} \frac{\partial^2}{\partial x_k \partial x_\ell}$
$q_k q_\ell$	$ix_k x_\ell$

(7.9)

Now let X be an arbitrary metaplectic manifold and let

$$\begin{array}{c} Mp(X) \\ \downarrow \\ X \end{array}$$

be the bundle of metaplectic frames of X. Since $Mp(X)$ is a principal $Mp(n)$ bundle, any module for $Mp(n)$ will induce an associated bundle. In particular, if H denotes the Hilbert space of the metaplectic representation, we obtain a bundle of Hilbert spaces over X which we shall denote by $H(X)$. If $H_x = H(X)_x$

is the fiber at $x \in X$ then H_x is isomorphic to the space H, where the isomorphism depends on the choice of a metaplectic frame at x. Let u be a smooth (local) section of $H(X)$ and let $\mathbf{s}$ be a smooth section of the bundle of metaplectic frames. Then u corresponds to a (locally defined) smooth function

$$u_{\mathbf{s}}: X \to H.$$

If we think of H explicitly as $H = L^2(\mathbf{R}^n)$ then $u_{\mathbf{s}}$ is given as $u_{\mathbf{s}}(x, r)$, where for each $x \in X$ the function $u_{\mathbf{s}}(x, \cdot)$ lies in $L^2(\mathbf{R}^n)$. Let $\mathbf{s}'$ be another section of the bundle of metaplectic frames so that

$$\mathbf{s}'(x)g(x) = \mathbf{s}(x)$$

where $g(x) \in Mp(n)$ depends smoothly on x in the common domain of definition. Then, by the definition of an associated bundle,

$$u_{\mathbf{s}'}(x) = Q_{g(x)} u_{\mathbf{s}}(x). \tag{7.10}$$

Similarly from $\mathbb{S} \subset H \subset \mathbb{S}'$ we can construct the bundles $\mathbb{S}(X)$ and $\mathbb{S}'(X)$. Furthermore, since the Heisenberg algebra $\mathbf{R} \oplus V$ is also a module for $Mp(n)$ we can form the associated bundle which is just $\mathbf{R}(X) \oplus T(X)$ where $\mathbf{R}(X) = X \times \mathbf{R}$ is just the trivial bundle and TX is the tangent bundle. Furthermore the fiber $\mathbf{R} \oplus TX_x$ is a Heisenberg algebra, being the sum of $\mathbf{R}$ and the symplectic vector space, TX_x.

A choice of metaplectic frame at x gives an isomorphism of $\mathbf{R} \oplus TX_x$ with $\mathbf{R} \oplus V$. Let $\xi_x \in \mathbf{R} \oplus TX_x$ and $u_x \in \mathbb{S}'_x$, and let $\mathbf{s}_x$ be a frame at x, so that ξ_x corresponds to $\xi \in \mathbf{R} \oplus V$ and u_x to $u \in \mathbb{S}'$. Then we set

$$A_{\xi_x} u_x = v_x$$

where v_x corresponds to $A_\xi u$. Changing the frame amounts to replacing ξ by $a\xi$ and u by $Q(a)u$ where $a \in Mp(n)$. But then

$$\begin{aligned} A_{a\xi} Q(a)u &= Q(a)[Q(a)^{-1} A_{a\xi} Q(a)]u \\ &= Q(a)v \end{aligned}$$

so that v_x is independent of the choice of metaplectic frame. We thus see that the Heisenberg algebra $\mathbf{R} \oplus TX_x$ at each point acts on the Hilbert space H_x at each point. Notice that this action is intrinsic, coming from the metaplectic structure of X. Of course each TX_x has a metaplectic structure and the abstract definition of the metaplectic representation given at the beginning of this section shows that we get a corresponding metaplectic representation on each H_x.

Let W be a Lagrangian subspace of TX_x. Let

$$\mathbb{S}'_W = \{u_x \mid \xi_x u_x = 0 \text{ for all } \xi_x \in W\}.$$

Notice that dim $\mathbb{S}'_W = 1$. Indeed, let U be a complementary Lagrangian subspace and let $\zeta_1, \ldots, \zeta_n, \xi_1, \ldots, \xi_n$ be a symplectic basis at x with the ζ's $\in U$ and ξ's $\in W$. We can choose a metaplectic frame $\mathbf{s}$ covering this frame. With respect to $\mathbf{s}$ the basis corresponds to the basis

$$q_1, \ldots, q_n, p_1, \ldots, p_n$$

on $\mathbf{R}^n \oplus \mathbf{R}^{n*}$ and, identifying H_x with $L^2(\mathbf{R}^n)$, the operators A_{ξ_i} become $\partial/\partial x_i$. Thus the space $\mathbb{S}'_W$ corresponds to the constants. This proves that $\mathbb{S}'_W$ is one dimensional. We have actually proved more: We have established an isomorphism $\mathbb{S}'_W \to \mathbf{R}$ depending on our choice of metaplectic frame $\mathbf{s}$. Suppose we replace $\mathbf{s}$ by $\mathbf{s}a$ where $a \in ML(n) \subset Mp(n)$. According to (7.10) this has the effect of multiplying by $(\det A)^{-1/2}$ if

$$a = \begin{bmatrix} A & 0 \\ 0 & A^{t-1} \end{bmatrix},$$

when it is A^{t-1} that is acting on the ξ's. This shows that:

The space $\mathbb{S}'_W$ is canonically isomorphic to the space $\wedge^{1/2}(W)$ of half forms on W. (7.11)

In other words we can identify the space $\mathbb{S}'_W$ as the dual space to the space of negative half forms $\wedge^{-1/2} W$. Since we have a pairing between $\wedge^{-1/2}(U)$ and $\wedge^{-1/2} W$, if U and W are transverse Lagrangian subspaces of TX_x we expect that we should also get a pairing between $\mathbb{S}'_U$ and $\mathbb{S}'_W$. This is indeed the case. Notice that we have a Hilbert space, H_x at x, and so a well defined scalar product between elements of H_x. This scalar product will not extend to all pairs of elements of $\mathbb{S}'_x$ but might extend to certain pairs. We claim that if U and W are transversal then the scalar product between $\mathbb{S}'_U$ and $\mathbb{S}'_W$ is well defined. Indeed choosing a metaplectic frame whose first n elements are a metalinear basis of U and whose last n elements are a metalinear basis of W we have an identification of H_x with $L^2(\mathbf{R}^n)$ in such a way that $\mathbb{S}'_U$ corresponds to the Dirac measure at the origin, while $\mathbb{S}'_W$ corresponds to the constants. In this case it is clear that the scalar product extends to the evaluation map, and it is easy to check that this is indeed dual to the pairing on half forms we considered earlier.

We can actually get a vector bundle which will yield, by a similar process, the bundle of half forms, rather than the dual bundle. Indeed, consider $\wedge^n(TX) \otimes H(X)$ with the corresponding distribution bundle $\wedge^n(TX) \otimes \mathbb{S}'(X)$. If $W \subset TX_x$ is a Lagrangian subspace then $\wedge^n(W)$ is one dimensional and injects in the obvious way into the vector space $\wedge^n TX_x$. It is clear that $\wedge^n(W) \otimes \mathbb{S}'_W$ is precisely the bundle of negative half forms on W. The exterior product gives a pairing of $\wedge^n(TX_x) \times \wedge^n(TX_x) \to \wedge^{2n}(TX_x)$ and ω^n maps this last space into $\mathbf{R}$. Thus $\wedge^n(TX) \otimes H(X)$ is again a bundle of Hilbert spaces and $\wedge^n(TX) \otimes \mathcal{S}'(X)$ a bundle of distributions. This time the pairing between the lines $\wedge^n(W) \otimes \mathcal{S}'_W$ and $\wedge^n(U) \otimes \mathcal{S}'U$ is well defined for any pair of Lagrangian subspaces of TX_x. Indeed it is defined for transversal subspaces as before and tends to the limit zero as the subspaces become non-transverse.

Let V be a metaplectic vector space and $u \in \mathcal{S}(V)$. Then u defines a linear functional on $\mathcal{S}'(V)$ and hence, in particular, on any one dimensional subspace. If W is any Lagrangian subspace of V then u defines an element of the dual space of $\wedge^{1/2} W$, i.e., an element of $\wedge^{-1/2} W$. Thus u defines a smooth section of the bundle of negative half forms over Lagrangian subspaces. We can let $V = TX_x$ for some metaplectic manifold X, and conclude that a smooth section of $\mathcal{S}(X)$ determines a smooth section of the bundle of negative half forms on the Lagrangian subspaces of TX.

So far, we have been considering the metaplectic representation in terms of the Hilbert space associated with a real polarization. In his pioneering paper on the subject, Segal showed how to give a convenient form for this representation in terms of analytic functions of several complex variables. (In fact, Segal, [13], constructed the corresponding theory for infinitely many degrees of freedom.) Bargmann, [17], independently, gave a complex variable construction for the representation. In terms of our language, this amounts to expressing the representation relative to the Hilbert space associated to a complex polarization. We now sketch some of the ideas involved in setting up this version of the representation, referring to Bargmann's paper, [17], for an exposition of the details of the representation and many interesting applications. Suppose we set

$$z = (1/\sqrt{2})(p - iq) \quad \text{and} \quad \bar{z} = (1/\sqrt{2})(p + iq)$$

and consider the form

$$\beta = i\bar{z}\, dz$$

so that

$$d\beta = \omega = dp \wedge dq.$$

(Here, and in what follows, p stands for the n-vector $(p_1, \ldots, p_n)$, z for the complex n-vector $(z_1, \ldots, z_n)$ and summation is assumed where appropriate, so that, for example, the $dp \wedge dq$ in the last formula stands for $\sum dp_i \wedge dq_i$.) We wish to consider the complex polarization consisting of those complex tangent vectors ζ which satisfy $\zeta \lrcorner dz^i = 0$ for all i. This, in turn, picks out the subspace of the space of sections of $\mathbf{L}$ consisting of those sections which are covariant constant along the polarization. If s is one such section, then any other section is of the form fs, where f satisfies $\zeta f = 0$ for all ζ in the polarization. The polarization is spanned by the vectors $\partial/\partial\bar{z}_i$, and thus the condition on f is that $\partial f/\partial\bar{z} = 0$, i.e., that f is a holomorphic function of the complex variables z. Let us choose a section s so that the connection form α_s is β. Since β is not real, the section s will not have unit length; as we shall see, this fact will be of crucial importance to us in defining the Hilbert space structure. Let r denote the section of L (of unit length) with

$$\alpha_r = pdq.$$

Then

$$s = e^{\varphi} r$$

where φ is determined (up to an additive constant) by

$$\beta = \alpha_s = (1/i)\,d\varphi + pdq.$$

Let us fix the additive constant by setting $\varphi(0) = 0$. Then

$$\begin{aligned} d\varphi &= i\{(i/2)(p + iq)(dp - idq) - pdq\} \\ &= (-1/4)d(p^2 + q^2) - (i/2)d(pq) \end{aligned}$$

so

$$\begin{aligned} \varphi &= -(p^2 + q^2)/4 - (1/2)ipq \\ &= -|z|^2/2 - (1/2)ipq. \end{aligned}$$

This shows that the length of the section s is given by

$$\|s(q,p)\|^2 = e^{-|z|^2}.$$

We choose some non-zero constant half form normal to the polarization. This allows us to identify the space of sections associated with the polarization with the space of holomorphic functions, and the preceding computation shows that the scalar product on this space of entire functions should be given by

$$(f,g) = c \int f(z)\overline{g(z)}e^{-|z|^2}\,dp\,dq$$

where c is a constant. A computation shows that this constant is given by $c = (2\pi)^{-n}$. This is the starting point of the Segal-Bargmann representation: we let $\mathcal{F}_n$ denote the space of holomorphic functions of n complex variables for which the integral (f,f) converges, where

$$(f,g) = (2\pi)^{-n} \int f(z)\overline{g(z)}e^{-|z|^2}\,dp\,dq.$$

It is not hard to show that $\mathfrak{F}_n$ is a Hilbert space, and that the polynomials form a dense subspace, in fact, the monomials $u_m(z)$ defined by

$$u_m(z) = z^m/(m!)^{1/2}$$

form an orthonormal basis (where $m = (m_1, \ldots, m_n)$, $m! = m_1! \cdot \ldots \cdot m_n!$ and $z^m = z_1^{m_1} \cdots z_n^{m_n}$). For any $a \in \mathbf{C}^n$, the linear exponential function $e_a(z) = e^{\bar{a} \cdot z}$ belong to $\mathfrak{F}_n$ and has the expansion

$$e_a(z) = \sum \overline{u_m(a)} u_m(z)$$

in terms of the orthonormal basis, $\{u_m\}$. If $f \in \mathfrak{F}_n$ has the expansion $f = \sum c_m u_m$ relative to the orthonormal basis, then

$$(f, e_a) = \sum c_m u_m(a) \quad \text{or} \quad (f, e_a) = f(a).$$

Thus the evaluation functional is actually an element of the Hilbert space, as is typical in Hilbert spaces of analytic functions. We can also write the last equation as

$$f(a) = (2\pi)^{-n} \int e^{a \cdot \bar{z}} f(z) e^{-|z|^2} \, dp \, dq.$$

We shall use this integral formula, together with the pairing between the Hilbert space, H, associated with the real polarization $q =$ const and the above Hilbert space associated with the complex polarization to get an explicit expression for the unitary map relating the two Hilbert spaces. We choose a fixed constant half form normal to the real polarization. We can then think of H as $L^2(\mathbf{R}^n)$. The pairing between $\psi \in H$ and $f \in \mathfrak{F}_n$ is then given (up to a constant factor depending on the choice of half forms) by

$$\begin{aligned} \langle\langle \psi, f \rangle\rangle &= (2\pi)^{-n/2} \int \langle \psi r, fs \rangle \, dp \, dq \\ &= (2\pi)^{-n/2} \int \psi(q) \overline{f(z)} e^{-\frac{1}{2}(p^2+q^2)+ipq} \, dp \, dq. \end{aligned}$$

The map $U\colon H \to \mathfrak{F}_n$ that we are seeking is characterized by $(U\psi, f) = \langle\langle \psi, f \rangle\rangle$ for all $f \in \mathfrak{F}_n$. By the reproducing property of the functions e_a we know that

$$\begin{aligned} (U\psi)(a) &= (U\psi, e_a) \\ &= \langle\langle \psi, e_a \rangle\rangle \\ &= (2\pi)^{-n/2} \int \psi(q) e^{a \cdot (p - iq)/\sqrt{2}} \, e^{-\frac{1}{2}(p^2+q^2)+ipq} \, dp \, dq. \end{aligned}$$

We can perform the integration with respect to p, which just amounts to computing the Fourier transform of a Gaussian to conclude that

$$(U\psi)(a) = \int A(a,q)\psi(q)\,dq$$

where the kernel, A (with correct choice of constants to make it unitary) is given by

$$A(a,q) = \pi^{-n/4} e^{-\frac{1}{2}(a^2+q^2)+\sqrt{2}\,a\cdot q}.$$

This is the formula used by Bargmann [17] (cf. his equation (1.4)) to construct the unitary map from H to $\mathfrak{F}_n$. In our framework, we see that it is a consequence of the geometrical structure, in particular the pairing between Hilbert spaces associated with different polarizations.

Now

$$\frac{\partial}{\partial z_j}(f\bar{g}e^{-|z|^2}) = \frac{\partial f}{\partial z_j}\bar{g}e^{-|z|^2} - f\cdot\overline{z_j g}e^{-|z|^2}$$

if f and g are both holomorphic. Therefore, if f and g are both in $\mathfrak{F}_n$, integration by parts shows that the adjoint of the operator $\partial/\partial z_j$ is multiplication by z_j. Using the explicit form of the operator U as an integral transform, and applying integration by parts, it is easy to see that

$$U\frac{\partial}{\partial q_j}U^{-1} = (1/\sqrt{2})\left(z_j - \frac{\partial}{\partial z_j}\right)$$

and

$$Uq_jU^{-1} = (-i/\sqrt{2})\left(z_j + \frac{\partial}{\partial z_j}\right)$$

which gives the explicit form of the representation of the Heisenberg algebra on $\mathfrak{F}_n$, in terms of the "creation and annihilation operators", z_j and $\partial/\partial z_j$. In computing with the metaplectic representation, at least at the Lie algebra level, it is useful to know that the image of any element in $sp(V)$ can be expressed as a quadratic expression in the images of elements of the Heisenberg algebra (and hence in terms of the "creation" and "annihilation" operators). To formulate this fact correctly it is convenient to make use of the concept of a graded Lie algebra: A graded Lie algebra g is a $\mathbf{Z}_2$ graded vector space, $g = g_0 + g_1$ together with a bilinear map of $g \times g \longrightarrow g$ such that $[g_i, g_j] \subset g_{i+j}$ and which is graded antisymmetric in the sense that

$$[X, Y] = -(-1)^{ij}[Y, X] \quad \text{if } X \in g_i \text{ and } Y \in g_j,$$

and which satisfies the graded version of Jacobi's identity which is

$$[X, [Y, Z]] = [[X, Y], Z] + (-1)^{ij}[Y, [X, Z]] \quad \text{for } X \in g_i \text{ and } Y \in g_j.$$

We refer the reader to [18] for a discussion of graded Lie algebras and some of their applications. One of the examples described in [18] is the graded Lie algebra obtained by setting $g_0 = sp(V)$ and $g_1 = V$ where V is a symplectic vector space. The bracket of two elements of g_0 is defined to be the usual Lie bracket. The bracket of $X \in g_0$ with $Y \in g_1$ is defined by letting X, Y be the image of $Y \in V$ under the action of $X \in sp(V)$. Finally, the symmetric bracket of $g_1 g_1 \to g_0$ is defined by identifying $sp(V)$ with $S^2(V)$ and using ordinary (symmetric) multiplication. Put another way, we can think of $sp(V)$ as consisting of homogeneous quadratic polynomials under Poisson bracket, and of V as consisting of homogeneous linear polynomials. The symmetric bracket of $g_1 \times g_1 \to g_0$ is just the multiplication of two linear polynomials to obtain a quadratic polynomial. Let us fix some version of the metaplectic representation of $N \times M_p(V)$, and let A denote the infinitesimal representation of its Lie algebra. The Lie algebra of N can be identified with linear polynomials. Thus A_u denotes the image of any linear homogeneous polynomial, u, and similarly A_x the image of any quadratic polynomial, X. For example, the table (7.9) gives such a set of values of A_f, relative to a real polarization. If we set $\kappa(u) = i(i/2)^{1/2} A_u$ and $\kappa(X) = A_x$ *then* κ *gives a representation of the graded Lie algebra*, g. This fact, which is relatively easy to verify, provides a convenient method of analyzing how the metaplectic representation behaves when restricted to various interesting subgroups. We refer the reader to Sternberg and Wolf [18], for the details of this method.

§8. Some examples.

In this section we present, in rather sketchy form, various examples of the quantization procedure, indicating some possible directions in which the theory might have to be developed. It is clear that the main analytical theorems (such as establishing the convergence of the integrals entering into the pairing) and many geometrical questions (such as workable criteria for unitary equivalence of polarizations) are yet to be formulated and established.

Suppose that we have a metaplectic manifold X with Hermitian line bundle together with a chosen polarization, $\mathcal{F}$. Suppose that φ is a function such that ξ_φ is everywhere tangent to the polarization. Then the (infinitesimal) operator associated to φ on $H(\mathcal{F})$ is just multiplication by $i\varphi$. (For instance, if $X = T^*M$ and we choose the cotangent fibration then the functions on M are quantized by multiplication. This is the "standard" quantization procedure.) In a sense this is the simplest example of the quantization procedure, in that the issue of a transverse polarization and pairing doesn't even enter. This suggests that if we start with a φ, we should try to find a polarization $\mathfrak{F}$ with ξ_φ tangent to $\mathfrak{F}$. The trouble is the bundle $L \otimes \wedge^{-1/2} \mathfrak{F}$ may not have any smooth global sections which are covariant constant. A case in point, which we shall soon discuss in detail, occurs when we take $X = \mathbf{R}^2 - \{0\}$, with coordinates q, p and $\omega = dp \wedge dq$, and take $\varphi = \frac{1}{2}(p^2 + q^2)$. The polarization in this case consists of the tangents to the concentric circles, and, as we shall see below, there are no smooth global sections of $L \otimes \wedge^{-1/2} \mathfrak{F}$, covariant constant along each leaf. However there do exist "generalized sections" in the sense of distribution theory and these are supported along preferred circles. When we allow such "generalized sections", and idea due to Simms, we do get exactly the "correct" quantization of the harmonic oscillator. (Another approach, suggested by Kostant, and verified by Blattner, Rawnsley, Simms and Sniatycki, is to consider the sheaf of germs of locally constant sections, and use the higher cohomology of this sheaf, the zero cohomology consisting of the global sections. We shall not discuss this approach here.)

Our discussion follows, almost verbatim, the paper of Simms [7]. We let $X = \mathbf{R}^{2n} - \{0\}$, with coordinates $(p_1, \ldots, p_n, q_1, \ldots, q_n)$. We let

$$\omega = \sum dp_j \wedge dq_j$$

and

$$H = \frac{1}{2} \sum (p_j^2 + q_j^2).$$

We introduce complex coordinates

$$z_j = p_j - iq_j$$

so that

$$\omega = \frac{1}{2i} \sum dz_j \wedge d\bar{z}_j$$

and

$$\xi_H = i \sum \left(\bar{z}_j \frac{\partial}{\partial \bar{z}_j} - z_j \frac{\partial}{\partial z_j} \right).$$

(We are in the situation of the n dimensional harmonic oscillator with the same mass m, and frequency ν, both taken to be one, and in units where Planck's constant is also taken to be one. For general units, with Planck's constant, h, and general values of m and ν we would write

$$\omega = h^{-1} \sum dp_j \wedge dq_j,$$

$$H = \frac{1}{2m} \sum (p_j^2 + m^2 \nu^2 q_j^2),$$

and

$$z_j = p_j - imvq_j,$$

$$\xi_H = ivh \sum \left(\bar{z}_j \frac{\partial}{\partial \bar{z}_j} - z_j \frac{\partial}{\partial z_j} \right).$$

In what follows we shall carry out the computations for the case $h = m = \nu = 1$, and put the general values in only when we get to the final result, leaving the necessary adjustments in the argument to the reader.)

We now wish to embed ξ_H in a (complex) polarization. For the case $n = 1$ we simply take $\mathfrak{F}$ to be the space spanned by ξ_H. For the case $n > 1$ we take $\mathfrak{F}$ to be spanned by ξ_H and the vector fields

$$z_j \frac{\partial}{\partial \bar{z}_k} - z_k \frac{\partial}{\partial \bar{z}_j}.$$

Notice that these latter vector fields, at any point x, all lie in the n dimensional space spanned by

$$\frac{\partial}{\partial \bar{z}_1}, \ldots, \frac{\partial}{\partial \bar{z}_n}$$

and all satisfy the equation $\langle \xi, \sum z_j d\bar{z}_j \rangle = 0$. They span an $n - 1$ dimensional space and hence, together with ξ_H, span the n dimensional space $\mathfrak{F}_x$. To check that $\mathfrak{F}$ is a polarization, it suffices, for each coordinate neighborhood U, to find functions $\varphi_1, \ldots, \varphi_n$ such that $\xi_{\varphi_1}^{(x)}, \ldots, \xi_{\varphi_n}^{(x)}$ span $\mathfrak{F}_x$ at all $x \in U$. Let U_k denote the open subset of X defined by $z_k \neq 0$. Let $Z_{jk} = z_j/z_k$ so that Z_{jk} is defined on U_k and

$$\xi_{Z_{jk}} = \frac{2i}{z_k^2}\left(z_k \frac{\partial}{\partial \bar{z}_j} - z_j \frac{\partial}{\partial \bar{z}_k} \right)$$

so that the vector fields $\xi_{Z_{jk}}$ $(j = 1, \ldots, n; j \neq k)$ and ξ_H span $\mathfrak{F}$ on U_k. This shows that $\mathfrak{F}$ is indeed a polarization.

We let S_k denote the frame field

$$S_k = (\xi_{Z_{1k}}, \ldots, \widehat{\xi_{Z_{kk}}}, \ldots, \xi_{Z_{nk}}, \xi_H)$$

defined on U_k.

On $U_j \cap U_k$ we have $Z_{\ell j} = Z_{\ell k}/Z_{jk}$ and thus

$$\xi_{Z_{\ell j}} = \frac{1}{Z_{jk}} \xi_{Z_{\ell k}} - \frac{Z_{\ell k}}{Z_{jk}^2} \xi_{Z_{jk}}.$$

We can then explicitly compute the transition matrix $g(k,j)$ where

$$S_j = S_k g(k,j)$$

and a direct computation shows that

$$\det g(k,j) = \left(\frac{z_k}{z_j}\right)^n.$$

We now turn to the question of metaplectic structures on X. Since $H^2(X,\mathbf{Z}) = 0$, we know that X possesses a metaplectic structure. For $n = 1$, we have $H^1(X,\mathbf{Z}_2) = \mathbf{Z}_2$ so that X has two metaplectic structures, while for $n > 1$ we have $H^1(X,\mathbf{Z}_2) = \{0\}$ so that X has just one metaplectic structure, and so we must treat these two cases separately. In both cases, a metaplectic structure on X is equivalent to a (complex) metalinear structure on $\mathfrak{F}$. Let us now describe the metalinear structures: For $n = 1$ we can take as our first metalinear structure the one given by the section ξ_H: We set $MB_1(\mathfrak{F}) = X \times ML(1,\mathbf{C})$ with

$$\rho(x,\lambda) = \xi_H(x)r(\lambda), \qquad r\colon ML(1,\mathbf{C}) \to GL(1,\mathbf{C}).$$

For the other metalinear structure we let ϵ be the nontrivial element of the homomorphism $r\colon ML(1,\mathbf{C}) \to GL(1,\mathbf{C})$. We can regard the plane as $\mathbf{R}^+ \times \mathbf{R}/\mathbf{Z}$ where $\mathbf{Z}$ acts on $\mathbf{R}^+ \times \mathbf{R}$ sending $(r,\theta) \rightsquigarrow (r,\theta + 2n\pi)$. Let $\mathbf{Z}$ act on $\mathbf{R}^+ \times \mathbf{R} \times ML(1,\mathbf{C})$ by $(r,\theta,\lambda) \to (r,\theta + 2n\pi, \epsilon^n\lambda)$ and set

$$MB_2(\mathfrak{F}) = (\mathbf{R}^+ \times \mathbf{R} \times ML(1,\mathbf{C}))/\mathbf{Z}.$$

We have an obvious projection of $MB_2(\mathfrak{F})$ onto X and an obvious action of $ML(1,\mathbf{C})$ on $MB_2(\mathfrak{F})$ given by $[r,\theta,\lambda] \times \lambda' \to [r,\theta,\lambda\lambda']$ where $[\ \ ]$ denotes equivalence class mod $\mathbf{Z}$. We have the map $\rho\colon MB_2(\mathfrak{F}) \to B(\mathfrak{F})$ given by

$$\rho[s,\theta,\lambda] = \xi_H(se^{i\theta})r(\lambda).$$

It is clear that this makes $MB_2(\mathfrak{F})$ into a metalinear bundle over $\mathfrak{F}$. To see that it is the second metalinear structure on $\mathfrak{F}$, and hence induces the second metaplectic structure on X, we choose sections as follows: Let V_1 and V_2 be the subsets of X defined by $0 < \theta < 2\pi$ and $-\pi < \theta < \pi$ respectively. Let

$$u_1(re^{i\theta}) = [r,\theta,1] \qquad \text{on } V_1$$

and

$$u_2(re^{i\theta}) = [r,\theta,1] \qquad \text{on } V_2.$$

Then

$$u_1(re^{i\theta}) = u_2(re^{i\theta}) \qquad \text{for } 0 < \theta < \pi$$

while

$$u_1(re^{i\theta}) = [r, \theta, 1] = [r, \theta - 2\pi, \epsilon] = u_2(re^{i\theta})\epsilon \qquad \text{for } \pi < \theta < 2\pi.$$

Thus the transition functions are indeed multiplied by the nontrivial cocycle of X with values in $\mathbf{Z}_2$.

For $n > 1$ the unique metalinear structure on $\mathcal{F}$ lifts the sections s_k to sections $\tilde{s}_k$ where $\tilde{s}_j = \tilde{s}_k \tilde{g}(k,j)$ for suitable $\tilde{g}(k,j) \in M\ell(n, \mathbf{C})$ covering the $g(k,j) \in G\ell(n, \mathbf{C})$ introduced earlier.

We let $U_{\mathcal{F}}(X)$ denote the Lie algebra of complex vector fields everywhere tangent to $\mathcal{F}$. We let $\wedge^{-1/2}(\mathcal{F})$ denote the space of smooth inverse half forms on $\mathcal{F}$. The vector fields ξ in $U_{\mathcal{F}}$ act via "Lie derivative," D_ξ, on $\wedge^{-1/2}(\mathcal{F})$ satisfying the usual rules for Lie derivative. Furthermore if $\nu \in \wedge^{-1/2}(\mathcal{F})$ is such that $\nu(\mathbf{s}) \equiv 1$ when $\mathbf{s}$ is a metalinear frame such that $\rho\mathbf{s} = (\xi_{\varphi_1}, \ldots, \xi_{\varphi_n})$, then $D_\xi \varphi = 0$.

Let us set

$$\alpha = \frac{i}{2} \sum \bar{z}_j dz_j$$

so that

$$d\alpha = \omega.$$

Then the locally constant sections of $L \otimes \wedge^{-1/2}(\mathcal{F})$ can be identified with the set of all $\varphi \otimes \nu$ satisfying the differential equations

$$(D_\xi \varphi + 2\pi i \langle \xi, \alpha \rangle \varphi) \otimes \nu + \varphi \otimes D_\xi \nu = 0$$

for all $\xi \in U_{\mathcal{F}}$.

Let us begin by examining this equation for the case $n = 1$, relative to each of the two metaplectic structures. For the first metaplectic structure we have a globally defined metalinear frame covering ξ_H, and may choose ν to be identically one on this frame, so $D_\xi \nu = 0$. The equation for locally constant sections $\varphi \otimes \nu$ then becomes

$$D_\xi \varphi + 2\pi i \langle \xi, \alpha \rangle \varphi = 0.$$

We may take $\xi = \partial/\partial\theta$, and, using Fourier series, we look for φ of the form $\varphi = e^{iK\theta} f(r)$. Since

$$2\pi i \left\langle \frac{\partial}{\partial \theta}, \alpha \right\rangle = \frac{-2\pi}{2} \bar{z} \frac{\partial z}{\partial \theta} = \frac{-2\pi i r^2}{2}.$$

We obtain

$$\left(K - \frac{2\pi r^2}{2}\right)f = 0.$$

Thus $K/2\pi \geq 0$ is an integer and we see that f must be a generalized function: f must be some multiple of the Dirac delta function $\delta(r - \sqrt{K/\pi})$. The element

$$\hat{e}^{iK\theta}\delta(r - \sqrt{2K}) \otimes \nu$$

is an eigenvalue of the Hamiltonian $\frac{1}{2}r^2$ with eigenvalue $K/2\pi$.

For the second metalinear structure we introduce the sections u_1 and u_2 as above, and define ν_1 and ν_2 by $\nu_1(u_1) = 1$ and $\nu_2(u_2) = 1$. Now $u_1 = u_2$ if $0 < \theta < \pi$ and $u_1 = u_2\epsilon$ for $\pi < \theta < 2\pi$. Therefore

$$\nu_2 = \nu_1 \qquad \text{for } 0 < \theta < \pi,$$

$$\nu_2 = -\nu_1 \qquad \text{for } \pi < \theta < 2\pi.$$

Suppose that we look for a locally constant section of the form

$$e^{iK_1\theta}f_1(r) \otimes \nu_1 \qquad \text{on } V_1,$$

$$e^{iK_2\theta}f_2(r) \otimes \nu_2 \qquad \text{on } V_2.$$

As before, we concluded that $K_1 = K_2 \geq 0$ and that f_1 and f_2 are multiples of $\delta(r - \sqrt{2k/2\pi})$, with $2\pi K = K_1 = K_2$. Furthermore, we must have $e^{iK(3\pi/2)} = -e^{iK(-\pi/2)}$ so $e^{2\pi iK} = -1$ implying that $K = N + 1/2$. Thus the second metalinear structure gives a shift from N to $N + 1/2$. If we reintroduce the constants m, ν, and h we get the energy levels $\nu\hbar N$ for the first metaplectic structure and $\nu\hbar(N + 1/2)$ for the second metaplectic structure where $\hbar = h/2\pi$.

It is easy to see why the correct metalinear structure on each circle is the non-trivial one: If we identify $\mathbf{R}^2$ with $T^*\mathbf{R}^1$, then we get a metaplectic structure on $\mathbf{R}^2$ induced from that of $\mathbf{R}^1$. We know from our construction of the metaplectic representation that Exp $(\pi J/2) \in Mp(1)$ acts non-trivially in the metaplectic representation, although it acts as the identity in $\mathbf{R}^2$. Thus, as we rotate around once, we must pass to a different element of the metaplectic bundle and, hence, if we look at the induced metalinear bundle of an individual circle, we see that we get the non-trivial bundle.

For the case $n > 1$, we choose sections, $\tilde{s}_k$, of $MB(\mathfrak{F})$ as above, and this defines the half forms ν_k on U_k by the condition $\nu_k(\tilde{s}_k) = 1$, so that the ν_k are locally constant half forms defined on U_k with

$$\nu_k = \chi(\tilde{g}(k,j))\nu_j \qquad \text{on } U_j \cap U_k$$

where

$$\chi(\tilde{g}(k,j))^2 = z_k^n/z_j^n .$$

Let us introduce coordinates r, θ_k, and $Z_{\ell k} = z_\ell/z_k$ ($\ell \neq k$) on U_k, where $z_k = |z_k|e^{i\theta k}$ and $r^2 = \sum z_\ell \bar{z}_\ell$. On U_k we write our half form section as $\varphi_k \otimes \nu_k$ and the condition that it be locally constant can be written as

$$D_\xi \varphi_k + 2\pi i\langle \xi, \alpha\rangle \varphi_k = 0$$

where ξ ranges over the vector fields

$$\frac{\partial}{\partial \theta_k}, \frac{\partial}{\partial \bar{Z}_{1k}}, \ldots, \frac{\partial}{\partial \bar{Z}_{nk}} \qquad \left(\frac{\partial}{\partial \bar{Z}_{kk}} \text{ omitted}\right)$$

which span $U_{\mathfrak{F}}(X)$ at all points of U_k. We write

$$\varphi_k = e^{2\pi i K_k \theta_k} q_k(Z_{1k}, \ldots, \hat{Z}_{kk}, \ldots, Z_{nk}) f_k(r)$$

and, writing the equation as

$$\frac{\partial \varphi_k}{\partial \theta_k} = \frac{2\pi i r^2}{2} \varphi_k,$$

$$\frac{\partial \varphi_k}{\partial \bar{Z}_{jk}} = -\frac{Z_{jk}}{2 \sum Z_{\ell k} \bar{Z}_{\ell k}} \frac{2\pi r^2}{2} \varphi_k$$

we see that

$$f_k = c\delta(r - \sqrt{2K_k/2\pi})$$

$$\frac{\partial}{\partial \bar{Z}_{jk}} \log q_k = \frac{\partial}{\partial \bar{Z}_{jk}} \log\left(\sum Z_{\ell k} \bar{Z}_{\ell k}\right)^{-K_k/2}$$

so

$$q_k = \left(\frac{|z_k|}{r}\right)^{K_k} p_k$$

where p_k is holomorphic in the variables $Z_{\ell k}$ ($\ell \neq k$). Thus the local expression for $\varphi_k \otimes \nu_k$ on U_k is

$$z_k^{K_k} q_k(Z_{1k}, \ldots, \hat{Z}_{k,k}, \ldots, Z_{nk}) r^{-K_k} \delta(r - \sqrt{2m\nu\hbar K_k}) \otimes \nu_k$$

(where we have reinserted the parameters m, ν and $\hbar$). Since r is a well defined function on X, equating the singularities of $\varphi_k \otimes \nu_k$ shows that $K_k = K$ is independent of k. The transition laws imply that

$$z_k^{2K} p_k^2\left(\frac{z_1}{z_k}, \ldots, \frac{\hat{z}_k}{z_k}, \ldots, \frac{z_n}{z_k}\right) = \frac{z_k^n}{z_j^n} z_j^{2K} p_j^2\left(\frac{z_1}{z_j}, \ldots, \widehat{\frac{z_j}{z_j}}, \ldots, \frac{z_n}{z_j}\right).$$

Since p_k and p_j are holomorphic, we conclude that

$$2K - n = 2N, \qquad N = 0, 1, 2, \ldots,$$

and that p_k is a polynomial of degree at most N in its variables. These states are eigenvectors of $H = (\frac{1}{2}m) \sum z_j \bar{z}_j$ with eigenvalues

$$\nu h\left(N + \frac{n}{2}\right), \qquad N = 0, 1, 2, \ldots,$$

and multiplicity

$$\binom{N + n - 1}{N}.$$

We thus get the standard energy levels and multiplicity for the n dimensional harmonic oscillator.

In [6], Simms shows how to apply the quantization procedure to the hydrogen atom. On $S^2 \times S^2$ one must introduce the complex product polarization and then one gets the correct energy levels and multiplicity for the hydrogen atom, together with the standard representations of O(4). Actually, the group O(2,4) acts irreducibly on the space of bound states of the hydrogen atom, and acts transitively on the space on non-zero contagent vectors to S^3, but we shall not go into this fact here.

We now present some computations of Blattner [4] which illustrate an example of the pairing between the Hilbert spaces associated to two real polarizations which are unitarily related but not Heisenberg related. The polarizations in question are the two rulings of the single sheeted hyperboloid in R^3, as discussed in §3, where the hyperboloid is considered as an orbit of $SL(2,\mathbf{R})$ acting on the dual of its Lie algebra.

As we remarked in §3, we can identify $s\ell(2,\mathbf{R})^*$ with $s\ell(2,\mathbf{R})$, which can be considered as the algebra of all matrices of the form

$$A = \begin{pmatrix} a & b \\ c & -a \end{pmatrix}$$

with the Killing form

$$B(A, A') = aa' + \tfrac{1}{2}(bc' + cb').$$

We shall be interested in the single sheeted orbits $B = \lambda^2 > 0$. Let X be such an orbit and $\beta \in X$. Then the definition of ω gives

$$\begin{aligned}\omega(\hat{\xi}(\beta), \hat{\eta}(\beta)) &= B(\beta, [\eta, \xi]) \\ &= B([\xi, \beta], \eta).\end{aligned}$$

If $\beta = (a, b, c)$ with $b \neq 0$, and we choose $\xi = (0, 0, -b^{-1})$ so that $\hat{\xi}(\theta) = [\xi, \beta] = (1, 0, -2ab^{-1})$ and $\eta = ((2b)^{-1}, 0, 0)$ so that $\hat{\eta}(\theta) = [\eta, \beta] = (0, 1, -cb^{-1})$, we obtain

$$\omega(\hat{\xi}(\beta), \hat{\eta}(\beta)) = B((1, 0, -2ab^{-1}), (2b)^{-1}, 0, 0) = (2b)^{-1}$$

so that

$$\omega = (2b)^{-1} da \wedge db$$

for the region $b \neq 0$. Since $2ada + bdc + cdb = 0$ on X we conclude that the other local expressions for ω are

$$\begin{aligned}\omega &= (2b)^{-1} da \wedge db, \qquad & b \neq 0, \\ &= (4a)^{-1} db \wedge dc, \qquad & a \neq 0, \\ &= (2c)^{-1} dc \wedge da, \qquad & c \neq 0.\end{aligned}$$

As we have seen in §3, there are exactly two invariant lines at each point $\beta \in X$. At the point $\beta = (\lambda, 0, 0)$ we saw that these lines were given by $c = 0$ and $b = 0$. Group invariance (or a direct verification) then shows that the lines at any point are given by

$$u = \text{const.} \qquad \text{and} \qquad v = \text{const.}$$

where

$$u = \frac{a + \lambda}{b} = \frac{-c}{a - \lambda} \quad \text{and} \quad v = \frac{a - \lambda}{b} = \frac{-c}{a + \lambda}.$$

We can use u and v as local coordinates on the open set U given by $b \neq 0$. Solving for a, b, c in terms of u, v gives

$$a = \lambda \frac{u + v}{u - v}, \qquad b = \frac{2\lambda}{u - v}, \qquad c = \frac{-2\lambda uv}{u - v}$$

and

$$\omega = \lambda(u - v)^{-2} du \wedge dv.$$

We choose, as our polarizations, $\mathfrak{F}_1$, given by $u = \text{const.}$ and $\mathfrak{F}_2$ given by $v = \text{const.}$

If one takes G to be the universal covering group of $SL(2, \mathbf{R})$ then the set of all Hermitian line bundles with connection, whose curvature form is ω, and which are homogeneous under G, are parametrized by the characters of the group G_β, corresponding to the infinitesimal character β. The group G_β is the direct product of the line $\{\exp t\beta\}$ and an infinite cyclic group generated by the element covering $-I$ in G, i.e., by $k(\pi)$ where k is the one parameter group generated by

$$\begin{pmatrix} 0 & 1 \\ -1 & 0 \end{pmatrix} \in SL(2, \mathbf{R}).$$

Thus the characters $\chi(t, z)$ are given by $\chi(t, z) = e^{2\pi i\lambda a} s^z$ where s is any complex number of absolute value one, i.e., $\chi(t, z) = e^{2\pi i(\lambda a + rz)}$ where r is determined mod Z. Each r gives a line bundle and for each line bundle we can choose sections s_1 and s_2 which are covariant constant along the polarizations $\mathfrak{F}_1$ and $\mathfrak{F}_2$, giving corresponding forms α_1 and α_2. As Blattner [4] shows (by a fairly straightforward computation) one such choice is given in the local coordinates u, v, by

$$\alpha_1 = -\lambda[(u - v)^{-1} du - u(1 + u^2)^{-1} du] - r\pi^{-1}(1 + u^2)^{-1} du$$

and another by

$$\alpha_2 = -\lambda[(u - v)^{-1} dv - v(1 + v^2)^{-1} dv] - r\pi^{-1}(1 + v^2) dv.$$

Then a computation shows that if $g_2 = g_2(v)$ and $g_1 = g_1(u)$ are functions constant along the respective polarizations, we have

$$(g_2 s_2 \otimes dv^{1/2}, g_1 s_1 \otimes du^{1/2})$$

$$= \iint_{u \neq v} g_2(v) \overline{g_1(u)} \left[\frac{|v - u|}{(1 + u^2)^{1/2}(1 + v^2)^{1/2}} \right]^{2\pi i\lambda} \exp(-2\pi i r[\arctan v - \arctan u])$$

$$\times \exp(i\pi r \operatorname{sgn}(v - u)) |\lambda|^{-1/2} |v - u| \frac{|\lambda| \, |du \wedge dv|}{|v - u|^2}.$$

This pairing, as it stands, is a singular integral but can be given as an interpretation by a procedure of analytic continuation. The operator so obtained coincides (up to conventions and normalizations) with the (unitary) intertwining operator introduced by Knapp and Stein relating the unitary representations of G on each of the Hilbert spaces $H(\mathfrak{F}_1)$ and $H(\mathfrak{F}_2)$.

The interest of the above example, besides illustrating the necessity for care in defining the integrals entering into the pairing between polarization, is that it is case of two polarizations which are unitarily related but not Heisenberg related.

The question of giving a geometrical characterization of the condition that two transverse polarizations be unitarily related remains to be explored. We indicate some very partial results in this direction for real polarizations in the plane. Any fibration in two dimensions is automatically a polarization, so our strategy will be to fix two fibrations in the plane and allow the form ω to vary. The problem then becomes one of describing the forms ω for which the two given polarizations are unitarily related. We shall deal with an infinitesimal version of this question, namely: Given a curve of ω_t of symplectic forms which give unitarily related polarizations, what conditions are imposed on $\dot{\omega} = (d/dt)\omega_t \,|_{t=0}$?

Let

$$\omega = e^{\rho}\, dx \wedge dy$$

where ρ is some smooth function on the plane. We take the polarizations given by $x =$ const. and $y =$ const. We let α_x and α_y be the forms defined by

$$d\alpha_x = \omega, \qquad \alpha_x = f dx, \qquad f(x,0) = 0,$$

and

$$d\alpha_y = \omega, \qquad \alpha_y = g dy, \qquad g(0,y) = 0,$$

so that

$$\alpha_x - \alpha_y = d\varphi$$

where

$$\frac{\partial^2 \varphi}{\partial x \partial y} = e^{\rho}, \qquad \varphi(x,0) = \varphi(0,y) = 0.$$

Then the operator U mapping $L^2(\mathbf{R}) \to L^2(\mathbf{R})$ associated with the pairing is given by

$$(Uv)(y) = \frac{1}{\sqrt{2\pi i}} \int_{\mathbf{R}} e^{i\varphi(x,y)} e^{\rho(x,y)/2} v(x)\, dx.$$

If $\rho \equiv 0$ then $\varphi = xy$ and the operator U is essentially the Fourier transform. If we set

$$\omega_t = e^{t\rho}\, dx \wedge dy$$

then

$$\varphi_t = xy + t\psi(x,y) + \cdots$$

where

$$\frac{\partial^2 \psi}{\partial x \partial y} = \rho, \qquad \psi(x,0) = \psi(0,y) = 0,$$

If we set

$$\dot{U}_0 = \left.\frac{dU_t}{dt}\right|_{t=0}$$

we obtain

$$(\dot{U}_0 f)(y) = \frac{1}{\sqrt{2\pi i}} \int e^{ixy}(i\psi(x,y) + \tfrac{1}{2}\rho(x,y))f(x)\,dx$$

and hence

$$(U_0^{-1}\,\dot{U}_0 v)(z) = \frac{1}{2\pi}\int k(x,z)f(z)\,dz$$

where

$$k(x,z) = \int e^{i(x-z)y}(i\psi(x,y) + \tfrac{1}{2}\rho(x,y))\,dy.$$

If U_t is a one parameter family of unitary operators then $U_t^{-1}\,\dot{U}_t$ is a skew Hermitian operator and so the kernel k must satisfy

$$\overline{k(z,x)} = -k(x,z)$$

or

$$\int e^{i(x-z)y}(-i\psi(x,y) - \tfrac{1}{2}\rho(x,y))\,dy = \int e^{i(x-z)y}(-i\psi(z,y) + \tfrac{1}{2}\rho(z,y))\,dy.$$

Let $f(x,\eta)$ denote the Fourier transform of ψ with respect to y. Then the above equation can be written as

$$f(x+\eta,\eta) - f(x,\eta) = \frac{\eta}{2}\left(\frac{\partial f}{\partial x}(x,\eta) + \frac{\partial f}{\partial x}(x+\eta,\eta)\right).$$

Let w be the Fourier transform of f with respect to x. Then we get

$$e^{i\xi\eta}w(\xi,\eta) - w(\xi,\eta) = \frac{i\xi\eta}{2}(w(\xi,\eta) + e^{i\xi\eta}w(\xi,\eta))$$

or

$$w(\xi,\eta)(e^{i\xi\eta} - 1) = w(\xi,\eta)\frac{i\xi\eta}{2}(e^{i\xi\eta} + 1),$$

or

$$w(\xi,\eta)\left(\frac{\xi\eta}{2} - \tan\frac{\xi\eta}{2}\right) = 0.$$

Since the equation $x - \tan x = 0$ has a root of order 3 at the origin and a simple root, r_K, on each interval $(-\pi/2 + K\pi, \pi/2 + K\pi)$ for $K \neq 0$, w must be supported on the union of the cone $\xi\eta = 0$ and the hyperbolas $\xi\eta = r_K$. Suppose w was supported on the Kth hyperbola. Since r_K is a simple root

$$\left(\frac{\xi\eta}{2} - r_K\right)w = 0$$

or, taking the inverse Fourier transform

$$\frac{1}{2}\frac{\partial^2}{\partial x \partial y}\psi + r_K\psi = 0$$

together with the boundary conditions, $\psi(x, 0) = \psi(0, y) = 0$, this implies $\psi \equiv 0$. Next suppose w is supported on $\xi\eta = 0$. Since zero is a triple root of $x - \tan x = 0$,

$$(\xi\eta)^3 w = 0.$$

Thus

$$\left(\frac{\partial^2}{\partial x \partial y}\right)^3 \psi = 0.$$

Since

$$\frac{\partial^2}{\partial x \partial y}\psi = \rho,$$

we obtain the condition

$$\left(\frac{\partial^2}{\partial x \partial y}\right)^2 \rho = 0.$$

It is interesting to compare this retriction with the (infinitesimal) condition for the polarizations to be Heisenberg related, which is that

$$\left(\frac{\partial^2}{\partial x \partial y}\right)\rho = 0.$$

Thus the unitarity condition uses the square of the operator $(\partial^2/\partial x \partial y)$ while the Heisenberg condition was the operator itself. This suggests that the class of unitarily related polarizations is quite restricted. It would be very useful to put the above considerations on a more rigorous basis and to understand the geometrical implications of unitarity.

REFERENCES, CHAPTER V

1. B. Kostant, *Quantization and unitary representations*. I. *Prequantization*, Lectures in Modern Analysis and Applications, III, Lecture Notes in Math., vol. 170, Springer-Verlag, Berlin, 1970, pp. 87–208. MR **45** #3638.

2. J.-M. Souriau, *Structure des systèmes dynamiques*, Maîtrises de mathématiques, Dunod, Paris, 1970. MR **41** #4866.
3. L. Auslander and B. Kostant, Invent. Math. **14** (1971), 255–354. MR **45** #2092.
4. R. J. Blattner, *Quantization and representation theory*, Proc. Sympos. Pure Math., vol. 26, Amer. Math. Soc., Providence, R. I., 1973, pp. 147–165. MR **49** #6277.
5. B. Kostant, *Symplectic spinors*, Conv. di Geom. Simp. Fis. Mat., INDAM, Rome, 1973.
6. D. J. Simms, Proc. Cambridge Philos. Soc. **73** (1973), 489–491.
7. ———, *Metalinear structures and a geometric quantization of the harmonic oscillator*, Int. Coll. Sympos. Geom., Aix en Provence, 1974.
8. S. Sternberg, Amer. J. Math. **79** (1957), 809–824. MR **20** #3335.
9. P. Hartman, *Ordinary differential equations*, Wiley, New York and London, 1964. MR **30** #1270.
10. A. Newlander and L. Nirenberg, Ann. of Math. (2) **65** (1957), 391–404. MR **19**, 577.
11. P. Renouard, Thèse, Paris, 1972.
12. A. Weil, Acta Math. **111** (1964), 143–211. MR **29** #2324.
13. D. Shale, Trans. Amer. Math. Soc. **103** (1962), 149–167. MR **25** #956.
14. O. Loos, *Symmetric spaces*. I: *General theory*, II: *Compact spaces and classification*, Benjamin, New York and Amsterdam, 1969. MR **39** #365a, b.
15. J. H. Rawnsley, Math. Proc. Cambridge Philos. Soc. **78** (1975), 345–350.
16. J.-M. Souriau, *Construction explicite de l'indice de Maslov applications* (to appear).
17. V. Bargmann, in *Analytic methods in mathematical physics*, Gordon & Breach, (1970), 27–63.
18. S. Sternberg and J. Wolf (to appear).

Chapter VI. Geometric Aspects of Distributions

In this chapter we shall study the behavior of distributions with respect to various kinds of smooth maps between manifolds. We will begin with some elementary considerations involving the push forward and the pull back of distributions and use these results to derive some interesting formulas. We will then show (essentially using the Radon transform) how to decompose a distribution into a superposition of distributions of a simpler type and use this to introduce the notion of the wave front set, a subset of the cotangent bundle, which measures the singular codirections for distributions. By introducing a particular class of distributions on the line, and then extending it so as to be closed under functorial operations, we arrive at a class of distributions, whose wave front sets are Lagrangian submanifolds, and which are, essentially, the "Fourier integral operators" introduced by Hormander. We will develop the calculus of these operators and their associated symbol calculus, and give some applications of these results.

§1. Elementary functorial properties of distributions.

Let $f: X \to Y$ be a smooth map. If u is a C^∞ function on Y then $f^*u = u \circ f$ is a C^∞ function on X. Thus f induces a linear map, f^*, from $C^\infty(Y)$ to $C^\infty(X)$, where $C^\infty(X)$ denotes the space of all C^∞ functions on X. We can give $C^\infty(X)$ a topology—the topology of uniform convergence with each finite number of derivatives on each compact subset. (Here "derivatives" means with respect to some local coordinate system. One chooses a partition of unity subordinate to a coordinate cover and writes $u = \sum \varphi u$, then differentiates each φu with respect to the local coordinates associated with φ.) It is easy to check that $f^*: C^\infty(Y) \to C^\infty(X)$ is a continuous linear map.

If $C_0^\infty(X)$ denotes the space of C^∞ functions on X with compact support, then it carries a stronger topology—a sequence of $u \in C_0^\infty(X)$ converges if all the $\operatorname{supp} u$ lie in some *fixed* compact set K and the u converge in $C^\infty(X)$. If f is a proper map then $f^* : C_0^\infty(Y) \to C_0^\infty(X)$ and is continuous with respect to this topology.

A distribution on X is a continuous linear function on $C_0^\infty(X)$. It follows that if f is a proper smooth map from X to Y and if ν is a distribution on X then $f_* \nu$ is a distribution on Y where

$$\langle u, f_* \nu \rangle = \langle f^* u, \nu \rangle. \tag{1.1}$$

If we put the weak topology on the spaces of distributions then the map f_* is continuous.

The *support* of a distribution is defined as follows: A point x does *not* lie in $\operatorname{supp} \nu$ if there is some neighborhood U of x such that $\langle w, \nu \rangle = 0$ if $\operatorname{supp} w \subset U$. If ν is a distribution with compact support then it defines a linear functional on $C^\infty(X)$. Indeed let φ be some C^∞ function of compact support such that $\varphi \equiv 1$ in some neighborhood of $\operatorname{supp} \nu$. Then, for any $w \in C^\infty(X)$ we define

$$\langle w, \nu \rangle = \langle \varphi w, \nu \rangle$$

and observe that this is independent of the choice of φ. With this definition it is clear that $f_* \nu$ is defined (by (1.1)) for any smooth map f, proper or not, if ν has compact support. (More generally $f_* \nu$ will be defined if $f^{-1}(K) \cap \operatorname{supp} \nu$ is compact for each compact $K \subset Y$, i.e., if $f_{|\operatorname{supp} \nu}$ is proper.)

A nice example of a distribution on X is given by a smooth density ρ on X, the linear functional being given as

$$u \rightsquigarrow \int_X u\rho.$$

We shall therefore refer to a distribution as a "generalized density". This will be of use to us in keeping the variances straight. If $|\wedge|(X)$ denotes the line bundle of densities, then $C^\infty(|\wedge|(X))$ will denote the space of smooth densities and $C^{-\infty}(|\wedge|(X))$ will denote the space of generalized densities. The pairing between $\rho \in C^{-\infty}(|\wedge|(X))$ and $u \in C_0^\infty(X)$ we will write either as

$$\langle u, \rho \rangle \quad \text{or as} \quad \int_X u \cdot \rho,$$

the second notation having an interpretation as an integral only for ρ's which are actually densities.

More generally, if E is a vector bundle over X we will let $C^\infty(E)$ denote the space of smooth sections of E and $C^{-\infty}(E)$ the space of generalized sections of

E. A generalized section of E is a continuous linear functional on

$$C_0^\infty(E^* \otimes |\wedge|(X)).$$

For example, a generalized function would be a continuous linear functional on $C_0^\infty(|\wedge|(X))$, i.e., a continuous linear functional on the space of smooth densities of compact support. A generalized density is a continuous linear functional on $C_0^\infty(|\wedge|^*(X) \otimes |\wedge|(X))$ which is isomorphic to $C_0^\infty(X)$ since $L^* \otimes L$ is canonically trivial for any line bundle L. The notions of support, etc., continue to make sense as do the two notations for the pairing between $u \in C_0^\infty(E)$ and $\rho \in C^{-\infty}(E^* \otimes |\wedge|(X))$. Here, if ρ were an honest section of $E^* \otimes |\wedge|(X)$ then $u \cdot \rho$ would mean the density obtained from the map $u \otimes \rho \to u \cdot \rho$ given by the evaluation $E \otimes E^* \otimes |\wedge|(X) \to |\wedge|(X)$.

To define the notion of pull back and push forward for vector bundles we need to use the idea of a morphism of vector bundles: Let $E \to X$ and $F \to Y$ be vector bundles. Recall that $f: E \to F$ is a morphism if f defines a map of $X \to Y$ and a smooth section of Hom $(f^\# F, E)$, i.e., a linear map $r(x): F_{f(x)} \to E_x$ which depends smoothly on x. Then $f^*: C^\infty(F) \to C^\infty(E)$ is defined by $f^*u(x) = r(x)u(f(x))$, and, if f is proper, $f^*: C_0^\infty(F) \to C_0^\infty(E)$. Similarly,

$$f_*: C_0^{-\infty}(E^* \otimes |\wedge|(X)) \to C_0^{-\infty}(F^* \otimes |\wedge|(Y))$$

is defined by (1.1) and if f is proper then also

$$f_*: C^{-\infty}(E^* \otimes |\wedge|(X)) \to C^{-\infty}(F^* \otimes |\wedge|(Y))$$

is defined.

To summarize: Sections pull back and generalized sections push forward.

In order to avoid cluttering up the notation let us go back, for the moment, to the case of the trivial bundle, i.e., functions and generalized densities. Suppose that the generalized density ν is a measure. This means that ν extends to a functional on $C_0^0(X)$, the space of continuous functions of compact support. (Put another way, this means that ν is continuous with respect to the C_0^0 topology on $C_0^\infty(X)$.) Then (1.1) makes sense for continuous functions u and thus $f_*\nu$ is again a measure. Thus the push forward of a measure is a measure. We might ask if the push forward of a smooth density is again smooth. In general the answer is no. For instance if X were a point and $f(X) = y \in Y$ then $\langle u, f_*\nu\rangle = au(y)$ for some constant a, so $f_*\nu$ is not smooth (unless $a = 0$ or dim $Y = 0$).

On the other hand, if $f: X \to Y$ is a submersion, and ρ is a smooth density of compact support on X then $f_*\rho$ is a smooth density on Y, and, indeed, is given by integration over the fiber: Indeed, for each $y \in Y$ and each $x \in f^{-1}(y)$, we can identify the spaces $|\wedge|(TX_x)$ with $|\wedge|T(f^{-1}(y)) \otimes |\wedge|TY_y$ and thus the density ρ, when restricted to the fiber $f^{-1}(y)$, can be thought of as a density along the fiber, with values in the line of densities at y. We can integrate this along the fiber to get a density, σ, on Y which is smooth. If u is any smooth function on Y then

$$\int (f^* u)\rho = \int u\sigma$$

since u depends only on y and the integral over X is a double integral, first over the fiber and then over Y. (The same works if ρ does not necessarily have compact support but f is proper.)

(The same works for morphisms on vector bundles. This time $r(x)^* : E_x^* \to F_y^*$ so that if ρ is a section of $E^* \otimes |\wedge|(X)$ then $r(x)^* \rho(x)$ can be identified with a fiber density at x with values in the vector space $F_y^* \otimes |\wedge|(Y)_y$. Thus integration over the fiber gives an honest section of $F^* \otimes |\wedge|(Y)$.)

Thus *under submersions smooth densities push forward and hence generalized functions pull back.*

(Or, more generally, if $E \to F$ is a smooth morphism then smooth sections of compact support of $E^* \otimes |\wedge|(X)$ push forward and hence generalized sections of F pull back.)

As an illustration of an application of the above result, suppose that $f : X \to Y$ is not necessarily a submersion. Let $A_f \subset X$ be the set of critical points of f, i e., the set of x such that rank $df_x < \dim Y$. Let ρ be a smooth density of compact support on X so that $f_* \rho$ is a measure on Y. We can ask whether $f_* \rho$ is absolutely continuous relative to Lebesgue measure on Y. We claim that if A_f has measure zero on X then $f_* \rho$ is absolutely continuous on Y. Conversely, if A_f does not have measure zero then we can find a ρ such that $f_* \rho$ is not absolutely continuous. Indeed, let $C_f = f(A_f)$ be the set of critical values. By Sard's theorem we know that C_f has measure zero. Let $C_{f,\rho} = f(A_f \cap \operatorname{supp} \rho)$. Then $C_{f,\rho}$ is compact and has measure zero. We know that $f_* \rho$ is smooth on $Y - C_{f,\rho}$. On the other hand, if A_f has measure zero, it is clear that

$$\int_{C_{f,\rho}} f_* \rho = \int_{f^{-1}(C_{f,\rho})} \rho = \int_{A_f} \rho + \int_{(X-A_f)\cap f^{-1}(C_{f,\rho})} \rho.$$

The first integral vanishes since A_f has measure zero. The second integral vanishes since f is a submersion on $X - A_f$. Thus $f_* \rho$ vanishes when integrated over any set of measure zero. Conversely, if A_f has positive Lebesque measure then we can find a smooth ρ such that $\int_{A_f} \rho \neq 0$.

We have seen how to pull back generalized functions under submersions. If $X \xrightarrow{f} Y \xrightarrow{g} Z$ are two submersions then it is easy to check that

$$f^* \circ g^* = (g \circ f)^*$$

when applied to generalized functions on Z.

This suggests a way of defining pull backs for certain generalized sections even when f is not an immersion. Indeed, suppose that u is a generalized function on

Y of the form

$$u = g^* v$$

where v is a generalized function on Z and $g : Y \to Z$ is a submersion. Suppose that the map $f : X \to Y$ is not necessarily a submersion but that $g \circ f : X \to Z$ is a submersion. Then we would define $f^* u = (g \circ f)^* v$. Of course, we would have to show that this is independent of the particular representation, i.e., if $g' : Y \to Z'$ with $g'^* v' = u$ then $(g' \circ f)^* v' = (g \circ f)^* v$. This is indeed true. We shall prove a considerably more general result in §3. We will therefore postpone the proof until then, but shall do some computations involving this definition now.

Suppose that w is a smooth function on Y and $u = wg^* v$ where $g : Y \to Z$ is a submersion and v is a generalized function on Z. (The product of a smooth function w and a generalized function a is defined as $\langle w \cdot a, \rho \rangle = \langle a, w\rho \rangle$.) Then we can set

$$\begin{aligned} f^* u &= f^* w f^* g^* v \\ &= f^* w (g \circ f)^* v \end{aligned}$$

if $g \circ f$ is a submersion. Again, this is well defined as we shall see in §3.

Suppose that $Z = \mathbf{R}$ and $v = \delta$ is the delta function at the origin on $\mathbf{R}$. Thus $\langle \delta, a\,dt \rangle = a(0)$, if t is the standard coordinate on $\mathbf{R}$ and dt the standard density. The generalized function $g^* \delta$ can be described as follows: Let

$$W = g^{-1}(0)$$

which is a submanifold since g is a submersion. Any density ρ can be written on W as $\sigma \otimes dg$ where σ is a density along W. Then

$$\langle g^* \delta, \rho \rangle = \int_W \sigma.$$

If we consider $wg^* \delta$ then $\langle wg^* \delta, \rho \rangle = \int_W w\sigma$. We can replace the function g by a function g' having the same W, and get the same generalized function provided we change w appropriately. Indeed $g = h'g'$ for some h' which does not vanish along W. Then $dg = h'dg'$ on W and

$$\rho = \sigma \otimes dg = \sigma' \otimes dg'$$

where $\sigma' = \sigma h'$. So we must take $w = (h')^{-1} w'$. We can therefore think of w as the coefficient of a section of $|\wedge|^{-1}(NW)$ where NW is the normal bundle to W. Then $wg^* \delta$ defines a section of $|\wedge|(NW)^{-1}$. Since

$$|\wedge|(Y) = |\wedge|(W) \otimes |\wedge|(NW)$$

along W, we can pair a section of $|\wedge|(NW)^{-1}$ with a section of $|\wedge|(Y)$ (of compact support) to get a density along W which we then integrate to give a number.

If $Z = \mathbf{R}^k$ and δ is the δ-function at the origin of $\mathbf{R}^k$ all remains the same, except that W is now of codimension k. Also, if $F \to Y$ is a vector bundle and we take the trivial line bundle over $\mathbf{R}^k$ then a morphism from F to the trivial bundle over $\mathbf{R}^k$ gives a section $(r(y)1)$ of F along W.

Thus if F is a vector bundle over Y we define a δ-section along a submanifold W of Y to be a smooth section, u, of $F_{|W} \otimes |\wedge|^{-1}(NW)$, where

$$\langle u, \rho\rangle = \int_W u\rho_{|W}$$

for any section ρ of $F^* \otimes |\wedge|(Y)$. (This definition will make sense for a properly immersed submanifold as well as an embedded one.)

If $W = g^{-1}(0)$ where $g : Y \to \mathbf{R}^k$ and if $f : X \to Y$ then $g \circ f$ is a submersion if and only if f is transversal to W. In this event, $f^{-1}(W)$ is a submanifold of X and $df : TX_x \to TY_y$ carries $Tf^{-1}(W)_x \to TW_y$ for $x \in f^{-1}(W)$ and so induces an isomorphism

$$df^* : (NW)_y \to (Nf^{-1}W)_x$$

and hence of $|\wedge|^{-1}NW_y$ with $|\wedge|^{-1}N(f^{-1}W)_x$. Let $u = g^*\delta$ be a δ-section along W. Then $f^*u = (g \circ f)^*\delta$ is a δ-section along $f^{-1}W$. If we think of u as a section of $F \otimes |\wedge|^{-1}W$ then $r \otimes df^*$ gives a map from

$$F_y \otimes |\wedge|^{-1}NW_y \to E_x \otimes |\wedge|^{-1}Nf^{-1}W_x$$

for $x \in f^{-1}W$ and $y = f(x)$, and hence determines a map, f^*, of sections of $F \otimes |\wedge|^{-1}NW$ to $E \otimes |\wedge|^{-1}Nf^{-1}W$. It follows from the definitions that the two ways of defining f^*u, as a pull back of a section along W or as $(g \circ f)^*\delta$ coincide. (The definition in terms of sections is a little more general in that it makes sense for proper immersed submanifolds.)

Suppose that $x_1, \ldots, x_m$ are coordinates on $U \subset X$ such that $f^{-1}W \cap U$ is given by $x_1 = \cdots = x_k = 0$ and that $y_1, \ldots, y_n$ are coordinates on $V \subset Y$ (with $f(U) \subset V$) such that $y_1 = \cdots = y_k = 0$ describes $W \cap V \subset V$. Then a section of $F \otimes |\wedge|^{-1}W$ can be written as $u = s \otimes |dy_1 \cdots dy_k|^{-1}$ and f^*u will then be the section

$$f^*u = \frac{(rs)}{|J_f|} \otimes |dx_1 \cdots dx_k|^{-1} \tag{1.2}$$

where J_f is the Jacobian determinant

$$J_f = \det\begin{bmatrix} \partial f_1/\partial x_1 & \cdots & \partial f_k/\partial x_1 \\ \vdots & & \\ \partial f_1/\partial x_k & & \partial f_k/\partial x_k \end{bmatrix}.$$

This gives the local expression (along $f^{-1}W$) of the pull back of a δ-section along W.

Now let us examine the push forward of a δ-section. Let $f: X \to Y$ and let ρ be a δ-section of the density bundle along a submanifold $Z \subset X$. By definition, ρ is a section of $|\wedge|(X)_{|Z} \otimes |\wedge|^{-1}(NZ) \sim |\wedge|(Z)$, i.e., ρ defines a smooth density along Z. If v is a function on Y then $\langle f^* v, \rho\rangle = \int f^*_{|Z} v \cdot \rho$. In other words

$$f_* \rho = (f_{|Z})_* \rho \tag{1.3}$$

where, on the left, we are considering ρ as a δ-section of the density bundle of X and, on the right, we are considering ρ as a smooth density on Z. If $E \to X$ and $F \to Y$ are vector bundles with f a morphism and if ρ is a δ-section of $E^* \otimes |\wedge|(X)$ then ρ can be considered as a smooth section of $E^*_{|Z} \otimes |\wedge|(Z)$, and again (1.3) holds.

Notice the following consequence of (1.3): Suppose that $f_{|Z}$ is a submersion. Then $f_* \rho$ is a smooth density (or a smooth section of $F^* \otimes |\wedge|(Y)$). For example, suppose that $f: X \to Y$ is a submersion and $Z \subset X$ intersects each fiber transversally. Then $f_{|Z}$ is a submersion. Thus, in this case, the push forward of a δ-density along Z is smooth.

Let us give an example of an interesting δ-density. Let $h: W \to X$ be a differentiable map. Then $h^*: C^\infty(X) \to C^\infty(W)$ could be thought of as defined by a generalized kernel on $W \times X$, that is, we would like to write

$$h^* u = \int k(w, x) u(x)\, |dx|$$

so that k is to be considered as a generalized section of $|\wedge|(X)$ (where we write $|\wedge|(X)$ for short, instead of $\pi_X^*|\wedge|(X)$ as a bundle on $W \times X$). In local coordinates, x and w, we would write

$$k(w, x) = \delta(x - h(w))\, |dx| \qquad \text{or} \qquad k = (x - h)^* \delta \otimes |dx|$$

where δ is the δ-function on $\mathbf{R}^n$, with $n = \dim X$, and then write

$$u(h(w)) = \int \delta(x - h(w) u(x))\, |dx|.$$

Our problem is to give an interpretation of the expression $\delta(x - h(w))\,|dx|$ as the local expression for a globally defined δ-section of $|\wedge|(X)$ along $Z = \operatorname{graph} h$. Now the projection of $W \times X$ onto X gives a specific isomorphism of $N(Z)_z$ with $T^* X_x$ where $z = (w, x) \in Z$. We thus get an isomorphism of $|\wedge|(X)_x$ with $|\wedge|(N(Z))_z$ and hence a preferred section of

$$|\wedge|(X) \otimes |\wedge|^{-1}(NZ).$$

This gives a δ-section, k, of $|\wedge|(X)$. Let us show that this section coincides with $\delta(x - h(w))|dx|$. Let x, w be local coordinates. Then $d(x - h)$ span the normal bundle to Z and we are identifying dx with $d(x - h)$ so that our preferred section is $|dx| \otimes |d(x - h)|^{-1}$ along Z. But the δ-section $(x - h)^* \delta$ is precisely

$|d(x-h)|^{-1}$ (where, of course, we write $|d(x-h)|$ for $|d(x_1-h_1)\cdots d(x_n-h_n|$.) Let us now examine the effect of multiplying k by some function u and then pushing forward onto W. To push forward onto W (in our scheme) we would like to think of k as a δ-section of some density bundle on $W\times X$. We do this by writing $|\wedge|(W\times X) = |\wedge|(W)\otimes|\wedge|(X)$ so $|\wedge|(X) = |\wedge|^{-1}(W)\otimes|\wedge|(W\times X)$. Thus k is a δ-section density of $|\wedge|^{-1}(W)$, i.e., a section of $|\wedge|^{-1}(W)\otimes|\wedge|^{-1}(NZ)\otimes|\wedge|(W\times X) = |\wedge|^{-1}(W)\otimes|\wedge|^{-1}(NZ)\otimes|\wedge|(X)\otimes|\wedge|(W)$ and is, by construction, that δ-section density obtained by identifying $|\wedge|(NZ)$ with $|\wedge|(X)$ and thus gives the canonical section of $|\wedge|^{-1}(W)\otimes|\wedge|(W)$. Since π: graph $h\longrightarrow W$ is a diffeomorphism, integration over the fiber is a trivial operation and thus $\pi_* ku$ is a smooth function on W and indeed

$$\pi_*(ku)(w) = u(h(w)).$$

Suppose that $f: Y\to W\times X$ is transversal to graph h. Then we can form f^*k and obtain a δ-section of $f^{\#}|\wedge|(X)$ along $f^{-1}(\text{graph}\, h)$. Let us apply this to the case where $W = X$ and take $Y = X$ where $f = \Delta$ is the diagonal map so that $\Delta(x) = (x,x)$. Then $\pi\circ\Delta = \text{id}$ so that we may identify $\Delta^{\#}|\wedge|(X)$ with $|\wedge|(X)$. To say that Δ is transversal to graph h is to say that h is a Lefschetz map, i.e., that h has isolated fixed points and that id $-dh$ is invertible at each one of these fixed points. Then $\Delta^{-1}(\text{graph}\, h)$ consists exactly of the set of fixed points and $\Delta^* k$ will be a δ-section of $|\wedge|(X)$ at these points. To evaluate $\Delta^* k$ let $x_1,\ldots,x_n$ be coordinates about the fixed point, p, given by $x_1 = \cdots = x_n = 0$ and coordinates $y_1,\ldots,y_n, z_1 - h(y_1),\ldots, z_n - h(y_n)$ around graph h so that the vanishing of the *last* n coordinates describes graph h. Then

$$\Delta(x_1,\ldots,x_n) = (x_1,\ldots,x_n,x_1-h(x_1),\ldots,x_n-h(x_n))$$

and so, by (1.2), near p we have

$$\Delta^* k = \frac{1}{|\det(\text{id} - dh_p)|}|dx|\otimes\delta(x_1,\ldots,x_n).$$

If X is compact we can integrate this generalized section density (i.e., form $\pi_* k$ where now $\pi: X\to \text{pt.}$). This gives

$$\pi_*\Delta^* k = \sum_{h(p)=p}\frac{1}{|\det(\text{id} - dh_p)|}.$$

If k were a smooth kernel, $k = k(x,y)\,dx$, then the sequence of operations $\pi_*\Delta^* k$ would yield $\int k(x,x)\,dx$ so we will write the above equality in the more suggestive form

$$\int k(x,x) = \sum\frac{1}{|\det(\text{id} - dh_p)|}.$$

It is very easy to extend the above considerations to the case of a vector bundle morphism. Here h will also carry with it an element r of $\mathrm{Hom}(E_{h(x)}, E_x)$ and so k is to be thought of as a generalized section of the bundle $\mathrm{Hom}(E_2, E_1) \otimes |\wedge|(X)$ where E_1 is the bundle E, pulled up to $X \times X$ via projection on the first factor and E_2 is E pulled up via the second projection. Then $\Delta^* k$ will be a δ-section of $\mathrm{Hom}(E, E) \otimes |\wedge|(X)$ and so $\mathrm{tr}\, \Delta^* k$ will be a section of $|\wedge|(X)$. Then we get the formula

$$\pi_* \,\mathrm{tr}\, \Delta^* k = \sum_{h(p)=p} \frac{\mathrm{tr}\, r_p}{|\det(\mathrm{id} - dh_p)|}. \tag{1.4}$$

We will derive some interesting applications of this formula in the next section.

Let us examine the situation where $W = G \times X$ where G is a Lie group and where the map $h: G \times X \to X$ is a group action, so that $h(a, h(b, x)) = h(ab, x)$. Let us take $Y = G \times X$ and f to be the diagonal map Δ where $\Delta(a, x) = (a, x, x)$. Then

$$\Delta^{-1}(\text{graph } h) = \{(a, x) | h(a, x) = x\}.$$

Let us examine the condition that the map Δ be transversal to graph h at some point (a, x) where $h(a, x) = x$. The image of $d\Delta_{(a,x)}$ consists of all vectors of the form (ζ, ξ, ξ) where $\xi \in TX_x$ and $\zeta \in g$ where g is the Lie algebra of G, i.e. the tangent space to G at e, considered as a left invariant vector field on G, so that ζ determines a tangent vector $\zeta_a \in TG_a$. If $\exp t\zeta$ is the one parameter group generated by ζ then

$$h(a \exp t\zeta, x) = h(a, h(\exp t\zeta, x)).$$

Let $\bar{\zeta}$ be the vector field on X corresponding to the element ζ, so that $\bar{\zeta}$ is the infinitesimal generator of the one parameter group $h(\exp t\zeta, \cdot)$. It follows from the above equation that the tangent space to graph h consists of all vectors of the form $(\zeta, \xi, d\bar{a}_x(\bar{\zeta}_x + \xi))$ where $\xi \in TX_x$ and $\zeta \in g$, and where we let $\bar{a}$ denote the transformation $z = h(a, z)$. Notice that

$$\begin{aligned} h(a, h(\exp t\zeta, x)) &= h(a(\exp t\zeta), x) = h\big(a(\exp t\zeta)a^{-1}, h(a, x)\big) \\ &= h\big(a(\exp t\zeta)a^{-1}, x\big) \end{aligned}$$

since $h(a, x) = x$. Thus

$$da_x\big(\bar{\zeta}_x\big) = \overline{(\mathrm{Ad}_a \zeta)}_x$$

and therefore the tangent space to graph h consists of all vectors of the form

$$\big(\zeta, \xi, \overline{(\mathrm{Ad}_a \zeta)}_x + d\bar{a}_x \xi\big) \quad \text{where } \xi \in TX_x \text{ and } \zeta \in g.$$

The intersection of im $d\Delta_{(a,x)}$ with the tangent space to graph h consists of vectors of the form (ζ, ξ, ξ) where

$$\xi - da_x \xi = \overline{(\mathrm{Ad}_a \zeta)}_x.$$

Since dim graph h = dim $G \times X$ = dim G + dim X while dim $G \times X \times X =$

dim G + 2 dim X a necessary and sufficient condition for transversality is that the above intersection have dimension equal to dim G. Now the tangent space to the orbit through x gives rise to a subspace of the intersection which already has dimension equal to dim G. In fact, suppose that

$$\xi = \overline{\zeta'_x} \quad \text{where } \zeta' \in g \text{ and } \zeta - \mathrm{Ad}_{a^{-1}}(\zeta' - \mathrm{Ad}_a \zeta') \in g_x$$

where g_x is the isotropy algebra of x, i.e. the set of all $\eta \in g$ with $\overline{\eta}_x = 0$. Then it is clear that (ζ, ξ, ξ) lies in the intersection. For each fixed such ξ the space of corresponding ζ has dimension equal to dim g_x while the dimension of the ξ's is dim g − dim g_x. Thus transversality is equivalent to the condition that these be the only solutions. Let O_x denote the orbit through x. The map da_x of TX_x into itself preserves the tangent space to the orbit. Let $P_{a,x}$ denote the induced map on the quotient space TX_x/TO_x. Then we can summarize the preceding discussion by the assertion

> *The map* Δ *is transversal to* graph h *at* (a, x) *if and only if* id $- P_{a,x}$ *is bijective.*

Notice that if bx is some other point on the orbit O then $(bab^{-1})bx = bx$ and the maps $P_{a,x}$ and $P_{bab^{-1},bx}$ are conjugate. Therefore, Δ is transversal to graph h at (a, x) if and only if it is transversal at (bab^{-1}, bx).

Let $G^+ \subset G$ be an open subset of G such that Δ is transversal to graph h at all (a, x) with $ax = x$, and let $Z \subset G^+ \times X$ consist of all such pairs (a, x). We continue to let Δ denote the restriction of Δ to $G^+ \times X$. Then $Z = \Delta^{-1}$ graph h is a submanifold of $G^+ \times X$ and $\Delta^* k$ is a δ-section of $|\wedge|X$ along Z. If the projection $\pi: Z \to G$ is proper then $\pi_* \Delta^* k$ is a generalized function on G. Notice that if G acts transitively on X, in other words if all of X is a single orbit of G, then it follows from the preceding characterization of transversality that all fixed points are transversal since the normal bundle is trivial. In this case we can take $G^+ = G$ and $\pi_* \Delta^* k$ is a generalized function defined on G. We shall study this situation in detail in the next section, in connection with the theory of characters. For the present let us consider a case at the opposite extreme, where $G = \mathbf{R}$ so that we are given a one parameter group, $t \rightsquigarrow \exp t\xi$ of transformations on X whose infinitesimal generator is the vector field ξ. All points of X are fixed points for $t = 0$, and if X is not one dimensional, no point $(0, x)$ can be regular. Let us examine the condition that Δ is transverse to graph h at all fixed points (t, x) when $t \neq 0$. A fixed point at T can arise from either a zero of ξ or a periodic trajectory of period T. If x is a zero of ξ, then the transversality requirement is that $d(\exp T\xi)_x$ have no eigenvalue equal to one. Now ξ induces a linear transformation $\ell_x(\xi)$ on TX_x. (It is given by $\ell_x(\xi)\eta_x = [\xi, \eta]_x$ where $\eta_x \in TX_x$ and η is any vector field whose value at x is η_x. The value of $[\xi, \eta]_x$ is independent of the choice of extension.) Furthermore, $d(\exp T\xi)_x = \exp T\ell_x(\xi)$. Since $\exp T\ell_x(\xi)$ is to have no eigenvalue equal to one for any non-zero T we conclude that $\ell_x(\xi)$ can have no purely imaginary eigenvalue. In particular, $\ell_x(\xi)$

can not have zero as an eigenvalue and hence the zeros of ξ are isolated and, since X is compact, finite in number. If x is a periodic point with period T, then the orbit through x is precisely the periodic trajectory passing through x. All points of this trajectory are fixed under T and the map $P_{T,x}$ is called the Poincaré map. The transversality condition implies that there is no nearby periodic trajectory whose period is close to T. Thus, the compactness of X implies that there are only a finite number of periodic trajectories whose periods lie in any bounded interval of $\mathbf{R}$, and that for each of these periodic trajectories the Poincaré map have no eigenvalue equal to one.

Thus Z is a union of sets of the form $\{x\} \times \mathbf{R}^+$ where x is a zero of ξ and of sets of the form $\{(x, T)\}$ where T is a period of some closed trajectory and x lies on this trajectory. The sets of the form $\{x\} \times \mathbf{R}^+$ are transversal to the fibration of $\mathbf{R}^+ \times X$ over $\mathbf{R}^+$ and hence a δ function along $\{x\} \times \mathbf{R}^+$ gives rise to a smooth function on $\mathbf{R}^+$. In fact, it is clear that each zero of ξ contributes the term

$$\frac{1}{|\det(\mathrm{id} - d(\exp t\xi)_x)|}.$$

Let us now examine the contributions from the periodic trajectories.

In this case a subset $\{(x, T)\} = Z_T$ corresponding to a periodic trajectory is not transversal to π and hence the contribution coming from the periodic trajectories will not be smooth. On the other hand the map π does have constant rank on Z_T and we can factor Π as $\pi = \iota O \pi'$ where $\pi'\colon Z_T \to \mathrm{pt}$ is a submersion on $\iota(\mathrm{pt}) = T$ is an injection. We see that the image of the portion of $\Delta^* R$ supported along Z_T will be mapped onto a δ-function at T. A computation similar to the one we did for the fixed point formula, and which we shall leave to the reader, shows that the coefficient of this δ-function is

$$\frac{T^\#}{|I - P_T|}$$

where $T^\#$ is the length of the *primitive* periodic trajectory of which Z_T is the iterate. (For instance, if Z_T is a simple closed curve traversed 3 times, then $T = 3T^\#$.) The P_T above is the Poincaré map at any pt $x_0 \in Z_T$. Since the Poincaré map at distinct points x_0 and x_1 are conjugate, this determinant does not depend on the choice of x_0. Let us assume that all non-zero t are regular in the sense described above. Let us also extend the preceding arguments to the case of a one parameter group of morphisms of a vector bundle as in the fixed point case. We then get the formula

$$\pi^*\Delta^* R = \sum_{x|\xi_x = 0} \frac{\mathrm{tr}^r_{\exp t\xi, x}}{|\det(\mathrm{id} - d(\exp t\xi)_x)|} + \sum_{\text{periodic trajectories}} \frac{T^\# \mathrm{tr}^r_{\exp T\xi, x}}{|\det(\mathrm{id} - P_T)|}\,\delta(t - T)$$

valid as a generalization function on $\mathbf{R}^+$.

§2. Traces and characters.

In this section we will apply the elementary considerations of the previous section to derive some interesting formulas. We will consider a parametrized family of maps of some manifold X into itself. We thus assume that X and Y are differentiable manifolds and that we are given a differentiable map

$$f: Y \times X \to X.$$

Let us assume that this action is locally transitive, which means that for each (y, x) the map of $TY_y \to TX_{f(y,x)}$ given by

$$\eta \to df_{(y,x)}(\eta, 0)$$

is surjective.

For instance, the case which will be of most interest will be where $Y = G$ is a Lie group and $f: G \times X \to X$ is a Lie group action, so that

$$f(a, f(b, x)) = f(ab, x).$$

If L_a is left multiplication by a we can write any $\zeta \in TG_a$ as $dL_a\eta$ where $\eta \in TG_e$. Differentiating the above equation with respect to b, i.e., setting $b = b_t$ with $b_0 = e$ and $b'(0) = \eta$,

$$df_{(a,x)}(0, df_{(e,x)}(\eta, 0)) = df_{(a,x)}(dL_a\eta, 0).$$

Since the map $x \rightsquigarrow ax$ is a diffeomorphism for each fixed a, we conclude that for a group action G to be locally transitive it suffices to know that

$$df_{(e,x)}: TG_e \to TX_x$$

is surjective for each x.

Consider the map

$$F: Y \times X \to X \times X, \quad F(y, x) = (f(y, x), x).$$

If f is locally transitive then F is transversal to the diagonal $\mathbf{\Delta}$. Suppose that f is locally transitive. We can then form the *isotropy bundle* $Z \subset Y \times X$ where

$$Z = F^{-1}(\mathbf{\Delta}) = \Delta^{-1} \text{ graph } f$$

where $\Delta: Y \times X \to Y \times X \times X$ is the diagonal map. As a set,

$$Z = \{(y, x) | f(y, x) = x\}.$$

The tangent space to Z at $(y, x) \in Z$ consists of those (η, ξ) such that

$$df(\eta, \xi) = \xi.$$

The transitivity shows that given any ξ we can find an η satisfying the above equation. Thus $f_{|Z}$ is a submersion.

We recall the following elementary computation: Let $F : Y \times X \to W$ be transversal to some submanifold $W' \subset W$ and let $Z = F^{-1}W'$. Let $\pi : Z \to Y$ be the restriction to Z of the projection of $Y \times X$ onto Y, and let $F_y : X \to W$ be the map $F_y(x) = F(y, x)$. Then y is a regular value of π if and only if F_y is transversal to W'. Indeed, to say that $(y, x) \in Z$ is a regular point of π means that we can, for each $\boldsymbol{\eta}$, find a $\boldsymbol{\zeta}$ such that

$$dF(\boldsymbol{\eta}, \boldsymbol{\zeta}) \in TW'.$$

The transversality of F implies that $\{dF(\boldsymbol{\eta}, \boldsymbol{\xi})\} + TW'$ spans all of TW and so, π is regular at (y, x) if and only if the $\{dF(0, \boldsymbol{\xi} - \boldsymbol{\zeta})\} + TW'$ also span all of TW, i.e., if F_y is transversal to W' at (y, x).

Applying this to $F : Y \times X \to X \times X$ and letting $f_y : X \to X$ be defined by $f_y(x) = f(y, x)$, we see that y is a regular value of $\pi : Z \to Y$ if and only if

$$\text{graph } f_y \text{ is transversal to } \boldsymbol{\Delta} \subset X \times X.$$

Thus y is a regular value of π if and only if f_y is a Lefschetz map, i.e., if, for each fixed point x of f_y the map $(\mathrm{id} - df_y) : TX_x \to TX_x$ is an isomorphism.

If $E \to X$ is a vector bundle then a parameterized family of morphisms would be a smooth map $f : Y \times X \to X$ together with a smooth section r where $r(y, x) : E_{f(y,x)} \to E_x$. (Thus, if E_1 and E_2 denote the vector bundle E pulled back to $X \times X$ via the projection onto the first or second factors then r is a section of $F^{\#}\,\mathrm{Hom}(E_1, E_2)$.) For example, if G is a group acting as vector bundle automorphisms of E so that $g : E_x \to E_{f(g,x)}$ then $r(g, x) = g^{-1} : E_{f(g,x)} \to E_x$.

For each fixed y, the map f_y^* sends $C^\infty(E) \to C^\infty(E)$ and depends smoothly on y. Actually, let E' denote the vector bundle E pulled back to $Y \times X$ via projection onto the second factor. We thus have the diagram

$$\begin{array}{ccccc} E & & E' & & E \\ \downarrow & & \downarrow & & \downarrow \\ X & \xrightarrow{\quad \iota_y \quad} & Y \times X & \xrightarrow{\quad f \quad} & X \end{array}$$

with

$$f_y = f \cdot \iota_y$$

and $\iota_y(x) = (y, x)$. Thus f^* maps $C^\infty(E)$ to $C^\infty(E')$. By the discussion in §1 we know that f^* is given by a δ-section kernel k associated with graph f. To describe pedantically the bundle of k, let π_1, π_2, and π_3 be the projection of $Y \times X \times X$ onto each of the three factors, and let $E_2 = \pi_2^{\#} E$, $E_3 = \pi_3^{\#} E$ and continue to write $|\wedge|(X)$ for $\pi_3^{\#}|\wedge|(X)$. Then k is a generalized section of

$$\mathrm{Hom}(E_3, E_2) \otimes |\wedge|(X).$$

We are exactly in the situation of §1 with $W = Y \times X$. We can form $\Delta^* k$ since

$\Delta : Y \times X \to Y \times X \times X$ is transversal to graph f by assumption, and form tr $\Delta^* k$ which is a δ-section of $|\wedge|(X)$ along Z. Suppose that this section has compact support in the X direction. (For instance, suppose that X is compact.) Then, as a section of $|\wedge|(X)$ we can push it forward so as to obtain a generalized function

$$\pi_{Y*}(\text{tr } \Delta^* k)$$

on Y which we shall consider as the generalized function assigning tr f_y^* to y. If $k = Kdx$ were an honest kernel on $Y \times X$ then

$$u \rightsquigarrow \int K(y, x_1, x_2) u(x_2)\, dx_2$$

would be of trace class and would have, as its trace,

$$\int \text{tr } K(y, x, x)\, dx = \pi_{Y*}(\text{tr } \Delta^* k)(y).$$

As it is, the right hand side is defined as a generalized function. Notice that *at regular values of π_Y, the generalized function is an actual function and we have the formula*

$$\pi_{Y*}(\text{tr } \Delta^* k)(y) = \sum_{x | f_y(x) = x} \frac{\text{tr } r(y, x)}{|\det(\text{id} - df_y(x))|}. \tag{2.1}$$

Indeed, tr $\Delta^* k$ is a δ-section along Z. In fact, by (1.3) (with Z replaced by the preimage of the regular values) we know that $\pi_{Y*}(\text{tr } \Delta^* k)$ is a function and that the diagram

$$\begin{array}{ccc} X \supset (\Delta \circ \iota_y)^{-1} \text{ graph } f_y & \xrightarrow{\iota_y} & Z \subset Y \times X \\ \pi \downarrow & & \downarrow \pi_Y \\ y = \text{pt} & & Y \end{array}$$

induces the equation

$$\pi_* \text{ tr } (\Delta \circ \iota_y)^* k_y = \pi_* \iota_y^* (\text{tr } \Delta^* k) = \pi_{Y*}(\text{tr } \Delta^* k)(y)$$

and the result now follows from (1.4) applied to the Lefschetz map f_y.

We can actually interpret $\pi_{Y*}(\text{tr } \Delta^* k)$ as a "generalized trace" in the following manner: We first notice that graph $f \subset Y \times X \times X$ is transversal to the fibers of $\pi_2 \times \pi_3 : Y \times X \times X \to X \times X$, i.e., that graph f is fibered over $X \times X$. Indeed the tangent space to graph f contains all vectors of the form $(\eta, \xi, df\eta + df\xi)$ and the transitivity guarantees that the last two components are arbitrary. The tangents to the fiber of $\pi_2 \times \pi_3$ are all vectors $(\eta', 0, 0)$; together these clearly span $T(Y \times X \times X)_{(y, x, f(y, x))}$. Let $\pi_{X \times X}$ denote the projection $Y \times X \times X \to X \times X$

and let ρ be any smooth density of compact support on Y (considered as being defined on $Y \times X \times X$ as well). We can form $k_\rho = (\pi_{X \times X})_*(k\rho)$ which, by (1.3), will be a smooth section of $\operatorname{Hom}(E, E) \otimes |\wedge|(X)$ on $X \times X$, and hence defines an operator of trace class on $C^\infty(E)$. Its trace is, by definition,

$$\bar{\pi}_* \operatorname{tr} \Delta^* k_\rho, \qquad \bar{\pi} : X \to \mathrm{pt}\,.$$

If we write, symbolically, $\rho = \rho(y)dy$ and $k = K(y, x_1, x_2)dx_2$ then

$$K_\rho(x_1, x_2) = \int_Y K(y, x_1, x_2)\rho(y)\,dy \quad \text{and} \quad \operatorname{tr} k_\rho = \int_X \int_Y \operatorname{tr} K(y, x, x)\rho(y)\,dy\,dx.$$

Interchanging the order of integration suggests that we interpret

$$\operatorname{tr} k_\rho = \operatorname{tr} \int \rho(y) f_y^*\,dy.$$

Indeed $k\rho$ is a δ-section on graph f and (1.3) again implies that

$$\pi_{X*} \Delta^*(k\rho) = \Delta^*(\pi_{X \times X})_*(k\rho)$$

where $\pi_X : Z \to X$. The situation is described by the following diagram

$$\begin{array}{ccc}
Z & \xrightarrow{\quad \Delta \quad} & \operatorname{graph} f \\
\downarrow & & \downarrow \\
Y \times X & \xrightarrow{\quad \Delta \quad} & Y \times X \times X \\
\downarrow \pi_X & & \downarrow \pi_{X \times X} \\
X & \longrightarrow & X \times X \\
\downarrow \pi & & \\
\mathrm{pt} & &
\end{array}$$

Now $\pi_{X*}\Delta^*(k\rho)$ is clearly the kernel of $\int f_y^* \rho(y)\,dy$. We have thus proved:

PROPOSITION 2.1. *Let f be a transitive action of Y as morphisms of the vector bundle $E \to X$ where X is a compact manifold. Let $Z = \Delta^{-1}$ graph f, and $\pi : Z \to Y$. For each smooth density of compact support, ρ, on Y, the operator*

$$\int_Y \rho(y) f_y^*\,dy : C^\infty(E) \to C^\infty(E)$$

is compact, and its trace considered as a functional on ρ defines a generalized function on Y, which we shall denote by $\operatorname{tr} f^*$. *We have*

$$\operatorname{tr} f^* = \pi_* \operatorname{tr} \Delta^* k$$

where k is the kernel of f^*, a δ-section along graph f. At regular values of π the *generalized function* tr f^* *is actually a function and we have the formula*

$$\operatorname{tr} f_y^* = \sum_{x \mid f_y(x)=x} \frac{\operatorname{tr} r(y,x)}{|\det(\mathrm{id} - df_y(x))|}.$$

As an important case where this set up holds let G be a Lie group and let H be a closed subgroup such that $G/H = X$ is compact. Let there be given a (finite dimensional) representation, σ, of H on some vector space V and let χ_σ be its character

$$\chi_\sigma(h) = \operatorname{tr} \sigma(h).$$

Then, in the usual way, σ determines a vector bundle E over X on which G acts as a group of automorphisms. Then the representation, I_σ, of G on $C^\infty(E)$ given by

$$I_\sigma(g)u = f_{g^{-1}}^* u$$

is called the representation induced by σ on G. It then follows from Proposition 2.1 that for any smooth density ρ of compact support on G the operator

$$\int I_\sigma(g)\rho(g^{-1})\, dg$$

is compact and its trace, considered as a functional on ρ, is a generalized function on G which we shall call the character of the induced representation and denote by χ_{I_σ} (or simply by χ when there is no confusion), so that

$$\chi_{I_\sigma}(g) = \operatorname{tr} f_{g^{-1}}^*.$$

We have established that χ is a smooth function at regular values of g whose value is given by (2.1). In evaluating (2.1) we note that to say that $x \in X$ is a fixed point for $f_{g^{-1}}$ means that $x = aH$ for some $a \in G$ (determined up to right multiplication by $h \in H$) satisfying

$$g^{-1}aH = aH,$$

i.e.,

$$a^{-1}ga \in H.$$

Also $\operatorname{tr} r(g,x) = \chi_\sigma(a^{-1}ga)$ (which does not depend on the choice of a). Let us now examine the denominator of (2.1). We can write

$$f_{g^{-1}} = f_a \circ f_{a^{-1}g^{-1}a} \circ f_{a^{-1}}$$

so

$$\det(\mathrm{id} - df_{g^{-1}}) = \det(\mathrm{id} - df_{a^{-1}g^{-1}a}).$$

Let us write $h = a^{-1}g^{-1}a$. We wish to compute df_h acting on TX_H. We may identify TX_H with TG_e/TH_e. Let ξ be an element of TG_e, and let $(\exp t\xi)H$ be the curve in X that it generates. Then

$$h(\exp t\xi)H = [h(\exp t\xi)h^{-1}]H = (\exp t[\mathrm{Ad}\, h\xi])H.$$

Differentiating shows that

$$df_h = \mathrm{Ad}\, h \qquad \text{for } h \in H \text{ acting on } TG_e/TH_e \tag{2.2}$$

so that we get the formula

$$\chi_{I_\sigma}(g) = \sum_{\substack{x \mid f_g x = x \\ x = aH}} \frac{\chi_\sigma(a^{-1}ga)}{|\det(\mathrm{id} - \mathrm{Ad}\,(a^{-1}g^{-1}a))_{TG_e/TH_e}|}. \tag{2.3}$$

Notice that it follows from (2.2) that for $\eta \in TH_e$, and for small values of s the element $(\exp s\eta H)$ will be a regular element of $\pi_Y : Z \to Y$, i.e., the transformation id $-df_{\exp s\eta}$ will be invertible if ad η induces an isomorphism on TG_e/TH_e. Indeed

$$\mathrm{Ad}\exp s\eta = \mathrm{id} + s\,\mathrm{ad}\,\eta + O(s^2)$$

which implies, for small values of $|s|$, that id $-$ Ad exp $s\eta$ is invertible if ad η is. Notice that the set of singular points of Z is an analytic subvariety and will thus be a proper analytic subvariety (and in particular a set of measure zero) if there exists one regular point. If there are no regular points then all of Z is mapped by π_Y onto a set of singular values which must carry supp χ. We have thus proved (using the remarks in §1 on when a push forward is smooth):

PROPOSITION 2.2. *If there is some $\eta \in TH_e$ such that* ad η *induces an isomorphism on TG_e/TH_e then the character χ is represented by a locally integrable function on G and χ is smooth on a non-empty set of G. If no such η exists then χ is concentrated on a proper subset of G.*

Let us give an illustration of (2.3) in a special case. Suppose we take $G = SL(n, \mathbf{R})$. Any matrix, F, can be written uniquely as the product $F = ODT$ where O is orthogonal, D is diagonal with positive entries and T is upper triangular with ones on the diagonal, so that $G = K \cdot A \cdot N$ where K is the group of orthogonal matrices, etc.

[We recall the proof of this fact, which is a special case of what is called the Iwasawa decomposition. It is clear from elementary linear algebra that any matrix can be written as the sum of an antisymmetric matrix and an upper triangular matrix. Exponentiating gives the desired decomposition near the identity. To get it globally, we use the following lemma about groups: Let G be a connected topological group with G_1 and G_2 subgroups with G_1 compact and G_2 connected. Suppose there are neighborhoods U_1 in G_1 and U_2 in G_2 such that $U_1 U_2$ is a neighborhood of G. Then $G = G_1 \cdot G_2$.

PROOF. For any a in G_1 we can find a neighborhood W of e in G_2 so that $aWa^{-1} \subset U_1 U_2$. A slightly smaller W works for nearby a and thus, since G_1 is compact, we can find a W that works for all of G_1. Therefore $WG_1 \subset G_1 G_2$ and so $(W \cdot W)G_1 = W(WG_1) \subset WG_1 G_2 \subset G_1 G_2$, etc., and since G_2 is connected, the powers of W exhaust all of G_2 and so $G_2 G_1 \subset G_1 G_2$. Now $G_1(G_1 G_2) = G_1 G_2$ and $G_2(G_1 G_2) = (G_2 G_1)G_2 \subset G_1 G_2$ so $(G_1 G_2)(G_1 G_2) \subset G_1 G_2$ and $(G_1 G_2)^{-1} = G_2 G_1 \subset G_1 G_2$. Thus $G_1 G_2$ is an open subgroup, and, since G is connected, is all of G, completing the proof of the lemma. In the case of $SL(n)$, the uniqueness of the decomposition is obvious: If $ODT = I$ then O is simultaneously orthogonal and triangular with positive entries on the diagonal and hence $O = I$.]

Let us take $H = AN$. (In general for a real semi-simple group one takes $H \supset MAN$ where M is the centralizer of A in K.) Then if $A = (\lambda_1, \ldots, \lambda_n)$ it is clear that the eigenvalues of Ad A in TG_e/TH_e are precisely $\lambda_j \lambda_i^{-1}$ $(j > i)$. The denominator in (2.2) thus becomes, in this case $|\Pi(1 - \lambda_j \lambda_i^{-1})|$ and the formula reduces to the one given by Gelfand-Naimark.

Of course, in general, the transversality condition of Proposition 2.2 need not hold. For example, if G is a nilpotent group, this condition can never hold. Also, in interesting cases, the space X will not be compact so that the expression π_{G*} tr $\Delta^* k$ will not be defined directly as an absolutely convergent integral. (It will, in many cases, be defined as an oscillatory integral.)

Let us consider the case of an induced character of a compact group. Thus suppose that G is a compact group and H a closed subgroup so that $X = G/H$ is automatically compact. In this situation G and H have preferred densities which we shall write as dg and dh, the ones corresponding to Haar measure with total measure one. Let ψ be the character of some irreducible representation of G and let us compute $\langle \psi dg, \chi_{I_\sigma} \rangle$.

We have

$$\langle \psi dg, \chi_{I_\sigma} \rangle = \langle \psi dg, \pi_{\sigma *} \operatorname{tr} \Delta^* k \rangle = \langle (\pi^* \psi) dg, \operatorname{tr} \Delta^* k \rangle.$$

This is an integral over Z of a δ-section of $|\wedge|(G \times X)$. Using the notation

$$\delta(x - f(g, x)) = \Delta^* \delta(x_2 - f(g, x_1))$$

we can write, symbolically,

$$\langle \pi^* \psi dg, \operatorname{tr} \Delta^* k \rangle = \iint \psi(g)(\operatorname{tr} r_{g,x}) \delta(x - f(g^{-1}, x))\, dx\, dg.$$

This integral is on a fiber bundle over X, so we can first integrate over the fibers and then over X. The fiber over $x = aH$ is $G_x = aHa^{-1}$ and the integral over the fiber becomes

$$\int_{G_x} \psi(g)\chi_\sigma(a^{-1}ga)\,dg_x = \int_H \psi(aga^{-1})\chi_\sigma(h)\,dh = \int_H \psi(g)\chi_\sigma(h)\,dh$$

which is an integer, independent of x. Since the total volume of X is one, we conclude that

$$\int \psi(g)\chi_{I_\sigma}(g)\,dg = \int_H \psi_{|H}(h)\chi_\sigma(h)\,dh. \tag{2.4}$$

The left hand side of this equation is defined as the pairing of a generalized function with a smooth density. The right hand side is an honest integral and represents the intertwining number of $\psi_{|H}$ with σ. (If σ is irreducible this is the number of times that σ occurs in $\psi_{|H}$.) In a sense, (2.4) can be regarded as a version, in terms of characters, of the Frobenius reciprocity formula.

Let us sketch some further applications of the methods used in deriving Proposition 2.1. Suppose we have a sequence of differential operators on vector bundles

$$0 \to C^\infty(E_1) \xrightarrow{d_1} C^\infty(E_2) \xrightarrow{d_2} \cdots \xrightarrow{d_{N-1}} C^\infty(E_N) \to 0$$

which is a complex, i.e., satisfies $d^2 = 0$, i.e., $d_{i+1} \circ d_i = 0$, $i = 0, \ldots, N$.

Suppose that the homology groups, $H(E_i)$, are finite dimensional, e.g., that it is an elliptic complex. Also, suppose that $f : Y \times X \to X$ is a transitive family of morphisms of each E_i and that

$$f_y^* d = df_y^*$$

for each i. In other words we assume that for each y we are in the situation described by Atiyah-Bott [2]. For each smooth density, ρ, of compact support, the operator $f_\rho^* = \int_Y f_y^* \rho(y)\,dy$ is a compact operator on each $C^\infty(E_i)$ and

$$f_\rho^* d = df_\rho^*.$$

Now for compact operators it is an elementary algebraic exercise to verify (cf. Atiyah-Bott [2, §7]) that

$$\sum (-1)^i \operatorname{tr} f^*_{\rho_{|E_i}} = \sum (-1)^i \operatorname{tr} f^*_{\rho_{|H(E_i)}}.$$

Let k_{E_i} be the kernel of f^* on E_i and $k_{H(E_i)}$ be the kernel of f^* on $H(E_i)$. We can thus assert the following equation for generalized functions:

$$\sum (-1)^i \pi_{Y*}(\operatorname{tr} \Delta^* k_{E_i}) = \sum (-1)^i \pi_{Y*}(\operatorname{tr} \Delta^* k_{H(E_i)}). \tag{2.5}$$

In particular, using Proposition 2.1 for a regular value of y we obtain the Atiyah-Bott fixed point formula

$$\sum (-1)^i \pi_{Y*} \operatorname{tr} \Delta^* k_{H(E_i)}(y) = \sum_{x | f_y(x)=x} \frac{\operatorname{tr} r(y,x)}{|\det(\mathrm{id} - df_y(x))|}. \tag{2.6}$$

We have thus proved that if a Lefschetz morphism, f_y, can be embedded in a transitive family of morphisms, then the Atiyah-Bott formula, (2.6), holds. Of course, in general, a Lefschetz morphism can not be embedded in a transitive family so that this does not give a general proof of the Atiyah-Bott formula. We shall show in a later section how the ellipticity of the complex can be used to achieve the same effect as a transitive family. The point of the above discussion is that it relies only on the elementary considerations of §1. Also, for a number of applications of the Atiyah-Bott formula there do exist transitive families. For example, taking $E_i = \wedge^i (X)$ with the usual d would yield the usual Lefschetz fixed point theorem. In this case all that is required of $f: Y \times X \to X$ is that it be a transitive family of diffeomorphisms and it is easy, by purely local geometric means, to embed any diffeomorphism in a transitive family of diffeomorphisms. We have thus proved the Lefschetz fixed point theorem. Also, in deriving the Weyl character formula from (2.6) as in Atiyah-Bott [2], there is a natural transitive family in evidence, so again the formula follows from elementary considerations.

Finally Atiyah-Bott prove a very interesting residue formula for the zeroes of a collection of n polynomials $P_1, \ldots, P_n$ on $\mathbf{C}^n$. Without formulating their theorem here we note that the proof consists of associating to the P_i a map of complex projective n-space into itself, and this is embeddible in a transitive family just using the projective collineations.

§3. The wave front set.

In the previous two paragraphs we discussed the pull back and push forward operators for rather special kinds of maps and sections of vector bundles. We now wish to extend these notions. The basic idea is as follows: If $f: X \to Y$ is a submersion, then, as we saw in §1, the pull back, $f^* u$, is defined for any generalized function u. On the other hand, if u is smooth, $f^* u$ is always defined, whether or not f is a submersion. These two cases can be regarded as extreme instances of the following phenomenon: f is transversal to "the singularities of u" (where, of course, we must define what we mean by "the singularities of u"). In the first instance, f, being a submersion is transversal to everything; in the second, u has no singularities. Up until recently, the singularities of u were described by a subset of Y, called the singular support of u. A major advance, introduced by Sato [3] and Hormander [4], was the realization that the singularities of u are better described as a subset of $T^* Y$, called the wave front set. The purpose of this section is to introduce the wave front set and study its functorial properties.

We begin with the singular support. Let u be a generalized section of some vector bundle F over Y. We define a closed set, sing supp $u \subset Y$ by saying that $y \in Y$ does *not* belong to sing supp u if there is some C^∞ function φ of compact support such that $\varphi(y) \neq 0$ and φu is an honest section of F, i.e., if $\varphi u \in C^\infty(F) \subset C^{-\infty}(F)$. It is clear that if $f: X \to Y$ is a morphism from $E \to X$ to $F \to Y$ such that $f(X) \cap \text{sing supp } u = \varnothing$ then we can define $f^* u$ as follows: Each $x \in X$ has a compact neighborhood W such that $f(W) \cap \text{sing supp } u = \varnothing$. We can then find a C^∞ function φ such that

$$\varphi \equiv 0 \text{ on sing supp } u,$$

$$\varphi \equiv 1 \text{ on } f(W).$$

It is easy to check that φu is smooth and hence $f^*(\varphi u)$ is well defined on W and that the $f^*(\varphi u)$ so defined piece together to give a smooth section of E. It is also easy to check that the section so defined is independent of the various choices, by comparing what happens when we make two different choices. We sketch another way of establishing the uniqueness. Let K be some closed subset of Y. We let $C_K^{-\infty}(F)$ denote the subspace of $C^{-\infty}(F)$ consisting of those generalized sections whose singular support lies in K. On $C_K^{-\infty}(F)$ we can introduce a topology that is stronger than the weak topology inherited from $C^{-\infty}(F)$. Roughly speaking we put on a topology so that $\{u_\alpha\}$ converges if (i) $\{u_\alpha\}$ converges in the weak topology and (ii) for each compact set A with $A \cap K = \varnothing$, we have uniform convergence of $\{u_\alpha\}$ together with any finite number of derivatives. Then $C^\infty(F) \subset C_K^{-\infty}(F)$ and $C^\infty(F)$ is dense in $C_K^{-\infty}(F)$ with respect to this topology. One checks that $u \to f^* u \in C^{-\infty}(E)$ which is defined on $C^\infty(F)$ is continuous with respect to the topology inherited from $C_K^{-\infty}(F)$, and hence extends to a unique map $f^* : C_K^{-\infty}(F) \to C^{-\infty}(E)$.

It follows readily from the definitions that if $f: X \to Y$ is a submersion, so $f^* u$ is defined for any u, that we have

$$\text{sing supp } f^* u \subset f^{-1}(\text{sing supp } u).$$

If $f: Y \to Z$ is a submersion which is proper (or if u has compact support) so that $f_* u$ is defined, we have

$$\text{sing supp } f_* u \subset f(\text{sing supp } u).$$

Let us now turn to the definition of the wave front set. In what follows, we shall not constantly mention the vector bundles, but rather carry the discussion for functions and for densities, or perhaps talk of sections without specifically listing the vector bundles and morphisms, so as not to have excess notational baggage. All the notions carry over without complication to the vector bundle situation.

Let u be a generalized density on X, and let $\ell \in T^*X_x$ be a non-zero covector. We say that *u is smooth at ℓ* if the following holds:

Let S be any auxiliary manifold and let $f: S \times X \to \mathbf{R}$ be any smooth map such that $df_s(x) = \ell$, where $f_{s'}(x') = f(s', x')$. Let $F: S \times X \to S \times \mathbf{R}$ be defined by

$$F(s, x) = (s, f(s, x)).$$

Then there is a C^∞ function, b, of compact support defined on X (and hence on $S \times X$) such that

(i) $b(x) \neq 0$,

(ii) $F_*(bu)$ is smooth on $S \times \mathbf{R}$ near s.

Intuitively speaking, we are saying that u is smooth at ℓ if, for any function f on X with $df = \ell$, the generalized density $f_* u$ is smooth on $\mathbf{R}$. We throw in the blip function b because $f_* u$ need not be defined and because we don't want to consider singularities introduced by f far away from ℓ. We throw in S because we want smooth dependence on parameters.

Notice that $f_*(bu)$ is well defined because the "integration" is over the X direction and bu has compact support in the X direction. Notice also that if u is smooth at ℓ, it is also smooth at $t\ell$ for any $t \neq 0$ (since multiplication by t is a diffeomorphism of $\mathbf{R}$). We let $[\ell]$ denote the line through ℓ.

Let PT^*X denote the projectivized cotangent bundle, so that PT^*X_x consists of all lines through the origin in T^*X_x. We define the projective wave front set, $PWF(u) \subset PT^*X$, by

$$[\ell] \in PWF(u) \textit{ if } u \textit{ is not smooth at } \ell.$$

The definition, as it stands, is rather cumbersome. We will make some simplifications, but let us first give some preliminary results. If π denotes the projection of $PT^*X \to X$ then it is clear that

$$PWFu \subset \pi^{-1}(\text{sing supp } u).$$

To study what happens above sing supp u we begin with

PROPOSITION 3.1. *Let $g: X \to \mathbf{R}$ be a submersion and let v be a generalized function on $\mathbf{R}$ with* sing supp $v = \{0\}$. *Let $u = hg^*v$ where h is some smooth function. Then*

$$PWF(u) \subset [N(g^{-1}(0))]$$

where $[Ng^{-1}(0)]$ denotes the projectivized normal bundle of $g^{-1}(0)$, i.e., the set of all $[dg_x]$, $x \in g^{-1}(0)$. For any $v \in C^{-\infty}(\mathbf{R})$ we have

$$PWF(hg^*v) \subset \{[dg_x] | x \in g^{-1}(\text{sing supp } v)\}.$$

PROOF. Since sing supp $u \subset g^{-1}(0)$ we need only look at $x \in g^{-1}(0)$. If $[\ell] \neq [dg_x]$ and f is some parametrized family of functions with $df_s = \ell$, we can introduce local coordinates about x of the form $(x^1, x^2, \ldots, x^n)_{s'}$ with $x^1 = g$ and $x^2 = f_{s'}$. Let $bdx = b(x^1, x^2, \ldots, x^n)dx^1 \cdots dx^n$ be a smooth density of compact support and let $w = w(s', t)$ be a smooth function of compact support on $S \times \mathbf{R}$. We may write $F^* w$ as $w(x^2)$, where we leave the dependence on s' implicit; that is, we write $w(x^2)$ instead of $w(x^2, s')$. Then

$$\begin{aligned}\langle F_*[g^*(v)bdx], w\rangle &= \langle g^*(v)bdx, F^* w\rangle \\ &= \langle g^*(v), (F^* w)bdx\rangle \\ &= \langle v, g_*(F^* w)bdx\rangle \\ &= \left\langle v, \int w(x^2)b(\cdot, x^2, \ldots, x^n)\, dx^2 \cdots dx^n\right\rangle \\ &= \int w(x^2)\langle v, b(\cdot, x^2, \ldots, x^n)\rangle\, dx^2 \cdots dx^n.\end{aligned}$$

Thus $(F_* u) = \left[\int \langle v, hb(\cdot, x^2, \ldots, x^n)\rangle\, dx^3 \cdots dx^n\right] dx^2$ is smooth on $S \times \mathbf{R}$. Q.E.D.

Let wdt be a generalized density of compact support on $\mathbf{R}$. Its Fourier transform is, by definition,

$$\hat{w}(\tau) = \frac{1}{\sqrt{2\pi}} \langle wdt, e^{-i\tau \cdot t}\rangle$$

and it is a well known fact that wdt is a smooth density if and only if $\hat{w}(\tau)$ vanishes to all orders at $\pm\infty$. Thus, if u is a generalized density on X which is smooth at some $\ell \in TX_x$ then $(f_{s*} bu)\hat{}\,(\tau)$ vanishes to all orders as $\tau \to \pm\infty$, i.e.,

$$\langle bu, e^{-i\tau f_{s'}}\rangle = O(|\tau|^{-N}) \qquad \text{for all } N \tag{3.1}$$

as $\tau \to \pm\infty$. We can now introduce a slightly refined notion of wave front set by saying that u is *forward smooth* at ℓ if (3.1) holds as $\tau \to +\infty$. In (3.1) the left hand side is to be considered as an asymptotic function on S, and (3.1) asserts that this asymptotic function on S vanishes near s. Thus (3.1) holds uniformly in s' near s and the various partial derivatives satisfy (3.1) (uniformly) as well. We define the wave front set of u, denoted by $WF(u)$, to be the set of $\ell \neq 0$ such that u is *not* forward smooth at ℓ. Note that if $\ell \in WF(u)$ so is $a\ell$, for any positive real number. Thus, if $ST^* X$ denotes the cosphere bundle, i.e., the set of rays from the origin in T^*X, then the wave front set determines by projection from the origin a subset $SWFu \subset ST^* X$, and WFu is the inverse image of $SWFu$ under the projection of $T^* X - 0_X$ onto $ST^* X$.

We can now sharpen Proposition 3.1:

PROPOSITION 3.2. *Let v be a generalized function on $\mathbf{R}$ with $WFv = \{(0, \sigma) | \sigma > 0\} \subset T^*\mathbf{R} = \mathbf{R} \oplus \mathbf{R}$. Let $g : X \to \mathbf{R}$ be a submersion. Then*

$$WFg^* v \subset \{\lambda dg(x) | \lambda > 0, g(x) = 0\}.$$

By Proposition 3.1, all we need to check is that $-dg(x) \notin WFg^*v$. Thus we show that $\langle bg^*v, e^{i\tau f_{s'}}\rangle = O(|\tau|^{-N})$ for all N if $f: S \times X \to \mathbf{R}$ is some function with $df_s(x) = dg(x)$, and where b is a suitable density. We can introduce coordinates $x^1, \ldots, x^n$ on a neighborhood U in X with $g = x^n$ and write $S \times U = S \times V \times I$ where $V \subset \mathbf{R}^{n-1}$ and $I \subset \mathbf{R}$. We write $x = (x^1, \ldots, x^n) = (y, x^n)$ and set $\bar{S} = S \times V$. We may think of f as being defined on $\bar{S} \times I$ and then $f_{s',y}$ is a function of the single variable x^n. Suppose our point x has coordinates (0,0). Then $df_{s,0} = dx^n$ and therefore, by the definition of WFv,

$$\langle cv, e^{i\tau f_{s',y}}\rangle = O(|\tau|^{-N})$$

uniformly in (s', y) near $(s,0)$ where c is some suitable density of compact support on $\mathbf{R}$ which does not vanish at 0. Let $d(y)$ be any smooth density of compact support near zero in V. Then

$$\langle (d \otimes c)g^*v, e^{i\tau f_{s'}}\rangle = \int_V d(y)\langle cv, e^{i\tau f_{s',y}}\rangle$$

and the last expression is $O(|\tau|^{-N})$ for all N.

So far, our definitions of the various kinds of wave front sets are highly implicit, since they depend on arbitrary functions defined on auxiliary parameter spaces. We will now show that it suffices to look at linear functions, defined locally, with their coefficients as parameters. This will not only give a way of computing the wave front set, it will also allow us to extend the notion of pull back to more general classes of maps and generalized functions. Our principal tools will be the Radon transform, for discussing the projective wave front set and the Fourier transform, for discussing the spherical wave front set. Since the discussion of the projective wave front set is simpler and more intuitive, we shall first develop its theory, even though the theory of the spherical wave front set is more refined.

We begin by reviewing some facts about the Radon transform. We let $\mathbf{R}^n$ have its standard density, dx, so that we may identify functions, u, with densities udx. We let S^{n-1} denote the unit sphere and $d\Omega$ its induced volume density. For a point, $\Omega \in S^{n-1}$, we define $\rho_\Omega : \mathbf{R}^n \to \mathbf{R}$ by

$$\rho_\Omega(x) = x \cdot \Omega.$$

For any compactly supported (generalized) function, u, we form $\rho_{\Omega *}(udx)$, which is a compactly supported (generalized) density on $\mathbf{R}$. We claim that the following formula holds:

$$\hat{u}(t\Omega) = \left(\frac{1}{2\pi}\right)^{(n-1)/2} (\rho_{\Omega *}(udx))\hat{\ }(t). \tag{3.2}$$

Here $\hat{\ }$ on the left denotes the Fourier transform on $\mathbf{R}^n$ while the $\hat{\ }$ on the right denotes the Fourier transform on **R**. Indeed, let $d\beta_{c,\Omega}$ denote the induced density (Lebesgue measure) on the hyperplane $\Omega \cdot x = c$. Then

$$\begin{aligned}\hat{u}(t\Omega) &= \left(\frac{1}{2\pi}\right)^{n/2} \int e^{-it(\Omega\cdot x)} u(x)\,dx \\ &= \left(\frac{1}{2\pi}\right)^{n/2} \int_{-\infty}^{\infty} e^{-itc}\left(\int_{\Omega\cdot x=c} u(x)d\beta_{c,\Omega}\right) dc \\ &= \left(\frac{1}{2\pi}\right)^{(n-1)/2} [\rho_{\Omega *} udx]\hat{\ }(t).\end{aligned}$$

Or, repeating the same steps

$$\begin{aligned}\hat{u}(t\Omega) &= \left(\frac{1}{2\pi}\right)^{n/2} \langle udx, e^{-it\Omega\cdot x}\rangle = \left(\frac{1}{2\pi}\right)^{n/2} \langle udx, e^{-it\rho_\Omega(x)}\rangle \\ &= \left(\frac{1}{2\pi}\right)^{n/2} \langle \rho_{\Omega *}\, udx, e^{-it(\cdot)}\rangle = \left(\frac{1}{2\pi}\right)^{(n-1)/2} [\rho_{\Omega *}\, udx]\hat{\ }(t).\end{aligned}$$

Using (3.2) and basic facts about the Fourier transform, we shall now show how to recover u from a knowledge of all the $\rho_{\Omega *} u$. Let u be a smooth density of compact support. We have, from the Fourier inversion formula,

$$\begin{aligned}u(x) &= \left(\frac{1}{2\pi}\right)^{n/2} \int \hat{u}(k)e^{ik\cdot x}\,dk \\ &= \left(\frac{1}{2\pi}\right)^{n/2} \int_0^{\infty}\int_{S^{n-1}} \hat{u}(t\Omega)t^{n-1}e^{it\Omega\cdot x}\,dt\,d\Omega \\ &= \left(\frac{1}{2\pi}\right)^{n-1/2} \int_{S^{n-1}}\int_{-\infty}^{\infty} [\rho_{\Omega *}\, udx]\hat{\ }(t)t_+^{n-1}e^{it(\Omega\cdot x)}\,dt\,d(\Omega)\end{aligned}$$

where $t_+ = t$ for $t \geqslant 0$ and $t_+ = 0$ for $t < 0$. We shall have a great deal to say about the function t_+^{n-1} (and its inverse Fourier transform) later on. For the moment let us introduce the operator I^r on densities on $\mathbf{R}^1$ given by

$$[I^r v]\hat{\ }(t) = v\hat{\ }(t)t_+^r$$

or,

$$[I^r v](c) = \left(\frac{1}{2\pi}\right)^{1/2} \int_{-\infty}^{\infty} t_+^r v\hat{\ }(t)e^{itc}\,dt.$$

We can then rewrite the last expression for $u(x)$ as

$$u(x) = \left(\frac{1}{2\pi}\right)^{n-1} \int I^{n-1}[\rho_{\Omega *}(udx)](\Omega\cdot x)\,d\Omega$$

or

$$u = \left(\frac{1}{2\pi}\right)^{n-1} \int \rho_\Omega^*\{I^{n-1}[\rho_{\Omega_*}(udx)]\}\,d\Omega. \tag{3.3}$$

Formula (3.3) is known as the Radon inversion formula. Notice that each of the operations built into the formula makes sense for generalized functions or densities. Thus

$$\rho_{\Omega_*} : C_0^{-\infty}(|\wedge|\mathbf{R}^n) \to C_0^{-\infty}(|\wedge|\mathbf{R}),$$

$$I^{n-1} : C_0^{-\infty}(|\wedge|\mathbf{R}) \to C^{-\infty}(\mathbf{R})$$

and

$$\rho_\Omega^* : C^{-\infty}(\mathbf{R}) \to C^{-\infty}(\mathbf{R}^n)$$

all make sense, and are continuous, and the integral is also well defined. Thus, since the smooth sections are dense in the generalized sections, equation (3.3) holds for generalized sections. We shall discuss a more general setting for equations such as (3.3), one suggested by Gelfand, at the end of this section.

Now let X be a differentiable manifold and U be a coordinate chart on X. We regard U as a subset of $\mathbf{R}^n$ via the coordinates and define $\rho : U \times S^{n-1} \to \mathbf{R}$ by

$$\rho(x,\Omega) = \rho_\Omega(x) = \Omega \cdot x.$$

The sphere bundle, ST^*U, can be identified with $U \times S^{n-1}$ (and also can be regarded as a sub-bundle of T^*U rather than a quotient).

PROPOSITION 3.3. *The generalized function u is smooth at* (x_0, Ω_0) *if and only if there exists a smooth density* $b = hdx$ *of compact support with* $b(x_0) \neq 0$ *such that* $\rho_{\Omega_*}(bu)$ *is smooth for* Ω *near* Ω_0.

PROOF. The necessity is obvious: We may take $S = S^{n-1}$ and $f = \rho$, so that $d\rho_\Omega(x) = (x,\Omega)$. To prove the sufficiency, let $f : S \times U \to \mathbf{R}$ be any parametrized family with $df_{s_0}(x_0) = (x_0, \Omega_0)$. Now

$$bu = \left[\int \rho_\Omega^*[I^{n-1}\rho_{\Omega_*}(bu)]\,d\Omega\right]dx$$

so that

$$f_{s*}bu = \int f_{s*}\{\rho_\Omega^*[I^{n-1}\rho_{\Omega*}(bu)]\,dx\}\,d\Omega.$$

We can split the integral on the right into two parts. For Ω near $\pm\Omega_0$ the expression $\rho_\Omega^*[I^{n-1}\rho_{\Omega*}(bu)]$ is smooth by hypothesis. On the other hand, for Ω outside of some neighborhood of $\pm\Omega_0$, we know from Proposition 3.1 that the wave front set of $\rho_\Omega^* v$ is contained in the set of vectors $(x, t\Omega)$, where v is any generalized function on **R**. Hence, for s sufficiently close to s_0, we conclude that $f_{s*}\rho_\Omega^* v$ is smooth. Q.E.D.

We can now see that the choice of b in the definition of the wave front set is not important:

PROPOSITION 3.4. *Suppose that $\rho_{\Omega'*}(bu)$ is smooth. Then so is $\rho_{\Omega'*}(hbu)$ where h is any smooth function. In particular, by Proposition* 3.3,

$$PWF(hu) \subset PWFu. \tag{3.4}$$

Indeed

$$\rho_{\Omega'*} h(bu) = \rho_{\Omega*} \int h\rho_\Omega^* I^{n-1} \rho_{\Omega*}(bu)\, d\Omega$$

and we may again break the integral up into two parts, on one the integrand is smooth and, pulling $\rho_{\Omega'*}$ under the integral sign, we apply Proposition 3.1 to the other.

PROPOSITION 3.5. *The set $PWFu$ is a closed subset of PT^*X.*

PROOF. It follows immediately from Proposition 3.3 that the set where u is smooth is open.

Let π denote the projection $\pi : PT^*X \to X$. We have already seen that $\pi(PWFu) \subset$ sing supp u. We now claim that

$$\pi(PWFu) = \text{sing supp } u. \tag{3.5}$$

PROOF. Suppose that $x \notin \pi(PWFu)$. Then there is a coordinate chart around x with the property that for each $\Omega_0 \in S^{n-1}$ there is a b such that $\rho_{\Omega*}\, bu$ is smooth for Ω near Ω_0. We can cover S^{n-1} by a finite number of such neighborhoods and, by taking h of sufficiently small support, apply (3.4) to conclude that there is *one* b such that $\rho_{\Omega*} bu$ is smooth for all Ω. We may now apply the Radon inversion formula to conclude that bu is smooth, proving (3.5).

Let $f : X \to Y$ be a smooth map. We know how to map subsets $A \subset T^*X$ into subsets $df_* A \subset T^*Y$ by setting $df_* A = \{\ell \in T^*Y | \ell \in T^*Y_{f(x)}, df_x^*\ell \in A \text{ for some } x\}$.

We want a similar definition for subsets of PT^* except that we now must take into account the possibility that df_x^* may send non-zero vectors into zero. Accordingly, if A is a subset of PT^*X we set

$$df_* A = \{[\ell] \in PT^*Y | \ell \in T^*Y_{f(x)} \text{ for some } x \\ \text{and either } df_x^*\ell = 0 \text{ or } [df_x^*\ell] \in A\}.$$

It is easy to check that if A is a compact subset of PT^*X then $df_* A$ is a compact subset of $PT^* Y$. It is also easy to check that if $g : Y \to Z$ is a second smooth map then

$$dg_* \circ df_* = d(g \circ f)_*.$$

Now let u be a generalized density of compact support on X. We claim that

$$PWF(f_* u) \subset df_* PWF(u). \tag{3.6}$$

Indeed, suppose that $[\ell] \notin df_* PWF(u)$. Let $g : S \times Y \to \mathbf{R}$ have $dg_{s_0}(y_0) = \ell$. Then $g \circ (\mathrm{id} \times f) : S \times X \to \mathbf{R}$ is a submersion for s near s_0 and for any $x_0 \in f^{-1}(y_0)$. Furthermore, for any $x_0 \in f^{-1}(y_0) \cap \operatorname{supp} u$ we can find a b of small support near x_0 such that $(g_s \circ f)_*(bu)$ is smooth. We can cover $f^{-1}(y_0) \cap \operatorname{supp} u$ by a finite number of neighborhoods on which the b's are defined and non-zero. By restricting a little further and multiplying by b^{-1} and using a partition of unity we can arrange that the b are $\equiv 1$ near $f^{-1}(y_0) \cap \operatorname{supp} u$. We thus may assume that $(g_s \circ f)_*(bu)$ is smooth where $b = f^* c$ for some C_0^∞ function defined near y_0 with $c(y_0) = 1$. But then

$$g_{s*}\, c f_*(u) = g_{s*} f_*(bu) = (g_s \circ f)_*(bu)$$

is smooth, for s near s_0, y near y_0.

We shall see that (3.6), together with the remarks at the end of §4 of Chapter IV, can be viewed as an assertion that singularities tend to accumulate at envelopes—a version of Huygens' principle. We shall have more to say about this later on. Before doing so, let us now turn to the problem of extending the notion of pull back.

Let W be a subset of $PT^* Y$ and let $f : X \to Y$ be a differentiable map. We say that f *is transversal to* W if, for any ℓ with $[\ell] \in W$ and $\ell \in T^* Y_{f(x)}$ we have $df_x^* \ell \neq 0$. For example, if Z is a submanifold of Y and $W = PN(Z)$ consists of the normal directions to Z, then to say that f is transversal to W is the same as saying that f is transversal to Z. Indeed, for f to be transversal to Z we must have, for each $x \in X$ with $f(x) \in Z$, that $df_x TX_x + TZ_{f(x)} = TY_{f(x)}$. But this is the same as saying that there is no $\ell \neq 0$ in $T^* Y_{f(x)}$ which vanishes both on $TZ_{f(x)}$ and $df_x TX_x$, i.e., that f is transversal to $PN(Z)$.

Suppose f is transversal to W. Then for any $\ell \in W$ with $\ell \in T^* X_{f(x)}$ the element $[df_x^* \ell]$ is well defined and we set

$$df^* W = \{[df_x^* \ell] | \ell \in T^* Y_{f(x)}, [\ell] \in W\}.$$

We have seen in §1 that if f is transversal to a submanifold Z then $f^* v$ can be defined if v is a δ-section along Z. We now extend this by defining $f^* v$ whenever $PWF[v] \subset W$ and f is transversal to W where W is some closed subset of $PT^* Y$. We first describe $f^* v$ by making some choices, and then show independence of these choices by describing $f^* v$ as a limit, as in the beginning of this section.

Given any $x \in X$ we can find coordinate neighborhoods U of x and V of $f(x)$ with the following properties:

(i) $f(U) \subset V \subset \mathbf{R}^n$,

(ii) there are disjoint closed subsets S_1 and S_2 of S^{n-1} such that $\pi^{-1} V \cap W \subset V \times S_1$ and $\{[\ell] | \ell \in T^* Y_{f(x')}, x' \in U, df^*_{x'} \ell = 0\} \subset V \times S_2$.

Now for any $u \in C_0^\infty(|\wedge|X)$ we can cover supp u by finitely many such U's with corresponding V's covering $f(\text{supp } u)$. By a partition of unity on X and Y we may replace, in trying to define $\langle u, f^* v\rangle$, the X and Y by U and V satisfying (i) and (ii) and assume that v has compact support. Let ψ_1 and ψ_2 be a partition of unity on the sphere with $S_1 \subset \text{supp } \psi_1$ and $S_2 \cap \text{supp } \psi_1 = \varnothing$ and $S_1 \cap \text{supp } \psi_2 = \varnothing$. Then

$$v = \int \rho_\Omega^* [I^{n-1} \rho_{\Omega *} v \, dy] \, d\Omega$$

$$= \int \psi_1 \rho_\Omega^* [I^{n-1} \rho_{\Omega *} v \, dy] \, d\Omega + \int \psi_2 \rho_\Omega^* [I^{n-1} \rho_{\Omega *} v \, dy] \, d\Omega.$$

The second integral is a smooth function and so its pull back is well defined. For the first integral we have that f is transversal to each of the surfaces $\rho_\Omega^{-1}(c)$, $c \in \mathbf{R}$ and thus

$$f^* \psi_1 \rho_\Omega^* [I^{n-1} \rho_{\Omega *} v \, dy]$$

is defined as in §1. We can thus define

$$f^* v = \int f^* \psi_1 \rho_\Omega^* [I^{n-1} \rho_{\Omega *} v \, dy] \, d\Omega + f^* \int \psi_2 \rho_\Omega^* [I^{n-1} \rho_{\Omega *} v \, dy] \, d\Omega.$$

To check that this definition is independent of all the choices, we proceed as follows: We introduce a topology on the space of all generalized functions having their wave front sets contained in W. We show that the smooth functions are dense in this topology, and thus there is at most one way of extending f^* by continuity from the smooth densities. For v of sufficiently small support, the above definition will prove to be continuous, establishing the global existence and uniqueness of f^*.

Before doing so, let us observe a result that follows directly from this local description of $f^* v$. For each Ω we know from Proposition 3.1 that the wave front set of $f^* \psi_1 \rho_\Omega^* [I^{n-1} \rho_{\Omega *} v dy]$ is contained in the set of all vectors of the form $df^* \ell$ where $\ell = (y, \Omega)$. From this it follows, by integrating over S_1 (and by choosing more and more refined product type covers of $PWFv$), that

$$PWF(f^* v) \subset df^* (PWFv). \tag{3.7}$$

Let Y be a manifold and let Z be a closed subset of $PT^* Y$. We shall let $C_Z^{-\infty}(Y) \subset C^{-\infty}(Y)$ consist of those generalized functions, v, on Y such that $PWFv \subset Z$. We will now put a topology on the space $C_Z^{-\infty}(Y)$. We wish to describe the topology in terms of a family of semi-norms. We first of all want the topology to be stronger than the weak topology, so for each $\rho \in C_0^\infty(|\wedge|X)$ we throw in the semi-norm $|\ |_\rho$ given by

$$|v|_\rho = |\langle v, \rho\rangle|.$$

Next let U be any coordinate patch and S any auxiliary space with $f: S \times U \to \mathbf{R}$ such that $df_s(x) \neq 0$ and $[df_s(x)] \notin Z$ for all $x \in U$ and $s \in S$. Let b be a density of compact support in U so that $f_{s*}(bu)$ is smooth on $S \times \mathbf{R}$. Let $c \in C_0^\infty(S)$ so that $c(s) f_{s*}(bu)$ is compactly supported and smooth on $S \times \mathbf{R}$. For each k we can introduce the C^k norm, $\|\ \|_k$, relative to some choices of coordinates on S and $\mathbf{R}$. Then to the quadruple f,b,c,k we introduce the semi-norm

$$|v|_{f,b,c,k} = \|cf_*(bv)\|_k.$$

We claim that

PROPOSITION 3.6. *$C^\infty(Y)$ is dense in $C_Z^{-\infty}(Y)$.*

Indeed, we first observe, that the proof of Proposition 3.2 shows that in the definition of the topology it suffices to restrict attention to f arising as the ρ in the Radon transform. Thus the proof of Proposition 3.6 can be reduced to the case where $Y \subset \mathbf{R}^n$ and Z is replaced by a set of the form $Y \times C$ where C is a closed subset of $\mathbf{R}^n$.

Now if v is any generalized function of compact support on $\mathbf{R}^n$ and h is any smooth function, then the convolution is a smooth function

$$(v \star h)(x) = \langle v, h_x\, dy\rangle$$

where $h_x(y) = h(x - y)$. As usual, we will sometimes write

$$v \star h(x) = \int h(x-y)v(y)\, dy.$$

The right hand side is defined for $v \in C_0^\infty$ and $h \in C^\infty$, and coincides with the left hand side. For $v \in C_0^{-\infty}$ we take the left hand side as the definition, carrying along the right hand side as a convenient formal expression. Notice that

$$\begin{aligned}\langle v \star h_1, h_2\rangle &= \iint h_1(x-y)v(y)\, dy h_2(x)\, dx\\ &= \int v(y)\Big(\int h_1(-(y-x))h_2(x)\, dx\Big)\, dy\\ &= \langle v, h_1^r \star h_2\rangle\end{aligned}$$

where $h_1^r(x) = h_1(-x)$. From this it is clear that $v \star h_1 \to v$ as $h_1 \to \delta$ in $C^{-\infty}$. Notice also that for h_1 and h_2 smooth functions of compact support and hence also for generalized functions, we have

$$\rho_{\Omega*}(h_1 \star h_2) = (\rho_{\Omega*} h_1) \star (\rho_{\Omega*} h_2).$$

Indeed, with no loss of generality we assume that $\Omega = (1, 0, \ldots, 0)$. The equation then becomes

$$\begin{aligned}&\int [h_1(x_1 - y_1, \ldots, x_n - y_n) h_2(y_1, \ldots, y_n)\, dy_1 \cdots dy_n]\, dx_2 \cdots dx_n \\ &= \Big[\int h_1(x_1 - y_1, \ldots, x_n - y_n)\, dx_2 \cdots dx_n\Big] h_2(y_1, \ldots, y_n)\, dy_1 \cdots dy_n \\ &= \int \Big[\int h_1(x_1 - y_1, \ldots, x_n - y_n)\, dx_2 \cdots dx_n\Big][h_1(y_2, \ldots, y_n)\, dy_2 \cdots dy_n]\, dy_1 \\ &= \rho_{\Omega *} h_1 \star \rho_{\Omega *} h_2(x_1).\end{aligned}$$

Now let h be any C^∞ function of compact support on $\mathbf{R}^n$ with $\int h\,dx = 1$, and set

$$h_R = R^{-n} h(Rx).$$

It is clear that $h_R \to \delta$ in $C^{-\infty}(\mathbf{R}^n)$ and that

$$\rho_{\Omega *} h_R(x) = R^{-1}(\rho_{\Omega *} h)(Rx) \qquad \text{or} \qquad \rho_{\Omega *}(h_R) = (\rho_{\Omega *} h)_R$$

so that

$$\rho_{\Omega *} h_R \to \delta \qquad \text{in } C^{-\infty}(\mathbf{R}^1).$$

Now let Ω range over some open set on the sphere for which $\rho_{\Omega *} v$ is smooth. Then

$$\rho_{\Omega *}(h_R \star v) = (\rho_{\Omega *} h)_R \star \rho_{\Omega *} v$$

approaches $\rho_{\Omega *} v$ in the C^∞ topology of $\mathbf{R}^1$, and all this depending smoothly on parameters. Thus we have completed the proof of Proposition 3.5. It is clear that the local definition of f^* is smooth in the $C_Z^{-\infty}$ topology. We have thus proved:

PROPOSITION 3.7. *Let Z be a closed subset of PT^*Y and let $f: X \to Y$ be a smooth map transversal to Z. If $v \in C_Z^{-\infty} Y$ and if $v_i \in C^\infty(Y)$ with $v_i \to v$ in $C_Z^{-\infty}(Y)$ then the $f^* v_i$ converge in $C^{-\infty}(X)$ to an element, $f^* v$, which is independent of the particular choice of v_i. Furthermore*

$$PWF(f^* v) \subset df^* PWF(v). \tag{3.7}$$

Let us now go back to equation (3.6) and point out that it does, in fact, represent Huygens' principle. Suppose that $X = Y \times S$ where S is some parameter space. Let $Z \subset Y \times S$ be a hypersurface which intersects each $Y \times \{s\}$ transversally and so defines a family of hypersurfaces, which we denote by Z_s. Suppose that k is a generalized density on X whose wave front set is contained in the non-zero normal bundle, $\mathfrak{N}(Z)$, to Z. Let us also assume that $\mathfrak{N}(Z)$ is transversal to the map f in the sense of Chapter IV. That is, it intersects

$H \subset T^*X$ transversally where $H \subset T^*X$ consists of all covectors lying in df^*T^*Y. In local coordinates $H \subset T^*X$ consists of points of the form $(y, s, \xi, 0)$ since $f(y,s) = y$ and therefore $df^*_{(y,s)}(y;\xi) = (y,s;\xi,0)$. Then $\mathfrak{N}(Z) \cap H$ consists precisely of those covectors normal to Z of the form $(y,s;\xi,0)$ and $df_* \mathfrak{N}(Z)$ consists precisely of the normals to the envelope of the surfaces Z_s. Thus (3.6) asserts that the singularities of $f_* k$ accumulate along the envelope of the surfaces Z_s. If we think of the operation f_* as a sort of continuous superposition then this is Huyghens' principle in the context of the theory of distributions. Notice that (1.3) was really the special case where there is no envelope.

Let us now look at the multiplicative properties of wave front sets. If X and Y are smooth manifolds then we have obvious maps

$$C^\infty(X) \otimes C^\infty(Y) \to C^\infty(X \times Y) \quad \text{and}$$

$$C^\infty(|\wedge|X) \otimes C^\infty(|\wedge|Y) \to C^\infty(|\wedge|(X \times Y)),$$

etc. with all the images dense in their range. We have corresponding maps for generalized sections and densities. Indeed, for $u \in C^{-\infty}(|\wedge|X)$ and $v \in C^{-\infty}(|\wedge|Y)$ we define $u \boxtimes v \in C^{-\infty}(|\wedge|(X \times Y))$ by observing that for any $h \in C_0^\infty(X \times Y)$ the function h_v lies in $C_0^\infty(X)$ where

$$h_v(x) = \langle v, h(x, \cdot)\rangle.$$

We thus may set

$$\langle u \boxtimes v, h\rangle = \langle u, h_v\rangle.$$

If we take $h(x,y) = h_1(x)h_2(y)$, then it is clear that

$$\langle u \boxtimes v, h\rangle = \langle u, h_1\rangle\langle v, h_2\rangle$$

and this suffices to show that we equally well could have defined

$$\langle u \boxtimes v, h\rangle = \langle v, h_u\rangle \qquad \text{where} \qquad h_u(y) = \langle u, h(\cdot, y)\rangle.$$

Notice that we do *not* have $PT^*(X \times Y) = PT^*X \times PT^*Y$ due to the existence of non-zero covectors in $T^*(X \times Y) = T^*X \times T^*Y$ of the form $(\ell_1, 0)$ and $(0, \ell_2)$. We do clearly have (as a set)

$$PT^*(X \times Y) = PT^*X \times PT^*Y \cup PT^*X \times 0_Y \cup 0_X \times PT^*Y$$

where 0_Y denotes the zero covectors of Y (which we could identify with Y) and similarly 0_X. We claim that

$$PWF(u \boxtimes v) \subset PWFu \times PWFv \cup PWFu \times 0_Y \cup 0_X \times PWFv. \tag{3.8}$$

Indeed, this is a purely local question and, by Proposition 3.3, we need only show that if ξ, η is such that $[(x, y, \xi, \eta)]$ does not lie in the right hand side of (3.8) then

$$\langle h(u \boxtimes v), e^{i(r\xi\cdot x+s\eta\cdot x)}\rangle = O((r^2+s^2)^{-N})$$

for all N, for some suitable blip function h. We may choose $h(x,y) = h_1(x)h_2(y)$. But then

$$\langle h_1 h_2 u \boxtimes v, e^{i(r\xi\cdot x+s\eta\cdot y)}\rangle = \langle h_1 u, e^{ir\xi\cdot x}\rangle\langle h_2 v, e^{is\eta\cdot y}\rangle$$

and the right hand vanishes to infinite order in $(r^2 + s^2)$.

So far, the entire discussion has been carried out for the projective wave front set. Let us now describe how the analogous results can be established using the spherical wave front sets. One way would be to show directly, using integration by parts and the Fourier transform, that it suffices to use linear functions f in (3.1). This is the original approach taken by Hormander [4] (he *defines* the wave front set, locally, via the Fourier transform, and shows that the local definiton is independent of the coordinate system). We will not repeat his arguments here. Let us sketch an alternative approach. If we go back and examine Proposition 3.2 we see that it doesn't really have much content until we have an effective way of describing the wave front set on $\mathbf{R}^1$. This defect would obviously be rectified once we prove

PROPOSITION 3.8 *Let u be a generalized function of compact support on $\mathbf{R}^1$ and suppose that $\hat{u}(\tau) = O(\tau^{-N})$ as $\tau \to +\infty$. Then u is forward smooth at all x, i.e., $WFu \subset \{(x,\xi)|\xi < 0\}$. Furthermore, if u is forward smooth at 0 so is hu for any $h \in C_0^\infty(\mathbf{R})$.*

PROOF. We use the Hilbert transform. Recall that for any $v \in C_0^\infty(\mathbf{R})$ we define its Hilbert transform, Hv, by the formula

$$(Hv)\hat{}\,(\tau) = \theta(\tau)\hat{}\,v(\tau)$$

where θ is the Heaveside function, $\theta(\tau) = 1$ for $\tau > 0$ and $\theta(\tau) = 0$ for $\tau \le 0$. The right hand side is a tempered distribution (i.e., an element of $\mathcal{S}'(\mathbf{R})$) which vanishes rapidly at infinity. Thus $Hv \in C^\infty(\mathbf{R})$. Similarly, if $v \in C_0^{-\infty}(\mathbf{R})$ then $Hv \in C^{-\infty}(\mathbf{R})$. Thus H: $C_0^{-\infty}(\mathbf{R}) \to C^{-\infty}(\mathbf{R})$ is a continuous linear map, and as such, is determined by its restriction to the dense subspace $C_0^\infty(\mathbf{R})$ which it maps into $C^\infty(\mathbf{R}) \subset C^{-\infty}(\mathbf{R})$. For $v \in C^\infty(\mathbf{R})$ we have

$$Hv(x) = \frac{1}{i}\int_{-\infty}^{\infty} \frac{v(y)}{x-y}\,dy$$

where the integral is taken as the Cauchy principal value, i.e.,

$$Hv(x) = \lim_{\varepsilon\to 0}\left(\int_{-\infty}^{-\epsilon} + \int_{\epsilon}^{\infty}\right)\frac{v(y)}{x-y}\,dy.$$

Notice that $\hat{u}(\tau) = O(\tau^{-N})$ for all N as $\tau \to +\infty$ if and only if $(Hu)\hat{}\,(\tau)$

$= O(\tau^{-N})$ as $\tau \to \pm\infty$, i.e., if and only if Hu is smooth.

We will have established Proposition 3.8 if we prove the following:

(i) Let f_s be an orientation preserving diffeomorphism of $\mathbf{R}^1$ depending smoothly on some parameter s. Then

$$f_s^* \circ H = H \circ f_s^* + k_s$$

where k_s is a smoothing operator (an integral operator with C^∞ kernel) depending smoothly on s.

(ii) Let $h_s \in C^\infty(\mathbf{R})$ depend smoothly on s. Then $h_s Hu = H(h_s u) + k_s u$ where k_s has a C^∞ kernel depending smoothly on s.

In the proof of (i) and (ii) we will suppress the explicit dependence on s. Also, it suffices to prove the desired equalities as operators on C_0^∞ since C_0^∞ is dense in $C_0^{-\infty}$. To prove (i) observe that

$$f^* Hv(x) = \frac{1}{i}\int \frac{v(y)}{f(x) - y}\,dy = \frac{1}{i}\int \frac{v(f(y))}{f(x) - f(y)} f'(y)\,dy$$

and

$$Hf^* v(x) = \frac{1}{i}\int \frac{v(f(y))}{x - y}\,dy.$$

Now

$$\frac{1}{f(x) - f(y)} = \frac{1}{(x - y)f'(y)[1 + g(x,y)(x - y)]}$$

where $g(x,y)$ depends smoothly on x and y. We can write

$$\frac{1}{1 + g(x,y)(x - y)} = 1 + (x - y)h(x,y)$$

where h is smooth. Then

$$f^* Hv = \frac{1}{i}\int \frac{v(f(y))}{f(x) - f(y)} f'(y)\,dy = \frac{1}{i}\int \left[\frac{v(f(y))}{x - y} + v(f(y))h(x,y)\right] dy$$

so that

$$(f^* H - H^* f)v = \frac{1}{i}\int v(f(y))h(x,y)\,dy = \frac{1}{i}\int v(y)h(x,f^{-1}(y))(f^{-1})'(y)\,dy$$

proving (i). Similarly, for (ii)

$$[(hH - Hh)v](x) = \frac{1}{i}\int \frac{[h(x) - h(y)]v(y)}{x - y}\,dy = \frac{1}{i}\int r(x,y)v(y)\,dy$$

where $r(x,y) = [h(x) - h(y)]/(x - y)$ is a smooth function of x and y. This proves Proposition 3.8. We can now state the analogues of Propositions 3.3–3.7

using the wave front set or spherical wave front set instead of the projective wave front set:

PROPOSITION 3.9 *Let u be a generalized density defined on X. Let U be a coordinate system on X with $U \times S^{n-1}$ giving the corresponding trivialization of $ST^* X$ over U. Then $(x_0, \Omega_0) \notin SWFu$ if and only if there is some blip function b, not vanishing at x_0, such that $H\rho_{\Omega *}(bu)$ is smooth on $\mathbf{R}$, where H is the Hilbert transform. This is the same as saying that $\langle bu, e^{i\tau\Omega\cdot x}\rangle$ is asymptotically zero as $\tau \to \infty$ for Ω near Ω_0. In particular $SWFu$ is closed in $ST^* X$ and WFX is closed in $T^* X - 0_X$. If $f: X \to Y$ is a smooth map and u has compact support then*

$$WF(f_* u) \subset \{\ell \in T^* Y | 0 \neq \ell \in T^* Y_{f(x)} \text{ and } df_x^* \ell \subset WFu \cup 0_X\}. \quad (3.9)$$

If v is a generalized function on Y with f transversal to WFv then $f^* v$ is defined by continuity and

$$WFf^* v \subset df^* WFv. \quad (3.10)$$

If u and v are generalized functions on X and Y then

$$WF(u \boxtimes v) \subset WFu \times WFv \cup WFu \times 0_Y \cup 0_X \times WFv. \quad (3.11)$$

There are a number of consequences of (3.9)—(3.11) (or their weaker projective versions) that are worthwhile recording.

i) Multiplication of generalized functions: Let u and v be honest functions. Then their product, uv, defined by $uv(x) = u(x)v(x)$ can also be described more abstractly as

$$uv = \Delta^* (u \boxtimes v)$$

where $\Delta : X \to X \times X$ is the diagonal map, $\Delta(x) = (x, x)$. To make sense of this expression for generalized sections we need to know that Δ is transversal to $WF(u \times v)$. Now $d\Delta^* (x, x, \xi, \eta) = (x, \xi + \eta)$ so that the normal bundle to Δ consists of all covectors of the form $(x, x, \xi, -\xi)$. We apply (3.11) and conclude

PROPOSITION 3.10 *Let u and v be generalized functions. Suppose that $0 \neq \xi + \eta$ for any $\xi \in WFu$ and $\eta \in WFv$ (where $\xi, \eta \in T^* X_x$). Then uv is defined by continuity and*

$$WF(uv) \subset (WFu + WFv) \cup WFu \cup WFv$$

(where $WFu + WFv$ denotes the set of all covectors of the form $\xi + \eta$ with $\xi \in WFu$ and $\eta \in WFv$).

Let X, Y and Z be differentiable manifolds and let K be a section of $|\wedge|Y$ over $X \times Y$ while L is a section of $|\wedge|Z$ over $Y \times Z$. For smooth sections (compactly supported in Y) the composition of these two kernels $K \circ L$, a section of $|\wedge|Z$ over $X \times Z$, is given as

$$K \circ L(x, z) = \int_Y K(x, y) L(y, z)$$

or, more abstractly,

$$K \circ L = \pi_* \Delta^* K \boxtimes L$$

where $K \boxtimes L$ is a section of $|\wedge|Y \otimes |\wedge|Z$ over $X \times Y \times Y \times Z$, where Δ : $X \times Y \times Z \to X \times Y \times Y \times Z$ is the diagonal map $\Delta(x, y, z) = (x, y, y, z)$ and π: $X \times Y \times Z \to X \times Z$ is projection onto the first and third factors. For $K \circ L$ to be defined for generalized sections we apply our criteria to obtain:

PROPOSITION 3.11 *Suppose K is a generalized section of $|\wedge|Y$ over $X \times Y$ and L is a generalized section of $|\wedge|Z$ over $Y \times Z$. Suppose there is no $\eta \neq 0$ in any $T^* Y_y$ such that $(0, \eta)_{(x,y)} \in WFK$ and $(-\eta, 0)_{(y,z)} \in WFL$. Then the composition $K \circ L$ is defined by continuity and*

$$WF(K \circ L) \subset \{(x, z; \xi, \zeta) | \textit{there is some } (y, \eta)$$
$$\textit{with } (x, y; \xi, \eta) \in WFK \textit{ and } (y, z; -\eta, \zeta) \in WFL\}.$$

Let us close this section by taking another look at the Radon transform from a point of view suggested by Gelfand [5]. Let X and Y be differentiable manifolds and let Z be a manifold which is fibered of X and over Y so that we have a diagram

$$\begin{array}{ccc} & Z & \\ \pi \swarrow & & \searrow \rho \\ X & & Y \end{array}$$

Assume, for simplicity, that X, Y and Z come equipped with nowhere zero densities so that we may identify functions with densities. Then, starting with any generalized function, u on X, we may form $\pi^* u$, and if appropriate compactness conditions are satisfied, then form $\rho_* \pi^* u$. This gives a way of going from $C_0^\infty(X)$ to $C^\infty(Y)$ and from $C_0^{-\infty}(X)$ to $C^{-\infty}(Y)$. Similarly, we may be able to go the other way, forming $\pi_* \rho^* v$ for $v \in C_0^\infty(Y)$ or $C_0^{-\infty}(Y)$. For example, we may let $X = \mathbf{R}^n$ and $Y = S^{n-1} \times \mathbf{R}$ and let $Z \subset X \times Y$ consist of those $(x; \Omega, t)$ such that $x \cdot \Omega = t$. We can think of $S^{n-1} \times \mathbf{R}$ as consisting of the oriented hyperplanes and then Z consists of those pairs (x, h) where x is a point of $\mathbf{R}^n$, h is an oriented hyperplane, and $x \in h$. Both X and Y have natural densities and the map $X \times Y \to \mathbf{R}$ sending $(x; \Omega, t) \rightsquigarrow x \cdot \Omega - t$ induces a density on Z. (In fact we are considering the generalized density $\delta(x \cdot \Omega - t)$ on $X \times Y$.) In this case if $u \in C_0^\infty(X)$, then

$$\rho_* \pi^* u(\Omega, t) = \rho_{\Omega *} u(t) = \int_{x \cdot \Omega = t} u$$

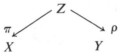

while

$$\pi_* \rho^* v(x) = \int_{S^{n-1}} v(\Omega, \Omega \cdot x) d\Omega.$$

Notice that whenever Z is a submanifold of $X \times Y$ then $\rho^{-1}(y) \cap \operatorname{supp} \pi^* u \subset \operatorname{supp} u \times y$ is compact and so there is no trouble in forming $\rho_* \pi^* u$ if u has compact support. (Of course, $\rho_* \pi^* u$ need not have compact support. For example, in the classical Radon transform $\pi_* \rho^* v$ will not, in general, have compact support for compactly supported v on $S^{n-1} \times \mathbf{R}$.)

As other examples, take $X = \mathbf{P}^n$ (= {lines through the origin in $\mathbf{R}^n$}) and $Y = \mathbf{P}^{n*}$ (= {hyperplanes through the origin in $\mathbf{R}^n$}); cf. Gelfand [5]. We can replace real projective n-space by complex or quaternionic projective n-space or by the Cayley projective plane; cf. Helgason [6]. In all these cases an inversion formula holds. Or, take $X = G_n(k) = \{k$ dimensional subspaces of $\mathbf{R}^n\}$ and $Y = G_{n-k}(k)$ where $k < n - k$ and $Z = \{(x,y) | x \in y\}$; cf. Helgason [6].

Let us examine the general situation where Z is a submanifold of $X \times Y$ and ask when does Z determine a double fibration. Let π denote the restriction of the projection $\pi_X : X \times Y \to X$ to Z. To say that π is a submersion means that $d\pi^*$ is injective. Now for any $\xi \in T^* X$ we have $d\pi_X^* \xi = (\xi, 0)$ and $d\pi^* \xi$ is just the restriction of $(\xi, 0)$ to Z. We thus want to be sure that the normal bundle, NZ, *does not contain any vectors of the form* $(\xi, 0)$ *with* $\xi \neq 0$ *or* $(0, \eta)$ *with* $\eta \neq 0$.

Since the operator $\rho_* \pi^* : C_0^\infty(X) \to C^\infty(Y)$ is given by a δ-section along Z we know that its wave front set consists of non-zero vectors in NZ. Suppose that X and Y are both compact so that we can form the composite operator

$$(\pi_* \rho^*) \circ (\rho_* \pi^*) : C^\infty(X) \to C^\infty(X).$$

By formula (3.9) the wave front set of this operator consists of those (ξ_1, ξ_2) for which there exists an η with $(\xi_1, \eta) \in WF(\rho_* \pi^*)$ and $(-\eta, \xi_2) \in WF(\pi_* \rho^*)$ (where by abuse of language we denote by $WF(\rho_* \pi^*)$ the wave front set of the distribution kernel associated to $\rho_* \pi^*$ etc.). In particular, suppose that the maps $SN^* Z \to ST^* X$ and $SN^* Z \to ST^* Y$ are both bijective. Then $\xi_2 = -\xi_1$ and $WF(\pi_* \rho^* \rho_* \pi^*) \subset N\Delta$ where $\Delta \subset X \times X$ is the diagonal. Actually, as we shall see later on, this is enough to tell us that $\pi_* \rho^* \rho_* \pi^*$ is an elliptic pseudo-differential operator, and this gives rise to an inversion formula for the operator $\rho_* \pi^*$. The bijectivity condition on the maps $SN^* Z \to ST^* X$ and $SN^* Z \to ST^* Y$ holds in the projective examples given above.

By use of results in differential topology one can show that if a double fibration satisfies the above bijectivity condition and X and Y are compact, then it bears a striking resemblance to the projective examples. For example $Z \subset X \times Y$ is of codimension 1, 2, 4 or 8 and the fibers are diffeomorphic to real, complex, quaternionic projective space or the Cayley plane.

§4. Lagrangian distributions.

Our procedure for the rest of this chapter will be to pick out a nice family of generalized sections—one that behaves nicely under functorial operations—and to study various properties of these generalized sections. In a certain sense, the generalized sections we will consider are generalizations of δ-sections, and arise when one wants to close the class of δ-sections under the push forward operation. Thus, for example, we will want to consider a generalized section of the form $\pi_* ag^* \delta$ on the manifold X where

$$\begin{array}{ccc} W & \xrightarrow{g} & \mathbf{R} \\ \pi \downarrow & & \\ X & & \end{array}$$

π and f are submersions and a is a smooth function (or density) defined on W. To illustrate what kind of sections will come up, let us consider the situation where $X = \mathbf{R}$, $W = \mathbf{R}^2$ with $g_\pm(x,t) = x \pm t^2$ and $\pi(x,t) = x$. We wish to consider $\pi_* ag^* \delta$ where $a = a(x,t)$ is some function which is compactly supported in t. As a warm up exercise let us replace δ by the Heaviside function x_+^0 where $x_+^0(x) = 1$ for $x > 0$ and $x_+^0 = 0$ for $x < 0$. (More generally we define $x_+^\lambda = x^\lambda$ for $x > 0$ and $x_+^\lambda(x) = 0$ for $x < 0$. We follow the notation of Gelfand-Shilov [1]. Recall that the derivative of x_+^0 at the origin is exactly δ.) Let the morphism associated to g and π be determined by the choice of densities $|dx|$ on $\mathbf{R}$ and $|dx \wedge dt|$ on $\mathbf{R}^2$. Also, in our preliminary computation, suppose that $a = a(t)$ is identically one for t near zero and that $\int_{-\infty}^{\infty} a(t)\,dt = A$. Then

$$g^* x_+^0 = \begin{cases} 1 \text{ on } g^{-1}(\mathbf{R}^+), \\ 0 \text{ on } g^{-1}(\mathbf{R}^-), \end{cases}$$

and so we obtain the pictures

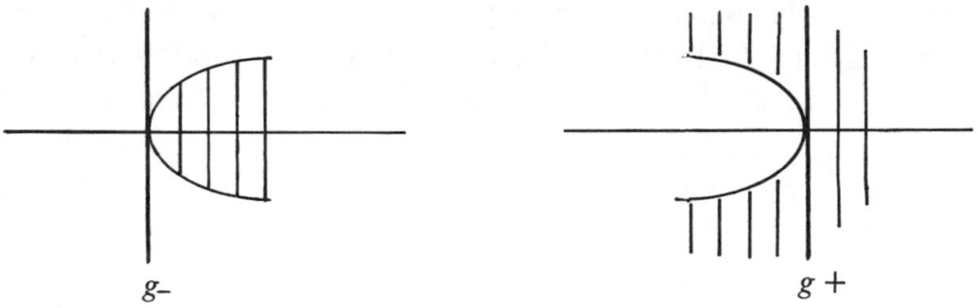

where the shaded region indicates where $g^* x_+^0 = 1$. Thus, for g_- we obtain the formula

$$\pi_* ag_-^* x_+^0 = \begin{cases} 0 \text{ for } x < 0 \\ \int_{-x^{1/2}}^{x^{1/2}} a(t)\,dt = 2x_+^{1/2} + \gamma(x) \end{cases}$$

where γ is a smooth function of x which vanishes near $x = 0$ and $\pi_* ag_-^* x_+^0 = A$ for large values of x. Similarly, for g_+, we obtain

$$\pi_* ag_+^* x_+^0 = \begin{cases} A \text{ for } x > 0 \\ A - \int_{-|x|^{1/2}}^{|x|^{1/2}} a(t)\, dt = A - 2x_-^{1/2} - \gamma(-x). \end{cases}$$

(Here $x_-^\lambda(x) = x_+^\lambda(-x)$) . Thus we can write

$$\pi_* ag_-^* x_+^0 \equiv 2x_+^{1/2} \text{ and } \pi_* ag_+^* x_+^0 \equiv -2x_-^{1/2} \tag{4.1}$$

where $\equiv$ means equality mod smooth functions.

Let us now do a slightly different computation, replacing x_+^0 by δ, taking $g = g_-$ and letting $a = a(x, t)$. (This computation will not only be illustrative, it will prove to be the crucial computation in our general considerations.)

Let $h \in C_0^\infty(\mathbf{R})$. Then

$$\langle \pi_* ag^* \delta, h \rangle = \int_Z a(x, t) h(x) \nu$$

where $Z \subset \mathbf{R}^2$ is the submanifold $x = t^2$ and ν is the induced density. If we use t to parametrize Z then $d(x - t^2) \wedge dt = dx \wedge dt$ so $\nu = dt$ and the above integral becomes

$$\int_{\mathbf{R}} a(t^2, t) h(t^2)\, dt.$$

Let us first assume that $a(x, t) = a(t)$, i.e., that a depends only on the second variable. Now we can write $a = a_e + a_o$ where

$$a_e(t) = \tfrac{1}{2}[a(t) + a(-t)] \quad \text{is even and} \quad a_o(t) = \tfrac{1}{2}[a(t) - a(-t)] \text{ is odd}$$

Since $h(t^2)$ is even, we have

$$\langle \pi_* ag^* \delta, h \rangle = \int_{\mathbf{R}} a_e(t) h(t^2)\, dt = 2 \int_0^\infty a_e(t) h(t^2)\, dt.$$

We now apply the well known fact, due to Whitney, that any even C^∞-function a_e can be written as $a_e(t) = b(t^2)$ where $b \in C^\infty$.[1] Then

$$2 \int_0^\infty a_e(t) h(t^2)\, dt = 2 \int_0^\infty b(t^2) h(t^2)\, dt = \int_0^\infty b(u) h(u) u^{-1/2}\, du$$

---

[1] *Proof of this fact*. Let $\Sigma a_n t^n$ be the Taylor expansion of a_e at 0. Since a_e is even it can involve only even powers of t. Thus $\Sigma a_n t^n = \Sigma c_n t^{2n}$. By Borel's theorem (cf. §2 of Chapter II) we can find a C^∞ function c with Taylor expansion $\Sigma c_n t^n$. Then $a_e(t) - c(t^2) = d(t)$ vanishes with all its derivatives at 0, hence $d(t^{1/2})$ is C^∞. We set $b(t) = c(t) + d(t^{1/2})$.

so that

$$\pi_* ag^* \delta = bx_+^{-1/2} \tag{4.2}$$

where $x_+ = x$ for $x > 0$ and $x_+ = 0$ for $x < 0$.

We should point out that if a depends smoothly on some auxiliary parameters then b will also depend smoothly on these parameters. If we take $a \equiv 1$ near the origin in (4.2) we see that our collection of generalized functions had better include $x_+^{-1/2}$. We can strengthen the assertion of (4.2) by the following trick: Let T_c denote translation by c of the line or plane in the x-direction, so $T_c(x,t) = (x+c,t)$. Then $T_c \circ g(x,t) = x + t^2 + c = g \circ T_c(x,t)$, so that

$$g^* \circ T_c^* = T_c^* \circ g^*$$

while multiplication by a commutes with T_c since a depends only on t. Finally $\pi \circ T_c = T_c \circ \pi$ and therefore $T_c^* \pi_* = \pi_* T_c^*$. Thus

$$(\pi_* ag^*)T_c^* u = T_c^* \pi_* ag^* u$$

for any $u \in C^{-\infty}(\mathbf{R})$. This implies that

$$v \star \pi_* ag^* u = \pi_* ag^*(v \star u)$$

for any $v \in C_0^\infty(\mathbf{R})$ and hence for any $v \in C_0^{-\infty}(\mathbf{R})$. In particular, taking $u = \delta$ we see that

$$\pi_* ag^* v = (b(x)x_+^{-1/2}) \star v$$

if either a or v has compact support. Here a, and hence b, can depend on auxiliary parameters. In particular, consider the diagram

$$\begin{array}{ccccc} & & \mathbf{R} \times \mathbf{R} \times \mathbf{R} \xrightarrow{G} \mathbf{R} \times \mathbf{R} & & (s,x,t) \rightsquigarrow (s, x+t^2) \\ & & \downarrow \pi_2 & & \downarrow \\ \mathbf{R} & \xrightarrow{\Delta} & \mathbf{R} \times \mathbf{R} & & (s,x) \end{array}$$

with $\Delta x = (x,x)$. Let $A(s,x,t) = a(s,t)$. Then

$$\Delta^* \pi_{2_*} AG^* = \pi_* ag^* \quad \text{and} \quad \pi_{2_*} AG^* v = \int B(x,y) y_+^{-1/2} v(x-y)\,dy.$$

Thus

$$\pi_* ag_-^* v(x) = \int b(x,y) y_+^{-1/2} v(x-y)\,dy \tag{4.3$_-$}$$

in the sense of generalized functions, where $b(x,t^2) = \frac{1}{2}[a(x,t) + a(x,-t)]$. Exactly the same computation yields

$$\pi_* a g_+^* v(x) = \int b(x,y) y_-^{-1/2} v(x-y)\, dy. \qquad (4.3)_+$$

Let us get back to our original discussion. We have seen that starting with δ and applying $\pi_* g^*$ we get $x_+^{-1/2}$. Applying $\pi_* g^*$ again we will get $x_+^{-1/2} \star x_+^{-1/2}$. We shall give a general recipe later on in the section which will allow us to compute (via the Fourier transform) such a convolution. Let us compute it now, directly (at least up to a constant factor). Clearly $x_+^{-1/2} \star x_+^{-1/2} = 0$ for $x < 0$. For $x > 0$ we have

$$x_+^{-1/2} \star x_+^{-1/2}(r) = \int (r-y)_+^{-1/2} y_+^{-1/2}\, dy = \int (r - rz)_+^{-1/2} (rz_+)^{-1/2} r\, dz$$
$$= \int (1-z)_+^{-1/2} z_+^{-1/2}\, dz = c$$

is thus some positive number independent of r. In other words

$$x_+^{-1/2} \star x_+^{-1/2} = c x_+^0.$$

The same argument shows that, for any a, $b > -1$,

$$x_+^a \star x_+^b = c(a,b) x_+^{a+b+1}, \qquad c(a,b) > 0.$$

Our collection of basic distributions on the line should therefore include $x_+^{n/2}$ for all $n \geq -1$. Now the most general distribution supported at the origin is a linear combination of derivatives of the δ-function. Thus if we wish to start out with derivatives of the δ-function then we must throw in $x_+^{n-1/2}$ for all values of n. If we allow the map $x \rightsquigarrow -x$ then we must also consider x_-^λ for various half integral values of λ. All these are examples of *homogeneous* generalized functions, e.g., if M_c denotes multiplication by c then

$$M_c^* x_+^\lambda = c^\lambda x_+^\lambda \quad \text{and} \quad M_c^* \delta = c^{-1} \delta,$$

etc. The subject of homogeneous generalized functions is given a beautiful and detailed presentation in Gelfand-Shilov [1]. We shall develop a part of the theory here from a slightly different point of view.

A function, f, defined on $\mathbf{R}^+$, is called a symbol of order r, if for each $n \geq 0$ there is a constant, c_n, such that

$$|f^{(n)}(\tau)| \leq c_n \tau^{r-n} \qquad \text{for } \tau > 1.$$

(In what follows immediately we are interested in the case where f is a real valued function. If f takes values in a vector space then we would require a set of such inequalities for each n corresponding to various semi-norms defining a topology on the vector space, in particular for the situation where f takes values in some function space.) Notice that if f is a symbol of order r and g is a symbol of order

s, then fg is a symbol of order $r + s$. Notice also that if h is a symbol of order r such that $|h(\tau)| \geq d\tau^r$ for some $d > 0$ and all $\tau > 1$, then $1/h$ is a symbol of order $-r$.

Let φ be a symbol of order 1 such that $|\varphi'| \geq d > 0$. Let f be a symbol of order r. Then

$$e^{i\varphi} f = \frac{1}{i\varphi'}(e^{i\varphi})' f = \left(\frac{e^{i\varphi} f}{i\varphi'}\right)' + e^{i\varphi}\left(\frac{if}{\varphi'}\right)',$$

where $g = (if/\varphi')'$ is a symbol of order $r - 1$. Integration then gives (for some $a > 0$)

$$\int_a^\tau e^{i\varphi} f = \frac{e^{i\varphi} f}{i\varphi'}(\tau) - \frac{e^{i\varphi} f}{i\varphi'}(a) + \int_a^\tau e^{i\varphi} g.$$

We will be interested in giving meaning to the expression $\int_a^\infty e^{i\varphi} f$. If we could ignore the upper boundary term

$$\lim_{\tau \to \infty} [e^{i\varphi} f / i\varphi'](\tau),$$

we would then reduce an integral $\int e^{i\varphi} f$, where f is a symbol of order r to an integral $\int e^{i\varphi} g$ where g is a symbol of order $r - 1$. If f were a function of compact support (or a symbol of degree less than -1) such an integration by parts would always be legitimate, since there would be no boundary term at infinity. For a general symbol of degree r we *define* the integral $\int^\infty e^{i\varphi} f$ by n successive integrations by parts so that $r - n < -1$. This gives a procedure for regularizing the integral. One can justify this regularization procedure by showing that the symbols of compact support are dense, in the appropriate topology, and that this procedure is an extension by continuity; cf. Hormander [4].

Thus, for example the integral

$$\int_0^\infty e^{ixt} t^{\lambda - 1}\, dt$$

makes sense for $x \neq 0$ and any value of λ. (Here an integration by parts might be necessary at $t = 0$ as well.)

To evaluate this integral we notice that for $x > 0$ a change of variables shows that

$$\int_0^\infty e^{ixt} t^{\lambda - 1}\, dt = x^{-\lambda} \int_0^\infty e^{is} s^{\lambda - 1}\, ds = c_\lambda x^{-\lambda}$$

where

$$c_\lambda = \int_0^\infty e^{is} s^{\lambda - 1}\, ds = \int e^{iz + (\lambda - 1)\log z}\, dz$$

where the integration is carried out along the positive real axis. By Cauchy's theorem, if $\lambda > 0$ we may replace the positive real axis by the positive imaginary axis. Thus

$$c_\lambda = e^{\pi\lambda i/2}\int_0^\infty e^{-s}s^{\lambda-1}\,ds = e^{\pi\lambda i/2}\Gamma(\lambda),$$

if $\lambda > 0$. For complex values of λ (other than $\lambda = -1, -2, \ldots$) this still holds by analytic continuation.

For $x < 0$,

$$\int_0^\infty e^{ixt}t^{\lambda-1}\,dt = |x|^{-\lambda}\int_0^\infty e^{-is}s^{\lambda-1}\,ds$$

and we must analytically continue to the negative real axis to obtain

$$|x|^{-\lambda}e^{-\pi\lambda i/2}\Gamma(\lambda) = c_\lambda e^{-\pi\lambda i}|x|^{-\lambda}.$$

Notice that

$$\lim_{y\to 0^+}(x+iy)^\mu = \begin{cases} x^\mu, & x > 0,\\ e^{\pi\mu i}|x|^\mu, & x < 0\end{cases}.$$

If we write $(x + i0)^\mu$ for this limit then we have

$$\int_0^\infty e^{ixt}t^{\lambda-1}\,dt = e^{\pi\lambda i/2}\Gamma(\lambda)(x+i0)^{-\lambda} = \Gamma(\lambda)[e^{\pi i\lambda/2}x_+^{-\lambda} + e^{-\pi\lambda i/2}x_-^{-\lambda}]$$

for $x \neq 0$. Notice that as a *distribution* $(x + i0)^\mu$ makes sense even near $x = 0$. Indeed, for $\mu > 0$ the function

$$(x+i0)^\mu = \begin{cases} x^\mu, & x > 0,\\ e^{\pi\mu i}|x|^\mu, & x < 0\end{cases}$$

is integrable near $x = 0$ so that

$$\int_{-\infty}^\infty a(x)(x+i0)^\mu\,dx = \langle a, (x+i0)^\mu\rangle$$

is well defined for any $a \in C_0^\infty(\mathbf{R}^1)$.

For $\mu < 0$ the integral

$$\langle a, (x+i0)^\mu\rangle = \frac{e^{\pi i\mu}}{\Gamma(-\mu)}\int_{-\infty}^\infty\int_0^\infty e^{ixt}a(x)t^{-\mu-1}\,dx\,dt$$

causes no trouble for t near zero, and can be regularized for large values of $|t|$ by integration by parts with respect to x.

Thus $(x + i0)^\lambda$ is a generalized function which depends holomorphically on λ. Its wave front set consists of $(0, \xi)$, $\xi > 0$. Also, it is homogeneous of degree λ in the sense that

$$M_a^*(x+i0)^\lambda = a^\lambda (x+i0)^\lambda, \qquad a>0,$$

where $M_a : \mathbf{R} \to \mathbf{R}$ denotes multiplication by a. We refer the reader to Gelfand-Shilov [1] for a detailed study of all the homogeneous distributions.

Since the wave front set of $(x+i0)^\lambda$ is the simplest possible, we shall use *these* distributions, or ones quite close to them, as our basic distributions on $\mathbf{R}^1$ rather than δ and its derivatives. Actually, we will be interested in the behavior of generalized functions, u, modulo smooth functions, and hence it is the behavior of $\hat{u}$ at infinity rather than at zero which will be of interest to us. We therefore proceed as follows: Let χ_+ be a C^∞-function of a real variable such that $\chi_+(\tau) = 0$ for $\tau < \epsilon$ ($\epsilon > 0$) and $\chi_+(\tau) = 1$ for all large values of τ. We shall use χ_+ to denote any function of this type. Similarly we define χ_- by $\chi_-(\tau) = \chi_+(-\tau)$. For each m we introduce the (class of) generalized functions

$$v_m^\pm(x) = \int_{\mathbf{R}} \chi_\pm(\tau) e^{i\tau x} |\tau|^m \, d\tau. \tag{4.5}$$

Thus $WF(v_m^\pm) = \Lambda_0^\pm$ where Λ_0^+ is the set of positive covectors in $T^*\mathbf{R}_0$ and Λ_0^- the set of negative covectors.

The generalized functions $v_m^\pm$ are homogeneous of degree $-m-1$ modulo smooth functions

$$M_a^* v_m^\pm \equiv a^{-m-1} v_m^\pm \qquad a>0$$

where $\equiv$ means that the two sides differ by a smooth function. Notice that

$$v_m^+ \star v_n^+ \equiv v_{m+n}^+, \quad v_m^- \star v_n^- \equiv v_{m+n}^- \quad \text{and} \quad v_m^+ \star v_n^- \equiv 0 \tag{4.6}$$

as can be read off directly from the fact that the Fourier transform carries convolution into multiplication. Observe also that we have the formulas (for $m \neq$ negative integer)

$$v_m^+ \equiv e^{\pi i(m+1)/2}(x+i0)^{-m-1} = i\Gamma(m+1)[e^{\pi i m/2} x_+^{-m-1} - e^{-\pi i m/2} x_-^{-m-1}] \tag{4.7}$$

and

$$\begin{aligned} v_m^- &\equiv e^{-\pi i(m+1)/2}(x-i0)^{-m-1} \\ &= i\Gamma(m+1)[-e^{-\pi i m/2} x_+^{-m-1} + e^{\pi i m/2} x_-^{-m-1}]. \end{aligned} \tag{4.8}$$

Solving for $x_+^{-1/2}$ and $x_-^{-1/2}$ gives

$$x_+^{-1/2} = \tfrac{1}{2}[(x+i0)^{-1/2} + (x-i0)^{-1/2}] \equiv \frac{1}{2\Gamma(\frac{1}{2})}[e^{-\pi i/4} v_{-1/2}^+ + e^{\pi i/4} v_{-1/2}^-] \tag{4.9}$$

and

$$\begin{aligned} x_-^{-1/2} &= \frac{i}{2}[(x+i0)^{-1/2} - (x-i0)^{-1/2}] \\ &\equiv \frac{i}{2}\frac{1}{\Gamma(\frac{1}{2})}[e^{-\pi i/4}v_{-1/2}^{+} - e^{i\pi/4}v_{-1/2}^{-}] \\ &= \frac{1}{2\Gamma(\frac{1}{2})}[e^{\pi i/4}v_{-1/2}^{+} + e^{-\pi i/4}v_{-1/2}^{-}]. \end{aligned} \tag{4.10}$$

Notice that if $h : \mathbf{R} \to \mathbf{R}$ is inversion,

$$h(x) = -x,$$

then $h^* x_+ = x_-$ so it follows from (4.7) and (4.8) (for non-integral m–then by analytic continuation for all m) that

$$h^* v_m^+ = v_m^-, \qquad h^* v_m^- = v_m^+. \tag{4.11}$$

Now let X be an arbitrary differentiable manifold and let Λ be a homogeneous Lagrangian submanifold of T^*X. We wish to associate a class of distributions on X to Λ. These distributions will all have their wave fronts on Λ and will be given in terms of local descriptions; that is, we will allow sums Σu where each u is compactly supported in an open set on X over which we are given a local parameterization of Λ:

$$\begin{array}{ccc} Z & \xrightarrow{\ f\ } & \mathbf{R} \\ \pi \downarrow & & \\ X & & \end{array} \qquad \begin{aligned} \Lambda &= \pi_* f^* T^* \mathbf{R}_0 \\ &= \pi_* N(f^{-1}(0)) \end{aligned}$$

such that

$$u = \pi_* a f^* v_m^{\pm}.$$

Let us use (4.1) and (4.2) to discuss the independence of this class of the choice of local parametrization.

We know from Chapter IV that any two parametrizations of the same Lagrangian submanifold can be obtained from one another by applying (up to diffeomorphism) a succession of each of the following two operations:

(i) replacing f by $g = bf$ where $b : Z \to \mathbf{R}$ is smooth and does not vanish on $f^{-1}(0)$;

(ii) replacing Z by $Z \times \mathbf{R}$ and f by g where $g(z, w) = f(z) \pm w^2$.

Let us examine the effect of (i). Consider the diagram

$$\begin{array}{ccc} \mathbf{R} \times \mathbf{R}^+ & \xrightarrow{\ M\ } & \mathbf{R} \\ \pi \downarrow & & \\ \mathbf{R} & & \end{array} \qquad \begin{aligned} M(x, c) &= cx, \\ \pi(x, c) &= x, \end{aligned}$$

so that $\pi^* v_m^{\pm}(x) = v_m^{\pm}(x)$ and $M^* v_m^{\pm}(x) \equiv c^{-m-1} v_m^{\pm}(x)$ or

$$M^* v_m^{\pm} = c^{-m-1} M^* v_m^{\pm} + \text{smooth function} .$$

Now (assuming $b > 0$ and reversing the signs if $b < 0$)

$$g = M \circ (f, b) \qquad \text{where } (f, b) : Z \to \mathbf{R} \times \mathbf{R}^+$$

and hence

$$g^* v_m^{\pm} = (f, b)^* M^* v_m^{\pm} = b^{-m-1} f^* v_m^{\pm} + \text{smooth function} .$$

Thus, up to smooth functions, the class is preserved under (i).

Now let us examine (ii). Since $df \neq 0$ we may introduce coordinates $x^1, \cdots, x^n$ on Z with $x^1 = f$. We will set $x = x^1$ and let $x' = (x^2, \ldots, x^n)$ denote the remaining coordinates. We wish to compare

$$\begin{array}{ccc} Z \times \mathbf{R} & \xrightarrow{\quad g \quad} & \mathbf{R} \\ \pi \downarrow & & \\ X & & \end{array} \qquad \text{with} \qquad \begin{array}{ccc} Z & \xrightarrow{\quad f \quad} & \mathbf{R} \\ \pi \downarrow & & \\ X & & \end{array}$$

which we can combine as

$$\begin{array}{ccccc} Z \times \mathbf{R} & & \overset{(x',x,w)}{\mathbf{R}^{n-1} \times \mathbf{R} \times \mathbf{R}} & \longrightarrow & \overset{x \pm w^2}{\mathbf{R}} \\ \pi_Z \downarrow & & \downarrow & \to & \underset{x}{\mathbf{R}} \\ Z & \xrightarrow{\quad \cong \quad} & \underset{(x',x)}{\mathbf{R}^{n-1} \times \mathbf{R}} & & \\ \pi_X \downarrow & & & & \\ X & & & & \end{array}$$

and wish to examine $\pi_{X*} \pi_{Z*} a g^* v_m^{\pm}$. We are clearly back in the one dimensional situation with x' playing the role of a parameter. We can apply $(4.3)_{\pm}$ to conclude

$$\pi_* a g_{\mp}^* v_m^{\pm} = \pi_* \int b(x, x', y) y_{\pm}^{-1/2} v_m^{\pm}(x - y)\, dy \tag{4.12}$$

where $2b(x, x', y^2) = a(x, x', y) + a(x, x', -y)$. Let us expand b in a Taylor series about $y = 0$:

$$b(x, x', y) = b_0(x, x') + b_1(x, x')y + \cdots + b_N(x, x')y^N + b_{N+1}(x, x', y)y^{N+1} .$$

Now the remainder term $b_{N+1}(x, x', y)y^{N+1} y_{\pm}^{-1/2} = b(x, x', y)y_{\pm}^{N-1/2}$ can be made as smooth as we please by choosing N large. Each of the other terms will yield in (4.8) an expression of the form $\pi_* u$ where

$$u = b_k(x, x')[y_{\pm}^{k-1/2} \star v_m^{\pm}].$$

Since $y_{\pm}^{k-1/2}$ is a linear combination of $v_{k-1/2}^{\pm}$ we may apply (4.6) to conclude that $\pi_* u$ is of the desired type. Notice that we lose the 'gradation' but keep the associated 'filtration' at least shifted by $\pm\frac{1}{2}$ when we change our parametrization. This suggests that we replace finite sums locally by series which converge modulo one local filtration (and hence all). We may also replace the expression

$$v_m^+(x) = \int \chi_+(\tau)\tau^m e^{i\tau x} d\tau$$

by

$$r_m^+(x) = \int \chi_+(\tau)s(\tau)e^{i\tau x}\, d\tau$$

where s is an arbitrary symbol of degree τ. Then we can write the local expression for $\pi_* af^* r_m$ as follows: Let $y_1, \cdots, y_d$ be coordinates on X and $\theta_1, \cdots, \theta_q$ be fiber coordinates on Z so that $y_1, \cdots, y_d, \theta_1, \cdots, \theta_q$ are coordinates on Z. Then

$$\pi_* af^* r_m = \iint a(y,\theta)\chi_+(\tau)e^{i\tau f(y,\theta)}s(\tau)\, d\tau d\theta.$$

With minor changes (passing from the 'inhomogeneous' to the 'homogeneous' coordinates) these are the 'Fourier integral operators' of Hormander [4].

Finally notice that the constant term in the Taylor expansion of b is just $a(x',x,0)$. If we combine (4.12) with (4.9) and (4.10) and expand $a(x',x,0) = a(x',0,0) + x[\cdots]$ we see that

$$\pi_* ag_{\pm}^* v_m^+ = \pi_*\left[a(x',0,0)\frac{1}{2\sqrt{\pi}}e^{\pm i\pi/4}v_{m-1/2}^+\right] + \cdots \tag{4.13}$$

where $\pm$ corresponds to $\pm w^2$ in $g(z,w) = f(z) \pm w^2$, and

$$\pi_* ag_{\pm}^* v_m^- = \pi_*\left[a(x',0,0)\frac{1}{2\sqrt{\pi}}e^{\pm i\pi/4}v_{m-1/2}^-\right] + \cdots. \tag{4.14}$$

We will denote by $I(X,\Lambda)$ the space of all distributions, μ, on X which can be written, to an arbitrary order of smoothness, as a locally finite sum of distributions of the form $\pi_* a\varphi^* v_m^+$ where

$$\begin{array}{ccc} Z & \xrightarrow{\quad\varphi\quad} & \mathbf{R} \\ \pi\downarrow & & \\ X & & \end{array}$$

is a local parametrization of Λ. We let degree $\mu = k$ where $k + (d-1)/2$ is the largest of the integers m occurring in this sum, d being the fiber dimension of Z. (Notice that according to this convention v_m^+ has degree $m + 1/2$.) It is clear from above that the degree doesn't depend on the parametrization. Denote by $I_k(X,\Lambda)$ the space of all distributions of degree $\leq k$ so that $I(X,\Lambda)$ is filtered by k,

$$\cdots \supset I_k(X,\Lambda) \supset I_{k-1}(X,\Lambda) \supset \cdots.$$

The intersection of all the I_k's is just $C^\infty(X)$. We now establish that this filtration has the so called 'Mittag-Leffler' property:

PROPOSITION 4.1. *Given a sequence* $\mu_k, \mu_{k-1}, \ldots$ *with* $\mu_l \in I_l(X, \Lambda)$, *there exists a* $\mu \in I_k(X, \Lambda)$ *such that*

$$\mu - \sum_N^k \mu_l \in I_{N-1}(X, \Lambda)$$

for all $N \leq k$.

PROOF. The problem is local, so we can describe Λ by a parametrization

$$\begin{array}{ccc} X \times \mathbf{R}^N & \xrightarrow{\quad\varphi\quad} & \mathbf{R} \\ \pi \downarrow & & \\ X & & \end{array}$$

and, with $r = m + (N - 1/2)$ we can write

$$\mu_i = \pi_* a_i \varphi^* \nu_{r-i} = \int_0^\infty a_i(x, z) e^{is\varphi(x,z)} s^{r-i} \chi(s)\, ds dz$$

where $\chi = 0$ for s small and $\chi = 1$ for s large. Let

$$a(x, z, s) = \sum a_i(x, z) s^{r-i} \chi(s/M_i),$$

the M_i going to infinity very fast. Then

$$\mu = \int a(x, z, s) e^{is\varphi(x,z)}\, ds dz$$

has the required properties.

§5. The symbol calculus.

We have associated, to each Lagrangian submanifold of T^*X a class of generalized functions. This class comes equipped with a filtration associated to every local parametrization. When we shift from one parametrization to another the filtrations are preserved, and the top order terms transform according to the relatively simple law, (4.13). This suggests that we be able to associate a geometric object on Λ to each generalized function associated to Λ and that this association be functorial, i.e., behave consistently with respect to push forward and pull back. Our experience in Chapters II, IV and V has indicated that the geometric objects are likely to be half densities or half forms, and this turns out to be the case. We therefore pause briefly to describe some basic functorial properties of morphisms of these objects.

Recall that if $f : X \to Y$ is a smooth map then we say that f is a morphism on half densities if we are also given a smooth section

$$r(x) \in \operatorname{Hom}\,(|\wedge|^{1/2} TY_{f(x)}, |\wedge|^{1/2} TX_x).$$

Similarly if X and Y are metalinear manifolds then

$$r(x) \in \operatorname{Hom}\,(\wedge^{1/2}\, TY_{f(x)}, \wedge^{1/2}\, TX_x).$$

We now observe that giving such an r is the same as giving a certain kind of half density (or half form) on the normal bundle to graph f.

Indeed, a point of $\mathfrak{N}(\text{graph } f)$ can be identified as $z = (x, f(x), -df^*\eta, \eta)$ where $\eta \in T^* Y_{f(x)}$; and thus we have the exact sequence determined by the projection onto graph f and then onto X,

$$0 \to T^* Y_y \to T(\mathfrak{N}\text{ graph } f)_z \to TX_x \to 0.$$

(where we have identified $TT^* Y_y$ with $T^* Y_y$). Therefore

$$\begin{aligned} |\wedge|^{1/2}\mathfrak{N}(\text{graph } f)_z &\cong |\wedge|^{1/2} TX_x \otimes |\wedge|^{-1/2} TY_{f(x)} \\ &= \operatorname{Hom}\,(|\wedge|^{1/2} TY_{f(x)}, |\wedge|^{1/2} TX_x). \end{aligned}$$

Thus a morphism determines a half density on $\mathfrak{N}(\text{graph } f)$ one which is ‘ constant along the fibers ’ of $\mathfrak{N}(\text{graph } f)$ over graph f. Conversely, given such a half density which is constant in $\mathfrak{N}$ we get a morphism from X to Y. Similarly, if X and Y are metalinear manifolds then $T^*(X \times Y)$ is a metaplectic manifold and the Lagrangian submanifold $\mathfrak{N}(\text{graph } f)$ carries a metalinear structure. We have, as above,

$$\wedge^{1/2}(\mathfrak{N}\text{ graph } f)_z = \wedge^{1/2} TX_x \otimes \wedge^{-1/2} TY_{f(x)} = \operatorname{Hom}\,(\wedge^{1/2} TY_{f(x)}, \wedge^{1/2} TX_x).$$

We now wish to use this identification to show how a morphism $f : X \to Y$ induces a push forward operation, f_*, from half densities (forms) on $\Lambda_X \subset T^* X$ to half densities (forms) on $f_* \Lambda_X$ (if the transversality condition required for $f_* \Lambda_X$ to be a Lagrangian submanifold is satisfied) and a pull back operation, f^*, from half forms (densities) on $\Lambda_Y \subset T^* Y$ to half forms (densities) on $f^* \Lambda_Y \subset T^* X$. Recall the definition of push forward or pull back for Lagrangian submanifolds. They are given as fiber products in the following diagrams where

$$\Lambda = \mathfrak{N}(\text{graph } f) \subset T^*(X \times Y) = T^* X \times T^* Y$$

and the transversality requirement is that the maps be transversal.

$$\begin{array}{ccc} \Lambda_X & \dashleftarrow & f_* \Lambda_X \\ \downarrow & & \vdots\downarrow \\ T^* X & \leftarrow & \Lambda \end{array} \qquad\qquad \begin{array}{ccc} \Lambda_Y & \dashleftarrow & f^* \Lambda_Y \\ \downarrow & & \vdots\downarrow \\ T^* Y & \leftarrow & \Lambda \end{array}$$

push forward pull back

We claim the following: Let $A \to B$ and $C \to B$ be transversal maps and D their fiber product:

$$\begin{array}{ccc} A & \longleftarrow & D \\ \downarrow & & \downarrow \\ B & \longleftarrow & C \end{array}$$

Suppose we are given a half density on A and on C and a negative half density on B. Then we get a half density on D, and similarly for metalinear manifolds and half forms. In fact, to say that $A \to B$ and $C \to B$ are transversal means that $A \times C \to B \times B$ is transversal to the diagonal, Δ, and then D is the inverse image of the diagonal. We thus have for $d = (a, c) \in D$ the exact sequence

$$0 \to TD_d \to TA_a \oplus TC_c \to TB_b \oplus TB_b/T\Delta_b \to 0.$$

Since $TB_b \oplus TB_b/T\Delta_b \sim TB_b$ we conclude that

$$|\wedge|^{1/2} TD_d \cong |\wedge|^{1/2} TA_a \otimes |\wedge|^{1/2} TC_c \otimes |\wedge|^{-1/2} TB_b,$$

and similarly for $\wedge^{1/2}$. We now observe that T^*X (or T^*Y) carries a canonical volume form coming from the highest non-vanishing exterior power of the symplectic form. Thus there is a canonical half form, negative half density, etc. on T^*X. Thus given a morphism $f: X \to Y$ we have a half density (form) on Λ and a negative half density (form) on T^*X. Thus a half density (form) on Λ_X determines, in the push forward diagram, a half density (form) on $f_*\Lambda_X$. Actually we will be mainly interested in the case where f: $X \to Y$ is a submersion in which case it will be convenient, in view of our eventual use of (4.13), to multiply the half form (density) so obtained from a half form (density), σ, on Λ_X via the push forward diagram by $(e^{i\pi/4}/\sqrt{2\pi})^k$ where k is the fiber dimension and denote this form by $f_*\sigma$.

For a half form (density), μ, on Λ_Y we simply apply the pull back diagram (without any normalizing factors) to obtain $f^*\mu$, a half form (density) on $f^*\Lambda_Y$.

Let us illustrate the pull back operation by computing it in the special case where $Y = \mathbf{R}$ and $\Lambda_Y = \Lambda_0^+ = T^*\mathbf{R}_0^+$ consists of positive covectors at the origin. Let us take $\alpha = |d\xi|^{1/2}$ as the half density on Λ_Y. Here x is the standard linear coordinate on $\mathbf{R}$ and ξ is the dual coordinate. If $f: X \to \mathbf{R}$ is a submersion and $S = f^{-1}(0)$ then $f^*\Lambda_0^+$ consists of all positive normal covectors, ξdf, $\xi > 0$, to S. The morphism f maps $|dx|^{1/2}$ into a half density, ν, on X. At points of S we can write $\nu = \mu \otimes |df|^{1/2}$ where μ is a half density on S. We claim that

$$f^*|d\xi|^{1/2} = \mu \otimes |d\xi|^{1/2}$$

where we identify $f^*\Lambda_0^+$ with $S \times \Lambda_0^+$.

Indeed, let us introduce local coordinates (x, y) where

$$x = x^1 \quad \text{and} \quad y = (x^2, \ldots, x^n)$$

on X so that $f(x, y) = x$ and S is given by $x = 0$. Then $\mathfrak{N}(\text{graph } f)$ consists of all points of the form $(x, y;\ \xi, 0) \times (x, -\xi)$ where (ξ, η) are dual to (x, y). In terms of these coordinates $T(\mathfrak{N}(\text{graph } f))$ is spanned by all vectors of the form

$$u_{y,v} = (0,1;0,0) \times (0,0),$$

$$u_\xi = (0,0;\ 1,0) \times (0,-1)$$

and

$$u_x = (1,0;0,0) \times (1,0).$$

In the entire computation we may treat y as a parameter and so forget $u_{y,v}$ and regard $\mathfrak{N}(\text{graph } f)$ as consisting of all points of the form $(x, x; \xi, -\xi)$ in $\mathbf{R}^4 = T^*\mathbf{R} \times T^*\mathbf{R}$.

Then $\mathfrak{N}(T \text{ graph } f)_z = C$ consists of all vectors of the form $(a, -b, a, b) \in T^*\mathbf{R} \times T^*\mathbf{R}$ where $z \in \mathfrak{N}(\text{graph } f)$. Let $A = T\Lambda_0^+ \subset T^*\mathbf{R}$ so that $A \subset \mathbf{R}^2$ consists of all vectors of the form $(0, c)$. Then we have

$$0 \to D \to A \oplus C \to B \oplus B/\Delta \to 0$$

where $B = \mathbf{R}^2$ and the maps $C \to B$, $A \to B$ are given by $(a, -b, a, b) \to (a, b)$ and $(0, c) \to (0, c)$. Thus in $A \oplus C = \mathbf{R}^3$ we have

$$D = \{(0, c, c)\}.$$

Let $v = (0, 1, 1)$. Our general procedure then says that

$$\left\langle \frac{\partial}{\partial \xi}, f^* |d\xi|^{1/2} \right\rangle = \gamma \otimes \alpha(w_1, w_2, v)\beta(w_1, w_2)$$

where $\gamma \otimes \alpha = |dx|^{1/2} \otimes |d\xi|^{1/2} \otimes |d\xi|^{1/2}$ and $\beta = |dx|^{-1/2} \otimes |d\xi|^{-1/2}$ and w_1, w_2 are any vectors such that w_1, w_2, v span $A \oplus C$. We may choose $w_1 = (1,0,0)$ and $w_2 = (0,1,0)$ and conclude that the above expression is indeed 1. We have thus proved:

Let $f : X \to \mathbf{R}$ be a submersion and $\Lambda_0^+ = T^*\mathbf{R}_0^+$ so that $f^*\Lambda_0$ consists of positive multiples of df at $S = f^{-1}(0)$. We may identify $f^*\Lambda_0^+$ with $S \times \Lambda_0^+$. If f is a morphism of half densities we may write

$$f^*|dx|_s^{1/2} = \mu_s \otimes |df|_s^{1/2} \qquad \text{for } s \in S.$$

Then

$$f^* |d\xi|^{1/2} = \mu \otimes |d\xi|^{1/2} \tag{5.1}$$

where ξ is the dual coordinate to x. If X is metalinear and f is a morphism of half forms then we have

$$f^* d\xi^{1/2} = \mu \otimes d\xi^{1/2}, \qquad f^* \overline{d\xi^{1/2}} = \mu \otimes \overline{d\xi^{1/2}}. \tag{5.2}$$

We are now in a position to state the main theorem of this section. In what follows we let $I_m(X, \Lambda)$ denote the space of generalized half forms on X of degree m associated to Λ (the degree being defined as in the last section). We assume, for simplicity, that all manifolds X, Y, etc. are metalinear. We let $S^m(\Lambda)$ denote the space of homogeneous half forms on Λ of degree m. (Here if μ is a half form on Λ and $M_t : \Lambda \to \Lambda$ is (the restriction to Λ of) multiplication by t then we say that μ is homogeneous of degree m when $M_t^* \mu = t^m \mu$.)

THEOREM 5.1. *There is a unique functor* $\sigma : I_m(X, \Lambda)/I_{m-1}(X, \Lambda) \to S^m(\Lambda)$ *characterized by the following*:

(a) *If* $f : X \to Y$ *is transversal to* Λ_Y *so that* $f^* \nu \in I_m(X, f^* \Lambda_Y)$ *for* $\nu \in I_m(Y, \Lambda_Y)$ *then*

$$\sigma(f^* \nu) = f^* \sigma(\nu). \tag{5.3}$$

(b) *If* $f : X \to Y$ *is a submersion transversal to* $\Lambda_X \subset T^* X$ *so that* $f_* \mu \in I_{m-d/2}(Y, f_* \Lambda_X)$ *we have*

$$\sigma(f_* \mu) = f_* \sigma(\mu). \tag{5.4}$$

(c) *For* $\Lambda_0 = T^* \mathbf{R}_0^+$ *and with* x, ξ *the canonical coordinates on* $T^* \mathbf{R}$ *we have*

$$\sigma(v_m^+ dx^{1/2}) = \xi^m d\xi^{1/2}. \tag{5.5}$$

PROOF. We begin by showing that (5.5) is consistent with (5.3) and (5.4) for the special case of Λ_0^+. That is, suppose we consider the diagram

$$\begin{array}{ccc} \mathbf{R} \times \mathbf{R} & \xrightarrow{g} & \mathbf{R} \\ \pi \downarrow & & \\ \mathbf{R} & & \end{array} \qquad \begin{aligned} g &= g_\pm = x \pm w^2, \\ g^* dx^{1/2} &= (dx \wedge dw)^{1/2}. \end{aligned}$$

and compare $\sigma(\pi_* a g^* v_m^+)$ with $\pi_*(a g^* \sigma(v_m^+))$. By (4.13) and (4.14) we have

$$\begin{aligned} \sigma(\pi_* a g_\pm^* v_m^+) &= \sigma\left(\frac{a(0)}{2\sqrt{\pi}} e^{\pm i\pi/4} v_{m-1/2}^+\right) \\ &= \frac{a(0)}{2\sqrt{\pi}} e^{\pm i\pi/4} \xi^{m-1/2} d\xi^{1/2} \qquad \text{on } \Lambda_0^+. \end{aligned}$$

Let us now compute

$$\pi_* ag^* \sigma(v_m^+) = \pi_* ag^* \xi^m d\xi^{1/2} \qquad \text{on } \Lambda_0^+.$$

We first compute $g^* \xi^m d\xi^{1/2}$. By formula (5.2) this is given by $\mu \otimes \xi^m d\xi^{1/2}$ where μ is the induced half form on the parabola $x \pm w^2 = 0$, from the half form $(dx \wedge dw)^{1/2}$ on $\mathbf{R} \times \mathbf{R}$. Since $d(x \pm w^2) \wedge dw = dx \wedge dw$ and since we can use w as coordinates on the parabola we have

$$g^* \xi^m d\xi^{1/2} = dw^{1/2} \otimes \xi^m d\xi^{1/2}.$$

We now must multiply by a and apply the push forward operation corresponding to the diagram

$$\begin{array}{ccc} g^* \Lambda_0 & \leftarrow & \pi_* g^* \Lambda_0 = \Lambda_0 \\ \downarrow & & \downarrow \\ T^*(\mathbf{R} \times \mathbf{R}) & \leftarrow & \mathfrak{N}(\text{graph } \pi) \end{array}$$

The corresponding local coordinates and half forms are as follows:

On $T^*(\mathbf{R} \times \mathbf{R})$ — coordinates $(x, \xi; w, \eta)$,

— half form $(d\xi \wedge dx)^{1/2} \otimes (d\eta \wedge dw)^{1/2}$.

On $g^* \Lambda_0^+$ — coordinates w_1, ξ_1 where $g^* \Lambda_0 \subset T^*(\mathbf{R} \times \mathbf{R})$ consists of all points of the form $(\mp w_1^2, w_1, \xi_1, \pm 2 w_1 \xi_1)$

— half form $dw_1^{1/2} \otimes d\xi_1^{1/2}$.

On $\mathfrak{N}(\text{graph } \pi)$ — coordinates x_2, w_2, ξ_2.

Here we regard $\mathfrak{N}(\text{graph } \pi) \subset T^*(\mathbf{R} \times \mathbf{R}) \times T^*\mathbf{R} = T^*(\mathbf{R} \times \mathbf{R} \times \mathbf{R})$ as consisting of all points of the form $(x_2, w_2, x_2; \xi_2, 0, -\xi_2)$. The morphism π associates $(dx_2 \wedge dw_2)^{1/2}$ to $dx^{1/2}$ and hence its corresponding half form on $\mathfrak{N}(\text{graph } \pi)$ is $(dx_2 \wedge dw_2 \wedge d\xi_2)^{1/2}$.

By our general prescription we must proceed as follows: We observe that indeed the set of points in $g^* \Lambda_0^+ \times \mathfrak{N}(\text{graph } \pi)$ which map onto the same point of $T^*(\mathbf{R} \times \mathbf{R})$ consists exactly of those points with $w = 0$, $x = 0$, and the projection onto $T^*\mathbf{R}$ yields Λ_0^+.

Now let $z = (0, \xi)$ be a point of Λ_0^+, and let $\partial/\partial\xi$ be a tangent vector at z. Its image in $Tg^* \Lambda_0 \oplus T\mathfrak{N}(\text{graph } \pi)$ is

$$u_\xi = \left(0, \frac{\partial}{\partial \xi}\right) \oplus \left(0, 0, \frac{\partial}{\partial \xi}\right) = (0, 1) \oplus (0, 0, -1).$$

We adjoin the vectors

$$\begin{aligned} v_1 &= (1,0) \oplus (0,0,0), \\ v_2 &= (0,0) \oplus (-1,0,0), \\ v_3 &= (0,0) \oplus (0,-1,0) \end{aligned}$$

and

$$v_4 = (0,0) \oplus (0,0,-1).$$

The images of v_1, v_2, v_3, v_4 in $T^*(\mathbf{R}^2) \cong T^*(\mathbf{R}^2) \times T^*\mathbf{R}^2/\Delta$ are

$$w_1 = \pm 2w \frac{\partial}{\partial x} + \frac{\partial}{\partial w} \pm 2\xi \frac{\partial}{\partial \eta}, \quad w_2 = \frac{\partial}{\partial x}, \quad w_3 = \frac{\partial}{\partial w}, \quad w_4 = \frac{\partial}{\partial \xi},$$

so

$$[d\xi^{1/2} \otimes dw^{1/2} \otimes (dx \wedge dw \wedge d\xi)^{1/2}] \cdot (u_\xi, v_1, v_2, v_3, v_4)$$
$$\cdot\, [d\xi \wedge dx \wedge d\eta \wedge dw]^{-1/2} (w_1, w_2, w_3, w_4) = \frac{1}{\sqrt{\pm 2\xi}} \cdot$$

Thus

$$\pi_* a g_+^* \xi_+^m d\xi^{1/2} = \frac{e^{+i\pi/4}}{\sqrt{2\pi}} \frac{1}{\sqrt{2\xi}} \xi_+^m d\xi^{1/2} = \frac{e^{+i\pi/4}}{2\sqrt{\pi}} \xi_+^{m-1/2} d\xi^{1/2}$$

which is consistent.

For g_- we get the same answer multiplied by $(-1)^{-1/2}$ establishing the consistency of the formulas.

We have established that replacing the identity parametrization of Λ_0^+ with the one giving Λ_0^+ as $\pi_* g_\pm^* \Lambda_0^+$ does not change the symbol map and is consistent with (5.3), (5.4) and (5.5). We will now use this fact to show that replacing any local parametrization

$$\begin{array}{ccc} Z & \xrightarrow{f} & \mathbf{R} \\ \pi_1 \downarrow & & \\ X & & \end{array} \quad \text{by} \quad \begin{array}{ccc} Z \times \mathbf{R} & \xrightarrow{f'_\pm} & \mathbf{R} \\ \pi \downarrow & & \\ Z & & \\ \downarrow \pi_1 & \pi'_1 & \\ X & & \end{array} \qquad f_\pm(z,t) = f(z) \pm t^2$$

also is consistent. Indeed, if we introduce coordinates $z_1 = f, z_2, \ldots, z_n$, we are back to the previous argument with $z_2, \ldots, z_n$ playing the role of parameters. Since all parametrizations of Λ_0^+ itself can be obtained by a sequence of such operations this proves the consistency for Λ_0^+. Applied to any Λ this will then show that σ is well defined for any Λ. For this we will use the following lemma:

LEMMA 5.1. *Let $\Lambda \subset T^*X$ be a homogeneous Lagrangian submanifold and $0 \neq \lambda \in \Lambda$. Then there exists a map $f : X \to \mathbf{R}$ which is a submersion near $\pi\lambda$ and is transversal to Λ such that $f_* \Lambda = T^*\mathbf{R}_0^+$ and $f_*(\lambda) = (0, 1)$.*

Assume the lemma for the moment. Suppose that

$$\begin{array}{ccc} Z & \xrightarrow{g} & \mathbf{R} \\ \pi \downarrow & & \\ X & & \end{array}$$

is a local parametrization of Λ and that $u = \pi_* a g^* v$. Then

$$\begin{array}{ccc} Z & \xrightarrow{g} & \mathbf{R} \\ \pi \downarrow & & \\ X & & \\ f \downarrow & & \\ \mathbf{R} & & \end{array}$$

gives a parametrization of Λ_0 and thus

$$\sigma(f_* \pi_* a g^* v) = \sigma(f_* u)$$

depends only on u. On the other hand we know from the invariance on Λ_0 that

$$\begin{aligned} \sigma(f_* \pi_* a g^* v) &= \sigma((f \circ \pi)_* a g^* v) \\ &= (f \circ \pi)_* a g^* \sigma(v) \\ &= f_* \circ \pi_* a g^* \sigma(v). \end{aligned}$$

But this last expression determines the value of $\pi_* a g^* \sigma(v)$, is independent of the particular parametrization and is determined by u.

We have thus established that σ is well defined. Let us now examine behavior under pull backs. Let $\Lambda_Y \subset T^*Y$ be a homogeneous Lagrangian submanifold and let $g : X \to Y$ be a smooth map transversal to Λ_Y. Let

$$\begin{array}{ccc} Z & \xrightarrow{f} & \mathbf{R} \\ \pi \downarrow & & \\ Y & & \end{array}$$

be a local parametrization of Λ_Y. We can form the fiber product:

$$\begin{array}{ccc} Z' & \xrightarrow{g'} & Z \\ \pi' \downarrow & & \downarrow \pi \\ X & \xrightarrow[g]{} & Y \end{array}$$

Since g is transversal to Λ_Y it follows that $f' = f \circ g'$ is a submersion at 0 and that f' provides a parametrization for $g'^* \Lambda_Y$. Then for $v = v_m^+$ on $\mathbf{R}$,

$$g^* \pi_* af^* v = \pi'_* g'^* af^* v = \pi'_* (g'^* a) f'^* v$$

gives a parametrization of $g^* u$ where $u = \pi_* af^* v$. Thus

$$\sigma(g^* u) = \pi'_* (g'^* a)\sigma(f'^* v) = \pi'_* (g'^* a) g'^* \sigma(f^* v) = \pi'_* g'^* a\sigma(f^* v) = g^* \sigma(u).$$

This establishes the functoriality of σ under pull back.

If $g : X \to Y$ is a submersion transversal to Λ, and

$$\begin{array}{ccc} Z & \xrightarrow{f} & \mathbf{R} \\ \pi \downarrow & & \\ X & & \end{array}$$

is a local parametrization of Λ then we can simply extend the diagram to obtain

$$\begin{array}{ccc} Z & \xrightarrow{f} & \mathbf{R} \\ \pi \downarrow & & \\ X & & \\ g \downarrow & & \\ Y & & \end{array}$$

as a local parametrization of $g_* \Lambda$. Again, if

$$u = \pi_* af^* v$$

then

$$g_* u = (g \circ \pi)_* af^* v$$

so

$$\begin{aligned} \sigma(g_* u) &= \sigma[(g \circ \pi)_* af^* v] = \sigma[g_* \pi_* af^* v] \\ &= g_* \sigma[\pi_* af^* v] = g_* \sigma(u). \end{aligned}$$

We have thus proved Theorem 5.1 up to the proof of Lemma 5.1.

PROOF OF LEMMA 5.1. Choose a Lagrangian submanifold transversal to Λ and to the vertical and intersecting Λ at λ. We can locally write this submanifold as graph df, and choose f so that $f(\pi\lambda) = 0$. Then f is a submersion at $\pi\lambda$ since $df(\pi\lambda) = \lambda \neq 0$. The 'horizontal' bundle H for this submersion consists of all rdf, $r \neq 0$. Since Λ intersects graph df transversally we conclude that Λ intersects H transversally and $\Lambda \cap H$ consists of multiples of df. Thus, in a conic neighborhood of λ we have $f_* \Lambda = T\mathbf{R}_0^+$ and $f_*(\lambda) = (0, 1)$, proving Lemma 5.1 and hence the theorem.

Appendix to §5.

We will indicate briefly how to develop the symbol calculus without making use of the formalism of half-forms. This will give a symbol calculus valid for manifolds which don't admit metalinear structures (for instance real projective space). The ideas described here are due to Hormander [4].

We recall from Chapter IV, §§3 and 5 that a Lagrangian manifold $\Lambda \subset T^*X$ possesses a canonical real line bundle, $\mathfrak{M}_\Lambda$, the Maslov bundle. It is locally flat in the sense that it possesses a canonical flat connection; so one can speak of locally constant sections of Λ. It also has a number of interesting functorial properties which we discussed in Chapter IV. We recall these properties here.

Let $f: X \to Y$ be transversal to $\Lambda_Y \subset T^*Y$ and let $\Lambda_X = f^*\Lambda_Y$. By definition Λ_X comes equipped with a map into Λ_Y. We showed in Chapter IV that the Maslov bundle of Λ_X is the pull back with respect to this map of the Maslov bundle of Λ_Y. Therefore sections of $\mathfrak{M}_{\Lambda_Y}$ can be pulled back to sections of $\mathfrak{M}_{\Lambda_X}$.

Next let $\pi: Y \to X$ be a fiber mapping, and $\Lambda_X = \pi_*\Lambda_Y$. By definition Λ_X comes equipped with an immersion into Λ_Y, and $\mathfrak{M}_{\Lambda_X}$ is the pull back of $\mathfrak{M}_{\Lambda_Y}$ with respect to this immersion; so we get a push forward operation π_* (really a pull back operation) from sections of $\mathfrak{M}_{\Lambda_Y}$ to sections of $\mathfrak{M}_{\Lambda_X}$.

Finally, we recall that if Λ_X is the conormal bundle of a submanifold of X the associated Maslov bundle admits a canonical trivialization.

Hormander defines $I_k(X, \Lambda)$ as we do in §4 except that: (1) he has a convention about defining the degree, k, which differs (trivially) from ours, and (2) the $\mu \in I_k(X, \Lambda)$ are half densities rather than half forms. (Therefore, they make sense even when the manifold is not metalinear.) In Hormander's set up the symbol of $\mu \in I(X, \Lambda)$ is a section of $|\wedge|^{1/2}\Lambda \otimes \mathfrak{M}_\Lambda$. Since $\mathfrak{M}_\Lambda$ is locally constant one can speak of homogeneous symbols of this bundle of any prescribed degree of homogeneity; and, with the convention about degrees described at the end of §4, $\mu \in I_k(X, \Lambda)$ has a symbol which is homogeneous of degree k. (As we just remarked, this is *not* Hormander's degree convention.) To define this symbol we first define it for our basic distributions $v_n^+\sqrt{|dx|}$ by

$$\sigma(v_n^+\sqrt{|dx|}) = \xi_+^n\sqrt{|d\xi|}. \tag{5.6}$$

(Compare with (5.5).) The Lagrangian manifold in this example is a conormal bundle; so, as we pointed out above, its Maslov bundle has a canonical constant section. We can interpret (5.6) as a section of the half density bundle times the Maslov bundle by regarding the right hand side as tensored by this canonical constant section.

To define the symbol map in general we proceed as in the proof of Theorem 5.1. The requirement of functoriality forces the definition of the symbol map on us. We only have to check that this definition is unambiguous. This amounts to repeating the computation preceding Lemma 5.1 using the various canonical sections of the Maslov bundles in question rather than the half form formalism.

One concluding remark: On a metalinear manifold a half form, μ, which belongs to $I_k(X, \Lambda)$ determines a half form $\sigma(\mu)$ on Λ. In the Hormander set up a half density μ which belongs to $I_k(X, \Lambda)$ determines a half density times a section of the Maslov bundle. This strongly suggests that a section of the Maslov bundle is an object which converts a half density on X to a half form, takes its half form symbol on Λ and converts it back into a half density on Λ. This can be stated a little more precisely as follows. For any metalinear manifold let MZ be the line bundle $\wedge^{1/2} Z \otimes |\wedge|^{-1/2} Z$. Then for a metalinear X and a $\Lambda \subset T^* X$ the Maslov bundle ought to be:

$$\mathfrak{M}_\Lambda = M\Lambda \otimes \pi^*(MX)^{-1}$$

where $\pi : \Lambda \to X$ is the projection map. This turns out indeed to be the case, but we won't prove it here.

§6. Fourier integral operators.

Let X and Y be n dimensional manifolds and $\Phi : T^*X - 0 \to T^*Y - 0$ a homogeneous symplectic transformation. (To say that Φ is homogeneous means that $\Phi(az) = a\Phi(z)$ for $z \in T^*X - 0$.) We will describe how to associate to Φ a family of operators from $C_0^\infty(\wedge^{1/2} Y)$ to $C^\infty(\wedge^{1/2} X)$.

Let π and ρ be the projections of $X \times Y$ on X and Y respectively and let κ be a generalized section of the line bundle $\pi^* \wedge^{1/2} \otimes \rho^* \overline{\wedge}^{1/2}$. Then κ defines an operator K: $C_0^\infty(\wedge^{1/2} Y) \to C^\infty(\wedge^{1/2} X)$ by

$$\mu \to K\mu = \pi_* \kappa \rho^* \mu. \tag{6.1}$$

Notice that the product $\kappa\rho^*\mu$ is a section $\pi^* \wedge^{1/2} \otimes \rho^* |\wedge|$; so the fiber integral π_* converts it into a section of $\wedge^{1/2} X$. (Notice also that the fiber integral makes sense since $\kappa\rho^*\mu$ is compactly supported on the fibers of π.) Finally note that we can think of κ as a generalized half-form on $X \times Y$ providing we give $X \times Y$ the following metalinear structure: the product of the given metalinear structure on X and the conjugate of the given metalinear structure on Y. Since $WF(\kappa\rho^* \mu) \subset WF(\kappa)$ we get a simple criterion for the right hand side of (6.1) to be smooth, namely: $\pi_* WF(\kappa) = \emptyset$; and this is true if $WF(\kappa)$ contains no vector of the form $(x, \xi, y, 0)$.

Let Γ_Φ be the set of points $\{(x, \xi, y, \eta) | (x, \xi, y, -\eta) \in \text{graph } \Phi\}$. This is a closed homogeneous Lagrangian submanifold of $T^*(X \times Y) - 0$. We want to investigate the class of operators given by the following:

DEFINITION 6.1. *An operator of the form K above is called a Fourier integral operator of order k associated to Φ if $\kappa \in I_{k+n/2}(\Gamma_\Phi)$. The set of all such will be denoted by $(F.I.)^k \Phi$. A pseudodifferential operator is a Fourier integral operator associated to the identity symplectic transformation.*

Note that Γ_Φ contains no elements of the form $(x, \xi, y, 0)$; hence K maps smooth $\frac{1}{2}$ forms to smooth $\frac{1}{2}$ forms. The symbol of κ is a smooth $\frac{1}{2}$ form homogeneous of order $k + n/2$ on Γ_Φ. However there is a canonical projection $\beta : \Gamma_\Phi \xrightarrow{\cong} T^*X - 0$ and on T^*X we have a canonical half form $\Omega_X^{n/2}$ which is homogeneous of degree $n/2$ and is nowhere zero. Every half form on Γ_Φ can be written uniquely as the product of a function and the half form $\beta^* \Omega_X^{n/2}$. In particular this is the case for the symbol of κ in Definition 6.1. The function we get will be called the symbol of the operator K and denoted $\sigma(K)$. Note that it has the same order of homogeneity as K.

A closed subset C of $X \times Y$ is *proper* if the projections π and ρ restricted to C are proper. The distribution κ of (6.1) is *properly supported* if its support is proper. If κ is properly supported then it is clear from (6.1) that K maps compactly supported half forms to compactly supported half forms.

If a closed set $C \subset S^*X \times S^*Y$ is proper so is its projection on $X \times Y$. This is true in particular of the spherical image of the Γ_Φ above, the projection of which on $X \times Y$ we'll denote by Γ'_Φ. Let U be a neighborhood of Γ'_Φ whose closure is proper (it is clearly possible to find such a U) and let ρ be a smooth function with support in U and equal to 1 on a smaller neighborhood of Γ'_Φ. Then if $\kappa \in I(X \times Y, \Gamma_\Phi)$ the same is the case for $\rho\kappa$. Writing $\kappa = \rho\kappa + (1 - \rho)\kappa$ where $\rho\kappa$ is properly supported and $(1 - \rho)\kappa$ is smooth we have proved:

PROPOSITION 6.1. *Every Fourier integral operator can be written as the sum of an operator with smooth kernel and a properly supported operator.*

Notice that the composite of two properly supported operators is well defined and is again properly supported. Let $\Phi : T^*X - 0 \to T^*Y - 0$ and $\Psi : T^*Y - 0 \to T^*Z - 0$ be homogeneous symplectic transformations. Our immediate goal is to prove:

PROPOSITION 6.2. *If $K \in (F.I.)^k \Phi$ and $L \in (F.I.)^\ell \Psi$ are properly supported then $K \circ L$ is in $(F.I.)^{k+\ell} \Psi \circ \Phi$ and its symbol is $\tilde{\sigma}(K)\tilde{\sigma}(L)$ where the $\tilde{\sigma}$'s are the σ's transported to $\Gamma_{\Psi\circ\Phi}$ via the identifications $\Gamma_{\Psi\circ\Phi} \to \Gamma_\Phi$, $(x, \xi, z, \gamma) \to (x, \xi, \Phi(x, \xi))$; and $\Gamma_{\Psi\circ\Phi} \to \Gamma_\Psi$, $(x, \xi, z, \gamma) \to (\Phi(x, \xi), z, \gamma)$.*

The proof will first of all require our looking more carefully at the distributions $I_m(X, \Lambda)$ where Λ is the conormal bundle to a submanifold of X. It will be enough to do this locally; so let $\mathbf{R}^n = \mathbf{R}^k \times \mathbf{R}^\ell$ with coordinates $x = (t, y)$, $t = (t_1, \ldots, t_k)$, $y = (y_1, \ldots, y_\ell)$, and let Λ be the conormal bundle to the submanifold $t = 0$; i.e., if $\xi = (\tau, \eta)$ are the corresponding cotangent coordinates then in (x, ξ) space Λ is defined by $t = 0$, $\eta = 0$.

LEMMA 6.3. *If Λ is as above, and $\mu \in I_m(X, \Lambda)$ then*

$$\mu = \int b(y, \tau) e^{it\tau} \, d\tau \tag{6.2}$$

modulo distributions of lower order, where $b(y,\tau)$ is 0 for τ small and homogeneous of degree $m - k/2$ in τ for τ large.

PROOF. Now Λ is describable by the parametrization

$$\begin{array}{ccc} \mathbf{R}^n \times S^{k-1} & \xrightarrow{\varphi} & \mathbf{R} \\ \downarrow \pi & & \\ \mathbf{R}^n & & \end{array}$$

where $\varphi(x,\omega) = t \cdot \omega$, $\omega \in S^{k-1}$. Therefore, with $r = m - 1 + k/2$,

$$\pi_* a\varphi^* \nu_r^+ = \int a(x,\omega) \int_0^\infty e^{is\varphi(x,\omega)} s^r \chi(s)\, ds d\omega$$

where $\chi(s) = 0$ for s small and 1 for s large. Setting $\tau = s\omega$ and $\tilde{a}(x,\tau) = \int_0^\infty \chi(s)a(x,\omega)s^{r-(k-1)}\, ds$, and noting that $t \cdot \tau = s\varphi(x,\omega)$ we get

$$\pi_* a\varphi^* \nu_r^+ = \int \tilde{a}(x,\tau)e^{it\tau}\, d\tau.$$

Finally let $\tilde{a}(x,\tau) = b(y,\tau) + \sum t_j b_j(x,\tau)$. By an integration by parts we get

$$\pi_* a\varphi^* \nu_r^+ = \int b(y,\tau)e^{it\tau}\, d\tau - \int \Sigma \frac{1}{\sqrt{-1}} \frac{\partial b_j}{\partial \tau_j} e^{it\tau}\, d\tau.$$

Since the second term is of lower order of homogeneity than the first this proves (6.2). Q. E.D.

We will leave the following formula as an exercise. If μ is given by (6.2), then

$$\sigma(\mu) = b(y,\tau)\sqrt{dyd\tau}. \tag{6.3}$$

Getting back to the proof of Proposition 6.2 the composition of two integral operators can, as we showed in §3, be factored into a push forward, a pull back and an exterior tensor product operation, so we will first have to look at exterior products of Lagrangian distributions. Let $\mu_i \in I_{k_i}(X_i, \Lambda_i)$, $i = 1, 2$; and let C be an arbitrary conic neighborhood of $\Lambda_1 \times 0 \cup 0 \times \Lambda_2$ in $T^*(X_1 \times X_2)$. Then we will prove

PROPOSITION 6.4. *We can write $\mu_1 \boxtimes \mu_2 = \nu + \nu'$ where $WF(\nu) \subset C$ and $\nu' \in I_{k_1+k_2}(X_1 \times X_2, \Lambda_1 \times \Lambda_2)$. Moreover*

$$\sigma(\nu') = \sigma(\mu_1) \boxtimes \sigma(\mu_2)$$

on the complement of C.

PROOF. We first prove it for our fundamental distributions $\nu_{k_1}^+$ and $\nu_{k_2}^+$ on the real line. Their exterior tensor products can be written as

$$\left(\int \rho(\xi_1)\xi_1^{k_1-1/2} \rho(\xi_2)\xi_2^{k_2-1/2} e^{ix\cdot\xi}\, d\xi\right)\sqrt{dx}. \tag{6.4}$$

Let $\varphi(\theta)$ be a function which is zero for $\theta \leq 0$ and $\theta \geq \pi/2$ and is one for $\epsilon/2 < \theta < \pi/2 - \epsilon/2$. Let

$$v_{k_1}^+ \boxtimes v_{k_2}^+ = \nu + \nu' \tag{6.5}$$

where ν and ν' are obtained from (6.4) by multiplying the integrand by $1 - \varphi(\theta)$ and $\varphi(\theta)$ respectively. It is clear that $WF(\nu)$ is contained in the set:

$$\left\{(r,\theta), 0 < \theta < \epsilon \text{ or } \frac{\pi}{2} - \epsilon < \theta < \frac{\pi}{2}\right\} \tag{6.6}$$

and ν' has the desired symbol on the complement of this set by Lemma 6.3 and formula (6.3).

Now we proceed to the general case. Let Λ_i, $i = 1, 2$, be parametrized by the diagram

$$\begin{array}{ccc} Z_i & \xrightarrow{\varphi_i} & \mathbf{R} \\ \pi_i \downarrow & & \\ X_i & & \end{array}$$

and let $\mu_i = (\pi_i)_* a_i \varphi_i^* v_{k_i}^+$, $i = 1, 2$. Then

$$u_1 \boxtimes u_2 = (\pi_1 \times \pi_2)_* a_1 a_2 \varphi_1^* \times \varphi_2^* v_{k_1}^+ \boxtimes v_{k_2}^+.$$

Decomposing $v_{k_1}^+ \boxtimes v_{k_2}^+$ according to (6.5) and observing that for ϵ sufficiently small $(\pi_1 \times \pi_2)_* (\varphi_1 \times \varphi_2)^*$ applied to the set (6.6) is contained in an arbitrary conic neighborhood of $\Lambda_1 \times 0 \cup 0 \times \Lambda_2$, we're done. This concludes the proof of Proposition 6.4.

Let's take up again the proof of Proposition 6.2. If the *F. I.* operators K and L of Proposition 6.2 have $\kappa(x,y)\sqrt{dxdy}$ and $\lambda(y,z)\sqrt{dydz}$ as their Schwartz kernels the composite operator has the kernel

$$\left(\int \kappa(x,y)\lambda(y,z)\,dy\right)\sqrt{dxdz}$$

which as a distribution must be interpreted as $\pi^* \Delta^* \kappa \boxtimes \lambda$ where $\Delta\colon X \times Y \times Z \to X \times Y \times Y \times Z$ is the diagonal map and $\pi : X \times Y \times Z \to X \times Z$ the projection $(x,y,z) \to (x,z)$. We can write $\kappa \boxtimes \lambda = \nu + \nu'$, ν so chosen that its wave front set in $T^*(X \times Y \times Y \times Z)$ is disjoint from the set

$$\{(x,\xi,y,\eta,y,-\eta,z,\gamma)\} \tag{6.7}$$

and ν' being a Lagrangian distribution on $\Gamma_\Phi \times \Gamma_\Psi$ with the appropriate product symbol $\sigma(\kappa) \boxtimes \sigma(\lambda)$ in the vicinity of the point (6.7). Since $WF(\nu)$ is disjoint from (6.7), $WF(\Delta^* \nu)$ doesn't contain elements of the form $(x,\xi,y,0,z,\gamma)$ and therefore $\pi_* WF(\Delta^* \nu)$ is empty, i.e., $\pi_* \Delta^* \nu$ is smooth. As for $\pi_* \Delta^* \nu'$ this is a Lagrangian distribution associated with the Lagrangian manifold $\pi_* \Delta^* \Gamma_\Phi \times \Gamma_\Psi = \Gamma_{\Phi \circ \Psi}$.

What about the order of $\pi_* \Delta^* \nu'$? If $\dim X = n$ then κ is of order $k + n/2$ and λ of order $l + n/2$; so $\kappa \boxtimes \lambda$ and $\Delta^* \kappa \boxtimes \lambda$ are of order $k + l + n$. The operator π_* lowers order by $n/2$ since it is integration over a fiber of dimension n so the order of $\pi_* \Delta^* \kappa \boxtimes \lambda$ is $k + l + n/2$ as required.

Finally we have to check that $\pi_* \Delta^* \sigma(k) \boxtimes \sigma(l)$, which is the symbol of $\pi_* \Delta^* \nu'$, is the appropriate product symbol. Let $\beta : T^*Z \to T^*W$ be a symplectic transformation. Then it maps Lagrangian submanifolds of $T^*(X \times Y \times Y \times Z)$ onto Lagrangian submanifolds of $T^*(X \times Y \times Y \times W)$ by its action on the right, and it also maps Lagrangian submanifolds of $T^*(X \times Z)$ onto Lagrangian submanifolds of $T^*(X \times W)$ in the same way. It is easy to check that, for a fixed Lagrangian $\Lambda \subset T^*(X \times Y \times Y \times W)$,

$$\beta^* \pi_* \Delta^* \Lambda = \pi_* \Delta^* \beta^* \Lambda$$

when the respective pull backs and push forwards are defined. (This is because π and Δ only affect the Y factor in the product $X \times Y \times Y \times Z$.) Moreover, if σ is a half form on Λ,

$$\beta^* \pi_* \Delta^* \sigma = \pi_* \Delta^* \beta^* \sigma.$$

Therefore to check that $\pi_* \Delta^* \sigma(\kappa) \boxtimes \sigma(\lambda)$ is the appropriate product symbol we only have to check it for $\Psi = \mathrm{id}$. The same argument on the left shows that we only have to check it when both Φ and Ψ are the identity. It is clear moreover that if we multiply κ by a function f and λ by a function g, $\Delta^* \pi^* \kappa \boxtimes \lambda$ get multiplied by fg so we only have to check it when κ and λ are both equal to $\Omega_X^{n/2}$. We will see below however that when $\Phi =$ identity, $\Omega_X^{n/2}$ is the symbol of the Schwartz kernel of the identity operator. However the identity operator composed with itself is the identity operator so we are done. This concludes the proof of Proposition 6.2.

A Fourier integral operator is elliptic if its symbol is nowhere equal to zero. As a corollary of Proposition 6.2 we will prove

PROPOSITION 6.5. *Let $K \in (F.I.)^k \Phi$ be a properly supported elliptic F.I. operator. Then there exists a properly supported F.I. operator $L \in (F.I.)^{-k} \Phi^{-1}$ such that* $\mathrm{id} - K \circ L$ *and* $\mathrm{id} - L \circ K$ *are properly supported operators with smooth kernels.*

PROOF. First of all let's note that the identity operator is in $F.I.^0$ (identity) and has symbol 1. In fact it is enough to check this for $X = \mathbf{R}^n$. The Schwartz kernel of the identity operator on $\mathbf{R}^n \times \mathbf{R}^n$ is $\delta(x - y)$ so this assertion follows from Lemma 6.3 with $\Lambda = N^* \Delta$.

Now if $\mathrm{id} - KL$ is smoothing, $\tilde{\sigma}(K)\tilde{\sigma}(L) = 1$ where the $\tilde{\sigma}$'s are the pull backs to T^*X so we must have $\tilde{\sigma}(L) = 1/\tilde{\sigma}(K)$. Choose L_1 to be any operator in $(F.I.)^{-k}\Phi^{-1}$ with this symbol. Then $\mathrm{id} - KL_1$ is in $(F.I.)^{-1}$ (identity). Let ρ be its symbol and pick any L_2 in $(F.I.)^{-k-1}\Phi^{-1}$ such that $\tilde{\sigma}(L_2) = \rho/\tilde{\sigma}(K)$. Then $\mathrm{id} - K \circ (L_1 + L_2)$ is in $(F.I.)^{-2}\Phi^{-1}$. Continuing this way we can find L_1, L_2,

... such that

$$\mathrm{id} - K \circ (L_1 + \cdots + L_N) \in (F.I.)^{-N}(\text{identity}).$$

Finally by the Mittag-Leffler property of Proposition 4.1, choose $L \in (F.I.)^{-k}\Phi^{-1}$ such that

$$L - (L_1 + \cdots + L_N) \in (F.I.)^{-N-k}\Phi^{-1}.$$

Then $\mathrm{id} - K \circ L$ is smoothing. A similar argument produces an L' such that $\mathrm{id} - L' \circ K$ is smoothing, and the usual argument on the uniqueness of inverses shows that $L' - L$ is smoothing, so we can replace L' by L and we're done.

For pseudodifferential operators the underlying Lagrangian manifold Γ_{id} is just $N^*\Delta$; that is, the set $\{(x, \xi, x, -\xi), \xi \in T^*X_x\}$. We will identify it with T^*X via $(x, \xi, x, -\xi) \to (x, \xi)$. We will say that a pseudodifferential operator K is *smooth* at (x, ξ) if, for the corresponding Schwartz kernel κ, $WF(\kappa)$ does not contain (x, ξ). We will say that K is of order l at (x, ξ) if there exists a pseudodifferential operator K' of order l such that $K - K'$ is smooth at (x, ξ). By exactly the same argument as that above (induction on order) one proves:

PROPOSITION 6.6. *Let K be a kth order pseudodifferential operator for which $\sigma(K)(x, \xi) \neq 0$. Then there exists a $(-k)$th order pseudodifferential operator L such that* $\mathrm{id} - KL$ *and* $\mathrm{id} - LK$ *are smooth at* (x, ξ).

We will use this result to give a new definition of wave front set. Before we do so, however, we must make an observation about *F. I.* operators applied to distributions. Let K be a *F. I.* operator belonging, let's say, to $(F.I.)\Phi$, and let μ be a compactly supported distribution. We will show that $K\mu$ is well defined and that

$$WF(K\mu) \subset \Phi WF(\mu). \tag{6.8}$$

The right way to define $K\mu$ is, of course, as $\pi_* \kappa \rho^* \mu$ where κ is the Schwartz kernel of K. We will show this makes sense by a variant of the argument used in §3 to define the product of two distributions. Namely, $WF(\rho^*\mu) \subset \{(x, 0, y, \eta), (y, \eta) \in WF(\mu)\}$, and $WF(\kappa) \subset \{(x, \xi, y, -\eta), (y, \eta) = \Phi(x, \xi)\}$; so, by §3, the wave front set of $\kappa\rho^*\mu$ is contained in

$$\{(x, \xi, y, -\eta + \eta'), (y, \eta) = \Phi(x, \xi), (y, \eta') \in WF(\mu)\}. \tag{6.9}$$

Note that $\kappa\rho^*\mu$ *is* well defined since (6.9) contains no zero vectors. For a point (x, ξ) to be in the push forward of this set there must exist a y for which $(x, \xi, y, 0)$ is in the set, i.e., a point of the form (6.9) for which $\eta = \eta'$. Hence $WF(K\mu) \subset \Phi WF(\mu)$ is as claimed.

If a pseudodifferential operator K is smooth at (x, ξ) then $(x, \xi, x, -\xi)$ is not in the WF set of the corresponding κ so the argument we've just given shows that, for all μ, $(x, \xi) \notin WF(K\mu)$.

Let μ be a fixed distribution and K a pseudodifferential operator. Then if $\sigma(K)(x_0, \xi_0) \neq 0$ and $K\mu$ is smooth, μ is smooth at (x, ξ_0). For, by Proposition 6.6 we can choose L such that $M = \mathrm{id} - LK$ is smooth at (x_0, ξ_0) and write $\mu = LK\mu + M\mu$ both of which terms are smooth at (x_0, ξ_0). This proves one half of the following:

PROPOSITION 6.7. *The point* $(x_0, \xi_0) \notin WF(\mu)$ *if and only if there exists a pseudodifferential operator* K *with* $\sigma(K)(x_0, \xi_0) \neq 0$ *and* $K\mu$ *smooth.*

To prove the other half we need:

LEMMA 6.7. *Let* $w \in I(X, \Lambda)$, *let* $(x_0, \xi_0) \in \Lambda$ *and let* V *be a conical neighborhood of* (x_0, ξ_0) *in* T^*X. *Then there exists* $w' \in I(X, \Lambda)$ *such that* w' *is smooth outside* V *and* $w - w'$ *is smooth near* (x_0, ξ_0).

Assuming this lemma, to prove the proposition choose a pseudodifferential operator K such that $\mathrm{id} - K$ is smooth near (x_0, ξ_0) and K is smooth on the WF set of μ. Then $K\mu$ is smooth.

To prove the lemma choose a parametrization

$$\begin{array}{ccc} Z & \xrightarrow{\varphi} & \mathbf{R} \\ \downarrow \pi & & \\ X & & \end{array}$$

of Λ near (x_0, ξ_0) such that w is a sum of terms of the form $\pi_* a\varphi^* v_m^+$ and let w' be the sum of $\pi_* \rho a \varphi^* v_m^+$ where ρ is a blip function equal to 1 on a neighborhood of the critical point corresponding to (x_0, ξ_0).

We will now discuss some examples. First of all let's go back to the generalized Radon transform described in §3. We are given n dimensional manifolds X and Y and a codimension l submanifold $Z \subset X \times Y$, such that the restrictions to Z of the projection maps

$$\begin{array}{ccc} & Z & \\ {}^{\pi}\swarrow & & \searrow^{\rho} \\ X & & Y \end{array}$$

are fibrations. Given a smooth density κ on Z we define the 'Radon' transform

$$K : C_0^\infty(Y) \to C^\infty(|\wedge|X)$$

by the formula

$$\mu \to \pi_* \kappa \rho^* \mu. \tag{6.10}$$

Ignoring the problem of variances (by fixing nonzero half densities on X and Y we can convert K into an operator on half densities without changing it in any essential way) it is clear that the Schwartz kernel κ of K is a δ distribution concentrated on Z. Suppose the projections

$$\begin{array}{ccc} & NZ - 0 & \\ {}^{\pi}\swarrow & & \searrow^{\rho} \\ T^*X - 0 & & T^*Y - 0 \end{array}$$

are bijective. (This is the notion of ellipticity for the Radon transform we introduced in §3.) Then the map Φ,

$$\Phi(x, \xi) = (y, -\eta), \qquad (y, \eta) = \rho\pi^{-1}(x, \xi),$$

is a homogeneous symplectic transformation; so by Lemma 6.3, K is in $(F.I.)^{(l-n)/2}\Phi$. The symbol of K is essentially the density κ entering into (6.10); so if $\kappa \neq 0$ everywhere K is elliptic. The operator

$$C_0^\infty(X) \ni \mu \to \rho_* \kappa \pi^* \mu$$

is just the transpose K^t, and by the same token it is an elliptic $F.\ I.$ operator of type $(F.I.)^{(l-n)/2}\Phi^{-1}$. Hence KK^t and K^tK are elliptic pseudodifferential operators of order $l - n$. In the case of the classical rank one symmetric spaces, Helgason [14] has shown that they are of the form $P(\Delta)^{-1}$ where Δ is the Laplace-Beltrami operator and P a polynomial of order $(n - l)/2$. This is in fact the content of the Radon inversion formula for these spaces.

Before discussing our second example note that if K is a pseudodifferential operator on $\mathbf{R}^n$ and κ its Schwartz kernel, then by Lemma 6.3, κ can be written as the sum of a smooth function and an expression of the form

$$\int a(x, \xi) e^{i(x-y)\cdot\xi}\, d\xi, \qquad a(x, \xi) \sim \sum_{i=m}^{-\infty} a_i(x, \xi) \tag{6.11}$$

with $a_i(x, \xi)$ homogeneous of degree i. The expression $a(x, \xi)$ is usually called the *total symbol* of K. In §1 of Chapter 2 we showed that differential operators could also be written in this form. In fact we showed that the differential operator $P(x, D) = \sum p_\alpha(x) D^\alpha$ could be written in the form

$$\mu \to \int \sum p_\alpha(x) \xi^\alpha e^{ix\cdot\xi} \hat{\mu}(\xi)\, d\xi.$$

Replacing $\hat{\mu}$ by $\int e^{-iy\cdot\xi} \mu(y)\, dy$ we get (6.11) with $a(x, \xi) = \sum p_\alpha(x) \xi^\alpha$. Thus we've shown:

PROPOSITION 6.9. *A differential operator is a pseudodifferential operator whose total symbol is a polynomial function of* ξ.

Notice that the symbol of P as we've defined it in Chapter II, namely $\sum_{|\alpha|=m} p_\alpha(x)\xi^\alpha$, is identical with the symbol as we've defined it here.

As an application of the machinery of this section we will prove a couple of standard results about elliptic differential operators.

PROPOSITION 6.10 *Let* X *be a compact manifold and* $P : C^\infty(X) \to C^\infty(X)$ *an elliptic differential operator. Then the kernel of* P *and cokernel of* P *are finite dimensional.*

In the proof we will need an elementary lemma about smoothing operators. For the lemma X is arbitrary.

LEMMA 6.11. *Let* $K : C_0^\infty(X) \to C_0^\infty(X)$ *have smooth Schwartz kernel* $k(x,y)dy$ *and assume*

$$\sup_{x,y} |k| \int_{y,k(x,y)\neq 0} dy < \frac{1}{2}. \tag{6.12}$$

Then id $-K$ *is invertible and its inverse is of the form* id $-L$ *where* L *is a smoothing operator.*

Formally L is given by the Neumann series

$$L = -(K + K \circ K + K \circ K \circ K + \cdots). \tag{6.13}$$

We'll prove the lemma by showing that the right hand side of (6.13) converges to a smooth limit. The Schwartz kernel of the mth term in the series is

$$k_m(x,y) = \int k(x,z_1)k(z_1,z_2)\cdots k(z_{m-1},y)\,dz_1 \cdots dz_{m-1}$$

so

$$\sup \left| \frac{\partial}{\partial x^\alpha} \frac{\partial}{\partial y^\beta} k_m(x,y) \right|$$

can be majorized by a constant independent of m times the $(m-1)$st power of

$$\sup |k(z,z')| \int_{z,k(z,z')\neq 0} dz.$$

Therefore by (6.12) the mth term can be majorized by a constant times $(1/2)^{m-1}$ proving that the sum does converge in the C^∞ topology. Q.E.D.

To prove Proposition 6.10 choose a pseudodifferential operator Q such that $PQ = \mathrm{id} - A$ and $QP = \mathrm{id} - B$ where A and B are smoothing operators. Let $a(x,y)dy$ be the Schwartz kernel of A. By the Stone-Weierstrass theorem we can approximate a in the C^0 topology by a smooth function of the form

$$a' = \sum_{i=1}^{N} f_i(x) g_i(y),$$

i.e., for any $\epsilon > 0$ we can find a function a' of this form so that $|a - a'| < \epsilon$. Let $a'' = a - a'$ and write

$$A = A' + A''$$

where A' and A'' are the operators with kernels $a'dy$ and $a''dy$. Then A' is a finite rank smoothing operator and for ϵ sufficiently small $I - A''$ is invertible (by the lemma) with an inverse which is again of the form id plus a smoothing operator. Let $Q' = Q(\mathrm{id} - A'')^{-1}$. Then $\mathrm{id} - PQ'$ is a finite rank smoothing operator. By a similar argument one can find Q'' so that $\mathrm{id} - Q''P$ is a finite rank smoothing operator, so we've proved that *P is invertible up to finite rank smoothing operators* which is the same as the assertion of Proposition (6.10).

Finally we will show that an elliptic differential operator is always solvable locally.

PROPOSITION 6.12. *If P is an elliptic differential operator on $\mathbf{R}^n$, $\mu \in C_0^\infty(\mathbf{R}^n)$ and $x_0 \in \mathbf{R}^n$ there exists a $v \in C^\infty(\mathbf{R}^n)$ such that $Pv = \mu$ on a neighborhood of x_0.*

PROOF. By Proposition 6.5 there exists a pseudodifferential operator Q such that $PQ = \mathrm{id} - A$ where A is smoothing . Let $a(x,y)dy$ be the Schwartz kernel of A. Let ρ be a blip function which takes the value 1 near x_0 and is supported in a small neighborhood of x_0. Then $\rho A \rho$ has Schwartz kernel $\rho(x)a(x,y)\rho(y)dy$. By Lemma 6.10, $\mathrm{id} - \rho A \rho$ is invertible providing the support of ρ is small enough, and the inverse is again of the form: Identity plus a smoothing operator. Let $v = (\mathrm{id} - \rho A \rho)^{-1}\mu$. Then

$$\begin{aligned} \mu &= v - \rho A \rho v = \rho v - A \rho v + (1 - \rho)(v + A\rho v) \\ &= PQ\rho v + (1 - \rho)(v + A\rho v). \end{aligned}$$

Thus $\mu = PQ\rho v$ on a neighborhood of x_0.

§7. The transport equation.

Let P be an mth order differential operator on the half form bundle of X, i.e., $P: C^\infty(\wedge^{1/2}X) \to C^\infty(\wedge^{1/2}X)$. In this section we will describe what happens when we apply P to a generalized half form of the type discussed in §4. Before doing so however we will need to review some material from Chapter II. If X is an open subset of $\mathbf{R}^n$ we can trivialize the half form bundle by identifying

$f \in C^\infty(X)$ with $f\sqrt{dx}$. Then P can be written in coordinates as $\Sigma_{|\alpha|\leq m} a_\alpha(x) D^\alpha$ where $\alpha = (\alpha_1, \ldots, \alpha_n)$ and

$$D^\alpha = D_1^{\alpha_1} \cdots D_n^{\alpha_n}, \quad D_i = \frac{1}{\sqrt{-1}} \frac{\partial}{\partial x_i}.$$

Let $p(x,\xi) = \Sigma_{|\alpha|\leq m} a_\alpha(x)\xi^\alpha$ and $p_k(x,\xi) = \Sigma_{|\alpha|=k} a_\alpha(x)\xi^\alpha$. The *principal symbol* (or, simply, *symbol*) of P is

$$\sigma(P) = p_m(x,\xi) \tag{7.1}$$

and the subprincipal symbol

$$\sigma_{\text{sub}}(P) = p_{m-1} - \frac{1}{2\sqrt{-1}} \sum \frac{\partial^2 p_m}{\partial x_i \partial \xi_i}. \tag{7.2}$$

Both (7.1) and (7.2) are intrinsically defined as functions on T^*X (we identify T^*X with (x,ξ) space in the usual way) and make sense when X is a manifold, not just a subset of $\mathbf{R}^n$. For detail see Chapter II, §6.

Given two differential operators P and Q of order ℓ and m respectively we have the following formulas for the principal and subprincipal symbol of PQ:

$$\begin{aligned} &\text{I. } \sigma(PQ) = \sigma(P)\sigma(Q) \\ &\text{II. } \sigma_{\text{sub}}(PQ) = \sigma(P)\sigma_{\text{sub}}(Q) + \sigma(Q)\sigma_{\text{sub}}(P) + \frac{1}{2i}\{\sigma(P),\sigma(Q)\}. \end{aligned} \tag{7.3}$$

PROOF: Write $P\mu$ as $\int p(x,\xi)e^{ix\cdot\xi}\hat{\mu}(\xi)\,d\xi$, $Q\mu$ as $\int q(x,\xi)e^{ix\cdot\xi}\hat{\mu}(\xi)\,d\xi$ and $PQ\mu$ as $\int r(x,\xi)e^{ix\cdot\xi}\hat{\mu}(\xi)\,d\xi$. We will first prove

$$r(x,\xi) = \sum_{|\alpha|\leq \ell} \frac{1}{\alpha!} \frac{\partial^\alpha p}{\partial \xi^\alpha} D_x^\alpha q. \tag{7.4}$$

PROOF. Apply $P(x,D)$ to $\int q(x,\xi)e^{ix\cdot\xi}\hat{\mu}(\xi)\,d\xi$ and use

$$Pe^{ix\cdot\xi}q = e^{ix\cdot\xi} \sum \frac{1}{\alpha!} \frac{\partial p}{\partial \xi^\alpha} D_x^\alpha q.$$

See Chapter II, (1.4). Q.E.D.

By (7.4) the term of order $\ell + m$ in $r(x,\xi)$ is $p_\ell q_m$ which proves (7.3) I. The term of order $\ell + m - 1$ is

$$p_\ell q_{m-1} + p_{\ell-1} q_m + \frac{1}{i} \sum \frac{\partial p_\ell}{\partial \xi_j} \frac{\partial q_m}{\partial x_j};$$

so the subprincipal part of PQ is

$$p_\ell q_{m-1} + p_{\ell-1} q_m + \frac{1}{i} \sum \frac{\partial p_\ell}{\partial \xi_j} \frac{\partial q_m}{\partial x_j} - \frac{1}{2i} \sum \frac{\partial^2 p_\ell q_m}{\partial \xi_j \partial x_j},$$

which is identical with (7.3) II.

We will now prove:

THEOREM 7.1. *If* $\mu \in I_k(X, \Lambda)$ *then* $P\mu \in I_{k+m}(X, \Lambda)$ *and* $\sigma(P\mu) = \sigma(P)\sigma(\mu)$.

PROOF. It is enough to prove the theorem when X is an open subset of $\mathbf{R}^n$ and Λ is given by a parametrization:

$$\begin{array}{ccc} X \times \mathbf{R}^N & \xrightarrow{\varphi} & \mathbf{R} \\ \downarrow \pi & & \\ X & & \end{array}$$

Moreover, because of (7.3) I, we only have to prove it for zeroth and first order differential operators. For zeroth order operators it is trivial so we only have to check it for the one operator $(1/\sqrt{-1})(\partial/\partial x_i)$. Applying this operator to $\mu = \pi_* a\varphi^* v_r$, $r = k + (N-1)/2$, we get

$$\frac{1}{\sqrt{-1}} \frac{\partial}{\partial x_i}\mu = \pi_* a \frac{\partial \varphi}{\partial x_i} \varphi^* \frac{1}{\sqrt{-1}} \frac{d}{dt} v_r + \pi_* \frac{1}{\sqrt{-1}} \frac{\partial a}{\partial x_i} \varphi^* v_r \tag{7.5}$$

by the chain rule. Both terms are in $I_{k+1}(X, \Lambda)$ and in fact the second term is in $I_k(X, \Lambda)$; so, as far as the symbol is concerned, we can ignore it. Since

$$\frac{1}{\sqrt{-1}} \frac{d}{dt} v_r = \frac{1}{\sqrt{-1}} \frac{d}{dt} \int_0^\infty s^r e^{ist}\, ds = \int_0^\infty s^{r+1} e^{ist}\, ds = v_{r+1} \quad \text{and} \quad \frac{\partial \varphi}{\partial x_i} = \xi_i;$$

on the critical set of φ, the symbol of (7.5) is $\xi_i \sigma(\mu)$. But ξ_i is the symbol of $(1/\sqrt{-1})(\partial/\partial x_i)$ so we are done. Q.E.D.

One corollary of Theorem 7.1 is: If $\sigma(P) = 0$ on Λ then $P\mu \in I_{k+m-1}(X, \Lambda)$. When this happens the computation of $\sigma(P\mu)$ is not quite so easy. Previous experience (Chapter II, §6) indicates it should involve the transport equation and this is indeed the case.

Let $p_m = \sigma(P)$ and let η_{p_m} be the Hamiltonian vector field associated with p_m. Recall that if $p_m = 0$ on Λ, η_{p_m} is tangent to Λ; so η_{p_m} can be viewed as a vector field on Λ, and as such it acts by Lie differentiation on all intrinsically defined objects on Λ including half forms. Let us denote the Lie derivative of η_{p_m} by $D_{\eta_{p_m}}$. Given $\mu \in I_k(X, \Lambda)$ we will derive the following formula:

$$\sigma(P\mu) = \frac{1}{\sqrt{-1}} D_{\eta_{p_m}} \sigma(\mu) + \sigma_{\text{sub}}(P)\sigma(\mu) \tag{7.6}$$

for the symbol of $P\mu$. (Compare with Chapter II, (6.8).)

PROOF OF (7.6). The proof will be by a series of lemmas.

LEMMA 7.1. (7.6) *is true when P is an mth order operator whose principal symbol is identically zero* (*i.e., an mth order operator which is really an* $(m-1)$*st order operator*).

PROOF. This is just Theorem 7.1.

LEMMA 7.2. *If* (7.6) *is true for* Q *with* $\sigma(Q) = 0$ *on* Λ *then* (7.6) *is true for any operator of the form* PQ.

PROOF. Let p and q be the top symbols of P and Q. Since $q = 0$ on Λ, $\boldsymbol{\eta}_{pq} = p\boldsymbol{\eta}_q$ on Λ. Let $\sigma = \sigma(\mu)$. By Chapter II (see the displayed formula in the last paragraph of §6),

$$D_{p\eta_q}\sigma = pD_{\eta_q}\sigma + \tfrac{1}{2}\{q,p\}\sigma;$$

so, by (7.3) II,

$$\frac{1}{\sqrt{-1}}D_{\eta_{pq}}\sigma + \sigma_{\text{sub}}(PQ)\sigma$$

$$= p\frac{1}{\sqrt{-1}}D_{\eta_q}\sigma + \frac{1}{2\sqrt{-1}}\{q,p\}\sigma + p\sigma_{\text{sub}}(Q)\sigma + \frac{1}{2\sqrt{-1}}\{p,q\}\sigma$$

$$= p\left(\frac{1}{\sqrt{-1}}D_{\eta_q}\sigma + \sigma_{\text{sub}}(Q)\sigma\right).$$

But the last term is $\sigma(PQ\mu)$ by Theorem 7.1. Q.E.D.

LEMMA 7.3. *It is enough to prove* (7.6) *when* Λ *is the conormal bundle of a submanifold of* X.

PROOF. If the projection

$$\begin{array}{c} \Lambda \subset T^*X \\ \pi \downarrow \\ X \end{array}$$

is of constant rank in a neighborhood of (x_0, ξ_0) then Λ is a conormal bundle near (x_0, ξ_0) (see Chapter IV, §5). Since π is of constant rank on a dense open subset of Λ it is enough to prove (7.6) in this case. Q.E.D.

Let $\mathbf{R}^n = \mathbf{R}^k \times \mathbf{R}^\ell$ with coordinates $x = (t, y)$ and let Λ be the conormal bundle of the submanifold $t = 0$. Then by Lemma 6.3 of §6 every $\mu \in I_m(X, \Lambda)$ can be written in the form

$$\int b(y,\tau)e^{it\tau}\,d\tau \tag{7.7}$$

modulo lower order terms where $b(y,\tau)$ is zero for τ small and homogeneous of degree $m - k/2$ in σ for τ large. Moreover the symbol of μ is just

$$\sigma(\mu) = b(y,\tau)\sqrt{dyd\tau}. \tag{7.8}$$

Let us now prove (7.6). By Lemma 7.3 it is sufficient to prove (7.6) when Λ is defined in $\mathbf{R}^n = \mathbf{R}^k \times \mathbf{R}^\ell$ by the equation $t = \eta = 0$. Since $\sigma(P) = p_m(x, \xi)$

vanishes on Λ we can write

$$p_m(x,\xi) = \Sigma\alpha_i(x,\xi)t_i + \Sigma\beta_i(x,\xi)\eta_i.$$

Therefore by Lemmas 7.1 and 7.2 it is enough to prove (7.6) for operators having t_i and η_i as symbols; for example, the operators "multiplication by t_i" and $(1/\sqrt{-1})(\partial/\partial y_i)$. If μ is equal to (7.7),

$$t_i\mu = \int t_i b(y,\tau)e^{it\tau}\,d\tau = \frac{1}{\sqrt{-1}}\int b(y,\tau)\frac{\partial}{\partial\tau_i}e^{it\tau}\,d\tau$$
$$= \int\frac{1}{\sqrt{-1}}\left(-\frac{\partial b}{\partial\tau_i}(y,\tau)\right)e^{it\tau}\,d\tau$$

and

$$\frac{1}{\sqrt{-1}}\frac{\partial}{\partial y_i}\mu = \int\frac{1}{\sqrt{-1}}\frac{\partial}{\partial y_i}b(y,\tau)e^{it\tau}\,d\tau.$$

The Hamiltonian vector field associated with t_i is $-\partial/\partial\tau_i$ and with η_i is $\partial/\partial y_i$, and the subprincipal parts of both of these operators is zero by (7.2); so in view of (7.8) our proof is finished. Q.E.D.

Our main application of (7.6) will be the following:

PROPOSITION 7.1. *Suppose that for any m and any* $\beta \in S^{m+k-1}(\Lambda)$ *there exists an* $\alpha \in S^m$ *satisfying*

$$\frac{1}{\sqrt{-1}}D_{\eta_p}\alpha + \sigma_{\text{sub}}(P)\alpha = \beta. \tag{7.9}$$

Then given $\alpha_0 \in S^m(\Lambda)$ *satisfying the homogeneous equation* $(1/\sqrt{-1})D_{\eta_p}\alpha_0 + \sigma_{\text{sub}}(P)\alpha_0 = 0$ *there exists* $\mu \in I_m(X,\Lambda)$ *with* $\sigma(\mu) = \alpha_0$ *and* $P\mu \in C^\infty$.

PROOF. Choose $\mu_0 \in I_m$ such that $\sigma(\mu_0) = \alpha_0$. Then by (7.6), $P\mu_0 \in I_{m+k-2}$. Let β_{-1} be the symbol of $P\mu_0$ and solve (7.9) for α_{-1} in terms of β_{-1}. Choose μ_1 such that $\sigma(\mu_1) = \alpha_{-1}$. Then $P(\mu_0 + \mu_1) \in I_{m+k-3}$. Continuing this way we can construct $\mu_0, \mu_1, \mu_2, \ldots$ such that $\mu_i \in I_{m-i}$ and $P(\mu_0 + \cdots + \mu_{N-1}) \in I_{m+k-N-1}$. Now apply Proposition 4.1. Q.E.D.

The hypothesis of Proposition 7.1 will be satisfied if the η_p flow lines on Λ admit a global C^∞ cross section. For example, let X be a compact oriented Riemannian manifold and Δ the Laplace-Beltrami operator. (We define Δ as $dd^* + d^*d$, so that on $\mathbf{R}^n$ it is the negative of the usual Laplacian.) Let $\delta(x-y)$ be the δ-function of the diagonal in $X \times X$. We will show that, at least modulo C^∞, there exists a "progressing" fundamental solution of the wave equation $(d/dt^2 - \Delta)\mu = 0$, i.e., there exists a generalized function $\mu_+(x,y,t)$ on $X \times X \times \mathbf{R}$ such that for y fixed $\mu_+(x,y,t)$ satisfies $(d/dt^2 - \Delta)\mu_+ \in C^\infty$, $\mu_+(x,y,0) = \delta(x-y)$ and $WF(\mu_+)$ is contained in $\tau - |\xi|$.

PROOF. First note that $\delta(x-y) \in I_{n/2}$ (N^* diagonal). In fact, locally $\delta(x-y) = \int e^{i(x-y)\cdot\xi}\, d\xi + \cdots$ (the dots indicating lower order terms); so by Lemma 7.3, $\delta(x-y) \in I_{n/2}$ (N^* diagonal). Let $p = \tau - |\xi|$ and let C be the Lagrangian manifold in $T^*(X \times X \times \mathbf{R})$ consisting of integral curves $\gamma(t)$ of $\boldsymbol{\eta}_p$, which have the property

$$\gamma(0) \in \{(x, \xi, x, -\xi, 0, |\xi|)\}.$$

For fixed y_0 the projection of C on $X \times \mathbf{R}$ is just the light cone with vertex at y_0; that is it consists of all points (x, t) such that x can be joined to y_0 by a geodesic of length $|t|$. The submanifold $t = 0$ of C is a global cross section for the η_p flow, so the hypotheses of Proposition 7.1 are satisfied. Moreover, not only can we solve (7.9) globally, but we can prescribe what α has to be at $t = 0$. In other words we can solve

$$\frac{d^2}{dt^2}\mu_+ - \Delta\mu_+ \in C^\infty,$$

and at the same time insure that $\mu_+(x, y, t) = \delta(x-y)$ when $t = 0$. Since $\tau - |\xi|$ vanishes on C we are done.

REMARK. One can construct a "regressing" fundamental solution in the same way, i.e., fundamental solution μ with $Wf(\mu_-)$ contained in $\tau + |\xi| = 0$.

To conclude this section we will describe how the theory of the transport equation can be generalized to pseudodifferential operators. The generalization of Theorem 7.1 is straight forward:

THEOREM 7.3. *Let* P: $C_0^\infty(\wedge^{1/2}X) \to C^\infty(\wedge^{1/2}X)$ *be an mth order pseudodifferential operator. Then if* $\mu \in I_k(X, \Lambda)$, $P\mu \in I_{k+m}(X, \Lambda)$ *and* $\sigma(P\mu) = \sigma(P)\sigma(\mu)$.

As the proof is very similar to the proof of Proposition 6.2 we will just sketch it. We write $P\mu = \pi_* \Delta^* \kappa \boxtimes \mu$ where κ is the Schwartz kernel of P, Δ: $X \times X \to X \times X \times X$ is the diagonal map $\Delta(x,y) = (x,y,y)$ and π is the usual projection. By Proposition 6.4, $\kappa \boxtimes \mu$ can be written as $\nu + \nu'$ where $WF(\nu)$ is contained in a small conic neighborhood of the set of "edge" points: $\Gamma_{\text{id}} \times 0 \cup 0 \times \Lambda$, and ν' is a Lagrangian distribution with wave front set concentrated on $\Gamma_{\text{id}} \times \Lambda$ and symbol equal to $\sigma(\kappa) \boxtimes \sigma(\mu)$ on the complement of a small conic neighborhood of the "edge". From facts proved in §3 about pull backs and push forwards of wave front sets one easily infers that $WF(\pi_* \Delta^* \nu)$ is empty, hence $\pi_* \Delta^* \nu$ is smooth. Therefore $P\mu$ is a Lagrangian distribution with symbol $\pi_* \Delta^* \sigma(\kappa) \times \sigma(\mu)$. All that's left to check is that

$$\pi_* \Delta^* \sigma(\kappa) \boxtimes \sigma(\mu) = \sigma(P)\sigma(\mu). \tag{7.10}$$

If $\sigma(\kappa)$ is multiplied by a function f then the left hand side of (7.10) gets multiplied by f; so it's sufficient to check (7.10) for a single nowhere vanishing $\sigma(\kappa)$. We have, however, already checked (7.10) for a lot of such $\sigma(\kappa)$'s ((7.10) is true for differential operators by Theorem 7.1), so we're done.

To discuss the analog of (7.6) one first has to define the subprincipal symbol for pseudodifferential operators. Locally this is easy. If

$$P\mu = \int P(x,\xi)e^{ix\xi}\hat{\mu}(\xi)\,d\xi, \qquad P(x,\xi) \sim P_m(x,\xi) + P_{m-1}(x,\xi) + \cdots$$

then

$$\sigma_{\text{sub}}(P) = P_{m-1} - \frac{1}{2\sqrt{-1}}\sum \frac{\partial P_m}{\partial x_i\,\partial \xi_i}$$

just as in (7.2). The problem is to show that this expression is intrinsically defined as a function on the cotangent bundle. This is not too hard, but we won't bother to do it here. (See, for example, Duistermaat-Hormander [9] §5.2.) Anyway, having defined $\sigma_{\text{sub}}(P)$ the formulas (7.3) and (7.4) are rather easy to prove, and (7.6) follows from Theorem 7.3 just as (7.6) for differential operators follows from Theorem 7.1.

§8 Some applications to spectral theory.

As above let X be a compact oriented Riemannian manifold and let Δ be its Laplace-Beltrami operator. It is well-known (see, for example, Warner [15]) that Δ has a complete orthonormal family of eigenfunctions

$$e_1, e_2, \ldots$$

and that the corresponding eigenvalues $\lambda_1, \lambda_2, \ldots$ are non-negative and tend to $+\infty$. From the e_i's one can construct two fundamental solutions of the wave equation:

$$v_+(x,y,t) = \sum e^{it\sqrt{\lambda_\alpha}}e_\alpha(x)\bar{e}_\alpha(y) \quad \text{and} \quad v_-(x,y,t) = \sum e^{-it\sqrt{\lambda_\alpha}}e_\alpha(x)\bar{e}_\alpha(y).$$

It turns out that, modulo C^∞ functions, v_+ and v_- are identical with the functions μ_+ and μ_- constructed above. We will not prove this here, but the proof is not very hard. The idea is to show that both μ_+ and v_+ satisfy the first order differential equation

$$\left(\frac{\partial}{\partial t} - \sqrt{\Delta}\right)w = 0, \qquad \text{modulo } C^\infty, \tag{8.1}$$

with the same initial data at $t = 0$. For v_+ this is obvious. To show it for μ_+ we need to know a little more about the operator $\sqrt{\Delta}$. By a theorem of Seeley, $\sqrt{\Delta}$ is a pseudodifferential operator with symbol $\sqrt{\sigma(\Delta)} = |\xi|$. See [16] . Let

$$w_+ = \left(\frac{\partial}{\partial t} - \sqrt{\Delta}\right)\mu_+ .$$

Since

$$\left(\frac{\partial}{\partial t^2} - \Delta\right)\mu_+ \in C^\infty$$

we get

$$\left(\frac{\partial}{\partial t} + \sqrt{\Delta}\right)w_+ \in C^\infty. \tag{8.2}$$

By Theorem 7.3 the symbol of the left hand side of (8.2) is $(\tau + |\xi|)\sigma(w_+)$. However, μ_+ and w_+ belong to $I(C)$ (see §7) and on C, $|\xi| = \tau \neq 0$ so (8.2) implies $\sigma(w_+) = 0$. Suppose $w_+ \in I_k(C)$. The vanishing of its symbol implies it is in $I_{k-1}(C)$ and, inductively, that it is in $I_{k-\ell}(C)$ for all ℓ. Hence $w_+ \in C^\infty$.

We will now explore some consequences of the fact that $\mu_+ = v_+ \bmod C^\infty$. Consider

$$\Delta : X \times \mathbf{R} \to X \times X \times \mathbf{R}, \qquad (x,a) \to (x,x,a)$$

and

$$\pi : X \times \mathbf{R} \to \mathbf{R}, \qquad (x,a) \to a.$$

The wave front set of μ_+ is, by construction, the set

$$C = \{(x,\xi,y,\eta,t,\tau)\}$$

with $\tau = |\xi|$ and $(x,-\xi)$ joined to (y,η) by a geodesic of length $|t|$. Since $\tau \neq 0$ for any element of C, $\Delta^*\mu_+$ is well defined, and its wave front set is contained in $\Delta^* C$. Note that $\Delta^* C$ is the set of all $(x,\xi+\eta,t,\tau)$ with $(x,\xi,x,\eta,t,\sigma) \in C$. The wave front set of $\pi_*\Delta^*\mu_+$ is the set of all (t,τ) such that there exists $(x,0,t,\tau) \in \Delta^* C$. This means that if $(t,\tau) \in WF(\pi_*\Delta^*\mu_+)$ there exists a geodesic of length $|t|$ for which $(y,\eta) = (x,\xi)$, i.e., a closed geodesic of period t. On the other hand,

$$\pi_*\Delta^* v_+ = \sum e^{it\sqrt{\lambda_\alpha}} \int e_\alpha(x)\bar{e}_\alpha(x)\,dx = \sum e^{it\sqrt{\lambda_\alpha}}.$$

Since $\mu_+ - v_+ \in C^\infty$ we have proved:

PROPOSITION 8.1. *The sum, $\sum e^{it\sqrt{\lambda_\alpha}}$, is well defined as a generalized function on* $\mathbf{R}$. *Moreover, if T is in its singular support, there exists a closed geodesic on X of period T.*

If Δ and π satisfy transversality conditions of the type discussed in §5, we can use the technique of §5 to obtain sharper information about $\pi_*\Delta^*\mu_+$. Without going into details we will describe one result which can be obtained this way (due to Duistermaat-Guillemin [17]). Let η by the Hamiltonian vector field associated with the function $(x,\xi) \to |\xi|$. If (x,ξ) lies on a closed geodesic γ of period T then $\exp T\eta$ maps (x,ξ) onto itself. Let N be the tangent space to T^*X at (x,ξ) divided by the two dimensional subspace spanned by η and the tangent vector to the cone axis: $\{(x,\lambda\xi), \lambda \in \mathbf{R}^+\}$. The derivative of $\exp T\eta$ maps N onto itself. We will denote this map by P_γ. The geodesic γ is called *generic* if the Lefschetz condition: $\det(I - P_\gamma) \neq 0$ is satisfied. It turns out that if there is just one closed geodesic γ of period T, and γ is generic, then $\sum e^{it\sqrt{\lambda_\alpha}}$ has a simple pole at $t = T$ and the following residue formula is valid:

$$\lim_{t\to T}(t - T)\left(\sum e^{it\sqrt{\lambda_\alpha}}\right) = \frac{|\gamma|}{\pi} i^\sigma |\det(I - P_\gamma)|^{-1/2} \tag{8.3}$$

where σ is the Morse index of γ and $|\gamma|$ is the length of γ. ($|\gamma|$ need not equal $|T|$ since γ might circle several times around the same orbit.) This formula shows that if every closed geodesic of period $\neq 0$ is generic and no two closed geodesics have the same period, then *the periods of the closed geodesics are identical with the singularities of* $\sum e^{it\sqrt{\lambda_\alpha}}$.

Now $\sum e^{it\sqrt{\lambda_\alpha}}$ can be expected to have a bad singularity at $t = 0$ since every point $x \in X$ is a closed geodesic of period 0. Let us see exactly what the nature of the singularity at $t = 0$ is. Recall that for a compact Riemannian manifold there exists a positive number r such that, for all x, $\exp_x : T_x \to X$ is a diffeomorphism of the open ball of radius r in T_x onto the subset of X consisting of all points y with $d(x,y) < r$. Given $\omega \in T_x^*$ and $(x,y) \in X \times X$ with $d(x,y) < r$, let $\varphi(x,y,\omega) = \langle \omega, v\rangle$ where $v \in T_x$ satisfies $\exp_x v = y$. For example if $X = \mathbf{R}^n$ we can take $r = +\infty$ and $\varphi(x,y,\omega) = (x - y)\cdot\omega$. Let

$$Z = \{(x, y, \omega, t) | x,y \in X, t \in \mathbf{R}, \omega \in T_x^*, |\omega| = 1, d(x,y) < r\}$$

and consider the diagram

$$\begin{array}{ccc} Z & \xrightarrow{\psi} & \mathbf{R} \\ \downarrow \pi & & \\ X\times X\times \mathbf{R} & & \end{array} \tag{8.4}$$

where $\psi(x,y,\omega,t) = t + \varphi(x,y,\omega)$. We will show that, for $|t| < r$, (8.4) is a parametrization of C.

PROOF (VIA GAUSS'S LEMMA). The fiber above (x,y,t) is the cosphere $S_x^* = \{\omega \in T_x^*, |\omega| = 1\}$. Restricted to this fiber φ is just the linear function $\langle v, \omega\rangle$ where $v \in T_x$ satisfies $\exp_x v = y$. Therefore if (x,y,ω,t) belongs to the critical set of ψ (the set where ψ and its fiber derivative vanish) we must have $v = c\omega$, for some $c \in \mathbf{R}$ and since $\psi = 0$, $-t = c$. Therefore

$$(x,y,\omega,t) \in \text{critical set} \Leftrightarrow y = \exp_x(-t)\omega.$$

(Here we have identified $\omega \in T_x^*$ with the element corresponding to it in T_x via the isomorphism $T_x^* \cong T_x$ given by $\langle\ ,\ \rangle_x$.) Now we will compute $d_y\varphi$ at a point of the critical set. Identify T_y with T_y^* by $\langle\ ,\ \rangle_y$ and let η be the element of T_y^* corresponding to the unit tangent vector to the geodesic joining x to y. We will show that $d_y\varphi = \eta$. Let $v(s)$ be a curve in T_x with $v(0) = v$ and let $y(s) = \exp_x v(s)$. Then $\varphi(x, y(s), \omega) = \omega \cdot v(s)$. Suppose first that $v(s)$ lies on a sphere of fixed radius a, in T_x. Then $(d/ds)\varphi(x, y(s), \omega) = \omega \cdot \dot{v}(0) = (v(0)/a) \cdot \dot{v}(0) = 0$ at $s = 0$. On the other hand by Gauss's lemma dy/ds is perpendicular to the tangent vector to the geodesic joining x to y so $(dy/ds) \cdot \eta = 0$ at $s = 0$. Next we will vary y by just varying the length of v, leaving its direction fixed, i.e., we will set $y(s) = \exp_x s\omega$ with $y = y(a)$, $a = |v|$. Then

$$(d\varphi/ds)(x,\ \omega,\ y(s)) = 1 = \eta \cdot (dy/ds) \qquad \text{at } s = a.$$

This proves $d_y\varphi = \eta$. A similar argument shows that $d_x\varphi = -\omega$. Since $d_t\psi = 1$ this proves that (8.4) is a parametrization of C, as claimed.

Since $\mu_+ \in I_{n/2}(X \times X \times \mathbf{R}, C)$ there exists a function $a = a(x, y, \omega, t)$ on Z such that

$$\mu_+ = \pi_* a\psi^* v_{n-1}^+ = \iint_0^\infty a(x, y, \omega, t) s^{n-1} e^{is\psi}\, ds\, d\omega \tag{8.5}$$

modulo elements of $I_{(n/2)-1}(X \times X \times \mathbf{R}, C)$. For $t = 0$, $\mu_+ = \delta(x - y)$ and $\psi(x, y, \omega, 0) = \varphi(x, y, \omega) = (x - y) \cdot \omega + O(|x - y|^2)$ locally; so comparing (8.5) with the standard representation of the δ-function in terms of plane waves

$$\delta(x - y) = \frac{1}{(2\pi)^n} \iint_0^\infty e^{is(x-y)\cdot\omega} s^{n-1}\, ds\, d\omega$$

we see that

$$a(x, y, \omega, 0) = \frac{1}{(2\pi)^n} + O(|x - y|). \tag{8.6}$$

We can write $a(x, y, \omega, t)$ as $a(x, y, \omega, 0) + tb(x, y, \omega, t)$. The contribution of the second term to (8.5) is

$$\iint_0^\infty tb(x, y, \omega, t) s^{n-1} e^{ist} e^{is\varphi}\, ds\, d\omega. \tag{8.7}$$

Since $te^{ist} = (1/\sqrt{-1})(d/ds)e^{ist}$ the integrand of (8.7) can be replaced by an integrand of order $O(s^{n-2})$ by an integration by parts. Therefore modulo elements of $I_{(n/2)-1}$,

$$\mu_+ = \iint_0^\infty a(x, y, \omega, 0) s^{n-1} e^{ist} e^{is\varphi}\, ds\, d\omega. \tag{8.8}$$

If we set $x = y$ in the integrand of (8.8), then, by (8.6),

$$\mu_+(x, x, t) = \frac{1}{(2\pi)^n} \int d\omega \int_0^\infty e^{ist} s^{n-1}\, ds. \tag{8.9}$$

Finally integrating (8.9) over X we get, for $|t| < r$,

$$\sum e^{it\sqrt{\lambda_\alpha}} = \frac{\operatorname{vol}(S^* X)}{(2\pi)^n} \int_0^\infty s^{n-1} e^{ist}\, ds + \cdots \tag{8.10}$$

(the dots indicating similar terms involving lower powers of s).

To exploit this formula we will multiply the right hand side by a function with support in $|t| < r$ and Fourier transform backwards. The function we will use ought to have the following properties:

(a) $\hat{\rho} \geq 0$,
(b) $\rho(0) = 1$,
(c) $\hat{\rho} > c > 0$ on $-1 < s < 1$.

To construct such a function let ρ_0 be an even function which is non-negative and has support in $-r/2 < t < r/2$; and let $\rho_1 = \rho_0 \star \rho_0$. Then $\hat{\rho}_1 \geq 0$, and for appropriate $\alpha, \beta > 0$, $\rho = \alpha\rho_1(t/\beta)$ has properties (b) and (c).

From (b) we get the following:

LEMMA 8.1. *For $s \gg 0$,*

$$\hat{\rho} \star s_+^{n-1} = s_+^{n-1} + O(s^{n-2}).$$

PROOF.

$$\begin{aligned} \hat{\rho} \star s_+^{n-1} &= \int_{-\infty}^s (s - z)^{n-1} \hat{\rho}(z)\, dz \\ &= s^{n-1} \int_{-\infty}^s \hat{\rho}(z)\, dz + (1 - n) s^{n-2} \int_{-\infty}^s z\hat{\rho}(z)\, dz + \cdots. \end{aligned}$$

Since $\hat{\rho}(z)$ is rapidly decreasing each of the integrals $\int_{-\infty}^s z^k \hat{\rho}(z)\, dz$ can be written as $\int_{-\infty}^\infty z^k \hat{\rho}(z)\, dz$ with an error term of order $O(s^{-N})$, N arbitrary. Since $\int_{-\infty}^\infty \hat{\rho}(z)\, dz = \rho(0) = 1$ this proves the lemma. Q.E.D.

Multiplying (8.10) by $\rho(t)$ and taking its Fourier transform we get

$$\sum \hat{\rho}(s - \mu_i) = \frac{\operatorname{vol}(S^* X)}{(2\pi)^n} \hat{\rho} \star s_+^{n-1} + O(s^{n-2})$$

where $\mu_i = \sqrt{\lambda_i}$. Therefore from the lemma we get the following formula:

$$\sum \hat{\rho}(s - \mu_i) = \frac{\operatorname{vol}(S^* X)}{(2\pi)^n} s_+^{n-1} + O(s^{n-2}). \tag{8.11}$$

Our first use of this formula will be to obtain an upper bound on the number of eigenvalues in any unit interval:

LEMMA 8.2. *The number of eigenvalues of* $\sqrt{\Delta}$ *contained in the interval* $(s-1, s+1)$ *is bounded above by* (const) s^{n-1}.

PROOF. If μ_i is such an eigenvalue its contribution to (8.11) is $\geq 1/c$ because of property (c). Q.E.D.

From this we will deduce the following theorem on asymptotic distribution of eigenvalues.

PROPOSITION 8.2. *Let* $g(s)$ *be the number of eigenvalues of* $\sqrt{\Delta}$ *in the interval between* 0 *and* s. *Then*

$$g(s) = \frac{\text{vol}\,(S^*X)}{(2\pi)^n}\frac{s_+^n}{n} + O(s^{n-1}).$$

PROOF. The left hand side of (8.11) can be written as $\hat{\rho} * (dg/ds)$ or $(d/ds)(\hat{\rho} * g)$; so if we integrate it from 0 to s we get

$$\hat{\rho} \star g = \frac{\text{vol}\,(S^*X)}{(2\pi)^n}\frac{s_+^n}{n} + O(s^{n-1}). \tag{8.12}$$

Now

$$\begin{aligned}(g - \hat{\rho} \star g)(s) &= g(s) - \int g(s-z)\hat{\rho}(z)\,dz \\ &= \int (g(s) - g(s-z))\hat{\rho}(z)\,dz.\end{aligned}$$

By Lemma 8.2, $|g(s) - g(s-z)| \leq C(|s| + |z|)^{n-1}|z|$; hence

$$|(g - \hat{\rho} \star g)(s)| \leq C\int (|s| + |z|)^{n-1}|z|\hat{\rho}(z)\,dz.$$

Since $\hat{\rho}(z)$ is rapidly decreasing the right hand side can be estimated by $O(s^{n-1})$, so we can replace $\hat{\rho} \star g$ by g in (8.12). Q.E.D.

REMARK. The proof we have just given of Proposition 8.2 is due to Hormander. (Hormander actually proves Proposition 8.2 for a much larger class of operators than we have considered here.)

As a last application of the techniques of this section we will give a proof of the Atiyah-Bott fixed point formula. Let

$$0 \to C^\infty(E^1) \xrightarrow{d} C^\infty(E^2) \xrightarrow{d} \cdots \xrightarrow{d} C^\infty(E^N) \to 0 \tag{8.13}$$

be a complex of differential operators defined on a compact manifold X. We say that (8.13) is *elliptic* if the Laplacian

$$\square = dd^t + d^t d$$

is elliptic in the usual sense for differential operators on vector bundles (has a bijective symbol). Let (f, r) be a morphism of (8.13) in the sense of §2, that is $f: X \to X$ is a Lefschetz map, and $r: f^* E^i \to E^i$ is a morphism of vector bundles such that the induced $f^*: C^\infty(E^i) \to C^\infty(E^i)$ commute with d. The Atiyah-Bott formula says that

$$\sum (-1)^i \operatorname{tr} f^* : C^\infty(E^i) \to C^\infty(E^i) \tag{8.14}$$

is equal to the alternating sum of the traces on homology providing the traces in (8.14) are defined according to the recipe given in §2. In §2 we showed that an elementary proof could be given of this formula whenever the morphism (f, r) could be imbedded in a transitive family of such morphisms $\{(f_y, r_y), y \in Y\}$. The main step in that proof was to show that

$$\int f_y^* \rho(y)\, dy$$

is a smoothing operator for all $\rho \in C_0^\infty(Y)$.

It is instructive to see how to imbed f in a transitive family of morphisms for the most elementary of all elliptic complexes, the de Rham complex. Here f^* is just the "pull back" operation on i forms which is defined for any map $f: X \to X$ and automatically commutes with d. To imbed f in a transitive family, choose vector fields $v_1, \ldots, v_N$ on X such that $v_1(x), \ldots, v_N(x)$ span T_x for all $x \in X$. Then, for $y = (y_1, \ldots, y_N)$,

$$f_y = \exp y_1 v_1 \circ \exp y_2 v_2 \circ \cdots \circ \exp y_N v_N \circ f \tag{8.15}$$

is a transitive family.

Given an arbitrary elliptic complex there is, in some sense, a natural substitute for the construction just outlined. Let $\square$ be the Laplacian defined above. Then $d\,\square = dd^t d = \square\, d$, so $\square$ is also a morphism of (8.12), though not a morphism of the type considered above. Since the complex is elliptic, $\square$ is a positive self-adjoint elliptic differential operator and $\sqrt{\square}$ a positive self-adjoint elliptic pseudodifferential operator. Both have self-adjoint extensions to $L^2(E^i)$, so by the spectral theorem one can exponentiate $i\sqrt{\square}$ and get a one parameter group of unitary operators $\exp it\sqrt{\square}$. Clearly $(1/i)(d/dt)\exp it\sqrt{\square} = \sqrt{\square}\exp it\sqrt{\square}$; so if $k(x,y,t)$ is the Schwartz kernel of $\exp it\sqrt{\square}$ it satisfies the equation

$$\left(\frac{1}{\sqrt{-1}}\frac{\partial}{\partial t} - \sqrt{\square_x}\right)k(x,y,t) = \left(\frac{1}{\sqrt{-1}}\frac{\partial}{\partial t} - \sqrt{\square_y}\right)k(x,y,t) = 0,$$
$$k(x,y,0) = \delta(x-y); \tag{8.16}$$

so, for example, if $\square$ were the usual Laplace-Beltrami operator we would know a lot about $k(x,y,t)$ from the results of the last two sections. Unfortunately, it is not known at present to what extent these results extend to elliptic operators on vector bundles. There are however some trivial facts on wave front sets which do extend to vector bundles, and this is all we will need for our proof. First we will need to extend the definition of wave front sets to sections of vector bundles:

DEFINITION. Let μ be a generalized section of the vector bundle E. Then (x,ξ) is in the wave front set of μ if, for some smooth section s, of the dual bundle, (x,ξ) is in the wave front set of $\langle \mu, s\rangle$.

Now let $D: C_0^\infty \to C^\infty(E)$ be a pseudodifferential operator. We will say that D is elliptic at (x,ξ) if $\sigma(D)(x,\xi)$ is bijective. We claim:

LEMMA 8.3. *If $D\mu$ is smooth and D is elliptic at (x,ξ) then μ is smooth at (x,ξ); i.e., $(x,\xi) \notin WF(\mu)$.*

The proof is exactly like the proof of Proposition 6.6. Namely, because of the ellipticity, one can construct a pseudodifferential operator Q such that $QD - I$ is smooth at (x,ξ). Applying this lemma to the two equations (8.16) satisfied by k we get:

LEMMA 8.4. *If (x,ξ,y,η,t,τ) is in $WF(k)$ then*

$$\tau = \gamma_j(x,\xi) = \gamma_k(y,\eta) \tag{8.17}$$

where $\gamma_j(x,\xi)$ is an eigenvalue of $\sigma(\sqrt{\square})(x,\xi)$ and $\gamma_k(y,\eta)$ an eigenvalue of $\sigma(\sqrt{\square})(y,\eta)$.

Because of the ellipticity, $\gamma_j(x,\xi)$ and $\gamma_k(y,\eta)$ are greater than zero for $\xi \neq 0$ or $\eta \neq 0$; so from Lemma 8.4 we conclude:

PROPOSITION 8.3. *If (x,ξ,y,η,t,τ) is in $WF(k)$ then $\tau \neq 0$.*

As a corollary we get:

PROPOSITION 8.4. *If $\rho \in C_0^\infty(\mathbf{R})$ then the operator*

$$\int \rho(t) \exp \sqrt{-1}\, t \sqrt{\square}\, dt \tag{8.18}$$

is smoothing.

PROOF. The Schwartz kernel of (8.18) is the push forward with respect to $\pi: X \times X \times \mathbf{R} \to X \times X$ of $\rho(t)k(x,y,t)$. By §3 the wave front set of this push forward is contained in the set of (x,ξ,y,η) such that $(x,\xi,y,\eta,t,0)$ is in $WF(\rho k)$. However, this set is empty by Proposition 8.3; so $\pi_* \rho k$ is smooth.

Composing (8.18) with f^* we see that

$$\int \rho(t)(\exp \sqrt{-1}\, t\sqrt{\Box})f^*\, dt \tag{8.19}$$

is smoothing. The proof of Atiyah-Bott now follows the lines of §2; namely we view the trace of (8.17) as a distribution on $C_0^\infty(\mathbf{R})$ and observe that as in §2 the alternating sum of the traces of (8.19) equals the alternating sum of the traces on homology. We leave the details to the reader.

Reference, Chapter VI

1. I. M. Gel'fand and M. A. Naĭmark, *Unitary representations of the classical groups*, Trudy Mat. Inst. Steklov **36** (1950). MR **13**, 722.
2. M. F. Atiyah and R. Bott, *Ann. of Math.* (2) **86** (1967), 374–407. MR. **35** #3701.
3. M. Sato, *Conf. on Functional Analysis and related topics* (Tokyo, April 1969), Univ. of Tokyo Press, Tokyo, 1970, pp. 91–94. MR **41** #6642.
4. L. Hörmander, *Acta Math.* **127** (1971), 79–183.
5. I. M. Gel'fand, M. I. Graev, and Z. Ja. Šapiro, Funkcional. Anal. i Priložen **3** (1969), no. 2, 24–40 = Functional Anal. Appl. **3** (1969), 101–114. MR **39** #6232.
6. S. Helgason, *Bull. Amer. Math. Soc.* **70** (1964), 435–446. MR **29** #4068.
7. A. Gabor, *Trans. Amer. Math. Soc.* **170** (1972), 239–244. MR **49** #5826.
8. I. M. Gel'fand and G. E. Shilov, Fizmatgiz, Moscow, 1958; English transl., Academic Press, New York, 1964. MR **20** #4182; **29** #3869.
9. J. J. Duistermaat and L. Hörmander, *Acta. Math.* **128** (1972), 183–269.
10. S. Helgason, *Differential geometry and symmetric spaces*, Academic Press, New York, 1962. MR **26** #2986.
11. R. T. Seeley, *Singular integrals* (Proc. Sympos. Pure Math., Chicago, Ill., 1966), Amer. Math. Soc., Providence, R. I., 1967, pp. 288–307. MR **38** #6220.
12. F. W. Warner, *Foundations of differentiable manifolds and Lie groups*, Scott, Foresman, Glenview, Ill., 1971. MR **45** #4312.
13. J. J. Duistermaat and V. Guillemin, *Invent. Math.* **29** (1975), 39–80.
14. J. Chazarain, *Invent. Math.* **24** (1974), 65–82. MR **49** #8062.
15. Y. Colin de Verdeère, Compositio. Math. **27** (1973), 159–184. MR **50** #1293.
16. A. A. Kirillov, Funkcional. Anal. i Priložen **1** (1967), no. 4, 84–85 = Functional Anal. Appl. **1** (1967), no. 4, 330–331. MR **37** #347.
17. C. Chevalley, *Theory of Lie groups*. I, Princeton Univ. Press, Princeton, N. J., 1946. MR **7**, 412.
18. N. Wallach, *Harmonic analysis on homogeneous spaces*, Dekker, New York, 1973.

Appendix to Chapter VI

§1. Let G be a Lie group, dg its Haar measure, and f a compactly supported smooth function of G. Given a unitary representation U: $g \to U(g)$ of G on the Hilbert space H one defines a bounded operator U_f on H by

$$U_f = \int U(g) f(g)\, dg \tag{A1.1}$$

Let $\hat{G}$ be the set of irreducible unitary representations of G. The formula (A1.1) assigns to f an (operator valued) function $\hat{f}$ on $\hat{G}$ by the assignment $\hat{f}(U) = U_f$. The basic problem of harmonic analysis on Lie groups is the problem of determining f from $\hat{f}$, or in other words inverting the group-theoretic "Fourier transform" $f \to \hat{f}$. It turns out that an inversion formula, if it exists, has to be of a rather special form. To see just what this form has to be let's inspect a simple case, the case of G a *finite* group. Let $L(G)$ be the space of complex valued functions on G, and take $L(G)$ into an algebra by convolution. The left regular action of G on $L(G)$ leaves fixed the bilinear form

$$(f_1, f_2) = \sum_{g \in G} f_1(g) \bar{f}_2(g);$$

therefore, $L(G)$ can be decomposed into a direct sum of irreducible unitary representations of G

$$L(G) = \sum m_i V^i \tag{A1.2}$$

(*Notation.* G acts on V^i by the unitary representation U^i and m_i is the multiplicity with which U^i occurs in the regular representation.) Now let $f \in L(G)$ and let L_f be the action of f on $L(G)$ by convolution. From (A1.2) we get

$$\operatorname{tr} L_f = \sum (\operatorname{tr} L_f | V_i) m_i .$$

Let r be the cardinality of G. The normalized Haar measure on G is the counting measure divided by r, so $L_f | V_i = r U_f^i$ by (A1.1). Thus we can rewrite our formula as

$$\operatorname{tr} L_f = \sum r m_i \operatorname{tr} U_f^i = \sum r m_i \chi^i(f) \tag{A1.3}$$

where $\chi^i(f)$ is, by definition, the *character of the representation* U^i. On the other hand

$$L_f = \sum_{g \in G} f(g) L_g$$

and $\operatorname{tr} L_g = r$ or 0 depending on whether $g = e$ or $g \neq e$. Thus from (A1.3) we get

$$f(e) = \sum m_i \chi^i(f). \tag{A1.4}$$

This formula is the so-called *Plancherel formula* for the group G. If we apply it to the function $L_a^* f$, the left translate of f by $a \in G$, we get the equivalent formula

$$f(a) = \sum m_i \operatorname{tr} (U_f^i U^i(a)^*). \tag{A1.5}$$

This formula solves our Fourier inversion problem by giving f explicitly in terms of $\hat{f}$.

Inspired by the results of the preceding paragraph we are led to what is nowadays the standard form of the "Fourier inversion problem" for Lie groups: Find a measure μ on the space $\hat{G}$ of (equivalence classes of) irreducible unitary representations of G such that for every $f \in C_0^\infty(G)$

$$f(e) = \int_{\xi \in \hat{G}} \chi^\xi(f)\, d\mu(\xi) \tag{A1.6}$$

where $\chi^\xi(f) = \operatorname{tr} U_f^\xi$. The measure μ is called the *Plancherel measure* of the group G. As in the example above it represents the multiplicity with which the representation ξ occurs in the left regular representation. Let us inspect a few examples of the formula (A1.6) (some of which are well known to the reader under other guises).

Example 1. $G = \mathbf{R}^n$. The irreducible unitary representations of G are all one-dimensional and are of the form $x \to e^{-i\langle x,\xi\rangle}$ where $\xi \in (\mathbf{R}^n)^*$. The Fourier transform of f is just its usual Fourier transform

$$\hat{f}(\xi) = \int f(x) e^{-i\langle x,\xi\rangle}\, dx$$

and (A1.6) is just the usual Fourier inversion formula

$$f(0) = \frac{1}{(2\pi)^n} \int \hat{f}(\xi)\, d\xi$$

the Plancherel measure being $d\xi/(2\pi)^n$.

Example 2. $G = T^n = \mathbf{R}^n/Z^n$ the standard n torus. All the irreducible unitary representations of G are one-dimensional and of the form $x \to e^{-2\pi i (x,k)}$ where $k \in Z^n$. The Plancherel formula takes the form

$$f(0) = \sum_{k \in Z^n} \hat{f}(k), \qquad \hat{f}(k) = \int_{T^n} f e^{-2\pi i\langle x,k\rangle}\, dx,$$

which is the standard Fourier inversion formula for Fourier series. The Plancherel measure this time is the counting measure on Z^n.

Example 3. $G = H^n$, the $2n + 1$ dimensional Heisenberg group. The underlying group manifold for this group is $\mathbf{R}^{2n+1}$ with coordinates $x_1, \ldots, x_n, y_1, \ldots, y_n$ and t; and the left invariant vector fields on this group are the vector fields:

$$\xi_i = \frac{\partial}{\partial x_i} + y_i \frac{\partial}{\partial t}, \qquad \eta_i = \frac{\partial}{\partial y_i}, \qquad \zeta = \frac{\partial}{\partial t}. \tag{A1.7}$$

These satisfy the standard Heisenberg bracket relations: $[\eta_i, \xi_j] = \delta_{ij}\zeta$. The irreducible unitary representations of G are of two kinds.

I. The one-dimensional representations. These factor through the quotient group $H^n/\mathbf{R} \cong \mathbf{R}^{2n}$, $\mathbf{R} \cong \exp t\zeta$, and are identical with those of Example 1.

II. A one-parameter family of infinite-dimensional representations, one such representation for each $\tau \in \mathbf{R} - \{0\}$. The underlying space on which each of these representations act is $L^2(\mathbf{R}^n)$ and the representation itself is:

$$(x, y, t) \to e^{i\tau(x\cdot q + t + x\cdot y/2)} T_y^* . \tag{A1.8$^\tau$}$$

Here $q_1, \ldots, q_n$ are the standard coordinates in $\mathbf{R}^n$, and T_y is translation by y. ξ_i gets represented as multiplication by $\sqrt{-1}\,\tau q_i$, η_i as $\partial/\partial q_i$ and ζ as multiplication by $\sqrt{-1}\,\tau$. These representations (the Stone-von Neumann representations) have been discussed at some length in Chapter V. To obtain the Plancherel formula, we first note that if we perform a partial Fourier transform in the variables x, t, leaving y fixed, then by (A1.7) the vector fields ξ_i, η_i and ζ get transformed into the operators $\sqrt{-1}\,(\xi_i + \tau y_i)$, $\partial/\partial y_i$, $\sqrt{-1}\,\tau$, where τ and ξ are the dual variables to t and x. These operators are, except for constants, identical with the images of ξ_i, η_i and ζ in the τth Stone-von Neumann representation. Otherwise stated consider the unitary map $L^2(\mathbf{R}^{2n+1}) \to L^2(\mathbf{R}^{2n+1})$ given by partial Fourier transform in the x, t variables. This map decomposes $L^2(\mathbf{R}^{2n+1})$ into a "direct integral" of irreducible representations

$$L^2(\mathbf{R}^{2n+1}) \cong \int H_{\xi,\tau}\, d\xi\, d\tau. \tag{A1.9}$$

Here $H_{\xi,\tau}$ is a copy of $L^2(\mathbf{R}^n)$ and the representation of the Heisenberg group on it is isomorphic with the τth Stone-von Neumann representation. The decomposition (A1.9) is called the "Plancherel decomposition" of $L^2(H^n)$.

To get the Plancherel formula we write the usual Fourier inversion formula in the form

$$f(0) = \frac{1}{(2\pi)^{n+1}} \int \tilde{f}(\xi, 0, \tau)\, d\xi\, d\tau \tag{A1.10}$$

where $\tilde{f}(\xi, y, \tau)$ is the partial Fourier transform of f with respect to x and t. We will show that (A1.10) is in fact the Plancherel formula for H^n. To do so we must compute the trace of U_f^τ with respect to the representation (A1.8)$^\tau$. Integrating (A1.8)$^\tau$ we get for U_f^τ:

$$\int e^{i\tau(x\cdot q + t + x\cdot y/2)} T_y^* f(x, y, t)\, dx\, dy\, dt$$

or

$$\int \tilde{f}\left(\tau\left(q + \frac{y}{2}\right), y, \tau\right) T_y^* \, dy.$$

Let us show that this operator is an integral operator with a smooth kernel. By definition, this operator applied to a function $g = g(q)$ is equal to

$$\int \tilde{f}\left(\tau\left(q + \frac{y}{2}\right), y, \tau\right) g(q + y) \, dy$$

or

$$\int \tilde{f}\left(\tau\left(\frac{y + q}{2}\right), y - q, \tau\right) g(y) \, dy$$

which is, as asserted, an integral operator with the smooth kernel

$$\tilde{f}\left(\tau\left(\frac{y + q}{2}\right), y - q, \tau\right). \tag{A1.11}$$

To compute its trace we set $y = q$ and integrate over y in (A1.11). This gives us

$$\operatorname{tr} U_f^\tau = \int \tilde{f}(\tau y, 0, \tau) \, dy = \tau^{-n} \int \tilde{f}(y, 0, \tau) \, dy. \tag{A1.12}$$

Therefore (A1.10) can be rewritten in the form

$$f(0) = \frac{1}{(2\pi)^{n+1}} \int \operatorname{tr} U_f^\tau \tau^n \, d\tau, \tag{A1.13}$$

which is the Plancherel formula; and, in particular, the Plancherel measure for H^n is $\tau^n d\tau/(2\pi)^{n+1}$. Note that only the representations of type II enter into the Plancherel formula.

Example 4. $G =$ a compact Lie group. The Peter-Weyl theorem says that all irreducible representations of G are finite dimensional and that $L^2(G)$ decomposes as a Hilbert space direct sum of irreducibles

$$L^2(G) = \sum m_i H^i$$

each representation H^i occurring with multiplicity $m_i = \dim H^i$. (For a proof of the Peter-Weyl theorem, see [18].) The Plancherel formula is identical with the Plancherel formula for finite groups namely

$$f(e) = \sum m_i \chi^i(f)$$

$\chi^i(f)$ being the character of f in the V^i representation.

We conclude these general remarks by noting that a Plancherel formula for nilpotent Lie groups involving the "orbit picture" of Chapter V was obtained by Kirillov [16]. The remainder of this appendix will be devoted to proving the Plancherel formula for certain non-compact semi-simple groups. The results we will be describing were obtained independently by Harish-Chandra and Gelfand-Neumark in the early fifties. The approach outlined below is mostly due to Gelfand.

§2. We will need to fix a few conventions about notation and note a few elementary facts about semi-simple Lie groups. Let G be a semi-simple Lie group and K a maximal compact subgroup of G. Then there exists a vector group A and a nilpotent group N contained in G such that the product map $K \times A \times N \to G$ is a diffeomorphism. This is the *Iwasawa decomposition* of G. (See [18].) Let M be the centralizer of A in K and let $B = MAN$. B is called the *Borel* subgroup of G. The quotient $G/B = K/M$ is compact. An example to keep in mind is $G = SL(n, \mathbf{C})$. In this example $K = SU(n)$, N is the group of triangular matrices with ones on the diagonal, A is the group of diagonal matrices with positive real entries, M is the group of diagonal matrices with entries of modulus one, and B the group of triangular matrices.

Since MA is the product of a compact group and an abelian group all its irreducible unitary representations are finite dimensional. Let $\sigma \in \widehat{MA}$ be such a representation. σ can be extended to a representation of B by the homomorphism $B \to B/N = MA$, then induced up to G. (See §2 of Chapter VI.) The representations on G obtained this way are called the *principal series* representations. As we will see below, for the complex groups they are the only representations occurring in the Plancherel formula.

In §5 we will need an elementary fact concerning the Lie algebra of G. Let $\mathfrak{g}$, b, n, a etc. denote the Lie algebras of the various subgroups introduced above; and recall the $\mathfrak{g}$ is equipped with a non-degenerate symmetric quadratic form, $\langle\, ,\rangle$, its Killing form.

LEMMA 2.1. *n is a maximal isotropic subspace of $\mathfrak{g}$ with respect to $\langle\, ,\rangle$; and its annihilator space is equal to b.*

A proof of this can be found, for example, in [18]. Let's just check it for $SL(n, \mathbf{C})$. Here n is the algebra of upper triangular matrices with zeroes on the diagonal and b the algebra of upper triangular matrices; so if $A \in n$ and $B \in b$, $AB \in n$ and $\operatorname{tr} AB = 0$. This shows that n is isotropic and that $b \subset n^\perp$. Since $\dim \mathfrak{g}/b = \dim n$, $b = n^\perp$ as asserted.

§3. In this section we will describe the "F_f formalism" of Harish-Chandra, which seems to be an essential ingredient in all proofs of the Plancherel formula. Let us first of all recall the main facts we have so far established about characters of induced representations. For the moment G can be an arbitrary connected Lie group. Let dg be its Haar measure, B a closed cocompact subgroup of G and $X = G/B$.

1. Let U: $g \to U(g)$ be a representation of G induced from B. Then for $f \in C_0^\infty(G)$, $U_f = \int U(g)f(g)\,dg$ is of trace class and $f \to \operatorname{tr} U_f$ is a distribution on G.

This distribution is, by definition, the *character* of U.

2. Suppose B has the following property: There exists $\xi \in b$ such that $\operatorname{ad}(\xi)\colon \mathfrak{g}/b \xrightarrow{\cong} \mathfrak{g}/b$. Then for every U induced from B the character of U is locally L^1 summable function on G.[1]

3. Let $L_g\colon X \to X$ be the left action of g on X. g is called *regular* if L_g is a Lefshetz map. Let σ be a representation of B, let E^σ be the induced vector bundle on X, $L_g^{\#}$ the induced action of g on E^σ and χ the character of the corresponding induced representation. Then χ restricted to the set of regular points of G is a smooth function and is expressible by the formula

$$\chi(g) = \sum_{gx=x} \frac{\operatorname{tr} L_g^{\#}\colon E_x^\sigma \to E_x^\sigma}{|I - (dL_g)_x|}. \tag{A3.1}$$

4. Let $F\colon G \times X \to X \times X$ be the fiber mapping $(g,x) \to (x,gx)$. Let $Z = F^{-1}(\Delta)$ and let $\pi\colon Z \to G$ be the projection of $G \times X$ on G restricted to Z. Then given a representation σ of B there exists a canonical smooth measure μ_σ on Z depending only on σ such that

$$\chi_\sigma = \pi_* \mu_\sigma \tag{A3.2}$$

χ_σ being the character of the representation induced from σ. For $\sigma = 0$, μ_0 is constructed as follows: For $z = (g,x)$ to be in Z we must have $gx = x$. Thus dF_z maps $T_g \oplus T_x$ onto $T_x \oplus T_x$. Identifying the normal space to the diagonal at (x,x) with T_x by the map $(v,-v) \to v$ we get an exact sequence

$$0 \to T_z Z \to T_g \oplus T_x \to T_x \to 0$$

and by standard functorial nonsense this gives a canonical identification $|\wedge| T_z Z \cong |\wedge| T_g$. The density μ_0 at z is the density on the LHS corresponding to Haar measure on the RHS. For σ an arbitrary representation of B let h_σ be the function

$$h_\sigma(z) = h_\sigma(g,x) = \operatorname{trace} L_g^{\#}\colon E_x^\sigma \to E_x^\sigma \tag{A3.3}$$

at $z = (g,x)$. Then for μ_σ one has the formula

$$\mu_\sigma = h_\sigma \mu_0. \tag{A3.4}$$

For the proofs of 1–4 we refer to §2 of VI. Let's now go back to the case at hand. G is a semi-simple Lie group, B its Borel subgroup and K, A, M and N are as in §2. $\widehat{MA}$ is the set of all equivalence classes at irreducible unitary representations of MA, the elements of $\widehat{MA}$ being in one-one correspondence with the principal series representations. Consider now the following diagram of mappings

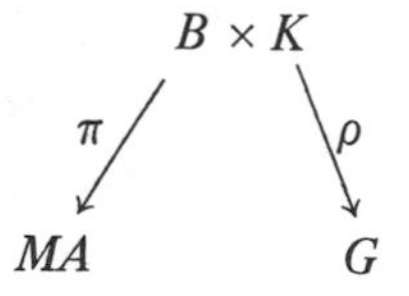

[1] The characters are, sui generis, generalized *densities*; however because of the existence of Haar measure, we are permitted to confuse functions and densities.

where π is the composite of the two projections $B \times K \to B$ and $B \to B/N = MA$, and ρ is the mapping $(b,k) \to kbk^{-1}$. Note that ρ is proper and π is a fiber mapping. The main result of this section is:

THEOREM 3.1. *There exists a smooth non-vanishing measure ν on $B \times K$ with the following properties.*

(i) ν is K invariant.

(ii) Given $f \in C_0^\infty(G)$, let $F_f = \pi_* \rho^* f\nu$.

Then the character, σ, of an element of MA and the character χ_σ of the corresponding induced representation are related by

$$\chi_\sigma(f) = \int_{MA} \sigma(ma) F_f. \tag{A3.6}$$

PROOF. Consider the principal M-bundle $K \xrightarrow{\kappa} K/M = X$. Let $F_1 \colon G \times K \to X \times X$ be the map

$$G \times K \xrightarrow{\kappa} G \times X \xrightarrow{F} X \times X$$

and let $Z_1 = \kappa^{-1}(Z) = F_1^{-1}(\Delta)$. Then $Z_1 \xrightarrow{\kappa} Z$ is a principal M-bundle over Z. Let μ_0 be the measure (A3.2), and let μ_1 be the unique M invariant measure on Z_1 such that $\kappa_* \mu_1 = \mu_0$. Let π_1 be the restriction to Z_1 of the projection of $G \times K$ onto G. Since $\pi_1 = \pi \circ \kappa$, (A3.2) can also be written

$$\chi_0 = (\pi_1)_* \mu_1 . \tag{A3.7}$$

We now make a trivial observation, which is essentially the main step in the proof of Theorem 3.1. Let Φ: $G \times K \to G \times K$ be the mapping $(g,k) \to (kgk^{-1},k)$. We claim that Φ maps $B \times K$ diffeomorphically onto Z_1. In fact, composing Φ with F_1 we get $F_1 \circ \Phi(g,k) = F_1(kgk^{-1},k) = (\bar{k},\overline{kg})$ where $\bar{k}$ and $\overline{kg}$ are the images of k and kg in G/B. $\bar{k} = \overline{kg}$ if and only if $g \in B$; so $(F_1 \circ \Phi)^{-1}(\Delta) = \Phi^{-1}(Z_1) = B \times K$. Thus Φ maps $B \times K$ onto Z_1 as asserted.

Note next that $\pi_1 \circ \Phi(b,k) = kbk^{-1} = \rho(b,k)$; so if we set $\nu = \Phi^* \mu_1$ we can rewrite (A3.7) as

$$\chi_0 = \rho_* \nu. \tag{A3.8}$$

Clearly ν is a smooth non-vanishing measure on $B \times K$. We claim it is K-invariant. To see this let $a \in K$. Associated with a are the following diffeomorphisms.

1. On G, $g \to aga^{-1}$.
2. On Z, $(g,x) \to (aga^{-1},ax)$.
3. On Z_1, $(g,k) \to (aga^{-1},ak)$.
4. On $B \times K$, $(b,k) \to (b,ak)$. We let the reader check that these diffeomor-

phisms commute with, π_1, κ, ρ etc. Since χ_0 is a character, and K acts on G by the adjoint representation, χ_0 is K-invariant. It follows that μ_0, μ_1 and ν are K-invariant as well. Q.E.D.

We next derive a formula analogous to (A3.8) for the induced characters associated with non-trivial $\sigma \in \widehat{MA}$. Consider the map $\kappa \circ \Phi\colon B \times K \to Z$. Let h_σ be the function (A3.3) and $\tilde{h}_\sigma$ its pull back to $B \times K$. We claim that $\tilde{h}_\sigma = \pi^* \sigma$ where π is the left-hand arrow in the diagram (A3.5). In fact $\tilde{h}_\sigma(b,k) = h_\sigma(kbk^{-1}, x)$, x being the coset of k in $X = G/B$. Thus,

$$\tilde{h}_\sigma(b,k) = \operatorname{trace}\,(L^{\#}_{kbk^{-1}}\colon E^\sigma_x \to E^\sigma_x) = \operatorname{trace}\,(L^{\#}_b\colon E^\sigma_0 \to E^\sigma_0) = \sigma(b)$$

as claimed. From (A3.4) and (A3.2) we get the formula

$$\chi_\sigma = \rho_*(\pi^* \sigma)\nu. \tag{A3.9}$$

Applying χ_σ to $f \in C_0^\infty(G)$ we get

$$\chi_\sigma(f) = \int_{B \times K} \pi^* \sigma \rho^* f \nu = \int_{MA} \sigma \pi_* \rho^* f \nu = \int_{MA} \sigma F_f$$

which is the formula (A3.6) we set out to prove.

§4. We now come to the main ingredient in our proof of the Plancherel formula. This is an inversion formula, due to Gelfand, for certain types of generalized Radon transforms. Let

$$\begin{array}{ccc} & Z & \\ {\scriptstyle\pi}\swarrow & & \searrow{\scriptstyle\rho} \\ Y & & X \end{array}$$

be a diagram of smooth manifolds and maps, π being a fiber mapping and ρ proper. Let ν be a smooth nowhere vanishing density on Z. Define

$$R\colon C_0^\infty(X) \to C_0^\infty(|\wedge|Y)$$

by $Rf = \pi_* \rho^* f \nu$. Given an $f \in C_0^\infty(X)$ we would like to be able to determine f from Rf, or, more geometrically, given $x_0 \in X$ we would like to determine the value of f at x_0 from the integrals of f over the fibers of $Z \xrightarrow{\pi} Y$. We will describe below some hypotheses which permit us to do this.

Assume there exist functions $\varphi \in C^\infty(X)$ and $\psi, \theta \in C^\infty(Y)$ such that the following is true:

Axiom I. $\pi^* \psi = \rho^* \varphi$.

Axiom II. $\rho_*(\pi^* \theta \nu)$ is in $C^\infty(|\wedge|X)$ and is non-zero at x_0.

Axiom III. φ has a non-degenerate critical point at x_0.

Then, setting $\mu = \rho_*(\pi^* \theta \nu)$, we get:

$$\begin{aligned}\int e^{i\lambda\varphi} f d\mu &= \int_Z e^{i\lambda\rho^*\varphi} p^* f \pi^* \theta\, d\nu \\ &= \int_Z e^{i\lambda\pi^*\psi} \pi^* \theta \rho^* f d\nu \\ &= \int_Y e^{i\lambda\psi} \theta \{\pi_* \rho^* f d\nu\}\end{aligned}$$

λ being a large real parameter. Comparing the two sides of this equation we get the formula:

$$\int_X e^{i\lambda\varphi} f d\mu = \int_Y e^{i\lambda\psi} \theta Rf. \tag{A4.1}$$

If the support of f is small then stationary phase applied to the right-hand side of (A4.1) determines f at x_0, and the left-hand side depends only on Rf, so we are done.

§5. We will derive the Plancherel formula by applying (A4.1) to the diagram (A3.5), ν being the ν occurring in Theorem 3.1. To do so, we need to make some rather strong assumptions about the regularity of the induced characters of G. Unfortunately these assumptions are only true for the complex semi-simple groups and a few very special types of real semi-simple groups.

Consider the character χ_0 of G induced from the trivial representation of B. We already know that χ_0 is a locally L^1-summable function, and that on the set of regular points of G it is smooth. Moreover, from the formula (A3.1) we know it is positive at the regular point, g, if $L_g\colon G/B \to G/B$ has at least one fixed point. Consider now the case of G a *complex* semi-simple group, for example $G = SL(n, \mathbf{C})$. Let $G^\#$ be the set of regular points of G. Then $G - G^\#$ is of real codimension 2 in G, so $G^\#$ is connected. Therefore, for g_1 and $g_2 \in G^\#$, L_{g_1} and L_{g_2} have the same number of fixed points; and χ_0 is positive on $G^\#$. In particular $\alpha = 1/\chi_0$ is a well-defined smooth function on $G^\#$. We will need the following fact.

LEMMA 5.1. *α extends to a continuous function on G, and the restriction of α to B is smooth and is the pullback of a smooth function θ on B/N.*

The proof of this lemma involves a certain amount of Lie group technology; so we will confine ourselves to proving it for $SL(n, \mathbf{C})$. Let A be an element of $SL(n, \mathbf{C})$. Let $\lambda_1, \lambda_2, \ldots, \lambda_n$ be the complex eigenvalues of A. A is in $G^\#$ if and only if all the λ_i's are distinct. By formula (A3.1) (see also formula 2.3 of VI, §2 and the discussion following Proposition 2.2), the value of the character χ_0 at A is:

$$1/\chi_0(A) = \frac{1}{n!} \prod_{i \neq j} |\lambda_i - \lambda_j|. \tag{A5.1}$$

The expression $\prod_{i\neq j}(\lambda_i - \lambda_j)$ is a symmetric polynomial in $\lambda_1, \ldots, \lambda_n$; so it can be written as a polynomial function of the coefficients of the characteristic polynomial of A and hence as a smooth function $p(A)$ of A. By (A5.1), $1/\chi_0(A) = |p(A)|$ which is well defined as a continuous function on G. On B each matrix can be written uniquely as a product of a nilpotent matrix and a diagonal matrix with entries $\lambda_1, \ldots, \lambda_n$ along the diagonal; so on B we can write (A5.1) as

$$\frac{1}{n!}\prod_{i<j}|\lambda_i - \lambda_j|^2$$

which is a smooth function of $(\lambda_1, \ldots, \lambda_n)$ defined on $B/N = MA$. This proves Lemma 5.1 for $SL(n, \mathbf{C})$. To apply the inversion formula of §4 to the diagram

$$\begin{array}{ccc} & B \times K & \\ {}^{\pi}\swarrow & & \searrow^{\rho} \\ MA & & G \end{array}$$

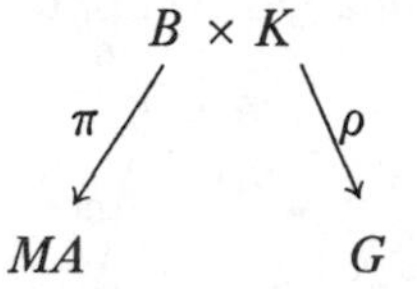

we must find functions $\varphi \in C^\infty(G)$ and ψ and $\theta \in C^\infty(MA)$ satisfying Axioms I, II, and III. For θ we can take the function θ of Lemma 5.1. In fact $\alpha = 1/\chi_0$ is a class function so $\alpha(kbk^{-1}) = \alpha(b)$; and, hence, $\pi^*\theta = \rho^*\alpha$ and

$$\rho_*(\pi^*\theta)\nu = \alpha\rho_*\nu = \alpha\chi_0\,dg = dg. \tag{A5.2}$$

We will construct φ and ψ as follows. Let $\mathfrak{g}$, b, n etc. be the Lie algebras of G, B, N etc. Let $\langle\ ,\ \rangle$ be the Killing form on $\mathfrak{g}$. This form is non-degenerate and invariant under the Ad action of G. Moreover, n is a maximal isotropic subspace of $\mathfrak{g}$ with respect to $\langle\ ,\ \rangle$ and its annihilator space is b by Lemma 3.1. Let $\varphi^0\colon \mathfrak{g} \to \mathbf{R}$ be the quadratic function $x \to \langle x, x\rangle$. The exponential mapping $\exp\colon \mathfrak{g} \to G$ maps a neighborhood of 0 diffeomorphically onto a neighborhood $\mathcal{O}$ of e. Let φ be the function on $\mathcal{O}$ corresponding to the function φ^0 on $\mathfrak{g}$. φ has a non-degenerate critical point at e, since φ^0 has a non-degenerate critical point at 0. Let us prove:

LEMMA. *$\rho^*\varphi$ is constant on the fibers of π.*

PROOF. K acts on $K \times B$ by left multiplication (on K) and acts on G by the adjoint action. ρ intertwines these two actions, and exp intertwines the adjoining actions on G and $\mathfrak{g}$; so $\rho^*\varphi$ takes the same values at (k, b) and (e, b). Therefore we only have to show that φ restricted to B is constant on the fibers of the fibration $B \to B/N$. The diagram

$$\begin{array}{ccc} b & \xrightarrow{\ \exp\ } & B \\ \downarrow & & \downarrow \\ b/n & \xrightarrow{\ \exp\ } & B/N \end{array}$$

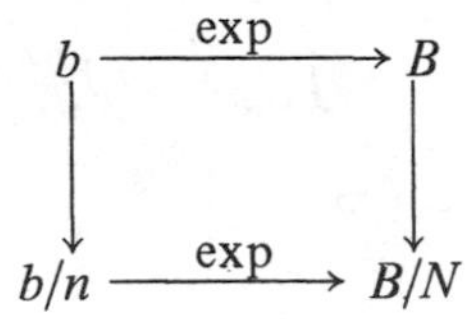

is commutative; so to show that φ is constant on the fibers of $B \to B/N$, it is enough to show that φ^0 is constant on the fibers of $b \to b/n$. However, $\varphi^0(b+n) = \langle b+n, b+n\rangle = \langle b, b\rangle = \varphi^0(b)$, since b is orthogonal to n. Q.E.D.

Since $\langle\ ,\ \rangle$ restricted to b has n as its annihilator space, $\langle\ ,\ \rangle$ induces a non-degenerate quadratic function ψ^0 on b/n. Let ψ be the extension of this function to B/N via the exponential map. We have proved

$$\pi^*\psi = \rho^*\varphi \tag{A5.3}$$

so φ and ψ satisfy Axiom I. Note that ψ also has a non-degenerate critical point at the identity, a fact which we will make use of in a moment.

Now let us consider the formula (A4.1) applied to the diagram

$$\begin{array}{ccc} & B \times K & \\ {\scriptstyle\pi}\swarrow & & \searrow{\scriptstyle\rho} \\ MA & & G \end{array}$$

with ν as our fixed density on $B \times K$ and φ, ψ and θ as in (A5.2) and (A5.3). Combining the formulas (A3.6) and (A4.1) we get

$$\int_G e^{i\lambda\varphi(g)} f(g)\, dg = \int_{MA} e^{i\lambda\psi}\theta(ma) F_f(ma). \tag{A5.4}$$

Let's apply stationary phase to both the left- and right-hand sides of (A5.4). The leading term on the left is, except for some multiplicative constants, equal to $\lambda^{-n/2} f(e)$ where n is the real dimension of G. The leading term on the right occurs with a coefficient of $\lambda^{-k/2}$ where k is the real dimension of MA. However, since $k < n$, this term must vanish, and so must all the terms in the stationary phase expansion up to order l where $l = (n-k)/2$. Thus on the right-hand side we get a stationary phase contribution of the form

$$\lambda^{-n/2} P(D) F_f(e)$$

where $P(D)$ is a constant coefficient differential operator homogeneous of order $2l$. From (A5.4) we deduce the formula

$$f(e) = (P(D)F_f)(e). \tag{A5.5}$$

By the Fourier inversion formula for the group MA and by the formula (A3.6) for the character, χ_σ, this can be rewritten in the form

$$f(e) = c \int_{MA} \chi_\sigma(f) P(\sigma)\, d\sigma. \tag{5.6}$$

This is the Plancherel formula for G.

Chapter VII. Compound Asymptotics

§0. Introduction.

Let V be a topological vector space. We recall from Chapter II that two functions, f_1 and f_2 from $\mathbf{R}^+ \to V$ are asymptotic if, for all N, we have $\tau^N(f_1(\tau) - f_2(\tau)) \to 0$ (in the topology of V) as $\tau \to \infty$. An equivalence class of maps with respect to this equivalence relation is called an asymptotic element of V. In particular, if $V = C^\infty(X)$, $C^\infty(|\wedge|^{1/2}X)$, $C^\infty(\wedge^{1/2}X)$, etc. one talks of asymptotic functions, asymptotic half densities, asymptotic half forms, etc. on a manifold, X. The goal in this chapter is to develop a theory of such asymptotic objects on a manifold, parallel to the theory of distributions that we have developed in Chapter VI. In particular, we will show that one can define a *frequency set* for asymptotics, which is analogous to the wave front set for distributions. We will first define this locally, using an asymptotic analogue of the Fourier transform, due to Leray [1], and then define it globally by integrating asymptotics against rapidly oscillating test functions. Just as in §3 of Chapter VI, we will show that the local and global definitions coincide.

Our analogue of the Lagrangian distributions of Chapter VI, §4, will be a special class of asymptotics called compound asymptotics. To define them we first recall our definition of Lagrangian distributions. Let $\Lambda \subset T^*X$ be a homogeneous Lagrangian submanifold parametrized by

$$\begin{array}{ccc} Z & \xrightarrow{\varphi} & \mathbf{R} \\ {\scriptstyle\pi}\downarrow & & \\ X & & \end{array}$$

so that

$$\Lambda = \pi_* \varphi^* \Lambda_0,$$

where Λ_0 is the cotangent space at the origin of $\mathbf{R}$. Then a Lagrangian distribution associated with Λ can be written as

$$\int a(x,\theta,\tau)e^{i\tau\varphi(x,\theta)}\,d\theta\,d\tau, \tag{0.1}$$

where $a(x,\theta,\tau)$ admits an asymptotic expansion of the form

$$a(x,\theta,\tau) \sim \sum_{-\infty}^{m} a_i(x,\theta)\tau^i \tag{0.2}$$

with x variables on X and θ fiber variables on Z. We refer to Lemma 4.1 of Chapter VI for a precise formulation. A compound asymptotic will be, roughly speaking, an asymptotic of the form

$$\int a(x,\theta,\tau)e^{i\tau\varphi}\,d\theta. \tag{0.3}$$

Thus (0.1) is obtained from (0.3) by integrating over the frequency variable, τ. It will turn out that the frequency set of (0.3) will lie in the inhomogeneous Lagrangian manifold $\Lambda' = \pi_* \varphi^* \Lambda_1$, where Λ_1 is the graph of dt in $T^*\mathbf{R}$, where t is the standard coordinate on $\mathbf{R}$.

There does not seem to be any simple analogue of the symbol calculus of §5 of Chapter VI for compound asymptotics, except when the canonical one form α, of T^*X, is exact when restricted to Λ. If it is not exact, but Λ satisfies the Bohr-Sommerfeld quantization conditions of Chapter II, §7, one can develop a symbol calculus for "$\mathbf{Z}$ asymptotics", where we look at sequences rather than functions, i.e., restrict the parameter τ to take on only integer values. The Bohr-Sommerfeld conditions themselves take on a simpler expression in the framework of the present chapter because we shall deal with half forms rather than with half densities.

The second half of the chapter is devoted to the local theory of compound asymptotics. In §5 we examine the pointwise behavior of compound asymptotics. In §§6 through 9 we examine the local behavior of asymptotics when the phase functions satisfy certain genericity assumptions. The results of §5 are essentially due to Bernstein, while §§6-9 represent joint work of Guillemin and Schaeffer, some of which overlaps with results developed independently by Duistermaat [2] and Arnold [3].

§1. The asymptotic Fourier transform.

Recall that the Schwartz space, $\mathcal{S}$, consists of those smooth functions, f, on $\mathbf{R}^n$, which satify

$$|x^\alpha D^\beta f| \leqslant C_{\alpha,\beta}, \qquad \text{for all } x \in \mathbf{R}^n,$$

where $C_{\alpha,\beta}$ is a constant, which may depend on f. We now want to consider asymptotic elements of the space $\mathbb{S}$ and, for this purpose, will consider those functions $f\colon \mathbf{R}^+ \to \mathbb{S}$ which satisfy

$$|x^\alpha D_\tau^\beta f(\tau)| \leqslant C_{\alpha,\beta}$$

for some $C_{\alpha,\beta}$ (depending on f). We recall that we are using the notation

$$(D_\tau)_j = \frac{1}{\tau\sqrt{-1}}\frac{\partial}{\partial x_j}$$

and

$$D_\tau^\beta = (D_\tau)_1^{\beta_1}\cdots(D_\tau)_n^{\beta_n}.$$

As usual, we shall regard $f_1\colon \mathbf{R}^+ \to \mathbb{S}$ and $f_2\colon \mathbf{R}^+ \to \mathbb{S}$ as equivalent if their difference is asymptotically zero as *an element of* $\mathbb{S}$; thus if

$$\lim_{\tau\to\infty} \tau^N \sup_x |x^\alpha D_\tau^\beta (f_1 - f_2)| = 0$$

for each fixed α, β and N. The space of equivalence classes we shall denote by $\mathcal{A}\mathbb{S}$ (or $\mathcal{A}\mathbb{S}(\mathbf{R}^n)$).

For any $f\colon \mathbf{R}^+ \to \mathbb{S}$ we define its asymptotic Fourier transform by

$$\hat{f}(\xi,\tau) = e^{-\pi i n/4}\left(\frac{\tau}{2\pi}\right)^{n/2} \int f(x,\tau)e^{-i\tau\xi\cdot x}\,dx. \tag{1.1}$$

Recall that the usual Fourier transform of a function $u \in \mathbb{S}$ is defined by

$$\tilde{u}(\xi) = \left(\frac{1}{2\pi}\right)^{n/2} \int u(x)e^{-i\xi\cdot x}\,dx.$$

Thus the relation between the asymptotic Fourier transform and the usual one is given by

$$\hat{f}(\xi,\tau) = e^{-\pi i n/4}\tau^{n/2}\tilde{f}(\tau\xi,\tau). \tag{1.2}$$

Notice that integration by parts shows that

$$\xi^\alpha D_\tau^\beta \hat{f}(\xi,\tau) = e^{-\pi i n/4}\left(\frac{\tau}{2\pi}\right)^{n/2} \int x^\beta D_\tau^\alpha f(x,\tau)e^{-i\tau\xi\cdot x}\,dx \tag{1.3}$$

and thus $\hat{f}(\cdot,\tau)$ belongs to $\mathbb{S}$.

We will define the inverse Fourier transform by

$$\check{g}(x,\tau) = e^{\pi i n/4}\left(\frac{\tau}{2\pi}\right)^{n/2}\int g(\xi,\tau)e^{i\tau\xi\cdot x}\,d\xi$$
$$= e^{\pi i n/4}\tau^{n/2}\tilde{g}(\tau x,\tau)$$

where

$$\tilde{g}(x,\tau) = \left(\frac{1}{2\pi}\right)^{n/2}\int g(\xi,\tau)e^{i\xi\cdot x}\,d\xi$$

is the inverse Fourier transform. Notice that if we set $g(\xi,\tau) = \tilde{f}(\tau\xi,\tau)$ then

$$\tilde{g}(x,\tau) = \left(\frac{1}{2\pi}\right)^{n/2}\int \tilde{f}(\tau\xi,\tau)e^{i\xi\cdot x}\,d\xi,$$

and setting $\zeta = \tau\xi$ this becomes

$$\tau^{-n}\left(\frac{1}{2\pi}\right)^{n/2}\int \tilde{f}(\zeta,\tau)e^{i\zeta\cdot\tau^{-1}x}\,d\zeta = \tau^{-n}f(\tau^{-1}x,\tau)$$

by the Fourier inversion formula. Thus we can write

$$f = (\hat{f})\check{}\,. \tag{1.4}$$

Notice that it follows from (1.3) that if $f \sim 0$ then $\hat{f} \sim 0$ and vice versa. Thus we can consider the asymptotic Fourier transform as transforming $\mathscr{AS}(\mathbf{R}^n) \to \mathscr{AS}(\mathbf{R}^{n*})$.

For any element $[f] \in \mathscr{AS}$ we will define its support as follows: a point $x \in \mathbf{R}^n$ does not lie in supp f if there is a C^∞ function b with $b(x) \neq 0$ such that $[bf] = 0$. For example, if a is a C^∞ function with $a \in \mathbb{S}$, $a \not\equiv 0$, then

$$f(x,\tau) = a(\tau x)$$

defines an element $[f]$ of $\mathscr{AS}$ with

$$\text{supp}\,[f] = \{(0)\}.$$

Similiarly if $a \in \mathbb{S}$ and $a(x,\tau) = a(x)$, $a \not\equiv 0$, then (1.2) shows that

$$\text{supp}\,\hat{a} = \{(0)\}.$$

Notice that it is possible for

$$\text{supp}\,[f] = \varnothing \quad\text{and}\quad [f] \neq 0.$$

For example, let $C(x)$ be a C^∞ function such that $C(x) = 0$ for $|x| \leqslant 1$ and $C(x) \equiv 1$ for $|x| \geqslant 2$ and let

$$f(x, \tau) = C\left(\frac{x}{(\log \tau)^{1/2}}\right)e^{-x^2}.$$

Then for any fixed x we have $f(x, \tau) = 0$ for $\log \tau \geqslant |x|^2$ so that $\operatorname{supp} f = \varnothing$. On the other hand, for $\log \tau > 4$,

$$\left|\sup_x f(x, \tau)\right| \geqslant e^{-\log \tau} = \frac{1}{\tau}$$

so $[f] \neq 0$.

Thus (applying the inverse asymptotic Fourier transform) it is possible that $\operatorname{supp} [\hat{f}] = \varnothing$ and $[f] \neq 0$. We shall describe a more restrictive function space in which this kind of pathology can not occur in §3.

Let us return to some further examples. By (1.2), $f(x, \tau) = a(x)e^{i\tau\xi\cdot x}$ has, as its asymptotic Fourier transform, $e^{-\pi in/4}\tau^{n/2}\tilde{a}(\tau(\cdot - \xi))$. Thus $\operatorname{supp} \hat{f} = \{\xi\}$. Suppose we consider

$$f(x, \tau) = a(x, \tau)e^{i\tau\varphi}, \qquad a(x, \tau) \sim \sum \frac{a_n(x)}{\tau^n}$$

where, in contrast to the linear case, we assume that the Hessian $d^2\varphi$ is non-degenerate at every x. We may then apply the method of stationary phase to the integral

$$\hat{f}(\xi, \tau) = e^{-\pi in/4}\left(\frac{\tau}{2\pi}\right)^{n/2} \int a(x, \tau)e^{i\tau(\varphi - \xi\cdot x)}\, dx. \tag{1.5}$$

The critical points of $\varphi - \xi \cdot x$ are the values of x such that

$$d_x\varphi = \xi. \tag{1.6}$$

To interpret this equation geometrically, we recall the notation of the Legendre transformation (cf. Sternberg [4, p.150]). If φ is any function defined on a vector space, V, then $d\varphi_x \in T^*V_x$ can be considered as an element of V^*, under the natural identification of TV_x with V and hence of T^*V_x with V^*. Thus φ determines a map, $\mathfrak{L}_\varphi$, of $V \to V^*$ called the *Legendre transformation* associated with φ. Suppose that $\mathfrak{L}_\varphi$ is a diffeomorphism (at least locally). Then $\mathfrak{L}_\varphi^{-1}$ is again a Legendre transformation associated with the function

$$\psi(\xi) = \xi \cdot x - \varphi, \qquad x = x(\xi),$$

where $x \in V$ is regarded as a function of ξ via $\mathfrak{L}_\varphi^{-1}$. Indeed

$$d_\xi\psi = x + \xi \cdot \frac{\partial x}{\partial \xi} - d_x\varphi \cdot \frac{\partial x}{\partial \xi} = x$$

since $d_x\varphi(x(\xi)) = \xi$. Now let us suppose, for the moment, that $\mathfrak{L}_\varphi$ is indeed a diffeomorphism. Then (1.6) has a unique solution and the method of stationary phase applies to the integral to yield

$$\hat{f}(\xi, \tau) = b(\xi, \tau)e^{-i\tau\psi(\xi)}.$$

Here b is given by the formula of stationary phase. For example $b(\xi, \tau) = \sum b_n(\xi)\tau^{-n}$ where $b_0(\xi) = e^{\pi i \operatorname{sgn} H\psi/2} a_0(x(\xi))$, etc. Notice that if all the $a_i(x)$'s have their support in some set $K \subset \mathbf{R}^n = V$ then all the b's will have their support in $\mathfrak{L}_\varphi(K)$. This suggests that in some sense the support be considered as lying on graph $\mathfrak{L}_\varphi \subset V + V^*$. That is we should be considering a subset of $V + V^*$ (in this case a subset of graph $\mathfrak{L}_\varphi$) whose projection onto V gives supp f and whose projection onto V^* gives supp $\hat{f}$. We shall do this (replacing $V + V^* = T^*V$ by T^*M where M is an arbitrary manifold, and replacing functions by half densities) in the next section. Notice also that our analysis also suggests a generalization of the notion of Legendre transformation. Indeed graph $\mathfrak{L}_\varphi = \Lambda$ is a Lagrangian submanifold of $V + V^*$ with the property that the projection of Λ onto V and onto V^* are diffeomorphisms. We can drop these latter conditions and define a *Legendre relation* to be any Lagrangian submanifold of $V + V^*$. Thus $a(x)e^{i\tau\xi_0 \cdot x}$ will be associated to the Lagrangian submanifold $\{\xi = \xi_0, x \text{ arb.}\}$ while $a(\tau(x - x_0))$ is associated to the Lagrangian submanifold $\{\xi \text{ arb.}, x = x_0\}$.

§2. The frequency set.

Let $[\nu]$ be an asymptotic half density (or half form) on the manifold M. We wish to associate to $[\nu]$ a subset $F([\nu])$ of T^*M called the frequency set of $[\nu]$. Intuitively, $F([\nu])$ should consist of those (relative) frequencies at which ν is oscillating. It will be easier for us to describe those covectors which do *not* belong to $F([\nu])$. Roughly speaking, a covector $\gamma \in T^*M_x$ does not belong to $F([\nu])$ if for any ψ defined near x with $d\psi(x) = \gamma$ there is some half density ρ with $\rho(x) \neq 0$ such that

$$[(\nu, \rho e^{i\tau\psi})] = 0.$$

Actually, we shall want this to hold not only for ψ such that $d\psi(x) = \gamma$ but also for all "nearby" ψ, so that the set of γ which do not belong $F([\nu])$ will be open. We shall formulate this in a somewhat roundabout way which will be useful for us when we study how $F([\nu])$ behaves under differentiable maps.

Let N be some other manifold and let $\pi\colon M \times N \to M$ and $\rho\colon M \times N \to N$ be the obvious projections. Actually, we want π and ρ to be morphisms so that $\pi^*[\nu]$ is an asymptotic half density (or half form) on $M \times N$. Let ψ be a smooth function on $M \times N$ such that $d_M\psi_{(x,y)} = \gamma$. Here d_M means differentiating in the M direction. We will demand that there exist some smooth function, b, of compact support, defined on M such that

$$[\rho_* e^{-i\tau\psi}\pi^* b\nu] = 0 \qquad \text{near } y. \tag{2.1}$$

If we let c be a function of small support with $c(y) \neq 0$ we can rephrase (2.1) as

$$[c\rho_* e^{-i\tau\psi}\pi^* b\nu] = 0. \tag{2.2}$$

Notice that at the moment the definition is highly complicated and difficult to check. It asserts that $\gamma \notin F([\nu])$ *if for any* N, π, ρ, *and* ψ *we can find* b *and* c *such that* (2.2) *holds.*

To illustrate the definition let us compute the frequency set of a simple asymptotic. Recall that a simple asymptotic half density is of the form $ae^{i\tau\varphi}$ where

$$a \sim \sum_{j}^{\infty} a_n \tau^{-n}$$

and the a_n's are half densities on M. We claim that:

PROPOSITION 2.1 *For a simple asymptotic* $\nu \sim ae^{i\tau\varphi}$ *we have*

$$F([\nu]) \subset \text{graph } d\varphi.$$

Indeed, let N, π, ρ and ψ be as in the definition where $d_M\psi_{(x_0,y_0)} = \gamma \notin \text{graph } d\varphi$. We may choose local coordinates about x_0 and y_0 and small neighborhoods about x_0 and y_0 such that $|d_M\psi(x, y) - d\varphi(x)| > c > 0$ where $|\ |$ denotes, say, Euclidean distance in terms of the local coordinates. If we now choose b and c to have sufficiently small supports it is clear that the expression (2.2) is asymptotic to a sum of terms of the form

$$\tau^{-n} \int d_n(x,y) e^{i\tau(\varphi(x)-\psi(x,y))}\, dx.$$

A repeated integration by parts shows that this is asymptotically zero.

To make the definition manageable we shall now show that it is sufficient to test with $N = \mathbf{R}^n$ and ψ linear. First of all, it is clear (since the morphism π is arbitrary and we have freedom of choice in b) that we may assume that ν has support in some coordinate chart, so that we may assume that M is an open subset of $\mathbf{R}^n$. Suppose that $\gamma = (x_0, \xi_0)$ when we identify T^*M with $\mathbf{R}^n \times \mathbf{R}^n$. Let us choose the standard half density $\sqrt{dxd\xi}$ on $\mathbf{R}^n \times \mathbf{R}^n$. This then determines morphisms:

$$\begin{array}{ccc} & M \times \mathbf{R}^n & \\ {}^{\pi}\swarrow & & \searrow^{\rho} \\ M & & \mathbf{R}^n \end{array}$$

Let us now take $\psi(x, \xi) = x \cdot \xi$, and let us write

$$\nu = a(x, \tau)dx^{1/2}.$$

Then we can find a b such that

$$\rho_* e^{-i\tau\psi}\pi^* b\nu = \int b(x)a(x,\tau)e^{-i\tau\xi\cdot x}\, dx$$

is asymptotic to zero for ξ near ξ_0. In other words

$$[\widehat{ba}(\xi,\tau)] = 0 \qquad \text{for } \xi \text{ near } \xi_0. \tag{2.3}$$

Observe, by the way, that if (2.3) holds, and if u is any smooth function of compact support then we also have

$$[\widehat{uba}(\xi,\tau)] = 0 \qquad \text{for } \xi \text{ near } \xi_0.$$

Indeed

$$\widehat{uba}(\xi,\tau) = \int \hat{u}(\eta,\tau)\widehat{ba}(\xi-\eta,\tau)\, d\eta.$$

Now by (1.3), we know that $\widehat{ba}(\zeta,\tau) = O(\tau^k)$ for some large k and, since $u \in \mathbb{S}$, that

$$\hat{u}(\eta,\tau) = O(\tau^{-N}) \qquad \text{for any } N,$$

uniformly in η for $|\eta| \geq \epsilon$ for any fixed $\epsilon > 0$. Now if (2.3) holds for $|\xi - \xi_0| < 2\epsilon$ then we can write

$$\int \hat{u}(\eta,\tau)ba(\xi-\eta,\tau)\, d\eta = \int_{|\eta|<\epsilon} + \int_{|\eta|\geq\epsilon}.$$

When $|\eta| < \epsilon$ and $|\xi - \xi_0| < \epsilon$ then the first integral is asymptotically zero by (2.3). The second vanishes asymptotically as the product of $O(\tau^{-N}) \cdot O(\tau^k)$.

We have seen that if $(x_0, \xi_0) \notin F([\nu])$ then (2.3) holds. The important result is that the converse is also true:

PROPOSITION 2.2. *Let M be an open submanifold of $\mathbf{R}^n$ and let $[\nu]$ be an asymptotic half density on M. Then $(x_0, \xi_0) \notin F([\nu])$ if and only if (2.3) holds.*

PROOF. Suppose that (2.3) holds. Let N be some second manifold and let

$$\begin{array}{ccc} & M \times N & \\ {}^{\pi}\swarrow & & \searrow^{\rho} \\ M & & N \end{array}$$

be a morphisms and ψ: $M \times N \to \mathbf{R}$ be such that $d_M\psi_{(x_0,y_0)} = (x_0, \xi_0)$. Let y be local coordinates on N. Since we are allowed to multiply by c we may assume that $\rho_* e^{-i\tau\psi}\pi^* b\nu$ has support in such a coordinate neighborhood and thus

$$\rho_* e^{-i\tau\psi}\pi^* b\nu = \int u(x,y,\tau)e^{-i\tau\psi(x,y)}b(x)a(x,\tau)\,dx,$$

where the u comes from the morphisms. Now

$$b(x)a(x,\tau) = e^{\pi in/2}\left(\frac{\tau}{2\pi}\right)^{n/2}\int \widehat{ba}(\xi,\tau)e^{i\tau\xi\cdot x}\,d\xi.$$

Substituting into the above integral gives

$$\iint u(x,y,\tau)\widehat{ba}(\xi,\tau)e^{i\tau[\xi\cdot x-\psi(x,y)]}\,d\xi\,dx.$$

Now $d_M\psi(x_0,y_0) = (x_0,\xi_0)$. Let us suppose that (2.3) holds for $|\xi-\xi_0| < 3\epsilon$. Now by further multiplication by another b and c (which in the integral can be absorbed into a new u) we can assume that, for all $(x,y) \in$ supp u,

$$d_M\psi(x,y) = (x,\zeta)$$

where

$$|\zeta-\xi_0| < \epsilon.$$

Now break the double integral up into two integrals, one over the range $|\xi-\xi_0| < 2\epsilon$ and the other for $|\xi-\xi_0| \geq 2\epsilon$. The first integral is asymptotically zero by (2.3). As to the second

$$|d_M[\xi\cdot x-\psi(x,y)]| > |\xi-\zeta| > \epsilon.$$

Since $\widehat{ba}(\xi,\tau)$ is at most of polynomial growth in τ a repeated integration by parts with respect to x shows that the integral is indeed asymptotically zero, proving the proposition.

Now condition (2.3) is clearly open. Therefore an immediate consequence of Proposition 2.2 is the following.

PROPOSITION 2.3. *For any manifold M the set $F([\nu])$ is a closed subset of T^*M.*

Let $f\colon M_1 \to M_2$ be a proper submersion. If $[\nu]$ is an asymptotic half density on M_1 then $f_*[\nu]$ is an asymptotic half density on M_2. We claim that

$$F(f_*[\nu]) \subset df_*\,F([\nu]). \tag{2.4}$$

Suppose that $\gamma \notin df_*\,F([\nu])$ so that $df^*\gamma \notin F([\nu])$. Let N be another manifold and $\psi\colon M_2\times N \to \mathbf{R}$ be such that $d\psi_{M_2}(x,y) = \gamma$. Let us define $\varphi\colon M_1\times N \to \mathbf{R}$ by $\varphi(z,y) = \psi(f(z),y)$. Then for any $z \in f^{-1}(x)$ we have $d_{M_1}\varphi(z,y) = df^*\gamma \notin F([\nu])$. For each such z and y we can find functions b on M_1 and c on N such that $[c\rho_* e^{-i\tau\varphi}\pi_1^* b\nu] = 0$. Near z we can write $M_1 = M_2 \times P$ and (by multiplying b by an appropriate function) we may assume that $b = b_2 r$ where b_2 is a function on M_2 with $b_2(x) \neq 0$ and r is a function on P with $r \equiv 1$ near y. Since the set of z's $\in \pi^{-1}x$ which come into consideration is compact, we can let $\bar{b}$ equal the product of the b_2's, and $\bar{c}$ equals the products of the c's. A partition of unity then shows that

$$[\bar{c}\rho_* e^{-i\tau\psi}\pi^* \bar{b} f_* \nu] = \sum [f_* \bar{c}\rho_* e^{-i\tau\varphi}\pi^*(\bar{b} \circ f)\nu] = 0,$$

proving (2.4).

Let $\Lambda \subset T^* M$ be a Lagrangian submanifold. We know that locally Λ can be represented by $d\pi_* d\varphi^* \Lambda_1$ where $\Lambda_1 = \{(x, 1) \in T^* \mathbf{R}^1\}$ is the standard Lagrangian submanifold of $T^* \mathbf{R}^1$, π: $N \to M$ is a submersion and φ: $N \to \mathbf{R}^1$ a map:

$$\begin{array}{ccc} N & \xrightarrow{\ \varphi\ } & \mathbf{R}^1 \\ {\scriptstyle\pi}\big\downarrow & & \\ M & & \end{array} .$$

If $[a(x, \tau)e^{i\tau\varphi(x)}]$ is a simple asymptotic on N then $d\pi_*[ae^{i\tau\varphi}]$ is an asymptotic whose frequency set lies in Λ. We say that $[\nu]$ is *a compound asymptotic* associated to Λ if

$$[\nu] = \sum [\nu_j]$$

where each $\nu_j = \pi_{j*}[a_j e^{i\tau\varphi_j}]$ and (π_j, φ_j) gives a local presentation of Λ, and where the sum is locally finite. In what follows, the compound asymptotics will be the asymptotics of principal interest to us. If $[\nu]$ is a compound asymptotic associated with Λ then it follows from (2.4) and Proposition 2.1 that

$$F([\nu]) \subset \Lambda. \tag{2.5}$$

Thus, if ψ is a function such that (graph $d\psi$) $\cap\, \Lambda = \varnothing$ then

$$[(\nu, \mu)] = 0 \tag{2.6}$$

if $\mu = be^{i\tau\psi}$ is any simple asymptotic half density associated to ψ. We can say somewhat more. Suppose that graph ψ intersects Λ transversally at a single point γ. We then claim that

$$(\nu, \mu) \sim e^{ic\tau} \sum d_n \tau^{-n} \tag{2.7}$$

where c and the d_n's depend only on ψ near $\pi_M \gamma$ where π_M: $T^* M \to M$. is the standard projection. Indeed, (ν, μ) can be written as a sum of terms of the form

$$(\pi_* ae^{i\tau\varphi}, be^{i\tau\psi}) = (ae^{i\tau\varphi}, \pi^* be^{i\tau\psi})$$

which will be an integral of the form

$$\int u(x,y)e^{i\tau[\varphi(x,y)-\psi(x)]}dx\,dy.$$

The assertion that graph $d\psi$ intersects Λ transversally means precisely that $\varphi(x,y) - \psi(x)$ has a non-degenerate critical point and thus we may apply the method of stationary phase to conclude (2.7).

Returning to the general case, notice that the formula for the frequency set for the pull back of an asymptotic is practically part of the definition, if f: $M_1 \to M_2$ is a submersion. Indeed, we claim that

$$F(f^*[\nu]) \subset df^* F([\nu]). \tag{2.8}$$

Indeed, since the considerations are entirely local, we may assume that $M_1 = M_2 \times Q$ and f is projection onto the first factor. Then $df^* F([\nu]) = F([\nu]) \times \{0\}$. If $\gamma_1 = (\gamma_2, \gamma_3) \notin df^* F([\nu])$ then either $\gamma_3 \neq 0$ or $\gamma_3 = 0$ and $\gamma_1 \notin F([\nu])_x$ where $x = f\pi_{M_1}\gamma_1$. In the first case we may write the integral in (2.1) as

$$\int u(x,\tau)a(y,z)e^{i\tau\psi(y,z)}dx\,dy$$

where $d_Q\psi \neq 0$. Here x is a variable on M_2, y is a variable on Q and z on N. Repeated integration by parts with respect to y shows that this integral vanishes asymptotically. In the second case, the integral entering into (2.1) takes the form

$$\int u(x,\tau)a(y,z)e^{-i\tau\psi(x,z)}dx\,dy$$

where $d_{M_2}\psi = \gamma \notin F([\nu])$ and, by assumption, the integral with respect to x already vanishes asymptotically.

We would like to be able to assert that (2.8) holds for all maps, not only for submersions. We would also like to be able to assert that if $F([\nu]) = \varnothing$ then $[\nu] = 0$. However, we know from the examples in §1 that this is not true. We must therefore restrict or modify the class of asymptotics we wish to consider. We will discuss the compound asymptotics, although a broader class of such asymptotics could be introduced. The discussion in the following section parallels, fairly closely, the corresponding discussion of §4 of Chapter VI.

§3. Functorial properties of compound asymptotics.

Let f: $M_1 \to M_2$ be a morphism and let $[\nu]$ be a compound asymptotic on M_2 associated with the Lagrangian manifold $\Lambda \subset T^*M_2$. Let π: $\Lambda \to M_2$ be the restriction to Λ of the projection of T^*M_2 onto M_2 and suppose that f: $M_1 \to M_2$ and π: $\Lambda \to M_2$ are transversal. Then we claim that:

PROPOSITION 3.1: *$f^*[\nu]$ is a compound asymptotic associated with $df^*\Lambda$.*

Indeed suppose, for the moment, that Λ has a presentation

$$\begin{array}{ccc} N_2 & \xrightarrow{\varphi} & \mathbf{R}^1 \\ {\scriptstyle g_2}\big\downarrow & & \\ M_2 & & \end{array}$$

so that

$$\Lambda = dg_{2*}\, d\varphi^* \Lambda_1 .$$

Then we can complete the diagram as

$$\begin{array}{ccccc} N_1 & \xrightarrow{\tilde{f}} & N_2 & \xrightarrow{\varphi} & \mathbf{R} \\ {\scriptstyle g_1}\big\downarrow & & \big\downarrow{\scriptstyle g_2} & & \\ M_1 & \xrightarrow{f} & M_2 & & \end{array}$$

where

$$N_1 = M_1 \underset{M_2}{\times} N_2 .$$

Now graph $d\varphi$ projects surjectively onto N_2 and hence pulls back to the Lagrangian manifold graph $d\psi \subset T^* N_1$ where $\psi = \varphi \circ \tilde{f}$. We make two assertions:

(i) *Let* $H_1 = dg_1^* T^* M_1$. *Then* graph $d\psi$ *intersects* H_1 *transversally and*

$$dg_{1*}(\text{graph } d\psi) = df^* \Lambda.$$

(ii) *Suppose that* f *and* g_2 *are morphisms. Then* $\tilde{f}$ *and* g_1 *can be equipped as morphisms so that*

$$f^* g_{2*} = g_{1*} \tilde{f}^* .$$

Granting (i) and (ii), if $\nu = g_{2*}\varphi^* \mu$ where μ is a half density on $\mathbf{R}^1$ then we can write $f^* \nu = g_{1*}\psi^* \mu$. Then the general case of Proposition 3.1 now follows since ν is a sum of terms of the form $g_{2*}\varphi^* \mu$.

Proof of (i). Since the computation is purely local we may introduce coordinates y, θ on N_2 with y as coordinates on M_2 (so that $g_2(y,\theta) = y$) and x, θ as coordinates on N_1 with x as coordinates on M_1. Let (y,θ,η,ζ) be the induced coordinates on $T^* N_2$ and (x,θ,ξ,ζ) the induced coordinates on $T^* N_1$. The fact that graph $d\varphi$ intersects $H = dg_2^* T^* M_2$ transversally means that given any vector z we can solve

$$z = \frac{\partial^2 \varphi}{\partial y \partial \theta} b + \frac{\partial^2 \varphi}{\partial \theta^2} c \tag{3.1}$$

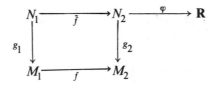

for b and c. The fact that f is transversal to Λ means that given any b we can find a', b' and c' such that

$$b = \frac{\partial f}{\partial x} a' + b' \qquad 0 = \frac{\partial^2 \varphi}{\partial y \partial \theta} b' + \frac{\partial^2 \varphi}{\partial \theta^2} c'. \tag{3.2}$$

To say that graph $d\psi =$ graph $d\varphi \circ \tilde{f}$ is transversal to $dg_1^* T^* M$ means that we should be able to solve

$$z = \frac{\partial^2 \varphi}{\partial y \partial \theta} \frac{\partial f}{\partial x} a'' + \frac{\partial^2 \varphi}{\partial \theta^2} c''$$

for a'' and c''. To do this choose b and c to solve (3.1), then a', b' and c' to solve (3.2) and set $a'' = a'$, $c'' = -c'$. This proves (i).

Proof of (ii). This reduces to the following:

Lemma 3.1. *Let*

$$\begin{array}{ccc} A & \xrightarrow{q'} & B \\ {\scriptstyle \ell'}\downarrow & & \downarrow{\scriptstyle \ell} \\ C & \xrightarrow[q]{} & D \end{array}$$

be a commutative diagram of linear maps describing A as a fiber product, i.e.,

$$A = B \underset{D}{\times} C.$$

Then

$$Hom\,(|\wedge|^{1/2} D, |\wedge|^{1/2} C) \sim Hom\,(|\wedge|^{1/2} B, |\wedge|^{1/2} A) \tag{3.3}$$

and

$$Hom\,(|\wedge|^{1/2} D, |\wedge|^{1/2} B) \sim Hom\,(|\wedge|^{1/2} C, |\wedge|^{1/2} A). \tag{3.4}$$

Applying the lemma to the differentials $d\tilde{f}$, df, dg_2, dg_1 shows that if f and g_2 are equipped with the structure of morphisms then so are $\tilde{f}$ and g_1, which is the assertion of (ii).

Proof of the lemma. The fact that A is a fiber product implies that in the diagram

$$\begin{array}{ccccccccccc} 0 & \longrightarrow & \ker q' & \longrightarrow & A & \xrightarrow{q'} & B & \longrightarrow & \operatorname{coker} q' & \longrightarrow & 0 \\ & & \downarrow & & {\scriptstyle \ell'}\downarrow & & \downarrow{\scriptstyle \ell} & & \downarrow & & \\ 0 & \longrightarrow & \ker q & \longrightarrow & C & \xrightarrow{q} & D & \longrightarrow & \operatorname{coker} q & \longrightarrow & 0 \end{array}$$

the induced vertical maps at the left and right are isomorphisms. But the exactness of the rows implies that

$$\mathrm{Hom}(|\wedge|^{1/2}D, |\wedge|^{1/2}C) \sim \mathrm{Hom}(|\wedge|^{1/2}\mathrm{coker}\ q, |\wedge|^{1/2}\ker q)$$

and

$$\mathrm{Hom}(|\wedge|^{1/2}B, |\wedge|^{1/2}A) \sim \mathrm{Hom}(|\wedge|^{1/2}\mathrm{coker}\ q', |\wedge|^{1/2}\ker q'),$$

proving (3.3).

Similarly, the fact that ℓ and ℓ' are surjective implies that

$$\mathrm{Hom}(|\wedge|^{1/2}D, |\wedge|^{1/2}B) \sim |\wedge|^{1/2}\ker \ell$$

and

$$\mathrm{Hom}(|\wedge|^{1/2}C, |\wedge|^{1/2}A) \sim |\wedge|^{1/2}\ker \ell'$$

and the fact that A is a fiber product implies that $\ker \ell \sim \ker \ell'$. This completes the proof of the lemma and hence of Proposition 3.1.

PROPOSITION 3.2. *Let $f\colon M_1 \to M_2$ be a submersion which is a proper morphism and let $\Lambda \subset T^* M_1$ be a Lagrangian submanifold which intersects $df^* T^* M_2$ transversally. Then if $[\nu]$ is a compound asymptotic associated with Λ then $f_*[\nu]$ is a compound asymptotic associated with $df_* \Lambda$.*

The proof is immediate and will be left to the reader.

Let Λ be a Lagrangian submanifold of $T^*(M_1 \times M_2)$ and Λ_2 a Lagrangian submanifold of $T^* M_2$. Suppose that Λ intersects $T^* M_1 \times \Lambda_2$ transversally so that $\rho_1(\Lambda \cap (T^* M_1 \times \Lambda_2)) = \Lambda_1$ is a Lagrangian submanifold of $T^* M_1$, where $\rho_1 : T^* M_1 \times T^* M_2 \to T^* M_1$ is the projection onto the first factor.

PROPOSITION 3.3. *Let $[\nu]$ be a compound asymptotic associated to Λ and $[\nu_2]$ a compound asymptotic of compact support associated with Λ_2. Then $[\int_{M_2} \nu\bar{\nu}_2]$ is a compound asymptotic associated with Λ_1.*

Again, it suffices to deal with local situations:

$$\begin{array}{lcll} N & \xrightarrow{\varphi} & \mathbf{R} & \nu = f_* \varphi^* \mu, \\ \downarrow f & & & \Lambda = df_* \text{ graph } d\varphi, \\ M_1 \times M_2 & & & \end{array}$$

$$\begin{array}{lcl} P & \xrightarrow{\psi} & \mathbf{R} \qquad \nu_2 = g_* \psi^* \mu_2, \\ \downarrow g & & \qquad \Lambda_2 = dg_* \text{ graph } d\psi. \\ M_2 & & \end{array}$$

Then we can form the fiber product $N \times_{M_2} P$ and consider φ and ψ as functions on $N \times_{M_2} P$ and consider $N \times_{M_2} P \xrightarrow{h} M_1$. We claim that graph $d(\varphi - \psi)$ intersects $dh^* T^* M_1$ transversally and $dh_*(\text{graph } d(\varphi - \psi)) = \Lambda_1$.

Indeed, let (x, y, θ) be local coordinates on N and (y, γ) be local coordinates on P. To say that graph $d(\varphi - \psi)$ intersects $dh^* T^* M_1$ transversally means that we can solve the equations

$$z_y = \frac{\partial^2 \varphi}{\partial x \partial y} a - \frac{\partial^2 \psi}{\partial \gamma \partial y} c + \frac{\partial^2 (\varphi - \psi)}{\partial y^2} b + \frac{\partial^2 \varphi}{\partial \theta \partial y} t,$$

$$z_\theta = \frac{\partial^2 \varphi}{\partial x \partial \theta} a + \frac{\partial^2 \varphi}{\partial \theta^2} t + \frac{\partial^2 \varphi}{\partial y \partial \theta} b,$$

$$z_\gamma = -\frac{\partial^2 \psi}{\partial y \partial \gamma} b - \frac{\partial^2 \psi}{\partial \gamma^2} c,$$

for a, b, t and c at the points where

$$\partial(\varphi - \psi)/\partial y = \partial(\varphi - \psi)/\partial \theta = \partial(\varphi - \psi)/\partial \gamma = 0.$$

Now since $T^* M_1 \times \Lambda_2$ is transversal to Λ, we can, for any b_3, η_3, find a_1, b_1, t, and b_2, c_2 such that

$$b_1 + b_2 = b_3,$$

$$\frac{\partial^2 \varphi}{\partial x \partial y} a_1 + \frac{\partial^2 \varphi}{\partial y^2} b_1 + \frac{\partial^2 \varphi}{\partial \theta \partial y} t_1 + \frac{\partial^2 \psi}{\partial y^2} b_2 + \frac{\partial^2 \psi}{\partial \gamma \partial y} c_2 = \eta_3,$$

$$\frac{\partial^2 \varphi}{\partial x \partial \theta} a_1 + \frac{\partial^2 \varphi}{\partial y \partial \theta} b_1 + \frac{\partial^2 \varphi}{\partial \theta^2} t_1 = 0,$$

$$\frac{\partial^2 \psi}{\partial y \partial \gamma} b_2 + \frac{\partial^2 \psi}{\partial \gamma^2} c_2 = 0.$$

The fact that $\varphi(x, y, \theta)$ gives a local presentation for Λ means that we can find a', b', t' solving the equation for z_θ and the fact that ψ gives a local presentation for Λ_2 means that we can find b'', c'' solving the equation for z_γ. Now take $b_3 = -(b' + b'')$ so that $b_1 + b' = -(b_2 + b'')$ and

$$\eta = z_y - \frac{\partial^2 \varphi}{\partial x \partial y} a' - \frac{\partial^2 \psi}{\partial \gamma \partial y} c' + \frac{\partial^2 \varphi}{\partial y^2} b' - \frac{\partial^2 \varphi}{\partial \theta \partial y} t'.$$

We now take $a = a_1 + a'$, etc. to solve our equations. The fact that

$$dh_*(\text{graph } d(\varphi - \psi)) = \Lambda_1$$

then follows directly from the definitions.

Now, in terms of local coordinates

$$\nu = \int e^{i\tau\varphi(x,y,\theta)} a(x, y, \theta; \tau)\, d\theta \quad \text{and} \quad \nu_2 = \int e^{i\tau\psi(y,\gamma)} b(y, \gamma; \tau)\, d\gamma$$

so that

$$\int_{M_2} \nu\bar{\nu}_2 = \int e^{i\tau[\varphi(x,y,\theta) - \psi(y,\gamma)]} c(x, y, \theta, \gamma; \tau)\, d\theta\, d\gamma\, dy$$

where $c = a\bar{b}$. We thus see that $[\int_{M_2} \nu\bar{\nu}_2]$ is a compound asymptotic associated with Λ_1, proving the proposition.

§4. The symbol calculus.

Let $\Lambda \subset T^*X$ be a Lagrangian manifold with a given parametrization

$$\begin{array}{ccc} Z & \xrightarrow{\ \varphi\ } & \mathbf{R} \\ {\scriptstyle \pi}\downarrow & & \\ X & & \end{array}$$

and let

$$\mu = \pi_* a e^{i\tau\varphi}$$

be a compound asymptotic defined by means of this parametrization. If $\psi \in C^\infty(X)$ has the property that graph $d\psi$ intersects Λ transversally at (x_0, ξ_0), then, by (2.7), the expression

$$\int_X \mu(x, \tau) e^{-i\tau\psi}\, dx \tag{4.1}$$

admits an asymptotic expansion of the form

$$e^{ic\tau} \sum a_i \tau^i \qquad \text{with } c = \varphi(x_0, \xi_0) - \psi(x_0). \tag{4.2}$$

Now a general compound asymptotic, ν, associated with Λ, will be a finite sum of such μ, each with a possible different φ. Then (4.1) will be given by a finite sum of expressions of the form (4.2), with differing c's. Now the "symbol" of ν, whatever it is, should determine the leading term in the asymptotic expansion of (4.1). Thus, for general ν, the symbol would have to consist of a large number of symbols, one for each c. Hence the symbol calculus for compound asymptotics can be expected to be rather complicated.

We can ask, however, the following question: When does there exist just one asymptotic series of the form (4.2) for each choice of ψ in (4.1)? Assuming that the a_i's are not all zero, *a necessary and sufficient condition is that the restriction of the fundamental one form*, α, *of* T^*X *to* Λ *be exact*. Indeed, the c in (4.2) determines a function, φ, on Λ by $\varphi(x_0, \xi_0) = c_\psi + \psi(x_0)$ and one easily checks that $d\varphi = \alpha$ on Λ. Conversely, given a φ on Λ, we can require an *admissable parametrization* of Λ to have the property that its phase function, restricted to the critical set, be equal to the pull back of φ to the critical set. If we only use such parametrizations, the asymptotic expansion of (4.1) will always consist of just one series of the form (4.2), and we can hope to construct a symbol calculus. We now proceed to develop these ideas.

A *Lagrangian pair* (Λ, φ) consists of a Lagrangian manifold, Λ, on which α is exact, together with a choice of function, φ, defined on Λ such that

$$d\varphi = \alpha.$$

Let us briefly indicate how the functorial properties of Lagrangian manifolds, described in §4 of Chapter IV, work for Lagrangian pairs: If (Λ, φ) is a Lagrangian pair in T^*Y and $f\colon X \to Y$ is a map which is transversal to Λ, then $f^*\Lambda$ comes equipped with a map into Λ and so φ pulls back to a function on $f^*\Lambda$ which we denote by $f^*\varphi$. One can easily check that $(f^*\Lambda, f^*\varphi)$ is a Lagrangian pair, which we call the pull back of (Λ, φ) and denote by $f^*(\Lambda, \varphi)$. Similarly, if $f\colon X \to Y$ is a submersion, the push forward Lagrangian submanifold, $f_*\Lambda$ is equipped with an injection into Λ, and hence φ pushes forward to a function, which we denote by $f_*\varphi$, on $f_*\Lambda$. Again one checks that $(f_*\Lambda, f_*\varphi)$ is a Lagrangian pair, which we denote by $f_*(\Lambda, \varphi)$ and call the push forward of (Λ, φ) under f.

As in Chapter VI, we will develop the symbol calculus for compound asymptotics within the metalinear setting; all manifolds will be metalinear, maps will be metalinear morphisms, and we shall be interested in asymptotic half forms. A half form, μ, will be said to be associated to the Lagrangian pair (Λ, φ) if it can be written as a sum of expressions of the form

$$\mu = \pi_* a\varphi_1^* e^{i\tau t}\sqrt{dt} \quad \text{where } \pi_*\varphi_1^*(\Lambda_1, t) \text{ is a local parametrization}$$
$$\text{of } (\Lambda, \varphi) \text{ with } t \text{ denoting the standard coordinate} \tag{4.3}$$
$$\text{on } \mathbf{R}, \text{ and } \Lambda_1 = \text{graph } dt \subset T^*\mathbf{R}.$$

For such a μ we define its symbol, $\sigma(\mu)$, as follows. Let a_0 be the leading term in the asymptotic expansion of the asymptotic function, a. Then $a_0\varphi_1^*\sqrt{dt}$ is a half form on Z, which pulls back to a half form on $\varphi_1^*(\Lambda_1) \subset T^*Z$, since $\varphi_1^*(\Lambda_1)$ projects diffeomorphically onto Z, under the projection of T^*Z onto Z. We now have a half form on $\varphi_1^*(\Lambda_1)$. By the method of §5 of Chapter VI, the push forward, π_*, takes this onto a half form on Λ. We multiply the resulting half form by $(2\pi i\tau)^{-k/2}$, where k is the fiber dimension of Z, and the expression so obtained is called the symbol of μ.

We must show that our definition of $\sigma(\mu)$ does not depend on the choice of parametrization. Let (x_0, ξ_0) be a point of Λ. Let z_0 be the critical point of φ_1 on Z corresponding to (x_0, ξ_0), and let (x, θ) be local coordinates near z_0 with θ being fiber coordinates. We also assume that the coordinates are so chosen that $\varphi_1^* \sqrt{dt} = \sqrt{dxd\theta}$ and $\pi_* \sqrt{dxd\theta} = \sqrt{dx}$. Then

$$\mu = \left(\int a(x, \theta, \tau) e^{i\tau\varphi_1(x,\theta)}\, d\theta\right)\sqrt{dx}. \tag{4.4}$$

Let us first assume that (x_0, ξ_0) is a regular point of the projection of $\Lambda \to X$, so that the fiber Hessian, $d_\theta^2 \varphi_1$ is non-degenerate at z_0. We can apply the method of stationary phase to (4.4) to obtain

$$\mu \sim e^{i\tau\varphi_1(x_0,\xi_0)}(2\pi\tau)^{-k/2} e^{(\pi i/4)\operatorname{sgn} d_\theta^2\varphi_1} |\det d_\theta^2 \varphi_1|^{-1/2}(a_0(x_0, \xi_0, \tau) + \cdots) \tag{4.5}$$

at x_0. Now sgn $d_\theta^2 \varphi_1$ is the number of positive eigenvalues minus the number of negative eigenvalues, while k is the number of positive eigenvalues plus the number of negative eigenvalues. If we define

$$(\det d_\theta^2 \varphi_1)^{1/2} = e^{(\pi i/2)(\operatorname{ind} d_\theta^2 \varphi_1)} |\det d_\theta^2 \varphi_1|^{1/2}$$

where ind $d_\theta^2 \varphi_1$ is the number of negative eigenvalues, then we can rewrite (4.5) as

$$\mu \sim e^{i\tau\varphi_1(x_0,\xi_0)}(2\pi i\tau)^{-k/2}(\det d_\theta^2 \varphi_1)^{-1/2} a_0(x_0, \xi_0, \tau) + \cdots. \tag{4.6}$$

Let us now compute $\sigma(\mu)$ at (x_0, ξ_0). Since $\varphi_1^* \Lambda_1$ projects diffeomorphically onto Z, we can use (x,θ) as coordinates on $\varphi_1^* \Lambda_1$. By definition, the symbol of μ is the push forward by π_* of $a_0(dxd\theta)^{1/2}$, considered as a half form on $\varphi_1^* \Lambda_1$, multiplied by $(2\pi i\tau)^{-k/2}$. To carry out the computation, we recall how to compute the push forward. Let $H \subset T^* Z$ be the bundle of horizontal covectors for the submersion π: $Z \to X$. Then $\Lambda = (\varphi_1^* \Lambda_1) \cap H$. Let p_0 be the point of $\varphi_1^* \Lambda_1$ corresponding to the critical point $z_0 \in Z$, and which then also corresponds to the point $(x_0, \xi_0) \in \Lambda$. Let

A denote the tangent space to Λ at (x_0, ξ_0),
B denote the tangent space to $\varphi_1^* \Lambda_1$ at p_0,
H_0 denote the tangent space to H at p_0,
T denote the tangent space to $T^* Z$ at p_0, and
F denote the tangent space to the fiber at z_0.

The transversality condition for push forward requires that

$$T = H_0 + B$$

and then

$$A = H_0 \cap B.$$

Thus

$$B/A \cong T/H_0 \cong F^*. \tag{4.7}$$

The fact that π is a morphism means that F is equipped with a half form $(d\theta^{1/2})$ and hence that F^* is equipped with a negative half form. By (4.7) this implies that B/A is equipped with a negative half form, and hence a half form on B determines a half form on A, i.e., a half form on $\varphi_1^* \Lambda_1$ determines a half form on Λ. Let us work this out explicitly in terms of local coordinates. We can consider T as the set of all vectors

$$(x_1, \ldots, x_n, \theta_1, \ldots, \theta_k, \xi_1, \ldots, \xi_n, \eta_1, \ldots, \eta_k).$$

Then B is the subspace defined by the equations

$$\xi_j = \sum x_i \frac{\partial^2 \varphi_1}{\partial x_i \partial x_j} + \sum \theta_\ell \frac{\partial^2 \varphi_1}{\partial \theta_\ell \partial x_j} \tag{4.8}$$

and

$$\eta_j = \sum x_i \frac{\partial^2 \varphi_1}{\partial x_i \partial \theta_j} + \sum \theta_\ell \frac{\partial^2 \varphi_1}{\partial \theta_j \partial \theta_\ell}. \tag{4.9}$$

The space A is defined by the further equations $\eta_1 = 0, \ldots, \eta_k = 0$.

We can thus use the variables $\eta_1, \ldots, \eta_k$ as coordinates on B/A and $\theta_1, \ldots, \theta_k$ as coordinates on F^*, the identification in (4.7) is then

$$\theta \rightsquigarrow \eta$$

where

$$\eta_i = \sum \frac{\partial^2 \varphi_1}{\partial \theta_i \partial \theta_j} \theta_j. \tag{4.10}$$

Thus the pull back of $(d\eta)^{-1/2}$ is $(d\theta)^{-1/2}(\det d_\theta^2 \varphi_1)^{-1/2}$. Multiplying this by $a_0(x_0, \theta_0, \tau)(dx d\theta)^{1/2}$ and then by the factor $(2\pi i \tau)^{-k/2}$ gives the leading term in (4.6). Since (4.6) is independent of the parametrization, we see that the symbol is well defined, at least at points where the projection is regular.

To rid ourselves of this assumption we have to refine the formula (4.2). We recall from Chapter V, that if V is a symplectic vector space and W_1 and W_2 are transversal Lagrangian subspaces, then there is a canonical pairing

$$\wedge^{1/2} W_1 \times \wedge^{1/2} W_2 \to \mathbf{C}.$$

Suppose that $\mu_1 = a_1 e^{i\tau\psi_1}$ and $\mu_2 = a_2 e^{i\tau\psi_2}$ are simple asymptotic half forms on X with compact support. By what we have just shown, the symbol of μ_i is the pull back to $\Lambda_i =$ graph $d\psi_i$ of a_i. We will prove the following:

LEMMA. *If Λ_1 and Λ_2 intersect transversally at (x_0, ξ_0), then*

$$\int_X \mu_1 \bar{\mu}_2 = (2\pi i\tau)^{-n/2} e^{i\tau c} \sum b_i \tau^i \tag{4.11}$$

where

$$c = \psi_1(x_0) - \psi_2(x_0)$$

and b_0 is the number obtained by pairing $\sigma(\mu_1)$ with $\sigma(\mu_2)$ at (x_0, ξ_0). Here we are taking V to be the tangent space to T^*X at (x_0, ξ_0) and W_i to be the tangent spaces to Λ_i.

PROOF. We have $d(\psi_1 - \psi_2) = 0$ at x_0 and the transversality condition implies that the Hessian of $\psi_1 - \psi_2$ is non-degenerate at this point, so that the method of stationary phase can be applied to the left side of (4.11). To evaluate the expressions entering into the formula of stationary phase we proceed as follows: Consider the map of T^*X into itself given by

$$f(x, \xi) = (x, \xi - d\psi_2(x)).$$

This is a canonical transformation carrying Λ_2 into the zero section and carrying Λ_1 into the graph of $d(\psi_1 - \psi_2)$. It commutes with the projection $T^*X \to X$, and so maps the preimage of a_i on Λ_i onto the corresponding preimage on $f(\Lambda_i)$, and preserves the pairing. The stationary phase formula is unchanged if we replace ψ_1 by $\psi_1 - \psi_2$ and ψ_2 by 0, in evaluating the left hand side of (4.11). We may thus assume that $\mu_1 = a_1 e^{i\tau\psi}$ and that $\mu_2 = a_2$ is independent of τ in evaluation of the highest order term in (4.11). But now we are reduced to the situation we have dealt with above. Namely, we can consider

$$\int_X a_1 \bar{a}_2 e^{i\tau\psi}$$

as a compound asymptotic over the one point manifold corresponding to the diagram

$$\begin{array}{l} X \to \mathbf{R} \\ \pi \downarrow \\ \text{pt} \end{array}$$

where π is given the structure of a morphism so that $\pi^*(1) = a_2$. This completes the proof of the lemma. We can regard the lemma, and formula (4.11), as the invariant formulation of the top order terms entering into the method of stationary phase when we use half forms.

We will now extend the lemma to compound asymptotics so as to complete the proof that the symbol is well defined, even at non-regular points. Let φ_1, π, and Z be as in (4.3), let x and θ be as above, and assume that the critical point, z_0, corresponding to the point $(x_0, \xi_0) \in \Lambda$ has coordinates $x = 0$, $\theta = 0$. Let $q(\theta)$ be a quadratic form which is non-degenerate on the null space of $d_\theta^2 \varphi_1$ at z_0, and is zero on a complementary space, and define the function $\varphi_r \colon Z \to \mathbf{R}$ by

$$\varphi_r(x,\theta) = \varphi_1(x,\theta) + rq(\theta).$$

We make φ_r into a morphism of half forms by setting

$$\varphi_r^*(dt)^{1/2} = (dx d\theta)^{1/2}.$$

Let us set

$$\Lambda_r = \pi_* \varphi_r^*(\Lambda_1) \qquad \text{and} \qquad \mu_r = \pi_* a \varphi_r^* e^{i\tau t}(dt)^{1/2}.$$

By construction, the point (x_0, ξ_0) is a regular point of Λ_r for small $r \neq 0$, so μ_r is a simple asymptotic at x_0 for small $r \neq 0$. Let

$$\nu = b e^{i\tau\psi}$$

be a compactly supported half form on X such that graph $d\psi$ intersects Λ_r transversally for all small values of r. Then

$$\int_X \mu\bar{\nu} = \int a\bar{b} e^{i\tau(\varphi_r(x,\theta) - \psi(x))} dx\, d\theta$$

admits an asymptotic expansion of type (4.2) for all small values of r, while for $r \neq 0$, the leading term in the expansion can be computed according to the recipe given in the lemma. By continuity, formula (4.11) is valid also at $r = 0$. This shows that the leading term in the asymptotic expansion of $\int_X \mu\bar{\nu}$ determines the symbol of μ at (x_0, ξ_0), and hence that the symbol is independent of the parametrization.

The order of a compound asymptotic of the form $\pi_* a e^{i\tau\varphi_1}$ is defined to be $r - (k/2)$ where r is the order of the leading term in the asymptotic expansion of a, and k is the fiber dimension of $Z \to X$. We have thus proved:

THEOREM 4.1 *Let μ be a compound asymptotic of order m associated with the Lagrangian pair (Λ, φ). Then the symbol of μ is well defined as a half form on Λ of the type $\tau^m\gamma$ where γ does not depend on τ. If two asymptotics have the same symbol, then they differ by an asymptotic of order one less. If $\nu = a e^{i\tau\psi}$ is a simple asymptotic of compact support such that* graph $d\psi$ *intersects Λ transversally at (x_0, ξ_0), then*

$$\int_X \mu\bar{\nu} = (2\pi i\tau)^{-n/2} e^{ic\tau} \sum b_i \tau^i \tag{4.11}$$

where $c = \varphi(x_0, \xi_0) - \psi(x_0)$, *and the leading coefficient*, b_m, *in* (4.11) *is obtained by pairing the half forms* $\sigma(\mu)$ *and* $\sigma(\nu)$ *at* (x_0, ξ_0).

The following statements are proved exactly as in Chapter VI for the corresponding statements for generalized half forms. We will let the reader consult §5 of Chapter VI for details.

THEOREM 4.2. *Let* (Λ, φ) *be a Lagrangian pair in* $T^* Y$ *and let* $f\colon X \to Y$ *be a morphism which satisfies the transversality condition for pull back. If* μ *is an asymptotic of order m associated with* (Λ, φ), *then* $f^* \mu$ *is an asymptotic of order m associated with* $(f^* \Lambda, f^* \varphi)$, *and*

$$\sigma(f^* \mu) = f^* \sigma(\mu). \tag{4.12}$$

THEOREM 4.3. *Let* (Λ, φ) *be a Lagrangian pair in* $T^* X$ *and let* $f\colon X \to Y$ *be a morphism satisfying the transversality condition for push forward, where f is a fiber map with fiber dimension k. If* μ *is a compound asymptotic of order m associated with* (Λ, φ), *then* $f_* \mu$ *is a compound asymptotic of order* $m - (k/2)$ *associated with* $(f_* \Lambda, f_* \varphi)$, *and*

$$\sigma(f_* \mu) = (2\pi i\tau)^{-k/2} f_* \sigma(\mu). \tag{4.13}$$

We now take up the transport equation for compound asymptotics. Let $P = P(x, D, \tau)$ be an asymptotic differential operator of order k acting on the asymptotic half forms of X. (See §2 of Chapter II for the definitions.) Let $p(x, \xi)$ be its principal symbol, and $\boldsymbol{\eta}_p$ the corresponding Hamiltonian vector field. Let $p_{\text{sub}}(x, \xi)$ be the subprincipal symbol of P. Finally, let (Λ, φ) be a Lagrangian pair such that $p = 0$ on Λ.

THEOREM 4.4. *If* μ *is a compound asymptotic of order m associated with* (Λ, φ) *then* $P\mu$ *is a compound asymptotic of order* $m - 1$, *and*

$$\sigma(P\mu) = \tau^{-1}(-iD_{\boldsymbol{\eta}_p} \sigma(\mu) + p_{\text{sub}} \sigma(\mu)). \tag{4.14}$$

PROOF. We follow the proof of the corresponding result in Chapter VI, §7. First of all, if the theorem is true for P and for Q, it is true for QP and for any operator with the same principal symbol as P. As we saw in Chapter VI, §7, this is a consequence of the composition rule for the subprincipal symbol. Next we note that it is sufficient to prove the theorem at points $(x, \xi) \in \Lambda$ for which the projection map $\Lambda \to X$ has constant rank in some neighborhood of (x,ξ), since these points form an open dense set. To conclude the proof we will need a characterization of Lagrangian manifolds in T^*X with the property that $\Lambda \to X$ has constant rank. For any Lagrangian manifold, Λ, and any smooth function, ψ, we set

$$\Lambda_\psi = \{(x, \xi + d\psi) \text{ where } (x,\xi) \in \Lambda\}.$$

As we have already seen, Λ_ψ is again a Lagrangian submanifold, since it differs from Λ by a canonical transformation. Suppose that $\Lambda \to X$ has constant rank, so that, locally, Λ is fibered over a submanifold, Y of X. On Λ, we can locally write $\alpha = d\varphi$, and since α is zero on vertical vectors, the function φ is constant along the fibers of $\Lambda \to X$, and thus can be regarded as the pull back of some function defined on Y. We extend this function, defined on Y, to be defined (locally) on X in some neighborhood of Y, and denote the extended function by ψ. Thus ψ is a function defined on X and $\alpha - \pi^* d\psi$ vanishes on Λ, where π denotes the projection of $T^*X \to X$. In other words, α vanishes on Λ_ψ, i. e., Λ_ψ is the normal bundle of Y. We have thus proved:

PROPOSITION 4.1. *Suppose that Λ is a Lagrangian submanifold of T^*X such that $\Lambda \to X$ has constant rank. Then, locally, there exists a smooth function ψ such that Λ_ψ is the normal bundle of $Y = \pi(\Lambda)$.*

We now prove Theorem 4.4. We introduce coordinates $(t,y) = (t_1, \ldots, t_k, y_1, \ldots, y_{n-k})$ near the Y specified by the proposition, so that Y is given by $t = 0$. We may assume that the function, ψ, of the proposition is a function of y alone. This implies that we can choose a phase function, φ, for Λ of the form

$$\varphi(x,\theta) = \psi(y) + t_1\theta_1 + \cdots + t_k\theta_k,$$

and thus an asymptotic associated to Λ can be written as

$$\mu = \left(\int a(x,\theta,\tau)e^{i\tau(\psi(y)+t_1\theta_1+\cdots+t_k\theta_k)}d\theta\right)(dx)^{1/2}$$

and then the symbol of μ at the point $d\psi + \theta_1\,dt_1 + \cdots + \theta_k\,dt_k$ is

$$\sigma(\mu) = (2\pi i\tau)^{-k/2}a_0(x,\theta,\tau)(dyd\theta)^{1/2}. \tag{4.15}$$

The ideal of functions vanishing on Λ is (locally) generated by the functions

$$t_1, \ldots, t_k \quad \text{and} \quad \eta_1 - (\partial\psi/\partial y_1), \ldots, \eta_{n-k} - (\partial\psi/\partial y_{n-k})$$

where the η's are the dual variables to the y's. It suffices to prove the formula (4.14) with P given by multiplication by t_j or by the differential operator $(i\tau)^{-1}(\partial/\partial y_j) - (\partial\psi/\partial y_j)$. For multiplication by t_j, formula (4.14) is obvious in view of the explicit form of μ and of $\sigma(\mu)$. The remaining operators involve only the y coordinates. Hence we can apply the proof of the transport equation given in §6 of Chapter II, regarding t and θ as parameters.

In many interesting situations the form α is not exact, but the cohomology class $(1/2\pi)[\alpha]$ is integral, in other words, the integral $(1/2\pi)\int_\gamma \alpha$ is an integer for any closed curve, γ, on Λ. In this case we will be able to develop a symbol calculus like that above for asymptotics, where τ ranges over integer values rather than over real numbers. Before developing this idea, let us first note that if $(1/2\pi)[\alpha]$ is integral on Λ, then we can write $\alpha = d\varphi$ on Λ, where φ is only determined up to additive multiples of 2π. In other words, φ is well defined as a map into $\mathbf{R}/\{2\pi\mathbf{Z}\}$ which can be identified with the unit circle, S^1. We have thus proved:

PROPOSITION 4.2. *If the periods of* $(1/2\pi)\alpha$ *are integral, then there exists a map* $\varphi\colon \Lambda \to S^1$, *uniquely determined up to a rotation of* S^1, *such that* $\alpha = \varphi^* d\theta$ *on* Λ *where* θ *is the angle variable on* S^1.

We now consider Lagrangian pairs (Λ, φ) where φ is a map of $\Lambda \to S^1$ and $\varphi^* d\theta = \alpha$ on Λ. It is easy to see that these new Lagrangian pairs have the same functorial properties as the Lagrangian pairs considered above: If $f\colon X \to Y$ satisfies the transversality condition for pull back then $f^*(\Lambda, \varphi) = (f^*\Lambda, f^*\varphi)$ is defined as above, where (Λ, φ) is a Lagrangian pair in T^*Y, and similarly for push forward. We similarly define admissible parametrizations of Lagrangian pairs as before.

We now consider asymptotics where the parameter τ is restricted to taking on integer values. Thus two sequences of half forms, $\{u_n\}$ and $\{v_n\}$, are said to be asymptotic if $n^k(u_n - v_n)$ tends to zero in the C^∞ topology. The equivalence class of such sequences will be called a $\mathbf{Z}$ asymptotic half form, or an integer asymptotic half form, or simply an asymptotic half form when it is clear from the context that we are working in the framework of integer asymptotics. There is no difficulty in working out the functorial properties as above for these objects.

The notion of an asymptotic expansion is the same as before, except that τ is restricted to take on integer values, and the coefficients, a_i, in an asymptotic expansion $\sum a_i n^i$ are uniquely determined.

Given a Lagrangian pair, (Λ, φ), we will say that an asymptotic half form $\{\mu_n\}$ is a compound asymptotic associated with (Λ, φ) if μ_n is a locally finite sum of (a fixed number of terms) of the form

$$\pi_* \nu_n e^{in\varphi_1}$$

where

$$\begin{array}{ccc} Z & \xrightarrow{\varphi_1} & \mathbf{R} \\ \downarrow{\scriptstyle \pi} & & \\ X & & \end{array}$$

is an admissible parametrization of (Λ, φ) and where $\{\nu_n\}$ admits an asymptotic expansion in powers of n. The order of this asymptotic is defined to be $m - k/2$ where m is the degree of the leading term in the expansion of $\{\nu_n\}$ and k is the fiber dimension.

If ψ is a smooth function on X and graph $d\psi$ intersects Λ transversally at (x_0, ξ_0), and a is a half form of compact support then the integral

$$\int_X \mu_n \bar{a} e^{-in\psi}$$

admits an asymptotic expansion of the form

$$e^{icn} \sum b_i n^i$$

with leading term of order $d = (\operatorname{ord} \mu) - (\dim X)/2$ and with $c = \varphi(x_0, \xi_0) - \psi(x_0)$. Here, of course, c is only determined up to multiples of 2π, but the expression e^{icn} is well defined since n is an integer. It is easy to check that the proof of Theorem 4.1 goes through virtually unchanged for integer asymptotics. Summarizing the contents of Theorems 4.1, 4.2 and 4.3 we have:

THEOREM 4.7. *Let (Λ, φ) be an integral Lagrangian pair in T^*X and μ an asymptotic half form associated with (Λ, φ). Then μ has a well defined symbol of the form $n^d \nu$ where d is the order of μ and ν is a half form on Λ which does not depend on n. The symbol determines μ up to asymptotics of lower order and commutes with the operations of pull back and push forward.*

By an asymptotic differential operator in the integer theory we will mean as in Chapter II an expression of the form

$$\sum n^{-j} P_j(x, D)$$

where $P_j(x, D)$ is a differential operator in the usual sense (i.e., not depending on n) and the order of P_j is less than or equal to j. The principal and subprincipal symbols of the above expression are defined as in Chapter II. The analogue of Theorem 4.4 is valid for integer asymptotics, the proof requiring only minor changes:

THEOREM 4.8. *Let $p = \sigma(P)$. If $p = 0$ on Λ and μ is a compound asymptotic associated with (Λ, φ), then $P\mu$ is a compound asymptotic of order one less, and*

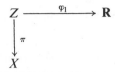

$$\sigma(P\mu) = (1/n)(i^{-1}\eta_p\mu + \sigma_{\text{sub}}(P)\mu).$$

As an application of this theorem we will rederive the results of Simms on the quantized energy levels of the one dimensional harmonic oscillator (Chapter V, §8) in the context of integer asymptotics. We are looking for solutions of the asymptotic differential equation

$$(1/2n^2)(d^2\psi/dx^2) + (E_0 + (1/n)E_1 - x^2/2)\psi = 0,$$

where E_0 and E_1 are constants to be determined. The principal symbol of the operator above is

$$E_0 - \tfrac{1}{2}(x^2 + \xi^2)$$

and the subprincipal symbol is just E_1. We must look for Langrangian submanifolds, Λ_{E_0}, on which the symbol vanishes and on which $(1/2\pi)\alpha$ has an integral period. Now Λ_{E_0} is clearly just a circle of radius $\sqrt{2}\,E_0$ and as for the integrality condition we have, by Stokes' theorem,

$$\frac{1}{2\pi}\int_{\Lambda_{E_0}} \alpha = \frac{1}{2\pi}\int_{x^2+\xi^2 \leqslant 2E_0} dx d\xi = E_0.$$

Therefore E_0 must take on integer values, and, in fact, if we want the period of $\alpha/2\pi$ to be exactly 1, then E_0 must equal 1 and Λ_{E_0} must be the unit circle S^1. We will determine E_1 by solving the transport equation on S^1; and in order to do this we must give S^1 a metalinear structure. Since $H^1(S^1, \mathbf{Z}_2) = \mathbf{Z}_2$ and $H^2(S^1, \mathbf{Z}_2) = 0$ there are precisely two metalinear structures on S^1 (see Chapter V, §8). For one of these, the Stiefel-Whitney class of the half form bundle is the zero element in $H^1(S^1, \mathbf{Z}_2)$ and this we call the *trivial* metalinear structure, while for the other, the Stiefel-Whitney class is the non-zero element, and this we will call the non-trivial metalinear structure. As we have seen in Chapter V, the metalinear structure on $\mathbf{R}$ induces a metaplectic structure on $T^*\mathbf{R} = \mathbf{R}^2$ and this metaplectic structure induces the non-trivial metalinear structure on each circle. We recall the definition of the two metalinear structures. If we take the basis bundle of S^1 and divide out by the action of $\mathbf{R}^+$, we just get two copies of S^1. A metaplectic covering of this "reduced" basis bundle of S^1 must therefore consist either of four copies of S^1 or two copies of S^1. In the latter case the covering map is given by $z \to z^2$. The first case is the trivial metalinear structure, the action of $\mathbf{Z}_4$ just permutes the four copies of S^1. In the second case, the generator of $\mathbf{Z}_4$ acts by changing z in the first copy to $-z$ in the second copy, then to $-z$ in the first copy and finally to z in the second copy.

Fix a half form ρ_0 at $\theta = 0$ in S^1 such that $\rho_0^2 = d\theta_0$. There is a unique continuous section, ρ, of the half form bundle over the interval $[0, 2\pi)$ such that $\rho(0) = \rho_0$ and $\rho^2 = d\theta$. We claim that:

For the trivial metalinear structure $\rho(2\pi) = \rho(0)$ *while for the non-trivial metalinear structure* $\rho(2\pi) = -\rho(0)$.

Indeed, since $\rho(2\pi)^2 = d\theta$, we must have $\rho(2\pi) = \pm\rho(0)$. If $\rho(2\pi) = \rho(0)$, then the half form bundle admits a nowhere zero section, and hence is trivial. Thus, for non-trivial metalinear structure we must have $\rho(2\pi) = -\rho(0)$. The transport equation on S^1 takes the form

$$-(1/i)(\partial\mu/\partial\theta) + E_1\mu = 0,$$

where μ is a half form on S^1. The general solution of this equation is

$$\mu = (\text{const.})e^{iE_1\theta}\rho,$$

where ρ is the half form introduced above. We must thus have

$$e^{iE_1 2\pi} = 1$$

for the trivial metalinear structure and

$$e^{iE_1 2\pi} = -1$$

for the non-trivial metalinear structure. Since we know that the metalinear structure on $\mathbf{R}$ induces the non-trivial metalinear structure on $\mathbf{R}^2$ and hence the non-trivial metalinear structure on each circle, this shows that the correct quantization of the energy levels is given as $n + \frac{1}{2}$.

§5. Pointwise behavior of compound asymptotics and Bernstein's theorem.

We will begin our study of compound asymptotics by studying their pointwise behavior: Given a compound asymptotic of the form

$$\int e^{i\tau\varphi(x,\theta)}\mu(x,\theta)\,d\theta,$$

we will study its behavior as a function of the frequency τ when x is held fixed. We will see later on that the asymptotic behavior in the frequency can vary drastically from point to point in x space, and that the geometric study of this variation is of great importance in physical applications, and can be quite complicated. Nevertheless, it turns out that for fixed x, the behavior is fairly simple. Note, that since x is held fixed, we need not write it in the expression for the integral. We are thus basically concerned in this section with the asymptotic evaluation of integrals of the form

$$I(\mu,\tau) = \int e^{i\tau\varphi(\theta)}\mu(\theta)\,d\theta, \tag{5.1}$$

φ being a smooth function defined on some compact subset, A, of $\mathbf{R}^n$, and μ a smooth function with support on A. Right off the bat we will make an assumption about φ which has the appearance of being quite restrictive; we will assume φ *is a polynomial.* This assumption is not as restrictive as it looks. For example, let φ be a smooth function on $\mathbf{R}^n$ with a singularity at 0. Suppose the first partials $(\partial\varphi/\partial\theta_1), \ldots, (\partial\varphi/\partial\theta_n)$ generate an ideal of finite codimension in $C_0^\infty(\mathbf{R}^n)$, $C_0^\infty(\mathbf{R}^n)$ being the ring of germs of smooth functions at the origin of $\mathbf{R}^n$. Then it is possible to make a C^∞ change of coordinates in a neighborhood of 0 such that in the new coordinates φ is a polynomial (see the Appendix to Chapter I). The condition just described turns out to hold for most functions φ; in fact the set for which it fails to hold is of "infinite codimension" in the space of all functions (see Tougeron [5]).

If φ is a polynomial then it has only a finite number of critical values. We will restrict A so that φ takes on only one of its critical values, say a_0, on A. The asymptotic behavior of (5.1) is completely described by the following theorem of I. N. Bernstein [6]:

THEOREM 5.1. *There exists a sequence of exponents $\epsilon_1, \ldots, \epsilon_k$ and non-negative integers $\ell_1, \ell_2, \ldots, \ell_k$ such that*

$$\int e^{i\tau\varphi}\mu\, d\theta \sim e^{i\tau a_0} \sum_{i=1}^{k} \sum_{\ell=0}^{\ell_i} \tau^{\epsilon_i}(\log \tau)^\ell \sum_{j=0}^{\infty} a_{ij\ell}(\mu)\tau^{-j} \tag{5.2}$$

the $a_{ij\ell}$'s being distributions, and $a_{ij\ell}(\mu)$ the values of these distributions on the smooth functions, μ.

The first step in the proof, which we will only give in rough outline, consists of reducing (5.2) to an equivalent but much more algebraic assertion.

Replacing φ by $\varphi - a_0$, which does not change the form of (5.2), we can assume that the critical value of φ is 0. Therefore, for $s \neq 0$ the set of points $\{x \in \mathbf{R}^n | \varphi(x) = s\}$ intersects a neighborhood of A in a smooth manifold; and so, for every $\mu \in C_0^\infty(A)$,

$$J(\mu, s) = \frac{1}{(2\pi)^n} \int \mu\, d\rho_s$$

is well defined for $s \neq 0$, where $d\rho_s$ is the intrinsic $n-1$ form on the surface $\varphi = s$ and is defined by $d\varphi \wedge d\rho_s = d\theta_1 \wedge \cdots \wedge \theta_n$.

Rewriting (5.2) in the form

$$(2\pi)^n \int e^{i\tau s} J(\mu, s)\, ds$$

we see that $J(\mu, s)$ is the Fourier transform of $I(\mu, \tau)$; so the asymptotic behavior of $I(\mu, \tau)$ at $\tau = \infty$ is completely determined by the singular behavior of $J(\mu, s)$ at $s = 0$ and vice versa. Let φ_+ be the function which is equal to φ for $\varphi \geqslant 0$ and equal to 0 for $\varphi < 0$. Then

$$\int \varphi_+^z \mu \, d\theta = \int_0^\infty s^z J(\mu, s) \, ds \tag{5.3}$$

when $\operatorname{Re} z \geqslant 0$, $z \in \mathbf{C}$. We have stipulated $\operatorname{Re} z \geqslant 0$ because otherwise the integrand in (5.3) will not be an integrable function. Nevertheless, (5.3) being a holomorphic function of z, it is reasonable to ask whether it can be extended by analytic continuation to a larger subset of $z \in \mathbf{C}$. For example, suppose $J(s) = s^{-\epsilon}\rho(s)$ for $s > 0$, $\rho(s)$ being compactly supported and smooth on the whole real line. Then

$$\int_0^\infty s^z J(s) \, ds = \int_0^\infty s^{z-\epsilon} \rho(s) \, ds$$

which is meromorphic function of z for all $z \in \mathbf{C}$ with poles at $\epsilon - 1$, $\epsilon - 2$, $\epsilon - 3$, etc. (See Gelfand-Shilov [7].) Bernstein proves that this example is far from atypical, namely he proves that *for all polynomials* φ, (5.3) *is a meromorphic function of* z *on the whole complex plane, and its poles lie on a finite sequence of arithmetic progressions*

$$\epsilon_i - 1, \epsilon_i - 2, \ldots, \tag{5.4$_i$}$$

$i = 1, \ldots, k$, the order of the pole being a fixed integer ℓ_i at each point of the sequence (5.4$_i$). Arguing backwards one can show that $J(s)$ admits an asymptotic expansion of the form

$$J(s) = \sum_{i=1}^{k} \sum_{\ell=0}^{\ell_i} s^{-\epsilon_i} (\log s)^\ell \sum a_{ij\ell} s^j \tag{5.5}$$

near $s = 0$. Taking the inverse Fourier transform of (5.5), we get (5.2).

To prove the assertions we have just made about (5.3), Bernstein shows that there exists an mth order partial differential operator, $P(x, \lambda, \partial/\partial x)$, whose coefficients are polynomial functions of $x \in \mathbf{C}^n$ and $\lambda \in \mathbf{C}$ such that, for $\operatorname{Re} \lambda > m$,

$$P\left(x, \lambda, \frac{\partial}{\partial x}\right) \varphi_+^\lambda = q(\lambda) \varphi_+^{\lambda - 1}, \tag{5.6}$$

$q(\lambda)$ being a polynomial function of λ. (Note that, for $\operatorname{Re} \lambda > m$, φ_+^λ is m times differentiable in x so (5.6) makes sense.) Let $\epsilon_1, \ldots, \epsilon_n$ be the roots of $q(\lambda)$. Using (5.6) one can define φ_+^λ as a distribution in x for all λ not equal to (5.4) by the following simple induction procedure. Having defined φ_+^z on a strip $|\operatorname{Re} z - \operatorname{Re} z_0| < \epsilon$ not containing any of the points (5.4) use the left hand side of (5.6) to define it on $|\operatorname{Re} z - \operatorname{Re} z_0 - 1| < \epsilon$. Since φ_+^z is already defined for $\operatorname{Re} z \geqslant 0$ this defines it for all z, excluding (5.4).

We claim that the assertion (5.2) has now been translated into an assertion that is entirely algebraic in character. Namely, given a polynomial $\varphi(x_1, \ldots, x_n)$

whose coefficients are complex numbers, consider the set consisting of all finite sums of expressions of the form: $p(\lambda, x)\varphi^{\lambda-k}$, $p(\lambda, x)$ being a polynomial in x whose coefficients are elements of the function field $\mathbf{C}(\lambda)$. Let M_φ be the quotient space obtained by identifying two such expressions, say $p_1\varphi^{\lambda-k_1}$ and $p_2\varphi^{\lambda-k_2}$, if $\varphi^{k_1}p_2 = \varphi^{k_2}p_1$. Let Ω_n be the ring consisting of all formal differential operators of the form:

$$\sum a_\alpha(x, \lambda)\frac{\partial}{\partial x^\alpha},$$

$a_\alpha(x, \lambda)$ being a polynomial in x with coefficients in the field $k = \mathbf{C}(\lambda)$. M_φ can be made into an Ω_n module by requiring that

$$\frac{\partial}{\partial x_i}(p\varphi^{\lambda-k}) = \left(\frac{\partial}{\partial x_i}p\right)\varphi^{\lambda-k} + (\lambda - k)p\varphi^{\lambda-k-1}\frac{\partial\varphi}{\partial x_i} \tag{5.7}$$

and

$$x_i(p\varphi^{\lambda-k}) = (x_i p)\varphi^{\lambda-k}.$$

It is clear that (5.6) is equivalent to the following purely algebraic assertion about the element φ^λ in M_φ:

$$P\left(x, \lambda, \frac{\partial}{\partial x}\right)\varphi^\lambda = q(\lambda)\varphi^{\lambda-1}. \tag{5.8}$$

One final observation: To prove there exists a $P \in \Omega_n$ satisfying (5.8) it is necessary and sufficient to prove the following assertion:

THEOREM 5.2. *M_φ is finitely generated as an Ω_n module.*

In fact let $p\varphi^{\lambda-k_i}$, $i = 1, \ldots, r$, be a set of generators and let $k = \max k_i$. From the fact that $\varphi^{\lambda-k-1}$ can be expressed as a linear combination of the $p_i\varphi^{\lambda-k_i}$ with coefficients in Ω_n one gets an equation of the form

$$p\left(x, \lambda, \frac{\partial}{\partial x}\right)\varphi^{\lambda-k} = \varphi^{\lambda-k-1}.$$

Replacing $\lambda - k$ by λ in this equation we get (5.8). Conversely (5.8) shows that M_φ is generated by φ^λ.

For the sake of completeness we have included a proof of Theorem 5.2 in the following appendix.

Appendix to §5 of Chapter VII.

To prove Theorem 5.2 we will need some classical results from commutative algebra. Let k be a field of characteristic zero and let $S = S_m$ be the polynomial ring $k[x_1, \ldots, x_m]$. S_m is a graded ring: $S_m = \sum S_m^i$ where S_m^i is the space of homogeneous polynomials of degree i. Let $M = \sum M^i$ be a finitely generated graded S_m module.

THEOREM A.5.1. *There exists a polynomial $p(t)$ with rational coefficients such that for all i sufficiently large* $\dim M^i = p(i)$.

REMARK. $p(t)$ is the so-called *Hilbert polynomial* of the module M.

PROOF (by induction on m). Let $x = x_m$. Consider the exact sequence

$$0 \to \mathrm{Ann}\,(x)^i \to M^i \to M^{i+1} \to (M/xM)^{i+1} \to 0$$

where Ann (x) is the set of $a \in M$ such that $xa = 0$, and the map in the middle is multiplication by x. Then

$$\dim M^{i+1} - \dim M^i = \dim (M/xM)^{i+1} - \dim \mathrm{Ann}\,(x)^i. \qquad \text{(A.5.1)}$$

Now M/xM and Ann (x) are finitely generated modules over

$$S_{m-1} = k[x_1, \ldots, x_{m-1}];$$

so for sufficiently large i the right hand side of (A.5.1) is a polynomial function of i. Summing the left hand side from 0 to i we see that the same is true of $\dim M^i$. Q. E. D.

Let $p(t) = \sum_{i=0}^d a_i t^i$, $a_d \neq 0$. Then $d = d(M)$ is called the *dimension* of the module M. The coefficients of $p(t)$ are clearly rational numbers, i.e., $a_i = (p_i/q_i)$ with p_i and q_i integers. For κ sufficiently large $p(\kappa q_1 q_2 \cdots q_d + i)$ is an integer for all i's so the $p(i)$'s are integers for all i's. It is well known that all polynomials with this property are linear combinations, with integer coefficients, of the polynomials $\binom{t}{r} = t(t-1)\cdots(t-r+1)/r!$. (See Zariski-Samuel [8 Chapter VII].) In particular $e(M) = a_d d!$ is an integer.

We are now going to define invariants analogous to $d(M)$ and $e(M)$ for a finitely generated module, M, over the ring $\Omega = \Omega_n$. Recall that Ω_n is the ring of formal differential operators of the form

$$a = \sum a_\alpha(x)\frac{\partial^{|\alpha|}}{\partial x^\alpha}, \qquad \alpha = (\alpha_1, \ldots, \alpha_n), \tag{A.5.2}$$

with $a_\alpha(x) \in k[x_1, \ldots, x_n]$ and

$$\frac{\partial^{|\alpha|}}{\partial x^\alpha} = \frac{\partial}{\partial x_1^{\alpha_1}} \cdots \frac{\partial}{\partial x_n^{\alpha_n}}.$$

Define

$$\Omega^\kappa = \{a \in \Omega, \text{ degree } a_\alpha + |\alpha| \leqslant \kappa \text{ in (A.5.2)}\}.$$

The Ω^κ's are a filtration of Ω in the sense of ring theory: $\Omega^\kappa \subset \Omega^{\kappa+1}$, $\bigcup \Omega^\kappa = \Omega$, and $\Omega^i \cdot \Omega^j \subset \Omega^{i+j}$. The graded ring gr $\Omega = \Omega^1/\Omega^0 + \cdots$ is isomorphic with S_m, $m = 2n$. Let M be an Ω_n module. Recall that a filtration of M is a sequence of finite dimensional subspaces (over k), $M^0 \subset M^1 \subset M^2$, etc. such that $\bigcup M^i = M$ and $\Omega^i M^j \subset M^{i+j}$. We will call a filtration admissible if $\Omega^i M^\kappa = M^{\kappa+i}$ for κ sufficiently large and i arbitrary. Since the M^i's are assumed to be finite dimensional over k, M clearly has an admissible filtration if and only if it is finitely generated. We will prove a couple of propositions about such filtrations:

Proposition A.5.2. *Let $\{M^i, i = 0, 1, 2, \ldots\}$ be an admissible filtration of M. Then there exists a polynomial $p(t)$ such that, for i sufficiently large,* dim $M^i = p(i)$. *Moreover if $p(t) = \sum_{r=0}^d a_r t^r$ with $a_d \neq 0$ then $e(M) = a_d d!$ is an integer; and $e(M)$ and $d(M) = d$ are independent of the filtration, i.e., are invariants of M alone.*

Proof. The first statement is a consequence of Theorem A.5.1. Just apply Theorem A.5.1 to gr Ω and gr M. Suppose $\{M^i\}$ and $\{M_1^i\}$ are two admissible filtrations. We can find r, s, t very large so that

$$M^r \subset M_1^s \subset M^t.$$

Since the filtrations are admissible this implies that

$$M^{r+i} \subset M_1^{s+i} \subset M^{t+i}$$

for all i, or that $p(r+i) \leqslant p_1(s+i) \leqslant p(t+i)$ for all i. This shows that the degree of p and its leading coefficient are equal to the degree of p_1 and its leading coefficient. Q. E. D.

Proposition A.5.3. *Let $0 \to M_1 \to M \to M_2 \to 0$ be an exact sequence of finitely generated Ω modules. Then $d(M) = \max(d(M_1), d(M_2))$. If $d(M_1) > d(M_2)$ (or $d(M_2) > d(M_1)$) then $e(M) = e(M_1)$ (or $e(M) = e(M_2)$). If $d(M_1) = d(M_2)$ then $e(M) = e(M_1) + e(M_2)$.*

Proof. Let $\{M^i\}$ be an admissible filtration of M. Let M_2^i be the image of M^i in M_2 and M_1^i be the preimage of M^i in M_1. It is clear that $\{M_2^i\}$ is an admissible filtration of M_2. We will show that $\{M_1^i\}$ is an admissible filtration of M_1. Note first that $\sum M_1^i/M_1^{i-1}$ is an S_m submodule of $\sum M^i/M^{i-1}$. Since the second module is Noetherian so is the first; in particular the first module is finitely generated; so for κ sufficiently large $\Omega^i M_1^\kappa = M_1^{\kappa+i}$ for all i. To conclude the proof of Proposition A.5.2 note that $\dim M^i = \dim M_1^i + \dim M_2^i$. Q. E. D.

The main result of Bernstein is the following:

Theorem A.5.4. *If M is a finitely generated Ω_n module then either $M = \{0\}$ or $d(M) \geqslant n$.*

Before discussing the proof of this theorem let us see how it implies Theorem 5.2. Since Theorem 5.2 has to do with modules which are not necessarily finitely generated we first need a substitute for the notion of admissible filtration for such modules.

Definition A.5.5. Let M be an Ω_n module and $\{M^i, i = 0, 1, 2, \ldots\}$ a filtration of M. We will say that $\{M^i\}$ is a (d, e) filtration if $\dim M^i \leqslant (e/d!)i^d + o(i^d)$ for all i.

For example, if M is a finitely generated Ω_n module then any admissible filtration of M is a $(d(M), e(M))$ filtration. It is clear that if M has a (d, e) filtration and L is any finitely generated submodule of M then $d(L) \leqslant d$ and, if $d(L) = d$, then $e(L) \leqslant e$.

Lemma A.5.6. *Let M be a Ω_n module having a (d, e) filtration. Then if $d < n$, $M = \{0\}$ and if $d = n$, M has finite length, not exceeding e. (In particular M is finitely generated.)*

Proof. If M is non-zero it contains a non-zero finitely generated submodule, so the first assertion is a direct consequence of Theorem 5.4. To prove the second assertion let $\{0\} = M_0 \subset M_1 \subset \cdots \subset M_\kappa \subset M$ be a chain of submodules with $M_i \neq M_{i+1}$. Let f_i be an element of $M_i \sim M_{i-1}$ and let $L_i = \Omega(f_1, \ldots, f_i)$. As we remarked above $d(L_i) \leqslant n$, so, by Theorem A.5.4, $d(L_i) = n$; therefore, again by Theorem A.5.4, $d(L_i/L_{i-1}) = n$. Since $e(L_i/L_{i-1}) \geqslant 1$ Proposition A.5.3 implies that $e(L_\kappa) \geqslant \kappa$. Since $e(L_\kappa) \leqslant e$ the length of M does not exceed e. Q.E.D.

Now consider the module M_φ in Theorem 5.2. If φ is of degree r this module has an $(n, (r + 1)n)$ filtration given by setting

$$M_\varphi^i = \{p\varphi^{\lambda-i}, \deg p \geqslant (r + 1)i\}.$$

The fact that this is a filtration is clear from (5.7), and that it is an $(n, (r + 1)n)$ filtration follows from the fact that the dimension of M^i is less than or equal to this dimension being of order

$$\binom{n + (r + 1)i}{n} = \frac{(r + 1)^n}{n!} i^n + O(i^{n-1})$$

By the lemma M_φ is finitely generated, so this proves Theorem 5.2.

Before proving Theorem A.5.4 it may be helpful to give a kind of heuristic proof of it suggested by the theory of group representations. To prove that $d(M) \geq n$, it is enough, by Proposition A.5.3, to prove there exists a submodule L with $d(L) \geq n$. Let e be a fixed non-zero element of M and let L be the submodule Ωe. This is isomorphic to Ω/I where I is the left ideal $\{a \in \Omega, ae = 0\}$. It is not hard to see that Ω is left Noetherian (this is an easy consequence of the fact that gr Ω is Noetherian), so I is contained in a maximal left ideal, J, of Ω. Therefore, by Proposition 5.3 it is enough to prove that $d(\Omega/J) \geq n$ for maximal left ideals J.

Now Ω_n can be thought of as the universal enveloping algebra of the $2n + 1$ dimensional Heisenberg algebra divided by $\Omega_n(z - 1)$, where z is the central element in the Heisenberg algebra. Suppose k were the field of real numbers. By the Stone-von Neumann theorem there is a unique irreducible unitary representation of the Heisenberg group for which z gets represented as the identity. The representation space of this representation is the space of L^2 functions on $\mathbf{R}^n$ and the representation itself is that for which x_i and $(\partial/\partial x_i)$ get represented in the usual way as a multiplication operator and a differentiation operator. If an analogous formal Stone-von Neumann theorem were true for the representation of Ω_n on Ω_n/J then $d(\Omega_n/J) = n$ would follow as an easy consequence. Such a theorem has in fact been proved by Philip Trauber; however, as it is a little complicated to state precisely, we will give instead Bernstein's original proof of Theorem A.5.4 (which is itself quite short and elegant).

The proof will be by induction on n. We will assume Theorem A.5.4 is true for Ω_{n-1} and prove it for Ω_n. We will also assume that the field k is algebraically closed and uncountable (if not, embed it in an overfield $\bar{k}$ with this property and replace Ω and M by $\Omega \otimes \bar{k}$ and $M \otimes \bar{k}$). Let $t = x_n \in \Omega_n$.

LEMMA A.5.7. *For some $\alpha \in k$, the map of M into M given by multiplication by $t - \alpha$ is not invertible (as a linear mapping over k).*

PROOF. If, for all $\alpha \in k$, $t - \alpha$ were invertible we would get a homomorphism of the field of rational functions, $k(t)$, into the ring of linear mappings of the vector space M. Choose $f \in M, f \neq 0$ and assign to each $Q \in k(t)$ the element $Qf \in M$. Regard this as a mapping of the vector space $k(t)$ into the vector space M (both are vector spaces over k). $k(t)$ has uncountable dimension over k (the elements $(t - \alpha)^{-1}$, $\alpha \in k$, are linearly independent) whereas M has countable dimension over k. It follows that $Qf = 0$ for some Q. But Q is a product of invertible linear mappings hence invertible. Contradiction. Q. E. D.

COROLLARY. *There exists* $\alpha \in k$ *such that one of the following is true*:
(a) $(t-\alpha)\colon M \to M$ *is injective and* $M/(t-\alpha)M \neq 0$.
(b) $\ker(t-\alpha) \neq 0$.

We will show that (a) cannot occur if $d(M) < n$. Consider $\tilde{M} = M/(t-\alpha)M$ as an Ω_{n-1} module. By assumption M has an admissible filtration $\{M^i\}$ such that $\dim M^i - \dim M^{i-1}$ is a polynomial of degree $< n-1$; so, for the quotient filtration $\tilde{M}$ we have

$$\dim \tilde{M}^i = \dim M^i - \dim(M^i) \cap (t-\alpha)M \leq \dim M^i - \dim(t-\alpha)M^{i-1}$$
$$= \dim M^i - \dim M^{i-1}.$$

Thus the filtration on $\tilde{M}$ is a (d, e) filtration with $d < n-1$. By induction $\tilde{M} = 0$.

Next consider (b). Making the change of variables $t - \alpha \to t$ we can assume $\ker t \neq 0$. Replacing M by the submodule $L = \{f \in M, t^r f = 0\}$ for large r we can assume that for all $f \in M$, $t^r f = 0$ for r large. We shall prove that the operator $\partial/\partial t - \alpha$ has trivial kernel on M for all $\alpha \in k$. Let

$$\left(\frac{\partial}{\partial t} - \alpha\right) f = 0 \quad \text{and} \quad t^n f = 0.$$

Then

$$\left(\frac{\partial}{\partial t} - \alpha\right) t^n f - t^n \left(\frac{\partial}{\partial t} - \alpha\right) f = n t^{n-1} f = 0, \qquad \text{i. e., } t^{n-1} f = 0.$$

Continuing this argument we get $t^{n-2} f = \cdots = tf = f = 0$.

Let ρ be the automorphism of Ω_n given by

$$\rho(x_i) = x_i,$$
$$\rho \frac{\partial}{\partial x_i} = \frac{\partial}{\partial x_i}, \qquad i = 1, \ldots, n-1,$$
$$\rho(t) = \frac{\partial}{\partial t},$$
$$\rho \frac{\partial}{\partial t} = -t.$$

Consider the Ω_n module M_ρ obtained from M by intertwining with ρ. It is clear that $d(M_\rho) = d(M)$; so if $d(M) < n$, $d(M_\rho) < n$. We have just shown that for M_ρ, $\ker(t-\alpha) = 0$ for all $\alpha \in k$, so we are back in situation (a) which we have shown to be impossible.

§6. Behavior near caustics.

Let $\nu = \int e^{i\tau\varphi(x,\theta)} a(x,\theta;\tau)\, d\theta$ be a compound asymptotic associated with the Lagrangian submanifold, Λ, of $T^*(M)$. If Λ projects diffeomorphically onto M then we can apply the method of stationary phase to eliminate the θ variables and obtain an asymptotic expansion for ν. In this section we wish to examine what happens near caustics, i. e., near values x where the projection is singular. As our discussion is entirely local, we may assume that M is some neighborhood of the origin in $\mathbf{R}^n$ and not worry about the distinction between functions, forms, half forms, etc. We let π denote the projection of Λ onto M (i.e., the restriction to Λ of the projection of $T^*(M) \to M$). Let us define

$$S_i(\Lambda) = \{\lambda \in \Lambda | d\pi_\lambda \text{ has rank } n - i\}. \tag{6.1}$$

We recall from Chapter IV, §5 that near $\lambda \in S_i(\Lambda)$ it is always possible to parametrize Λ with i phase variables. Furthermore, in *any* parametrization of Λ, we know that i is the nullity of the vertical Hessian, $d_\theta^2 \varphi$. We have two uses of this remark. First of all, by applying the method of stationary phase, we can, near $\lambda \in S_i(\Lambda)$, assume that there are exactly i θ variables in the expression for ν as an asymptotic integral. The key problem of the next few sections is to see what further simplifications can be made in the expression for ν by introducing suitable local coordinates. The possible choices of coordinates will be related to an analysis of the nature of the singularity. As a second application of the remark, we can compute the dimension of $S_i(\Lambda)$ for a "generic" Lagrangian submanifold. Indeed, we can consider $(x,\theta) \to d_\theta^2 \varphi(x,\theta)$ as a map of $\mathbf{R}^n \times \mathbf{R}^k$ into the space of $k \times k$ symmetric matrices, where k is the number of θ variables. Then $S_i(\Lambda)$ is the inverse image under this map of S_i, where S_i denotes the submanifold of matrices of corank i. By Proposition 3.6 of Chapter IV we know that S_i is a submanifold of codimension $i(i+1)/2$. By the Thom transversality theorem we know that by a slight perturbation of φ we can arrange that the map $(x,\theta) \rightsquigarrow d_\theta^2 \varphi(x,\theta)$ be transversal to all the S_i. A slight change in φ means a slight change in Λ. Thus, by slightly changing Λ we can arrange that all the $S_i(\Lambda)$ are submanifolds of codimension $i(i+1)/2$, in other words:

Generically, $S_i(\Lambda)$ is a submanifold of codimension $i(i+1)/2$. (6.2)

By the Thom-Boardman procedure (cf. Golubitsky-Guillemin [9]) one can refine this classification of singularities. For example, the next step is to examine $S_i(\Lambda)$ and let $S_{i,j}(\Lambda)$ consist of those points for which the restriction of $d\pi$ to $TS_i(\Lambda)_\lambda$ has a kernel of dimension j. For generic Λ this will again be a submanifold and it turns out that its codimension is a universal cubic polynomial in i and j.

Cf. Boardman [10] or Golubitsky-Guillemin [9]. Similarly, one constructs $S_{i,j,k}(\Lambda)$. Let us examine the $S_{1,\ldots,1}$ type singularities for Lagrangian submanifolds. Near a point of $S_1(\Lambda)$ we know that we can parametrize Λ by a phase function with a single phase variable, θ. Let φ: $M \times \mathbf{R} \to \mathbf{R}$ be the phase function and C its critical set, so that C is the set of points where $d_\theta \varphi = 0$. The map $(x,\theta) \rightsquigarrow d_x\varphi(x,\theta)$ is a fiber preserving diffeomorphism of C with Λ, and thus the singularities of $C \to M$ and $\Lambda \to M$ are the same. Let us examine the singularities of C. For $c \in C$ to be singular for the projection onto M means that at c the vertical vector $\partial/\partial\theta$ must be tangent to C, i. e., must annihilate $\partial\varphi/\partial\theta$. Thus $S_1(C)$ consists of those points where $d_\theta\varphi = d_\theta^2\varphi = 0$. Let us assume that these equations are independent so that $S_1(C)$ is indeed a submanifold of C (of codimension one). The tangent space of $S_1(C)$ consists of those tangent vectors of $M \times \mathbf{R}$ which annihilate both $\partial\varphi/\partial\theta$ and $\partial^2\varphi/\partial\theta^2$. For a tangent vector to $S_1(C)$ to be annihilated by the projection onto M means that it must be vertical, i. e., that $\partial/\partial\theta$ is tangent to $S_1(C)$, which means that $\partial^3\varphi/\partial\theta^3$ must vanish. This is one additional equation. If it is independent of the previous two we conclude that $S_{1,1}$ is of codimensions one in $S_1(C)$ and hence of codimension two in C. It is now clear how to proceed: Let $w(\varphi)$: $M \times \mathbf{R} \to \mathbf{R}^{n+1}$ be defined by

$$w(\varphi)(x,\theta) = \left(\frac{\partial\varphi}{\partial\theta}(x,\theta), \ldots, \frac{\partial^{n+1}\varphi}{\partial\theta^{n+1}}(x,\theta)\right).$$

Let $W_k \subset \mathbf{R}^{n+1}$ consist of the points whose first k components vanish. By perturbing φ we can arrange that $w(\varphi)$ is transversal to all the W_k. For such φ the preceding argument extends to show that

$$S_{\underbrace{1,\ldots,1}_{k\,\text{times}}}(C) = w(\varphi)^{-1}(W_k)$$

and is a submanifold of codimension k in C. Thus:

For generic Λ, $S_{\underbrace{1,\ldots,1}_{k\,\text{times}}}(\Lambda)$ is a submanifold of codimension k.

Thus, for example suppose that we are interested in generic singularities for $\dim M \leq 4$.

We have the table:

type	codimension
$S_{1,0}$	1
$S_{1,1,0}$	2
$S_{1,1,1,0}, S_{2,0}$	3
$S_{1,1,1,1}$	4

We claim that these are all. We know that S_3 has codimension 6. Let us examine $S_{2,1}$. We may assume $\varphi = \varphi(x, \theta_1, \theta_2)$. Then C is defined by

$$\frac{\partial\varphi}{\partial\theta_1} = \frac{\partial\varphi}{\partial\theta_2} = 0.$$

At $c \in S_2(C)$ both $\partial/(\partial\theta_1)$ and $\partial/(\partial\theta_2)$ must be tangent to S_2 so the points of $S_2(C)$ are described by

$$\frac{\partial^2\varphi}{\partial\theta_1^2} = \frac{\partial^2\varphi}{\partial\theta_1\,\partial\theta_2} = \frac{\partial^2\varphi}{\partial\theta_2^2} = 0$$

showing once again that $S_2(C)$ has codimension 3. Now at a point of $S_{2,1}(C)$ there must be some non-zero vector of the form

$$a\frac{\partial}{\partial\theta_1} + b\frac{\partial}{\partial\theta_2}$$

tangent to $S_2(C)$. Thus the matrix

$$\begin{pmatrix} \dfrac{\partial^3\varphi}{\partial\theta_1^3} & \dfrac{\partial^3\varphi}{\partial\theta_1^2\,\partial\theta_2} & \dfrac{\partial^3\varphi}{\partial\theta_1\,\partial\theta_2^2} \\ \dfrac{\partial^3\varphi}{\partial\theta_1^2\,\partial\theta_2} & \dfrac{\partial^3\varphi}{\partial\theta_1\,\partial\theta_2^2} & \dfrac{\partial^3\varphi}{\partial\theta_2^3} \end{pmatrix}$$

must have rank one. This is a matrix of the form

$$\begin{pmatrix} A & B & C \\ B & C & D \end{pmatrix}$$

and it is clear that it will have rank < 2 if the pair of equations $AC - B^2 = 0$ and $BD - C^2 = 0$ are satisfied if $C \neq 0$. Similarly, the rank < 2 condition is given by two independent equations on each of the open sets, $A \neq 0$, $B \neq 0$ or $D \neq 0$. Thus $S_{2,1}$ has codimension 2 in S_2 and hence codimension 5 in Λ.

It will turn out from the discussion below that all singularities of the S_1 type are equivalent in any given dimension. The $S_{2,0}$ singularities can be further refined. In three dimensions there are two kinds—the elliptic umbilic and the hyperbolic umbilic. In four dimensions there are three kinds of $S_{2,0}$ singularities—the elliptic, parabolic, and hyperbolic umbilics; cf. below. Thus there are seven distinct types of singularities for Lagrangian manifolds in dimension four. These are exactly the seven "Thom catastrophes"; cf. [11].

Let us begin our analysis by studying generic $S_{1,0}$ type singularities and associated asymptotics.

PROPOSITION 6.1: *For $\lambda_0 \in S_{1,0}(\Lambda)$ we can choose a phase function $\varphi(x, \theta)$ parametrizing Λ near λ_0 of the form*

$$\varphi(x,\theta) = \mu(x) + \rho(x)\theta - \frac{\theta^3}{3} \qquad \textit{with } d\rho \neq 0. \tag{6.3}$$

Note. For φ of the form above the critical set C where

$$\frac{\partial\varphi}{\partial\theta} = 0$$

is just the set $\{(x,\theta)|\theta^2 = \rho(x)\}$ and the caustic is the set $\rho = 0$. If we choose coordinates $x_1, \ldots, x_n$ on X such that ρ is x_1 and use $(\theta, x_2, \ldots, x_n)$ as a system of coordinates on Λ then the map $\Lambda \to X$ is given locally by

$$(\theta, x_2, \ldots, x_n) \to (\theta^2, x_2, \ldots, x_n).$$

In other words it is the map which folds the $\theta < 0$ plane onto the $\theta > 0$ plane with fold along the line $\theta = 0$. For this reason the elements of $S_{1,0}(\Lambda)$ are called *fold points* of Λ.

Assuming Proposition 6.1 for the moment we see that an asymptotic associated with Λ has the form

$$\int a(x,\theta)e^{i\tau(\mu+\rho\theta-(\theta^3/3))}\,d\theta \tag{6.4}$$

in the neighborhood of a fold point. We will simplify the right hand side of (6.4) as follows. By the Malgrange preparation theorem (cf. for example Golubitsky-Guillemin [9]) we can find functions $a_0(x)$, $a_1(x)$, and $h(x,\theta)$ such that

$$a(x,\theta) = a_0(x) + a_1(x)\theta + h(x,\theta)(\rho(x) - \theta^2)$$

where $\rho(x) - \theta^2 = \partial\varphi/\partial\theta$. Thus the above expression can be written as

$$a_0(x)\int e^{i\tau\varphi(x,\theta)}\,d\theta + a_1(x)\int \theta e^{i\tau\varphi}\,d\theta + \int h(x,\theta)\frac{\partial\varphi}{\partial\theta}e^{i\tau\varphi}\,d\theta.$$

The last term in this sum is

$$\int h(x,\theta)\frac{1}{i\tau}\frac{\partial}{\partial\theta}e^{i\tau\varphi}\,d\theta$$

which is of order $1/\tau$. Integrating by parts and then repeating the same argument over again we prove (with different a_0 and a_1 from the above):

PROPOSITION 6.2. *There exist asymptotic series $a_0(x,\tau)$ and $a_1(x,\tau)$ depending on $a(x,\theta)$, such that*

$$\int a(x,\theta)e^{i\tau\varphi}\,d\theta = a_0(x,\tau)\int e^{i\tau\varphi(x,\theta)}\,d\theta + a_1(x,\tau)\int \theta e^{i\tau\varphi}\,d\theta. \tag{6.5}$$

REMARK. It is rather complicated actually to write down the dependence of a_0 and a_1 on a. This involves looking at a special case of the following problem: Given a function $f(x,\theta)$ on $\mathbf{R}^n \times \mathbf{R}$ with

$$\frac{\partial^i f}{\partial \theta^i}(0) = 0 \qquad \text{for } i < n \qquad \text{and} \qquad \frac{\partial^n f}{\partial \theta^n}(0) \neq 0,$$

then the Malgrange preparation asserts that for every function $a = a(x,\theta)$ there exist functions $a_0(x), \ldots, a_{n-1}(x)$ and $h(x,\theta)$ such that $a(x,\theta) = \sum a_i(x)\theta^i + hf$. How do the a_i's and h depend on a? If f and a are real analytic then h and the a_i's are uniquely determined (by the uniqueness part of the Weierstrass preparation theorem) and one can show (see for example Arnold [3]) that the map

$$a \to (a_0, \ldots, a_{n-1}, h)$$

behaves in some ways like an $(n-1)$st order differential operator. If f and a are smooth then h and the a_i's may not even be uniquely determined (see, for example, Malgrange [17]).

To simplify (6.5) further we recall the definition of the Airy function

$$A(t) = \int e^{i(t\theta - (\theta^3/3))}\, d\theta, \qquad t \in \mathbf{R}, \quad \theta \in \mathbf{R}.$$

This is a Bessel function of type $\frac{1}{3}$. Its properties are exhaustively discussed in the Bureau of Standards tables [12]. Among other things it can be characterized as a solution of the ordinary differential equation $A'' + tA = 0$. This equation is the standard equation in one dimension which describes transitions from oscillatory behavior to exponentially damped behavior: Assuming t is approximately constant then in the region $t > 0$ the solutions of $A'' + tA = 0$ are approximately sine and cosine functions, and in the region $t < 0$ they are approximately exponentially increasing and exponentially decreasing.

Differentiating under the integral sign we get

$$A'(t) = \int i\theta e^{i(\theta t - (\theta^3/3))}\, d\theta.$$

Therefore from (6.3) we obtain the following:

THEOREM 6.1. *Let (x,ξ) be a fold point of Λ and (x,θ) the corresponding critical point of φ. There exist asymptotic series a_0 and a_1 on M such that*

$$\int a(x,\theta) e^{i\tau\varphi(x,\theta)}\, d\theta = e^{i\tau\mu}\left\{ \frac{a_0}{\tau^{1/3}} A(\tau^{2/3}\rho) + \frac{a_1}{\tau^{2/3}} A'(\tau^{2/3}\rho) \right\}.$$

REMARK. Airy functions play an important role in the asymptotic theory of the Schrödinger equation (see, for example, Messiah [13]), and of the reduced wave equation (see Ludwig [14]). The theorem above shows that this fact has nothing to do with the special properties of these equations but only with the presence of fold singularities. In the following we will use standard properties of Airy functions to describe what "illuminated regions" and "shaded regions" look like in the neighborhood of a simple caustic.

Finally we will prove Proposition 6.1. The idea of this proof is due to Chester, Friedman, and Ursell [15] (though they work with analytic rather than smooth data).

Our starting point is the following lemma, due to Whitney.

Let f be a smooth even function on the real line. Then there exists a smooth function g on the real line such that $f(x) = g(x^2)$. If f depends smoothly on a set of parameters, g can be chosen so that it depends smoothly on the same parameters.

See Chapter VI, §5, where we use this lemma in another context and give a proof.

Now let Λ be a Lagrangian manifold with a fold point at (x_0, ξ_0). Assume for simplicity that $M \subset \mathbf{R}^n$ and that x_0 is the origin. Let $\varphi = \varphi(x, \theta)$ on $M \times \mathbf{R}$ be a phase function parametrizing Λ in a neighborhood of x_0 and let C be its critical set. Assume that the point on C corresponding to (x_0, ξ_0) is the origin. We will prove:

LEMMA 6.1. *There exists smooth functions u_0 and ρ on M and ζ on $M \times \mathbf{R}$ such that restricted to C:*

$$\frac{\zeta^3}{3} - \rho\zeta + u_0 = \varphi, \qquad \frac{\partial \zeta}{\partial \theta} > 0, \qquad \textit{and} \qquad \zeta^2 - \rho = 0. \tag{6.6}$$

PROOF. First let us prove the assertion for the special case when the base manifold, M, is one dimensional. The assumption that the origin is a fold point of C means that

$$\frac{\partial \varphi}{\partial \theta} = \frac{\partial^2 \varphi}{\partial \theta^2} = 0 \qquad \text{and} \qquad \frac{\partial^2 \varphi}{\partial \theta \partial x} \neq 0, \quad \text{at } 0.$$

(See (6.3).) Since

$$\frac{\partial^2 \varphi}{\partial \theta \partial x} \neq 0$$

we can solve for x as a function of θ on C and we get $x = x(\theta)$. Since

$$\frac{\partial^2 \varphi}{\partial \theta^2}(x(\theta), \theta) + \frac{\partial^2 \varphi}{\partial x \partial \theta}(x(\theta), \theta) x'(\theta) = 0$$

on C we conclude that $x'(0) = 0$. Since

$$\frac{\partial^3 \varphi}{\partial^3 \theta} \neq 0$$

we conclude that $x''(\theta) \neq 0$ so by a change of coordinates on $\mathbf{R}$ we can assume $x = \theta^2$ on C. Let C^+ be the part of C where $\theta > 0$ and C^- the part where $\theta < 0$. By the last of the three equations (6.6) we must have $\zeta = +\sqrt{\rho}$ on C^-. So on C^+ we have

$$-\frac{2}{3}\rho^{3/2} + u_0 = \varphi(\theta)$$

and on C^- we have

$$\frac{2}{3}\rho^{3/2} + u_0 = \varphi(-\theta).$$

Since ρ and u_0 are function x alone we must have

$$u_0(x) = \frac{1}{2}(\varphi(\theta) + \varphi(-\theta)) \qquad \rho(x)^3 = \frac{9}{16}(\varphi(\theta) - \varphi(-\theta))^2 \tag{6.7}$$

with $x = \theta^2$. The expressions on the right are both even functions of θ, so u_0 and ρ^3 exist by the lemma on even functions. To show that the cube root of ρ^3 exists we note that since $\varphi'(\theta) = \varphi''(\theta) = 0$, and $\varphi'''(\theta) \neq 0$ the Taylor series for $(\varphi(\theta) - \varphi(-\theta))^2$ begins with a non-zero term of order 6. Thus ρ exists and is of order 2 with respect to θ and of order 1 with respect to x. In particular, $\zeta = +\sqrt{\rho}$ exists on C and $(\partial\zeta)/(\partial\theta) \neq 0$.

Now suppose dim $M > 1$. Choose coordinates $(x_1, \ldots, x_n)$ on M such that

$$\frac{\partial\varphi}{\partial\theta\partial x_1} \neq 0.$$

For $a = (a_2, \ldots, a_n)$ let C_a = the intersection of C with the line $x_2 = a_2, \ldots, x_n = a_n$. Applying the preceding argument to C_a we find functions u_0^a, ρ^a and ζ^a on C_a satisfying (6.6) and depending smoothly on a. We let u_0, ρ, and ζ be the corresponding functions on C.

Finally extend ζ from C to $M \times \mathbf{R}$ arbitrarily. This concludes the proof of our lemma.

To prove Proposition 6.1 let $\psi(x,\theta) = u_0(x) + \rho(x)\zeta(\theta) - (\zeta^3(\theta))/3$. From (6.6) one easily sees that the critical set of ψ equals the critical set of φ. Making the change of coordinates $x \to x$, $\theta \to \zeta(\theta, x)$ we get a phase function of the desired form.

REMARK. In §7 we will obtain another proof of Proposition 6.1 as a special case of a much more general result. (See Proposition 7.1.)

As an application of Theorem 6.1 we describe some results of Ludwig [14] on the reduced wave equation

$$\Delta u + \tau^2 u = 0 \tag{6.8}$$

with prescribed boundary data on an oriented hypersurface, $\mathcal{S}$, in $\mathbf{R}^n$ like the kind shown in Figure 1.

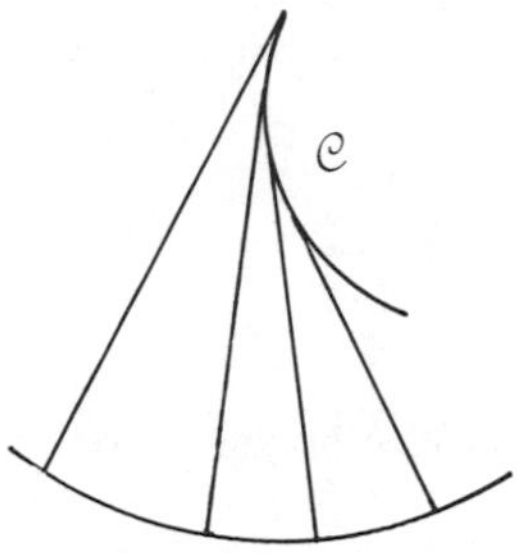

Suppose the family of normal lines to $\mathcal{S}$ have $\mathcal{C}$ as an envelope (i.e., they all lie on one side of $\mathcal{C}$ and are tangent to $\mathcal{C}$). For $\tau \gg 0$ geometric optics gives a very good approximate solution to (6.8) away from $\mathcal{C}$; cf. Chapter II. However geometrical optics makes some rather implausible assertions about what happens near $\mathcal{C}$, e.g., the light intensity at $\mathcal{C}$ is infinite and there is no illumination at all on the other side of $\mathcal{C}$. In the paper we mentioned above Ludwig works out the predictions of physical optics concerning what happens near $\mathcal{C}$. The picture he gets is roughly the following:

(a) At points on the illuminated side of $\mathcal{C}$ whose distance from $\mathcal{C}$ is large compared with $\tau^{-2/3}$ the approximation of geometric optics is correct to order $-\frac{1}{2}$.

(b) The light intensity on the caustic itself is large but finite (of order $\tau^{1/6}$) .

(c) On the dark side of $\mathcal{C}$ there is an illuminated strip of width approximately $\tau^{-2/3}$.

Ludwig's results are uniform in the sense that they are valid for all $\tau \gg 0$. In other words if we refract a monochromatic beam of light through a lens L producing an image on a screen we can predict from these results how the image changes as we change the frequency of the light (see Figure 2).

To solve equation (6.8) asymptotically with boundary data prescribed along $\mathcal{S}$ we proceed as in Chapter II. Consider, in the cotangent bundle of $\mathbf{R}^n$, the Lagrangian manifold consisting of all points, $(x + tn_x, n_x)$, where $x \in \mathcal{S}$ and n_x is the unit normal at x pointing in the direction of the orientation. The caustic $\mathcal{C}$ in Figure 1 is precisely the set of critical points of the map $\pi\colon \Lambda \to \mathbf{R}^n$. It is not hard to show that for $\mathcal{C}$ in general position most of these points are fold points; those which are not form a subset of codimension one. (See Figure 3. Also compare

The man varies the frequency of light by putting yellow, blue, red, etc., filters in front of the flashlight.

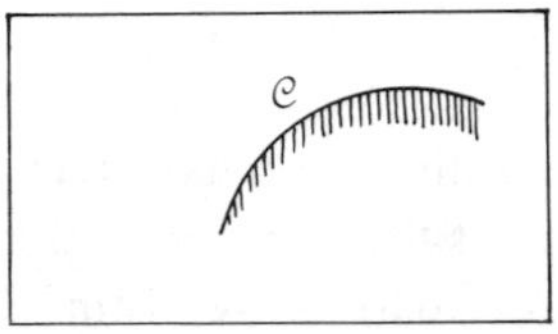

The region of illumination on the dark side of $\mathcal{C}$ gets thinner as one goes from the infrared to the ultraviolet part of the spectrum. For example the illuminated region for a red filter $\sim$ 1.4 times as wide as the illuminated region for a violet filter.

with Proposition 6.2)

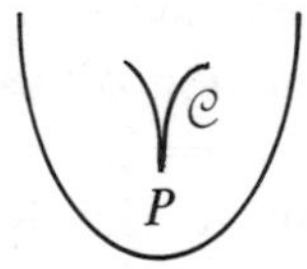

Reflection in a curved mirror produces a cusped caustic. All the points on $\mathcal{C}$ except P are fold points. (We will discuss the behavior of solutions of (6.8) in the neighborhood of the cusp, P, in §7.)

At a fold point Theorem 6.1 gives a solution of (6.8) which has the general form

$$e^{i\tau\sigma}\left\{\frac{g_0}{\tau^{1/3}}A(\tau^{2/3}\rho)+\frac{g_1}{i\tau^{2/3}}A'(\tau^{2/3}\rho)\right\}+O\left(\frac{1}{\tau}\right). \tag{6.9}$$

Here ρ, σ, g_0 and g_1 are functions of x alone not depending on τ, A is the Airy function and A' its derivative. We will worry about how to determine σ, ρ, g_0, and g_1 a little later on. However we already know that $d\rho \neq 0$ near $\mathcal{C}$ and $\rho = 0$ on $\mathcal{C}$, so ρ can be regarded as a normal coordinate for $\mathcal{C}$.

In order to discuss the qualitative behavior of the solution (6.9) we recall some basic facts about Airy functions.

For $t \gg 0$,

$$A(t) \sim \frac{1}{\sqrt{\pi}\, t^{1/4}} \cos\left(\frac{2}{3} t^{3/2} - \frac{\pi}{4}\right) \tag{6.10}$$

and

$$A'(t) \sim \frac{t^{1/4}}{\sqrt{\pi}} \left(-\sin\left(\frac{2}{3} t^{3/2} - \frac{\pi}{4}\right)\right). \tag{6.11}$$

For $t \ll 0$,

$$A(t) \sim \frac{1}{2\sqrt{\pi}\,(-t)^{1/4}} e^{-2/3(-t)^{3/2}} \tag{6.12}$$

and

$$A'(t) \sim \frac{1}{2\sqrt{\pi}} (-t)^{1/4} e^{-2/3(-t)^{3/2}}. \tag{6.13}$$

For t close to 0,

$$A(t) \sim c_1 + c_2 t \tag{6.14}$$

where $c_1 \doteq .355$ and $c_2 \doteq -.259$.

PROOF. (6.10) and (6.11) can be obtained by applying stationary phase to the integral form of the Airy function and (6.12), and (6.13) can be obtained from (6.10) and (6.11) by analytic continuation. For (6.14) see the Bureau of Standards tables [12], or Olver [17].

Combining these results with (6.9) we see that for $\tau^{2/3}\rho \gg 0$, the solution, u, of (6.8) satisfies

$$u \sim \frac{\tau^{-1/3} e^{i\tau\sigma}}{\sqrt{\pi}\,[\tau^{2/3}\rho]^{1/4}} \left(g_0 \cos\left(\frac{2\tau\rho^{2/3}}{3} - \frac{\pi}{4}\right) - g_1 \rho^{1/2} \sin\left(\frac{2\tau\rho^{2/3}}{3} - \frac{\pi}{4}\right)\right) \tag{6.15}$$

which is identical with the solution given by geometric optics. For $\tau^{2/3}\rho$ close to zero the first term of (6.9) dominates and we get

$$u \sim e^{i\tau\sigma} \frac{g_0}{\tau^{1/3}} A(\tau^{2/3}\rho) \tag{6.16}$$

where A is given by (6.14). Comparing (6.15) with (6.16) we see that the value of u on the caustic is approximately $\tau^{1/6}$ times its value for $\rho \sim 1$.

Finally when $\tau^{2/3}\rho \ll 0$ we get

$$u \sim \frac{\tau^{-1/3} e^{i\tau\sigma}}{\sqrt{\pi}\,[\tau^{2/3}(-\rho)]^{1/4}} \left(g_0 e^{-2/3\tau(-\rho)^{3/2}} + g_1(-\rho)^{1/2} e^{-2/3\tau(-\rho)^{3/2}} \right). \tag{6.17}$$

When ρ is comparable to $\tau^{2/3}$ we cannot simplify the solution (6.9) very much. This is a region of transition in which the approximation of geometric optics breaks down and the extremely simple solution (6.16) is not yet valid. Notice that there is still a significant amount of illumination on the dark side of the caustic if $-\tau^{2/3}\rho \cong 1$.

We leave it as an exercise for the reader to read off from the expressions above the rough qualitative behavior of u described earlier.

We will now discuss how one determines the functions σ, ρ, g_0, and g_1. The symbol of the equation (6.8) is the function $\xi^2 - 1$. Our general prescription for constructing asymptotic solutions of (6.8) is to find a Lagrangian manifold, Λ, on which $\xi^2 - 1 = 0$ and on Λ to find half densities which are invariant under the Hamiltonian flow associated with $\xi^2 - 1$; namely

$$2 \sum \xi_i \frac{\partial}{\partial x_i}.$$

Let us look at what the first condition involves.

Our phase function is of the form

$$\varphi = \varphi(x,\theta) = \sigma(x) + \rho(x)\theta - \frac{1}{3}\theta^3,$$

where ρ and σ are the same as the ρ and σ in the previous paragraph. The critical set, C, of φ is the set of all (x,θ) where $\rho(x) = \theta^2$ and the map $(x,\theta) \to (x,(d\varphi)_x)$ must map this set diffeomorphically onto our Lagrangian manifold. Therefore, we must have

$$\sum \left(\frac{\partial\varphi}{\partial x^i}\right)^2 = 1 \qquad \text{on } C. \tag{6.18}$$

Now

$$\frac{\partial\varphi}{\partial x^i} = \frac{\partial\sigma}{\partial x^i} + \frac{\partial\rho}{\partial x^i}\theta$$

and

$$\sum \left(\frac{\partial\varphi}{\partial x^i}\right)^2 = \sum \left(\frac{\partial\sigma}{\partial x^i}\right)^2 + \sum \rho\left(\frac{\partial\rho}{\partial x^i}\right)^2 + 2\sum \frac{\partial\sigma}{\partial x^i}\frac{\partial\rho}{\partial x^i}\theta.$$

Since $\theta^2 = \rho$ on C and since σ and θ are independent on C with respect to functions of x, the equation (6.18) breaks up into the pair of equations

$$(d\sigma)^2 + \rho(d\rho)^2 = 1 \tag{6.19}$$

and

$$d\sigma \cdot d\rho = 0. \tag{6.20}$$

These are the eikonal equations for ρ and σ which Ludwig derives in §1 of his paper. In two dimensions it is rather easy to analyze them geometrically. When $\rho = 0$, $d\rho$ is perpendicular to the caustic, so $d\sigma$ is tangent to the caustic by (6.20), and, by (6.19), σ is just the arc length variable along the caustic itself. To interpret (6.19) and (6.20) as equations for ρ, let us suppose first of all that $\mathcal{C}$ is a circle of radius a about the origin in $\mathbf{R}^2$. Let (r, θ) be the polar coordinates of any point in $\mathbf{R}^2$. Then on $\mathcal{C}$, $\sigma = a\theta$. It is clear that, for (6.19) and (6.20) to be satisfied with initial data $\rho = 0$ and $\sigma = a\theta$ on $\mathcal{C}$, ρ must be radially symmetric; so we can write $\rho = \rho(r)$. From (6.20) we deduce that σ is constant in the radial direction so $\sigma = a\theta$ in a whole neighborhood of $\mathcal{C}$. Thus (6.17) reduces to the equation

$$\left(\frac{a}{r}\right)^2 + \rho(r)(\rho'(r))^2 = 1 \quad \text{or} \quad \rho(r)(\rho'(r))^2 = \frac{(r-a)(r+a)}{r^2}.$$

Setting $\rho(r) = c(r-a) + O((r-a)^2)$ in the vicinity of $\mathcal{C}$ we get

$$c^3(r-a) = \frac{(r-a)(r+a)}{r^2}$$

at a or $c^3 = 2/a$. Thus:

$$\rho(r) \doteq \left(\frac{2}{a}\right)^{1/3}(r-a) \tag{6.21}$$

in the vicinity of $\mathcal{C}$.

Now we consider the case of a general caustic, $\mathcal{C}$. Let $x_0 \in \mathcal{C}$ and r distance along the normal line through $\mathcal{C}$ at x_0. Applying the argument above to the osculating circle to $\mathcal{C}$ at x_0 (which has third order contact with $\mathcal{C}$ at x_0) we get (6.21) holding along the normal line through $\mathcal{C}$ at x_0, where a is the radius of curvature at x_0.

Finally we will see what the analogue of the transport equation is in our set-up. Our asymptotic solution of (6.8) has the form

$$u(x) = g(x,\theta)e^{i\tau(\sigma+\rho\theta-\theta^3/3)}\,d\theta$$

where σ and ρ are in principle determinable by the equations (6.19) and (6.20). What about g? We will prove

Let φ *be the phase function above* (i.e., $\varphi(x,\theta) = \sigma + \rho\theta - \theta^3/3$ *on* $M \times \mathbf{R}$ *and let*

$$q = \frac{1 - (\nabla\varphi)^2}{\partial\varphi/\partial\theta}.$$

(*Since* $1 - (\nabla\varphi)^2$ *vanishes on the locus of points where*

$$\frac{\partial\varphi}{\partial\theta} = 0$$

and φ *is a generic phase function,* q *is a smooth function.*) *Then we have*

$$2(\nabla\varphi)\cdot\nabla g + (\nabla\varphi)g - \frac{\partial}{\partial\theta}(qg) = 0 \bmod\left(\frac{\partial\varphi}{\partial\theta}\right). \tag{6.22}$$

See Ludwig [14] §2.

PROOF. Equation (6.22) just asserts that the half density on C associated with g is invariant under the Hamiltonian flow. To see this we first of all prove:

LEMMA 1. *The vector field on* C *associated with the Hamiltonian vector field*

$$2\sum \xi_i \frac{\partial}{\partial x_i}$$

on Λ *is the vector field*

$$\boldsymbol{\eta} = 2\sum \frac{\partial\varphi}{\partial x_i}\frac{\partial}{\partial x_i} - q\frac{\partial}{\partial\theta}.$$

PROOF. The vector field is tangent to C since it kills φ and $(\partial\varphi)/(\partial\theta)$; and it and the Hamiltonian vector field project onto the same vector on the base at each point, namely:

$$2\sum \frac{\partial\varphi}{\partial x_i}\frac{\partial}{\partial x_i}.$$

Therefore, the two vector fields agree on the set where $d\pi$ is bijective. By continuity they agree everywhere. Q. E. D.

Now let Ω be an n form on $\mathbf{R}^{n+1}$ such that, on C,

$$\Omega \wedge d\left(\frac{\partial\varphi}{\partial\theta}\right) = dx \wedge d\theta.$$

Let Ω_0 be the restriction of Ω to C. Since Ω is determined up to a multiple of

$$d\left(\frac{\partial\varphi}{\partial\theta}\right),$$

Ω_0 is intrinsically defined on C and is the usual volume form there.

LEMMA 2.

$$D_\eta\sqrt{\Omega_0} = \left(\Delta\varphi - \frac{\partial q}{\partial\theta}\right)\sqrt{\Omega_0}.$$

PROOF. It is enough to show that

$$(D_\eta \Omega) \wedge d\frac{\partial \varphi}{\partial \theta} = 2\left(\Delta\varphi - \frac{\partial q}{\partial \theta}\right)\Omega \wedge d\left(\frac{\partial \varphi}{\partial \theta}\right), \tag{6.23}$$

for then we can get the equation above by restricting to C and taking square roots. To prove (6.23) we first of all note that

$$D_\eta \frac{\partial \varphi}{\partial \theta} = \frac{\partial q}{\partial \theta}\frac{\partial \varphi}{\partial \theta} \tag{6.24}$$

by a straightforward computation. Now the left hand side of (6.23) can be written

$$D_\eta\left(\Omega \wedge d\left(\frac{\partial \varphi}{\partial \theta}\right) - \Omega \wedge d\left(D_\eta \frac{\partial \varphi}{\partial \theta}\right)\right) = D_\eta(dx \wedge d\theta) - \frac{\partial q}{\partial \theta}\Omega \wedge d\frac{\partial \varphi}{\partial \theta}$$

on C by (6.24). Since

$$D_\eta(dx \wedge d\theta) = (\operatorname{div} \eta)dx \wedge d\theta = (\operatorname{div} \eta)\Omega \wedge d\frac{\partial \varphi}{\partial \theta},$$

$$(D_\eta \Omega) \wedge d\left(\frac{\partial \varphi}{\partial \theta}\right) = \left(\operatorname{div} \eta - \frac{\partial q}{\partial \theta}\right)\Omega \wedge d\left(\frac{\partial \varphi}{\partial \theta}\right),$$

which is essentially (6.23). Q.E.D.

Combining the lemmas we see that, for a half density $g\sqrt{\Omega_0}$ to be invariant under the transport equation, g must satisfy (6.22).

This gives an illustration of how the use of normal forms near singularities gives detailed qualitative information about the behavior of high frequency approximate solutions to partial differential equations near caustics.

We have only analyzed the simplest possible singularity, namely $S_{1,0}$. To proceed further we must use some deeper results in singularity theory. These will be discussed in the next few sections.

§7. Iterated S_1 and $S_{2,0}$ singularities, computations.

In order to analyze more complicated singularities we will need results analogous to Proposition 6.1. Our goal in this section will be to derive, at least formally, such results for the iterated S_1 singularities and, for the simplest kinds of S_2 singularities. We have the following result.

PROPOSITION 7.1. *Let X be a neighborhood of the origin in $\mathbf{R}^n$ and $\varphi = \varphi(x, \theta)$ a phase funcion on $X \times \mathbf{R}$. Suppose that at the origin*

$$\frac{\partial^i \varphi}{\partial \theta^i} = 0 \qquad \textit{for } i = 1, \ldots, k-1 \qquad \textit{and} \qquad \frac{\partial^k \varphi}{\partial \theta^k} \neq 0.$$

Then there exists a function $\theta_1 = \theta_1(x,\theta)$ *and functions of* x: $f_0(x), \ldots, f_{k-2}(x)$ *such that* $\theta(0,0) = 0$ *and* $f_1(0) = \cdots = f_{k-2}(0) = 0$,

$$\frac{\partial\theta_1}{\partial\theta} \neq 0 \qquad \text{at } 0$$

and

$$\varphi(x,\theta) = f_0 + f_1\theta_1 + \cdots + f_{k-2}\theta_1^{k-2} + \frac{\theta_1^k}{k} + \epsilon(x,\theta) \tag{7.1}$$

with $\epsilon(x,\theta)$ *vanishing to infinite order when* $x = 0$.

PROOF. Since

$$\frac{\partial^k\varphi}{\partial\theta^k} \neq 0$$

at 0, we can write $\varphi(0,\theta) = c + (\theta^k/k)h(\theta)$ where $h(\theta) \neq 0$ at 0. Making the coordinate change $\theta_1 = \theta(h(\theta))^{1/k}$, we get

$$\varphi(x,\theta) = c + \frac{\theta_1^k}{k} + \epsilon(x,\theta)$$

where $\epsilon(x,\theta)$ vanishes to order 1 in x when $x = 0$.

Now suppose by induction that we can write

$$\varphi(x,\theta) = f_0(x) + f_1(x)\theta + \cdots + f_{k-2}(x)\theta^{k-2} + \frac{\theta^k}{k} + \epsilon(x,\theta)$$

where the f_i's are polynomials of degree $< N$ in x and $\epsilon(x,\theta) = O(|x|^N)$. Let $\epsilon(x,\theta) = \sum_{|I|=N}\epsilon_I(\theta)x^I + O(|x|^{N+1})$. We will try to find homogeneous polynomials $f'_i = \sum_{|I|=N} f'_{i,I}x^I$ and $\theta_1 = \theta + \sum_{|I|=N}\tau_I(\theta)x^I$ such that

$$\sum_{i=0}^{k-2}(f_i + f'_i)\theta_1^i + \frac{\theta_1^k}{k} = \varphi(x,\theta) + \epsilon_1(x,\theta) \tag{7.2}$$

where $\epsilon_1(x,\theta)$ is $O(|x|^{N+1})$ when $x = 0$. Equating coefficients of x^I (and using the fact that $f_i(0) = 0$) we see that (7.2) reduces to the following system of equations:

$$f'_{0,I} + f'_{1,I}\theta + \cdots + f'_{k-2,I}\theta^{k-2} + \tau_I(\theta)\theta^{k-1} = \epsilon_I(\theta) \tag{$7.2)_I$}$$

which are easily solvable by Taylor's theorem with remainder. Q. E. D.

To study the S_2 singularities, we need a short digression on the theory of cubic binary forms over the reals. Let $\varphi = \varphi(\alpha,\beta)$ be a homogeneous polynomial of degree 3 in the variables α and β. We will say that φ is *degenerate* if the quadratic forms

$$\frac{\partial \varphi}{\partial \alpha} \quad \text{and} \quad \frac{\partial \varphi}{\partial \beta}$$

are multiples of each other. Suppose this the the case, i. e., suppose

$$\frac{\partial \varphi}{\partial \beta} = c \frac{\partial \varphi}{\partial \alpha}.$$

Making the coordinate change $\alpha_1 = \alpha$, $\beta_1 = \beta + c\alpha$, we get

$$\frac{\partial \varphi}{\partial \beta_1} = 0,$$

so $\varphi = a\alpha_1^3$. Relabelling $\alpha = \alpha_1 \sqrt[3]{a}$ we see that either $\varphi \equiv 0$ or $\varphi = \alpha^3$.

Now suppose φ is non-degenerate. Let L_φ be the one dimensional vector space consisting of quadratic forms in α and β modulo linear combinations of

$$\frac{\partial \varphi}{\partial \alpha} \quad \text{and} \quad \frac{\partial \varphi}{\partial \beta}.$$

Let K be the two dimensional vector space spanned by α and β. Multiplication gives us a bilinear map of K into L_φ; and if we choose a basis in L_φ this map can be viewed as a quadratic form on K. It is clear that this quadratic form is non-zero; otherwise α^2, β^2 and $\alpha\beta$ would all be multiples of

$$\frac{\partial \varphi}{\partial \alpha} \quad \text{and} \quad \frac{\partial \varphi}{\partial \beta}.$$

DEFINITION 7.1. *We will say φ is **hyperbolic** if the quadratic form described above is of rank 2 and index 1, **elliptic** if it is of rank 2 and index 2 or 0, and **parabolic** if it is of rank 1.*

PROPOSITION 7.2. (*Classification theorem for cubic binary forms over* **R**.) *If φ is hyperbolic, then by a linear change of coordinates it can be written in the form*:

$$\varphi(\alpha, \beta) = \frac{\alpha^3 + \beta^3}{3}. \tag{7.3}$$

If it is parabolic it can be written in the form:

$$\varphi(\alpha, \beta) = \alpha^2 \beta \tag{7.4}$$

and if it is elliptic it can be written in form:

$$\varphi(\alpha, \beta) = \alpha^3 - \alpha\beta^2. \tag{7.5}$$

PROOF. We will discuss the hyperbolic case. The other two cases are handled similarly. If $\varphi(\alpha, \beta)$ is hyperbolic we can make a linear change of coordinates such that the quadratic form just described sends the pair (α, α) to 0 and (β, β) to 0. This means α^2 and β^2 are both linear combinations of

$$\frac{\partial \varphi}{\partial \alpha} \quad \text{and} \quad \frac{\partial \varphi}{\partial \beta}.$$

Since α^2 and β^2 are linearly independent,

$$\frac{\partial \varphi}{\partial \alpha} \quad \text{and} \quad \frac{\partial \varphi}{\partial \beta}$$

must also be linear combinations of α^2 and β^2; i.e.,

$$\frac{\partial \varphi}{\partial \alpha} = a_1 \alpha^2 + a_2 \beta^2 \qquad \text{and} \qquad \frac{\partial \varphi}{\partial \beta} = b_1 \alpha^2 + b_2 \beta^2$$

for real numbers a_1, a_2, and b_1, b_2. Clearly, for the first of these equations to hold, the coefficient of $\alpha^2 \beta$ in φ must be zero, and for the second to hold the coefficient of $\alpha \beta^2$ must be zero. This shows that

$$\varphi = \frac{1}{3}\left(c_1 \alpha^3 + c_2 \beta^3\right).$$

Replacing α by $\sqrt[3]{c_1}\, \alpha$ and β by $\sqrt[3]{c_2}\, \beta$ we get φ in the desired form. Q.E.D.

COROLLARY. *Let $h = h(\alpha, \beta)$ be a smooth function on $\mathbf{R}^2$. Suppose that its first and second derivatives at 0 vanish and the cubic term in its Taylor series is a non-degenerate cubic binary form. If this form is hyperbolic we can find coordinates $\alpha_1 = \alpha_1(\alpha, \beta)$ and $\beta_1 = \beta_1(\alpha, \beta)$ such that*

$$h(\alpha, \beta) = c + \frac{\alpha_1^3 + \beta_1^3}{3} \qquad \textit{near } 0. \tag{7.6}$$

If it is elliptic we can find coordinates such that

$$h(\alpha, \beta) = c + \alpha_1^3 - \alpha_1 \beta_1^2 \qquad \textit{near } 0. \tag{7.7}$$

PROOF. We will discuss the hyperbolic case. The elliptic case is handled similarly. Clearly we can write

$$h(\alpha, \beta) = \frac{\alpha^3 + \beta^3}{3} + \sum_{i+j=4} f_{i,j}(\alpha, \beta) \alpha^i \beta^i.$$

Making the coordinate change $\alpha_1 = \alpha + f_{2,2} \beta^2$, $\beta_1 = \beta$ we can assume that in the expression on the right hand side $f_{2,2} = 0$. This means we can write

$$h(\alpha, \beta) = \frac{\alpha^3 + \beta^3}{3} + g_1 \alpha^3 + g_2 \beta^3$$

where $g_1(0) = g_2(0) = 0$. Letting $\alpha_1 = \alpha\sqrt[3]{1 + 3g_1}$ and $\beta_2 = \beta\sqrt[3]{1 + 3g_2}$, we get

$$h(\alpha, \beta) = \frac{\alpha_1^3 + \beta_1^3}{3}$$

as required. Q.E.D.

For the parabolic singularities the situation is a little more complicated. For example $\alpha^2\beta$ and $\alpha^2\beta + \beta^4$ are not equivalent. We will prove:

PROPOSITION 7.3. *Suppose $h = h(\alpha, \beta)$ vanishes together with its first and second derivatives at* 0. *Suppose the cubic term in its Taylor series at* 0 *is $\alpha^2\beta$. Then if*

$$\frac{\partial^4 h}{\partial \beta^4} \neq 0$$

we can find a change of coordinates $\alpha_1 = \alpha_1(\alpha, \beta)$ and $\beta_1 = \beta_1(\alpha, \beta)$ at 0 *such that*

$$\pm h = \alpha_1^2 \beta_1 + \beta_1^4. \tag{7.8}$$

PROOF. We can write

$$\pm h(\alpha, \beta) = \alpha^2\beta + \sum_{i+j=4} f_{i,j}\alpha^i\beta^j \qquad \text{with } f_{0,4} > 0$$

at the origin. Letting $\alpha_1 = \alpha/\sqrt[8]{f_{0,4}}$ and $\beta_1 = \sqrt[4]{f_{0,4}}\beta + f_{1,3}\alpha/4[f_{0,4}]^{3/4}$we get

$$\pm h(\alpha, \beta) = \alpha_1^2\beta_1 + \beta_1^4 + \rho\alpha_1^2\beta_1 + \tau\alpha_1^4$$

with ρ and τ appropriate functions of α and β, and $\rho(0) = 0$. Finally replacing α_1 by $\alpha_1\sqrt{1 + \rho}$, we can assume that $\rho = 0$. Thus we have reduced our problem to the case when $h(\alpha, \beta)$ has the form:

$$\pm h(\alpha, \beta) = \alpha^2\beta + \beta^4 + \tau\alpha^4. \tag{7.9}$$

We will now look for a coordinate change $\alpha_1 = \alpha$, $\beta_1 = \beta + u\alpha^2$, and a function ρ of α and β which vanishes at 0 such that

$$\pm h(\alpha, \beta) = \alpha_1^2\beta_1 + \beta_1^4 + \rho\alpha_1^2\beta_1.$$

Substituting the expressions for α_1 and β_1 in this expression we get $\pm h(\alpha, \beta)$ equal to the following mess:

$$\alpha^2\beta + \beta^4 + (u + u\rho + 6u^2\beta^2 + 4u^3\alpha^2\beta + u^4\alpha^4)\alpha^4 + (\rho + 4u\beta^2)\alpha^2\beta.$$

If we make the substitution $\rho = -4u\beta^2$ in the second bracketed expression, we get

$$\tau = u + 2u^2\beta^2 + 4u^3\alpha^2\beta + u^4\alpha^4.$$

Since $\tau = u$ when α and $\beta = 0$ this equation can be solved for u in terms of τ, α and β. Defining ρ to be $-4u\beta^2$ we get

$$\pm h(\alpha, \beta) = \alpha_1^2\beta_1 + \beta_1^4 + \rho\alpha_1^2\beta_1$$

as asserted. Finally if we replace α_1 by $(\sqrt{1+\rho})\alpha_1$ we can eliminate the ρ term. Q. E. D.

Our canonical form theorem for the $S_{2,0}$ singularities is the following:

PROPOSITION 7.4. *Let $\varphi = \varphi(x, \alpha, \beta)$ be a phase function on $X \times \mathbf{R}^2$. Suppose that at the origin φ and all its first and second derivatives with respect to α and β vanish. Suppose that the cubic term in the Taylor series expansion of $\varphi(0, \alpha, \beta)$ at the origin is non-degenerate. Then if it is hyperbolic we can find functions $\alpha_1 = \alpha_1(x, \alpha, \beta)$, $\beta_1 = \beta_1(\alpha, \beta, x)$ and $f_i = f_i(x)$, $i = 0, 1, 2, 3$, such that $\alpha_1(0) = \beta_1(0) = f_0(0) = \cdots = f_3(0) = 0$,*

$$\frac{\partial(\alpha_1, \beta_1)}{\partial(\alpha, \beta)} \neq 0,$$

and

$$\varphi = f_0 + f_1\alpha_1 + f_2\beta_1 + f_3\alpha_1\beta_1 + \frac{\alpha_1^3 + \beta_1^3}{3} + \epsilon(x, \alpha, \beta) \tag{7.10}$$

where $\epsilon(x, \alpha, \beta)$ vanishes to infinite order when $x = 0$.

If the cubic term is elliptic we can find $\alpha_1, \beta_1, f_0, \ldots, f_3$ as above such that

$$\varphi = f_0 + f_1\alpha_1 + f_2\beta_1 + f_3(\alpha_1^2 + \beta_1^2) + \alpha_1^3 - \alpha_1\beta_1^2 + \epsilon(x, \alpha, \beta) \tag{7.11}$$

where $\epsilon(x, \alpha, \beta)$ vanishes to infinite order when $x = 0$.

If the cubic term is parabolic and the hypothesis of Proposition 7.3 is satisfied by $\varphi(0, \alpha, \beta)$ we can find $\alpha_1, \beta_1, f_0, f_1, f_2, f_3, f_4$ as above such that

$$\pm\varphi = f_0 + f_1\alpha_1 + f_2\beta_1 + f_3\alpha_1^2 + f_4\beta_1^2 + \alpha_1^2\beta_1 + \beta_1^4 + \epsilon(x, \alpha, \beta) \tag{7.12}$$

where $\epsilon(x, \alpha, \beta)$ vanishes to infinite order when $x = 0$.

PROOF. We will just discuss the hyperbolic case, leaving the parabolic and elliptic cases as an exercise for the reader. We will prove (7.10) by the same kind of induction argument as that in the proof of Proposition 7.1. The case $N = 1$ is just the corollary to Proposition 7.2, so we will assume the case $N - 1$ and prove the case N. Our inductive assumption is that

$$\varphi = f_0 + f_1\alpha + f_2\beta + f_3\alpha\beta + \frac{\alpha^3+\beta^3}{3} + \epsilon(x,\alpha,\beta)$$

where the f_i's are polynomials of degree $< N$ in x, and $\epsilon(x,\alpha,\beta) = O(|x|^N)$. Let

$$\epsilon(x,\alpha,\beta) = \sum_{|I|=N} \epsilon_I(\alpha,\beta)x^I + O(|x|^{N+1}).$$

We will try to find homogeneous polynomials $f_i' = \sum_{|I|=N} f_{i,I}' x^I$ of degree N in x and functions

$$\alpha_1 = \alpha + \sum_{|I|=N} \tau_I(\alpha,\beta)x^I, \qquad \beta_1 = \beta + \sum_{|I|=N} \mu_I(\alpha,\beta)x^I$$

such that

$$\begin{aligned}\varphi = (f_0+f_0') + (f_1+f_1')\alpha_1 + (f_2+f_2')\beta_1 \\ +(f_3+f_3')\alpha_1\beta_1 + \frac{\alpha_1^3+\beta_1^3}{3} + \epsilon_1(x,\alpha,\beta)\end{aligned} \tag{7.13}$$

where $\epsilon_1(x,\alpha,\beta) = O(|x|^{N+1})$. Equating coefficients of x^I we get

$$f_{0,I}' + f_{1,I}'\alpha + f_{2,I}'\beta + f_{3,I}'\alpha\beta + \tau_I\alpha^2 + \mu_I\beta^2 = \epsilon_I(\alpha,\beta) \tag{7.14}$$

which can be easily solved for constants $f_{0,I}'$, etc. and functions $\tau_I(\alpha,\beta)$, $\mu_I(\alpha,\beta)$ using the integral form of the Taylor series with remainder. Q. E. D.

We will now discuss the S_2 singularities in a little more detail. Let Λ be a Lagrangian manifold in T_X^* and let (x_0,ξ_0) be an S_2 singularity on Λ. Let $\varphi = \varphi(x,\alpha,\beta)$ be a phase function on $X \times \mathbf{R}^2$ parametrizing Λ in a neighborhood of (x_0,ξ_0). We will assume that $(x_0,0,0)$ is the point on the critical set of φ corresponding to (x_0,ξ_0). Since (x_0,ξ_0) is an S_2 singularity the first and second derivatives of $\varphi(x,\alpha,\beta)$ with respect to α and β are zero at $(x_0,0,0)$. Let $\varphi_3(\alpha,\beta)$ be the cubic term in the Taylor series expansion of $\varphi(x_0,\alpha,\beta)$ at (0,0). It is easy to see that (x_0,ξ_0) is an $S_{2,1}$ singularity if and only if φ_3 is degenerate, and is an $S_{2,2}$ singularity if and only if $\varphi_3 \equiv 0$.

DEFINITION 7.2. *We will say that* (x_0,ξ_0) *is a hyperbolic (elliptic, parabolic)* S_2 *singularity if* φ_3 *is hyperbolic (elliptic, parabolic).*

We must check that this definition is independent of φ and is an intrinsic property of Λ:

Let R_C be the local ring of formal power series in the coordinates of C at $(x_0,0,0)$ and let R_Λ be the corresponding local ring at (x_0,ξ_0). Let $R_C^\#$ and $R_\Lambda^\#$ be the quotient rings $R_C/(x-x_0)$ and $R_\Lambda/(x-x_0)$. Since Λ and C are diffeomorphic R_Λ and R_C are isomorphic, and so are $R_\Lambda^\#$ and $R_C^\#$.

Now it is clear from Definition 7.2 that hyperbolicity, etc. is an algebraic property of the local ring of formal power series in α and β, modulo the ideal generated by

$$\frac{\partial\varphi}{\partial\alpha}(x_0, \alpha, \beta) \quad \text{and} \quad \frac{\partial\varphi}{\partial\beta}(x_0, \alpha, \beta).$$

However, this is just the ring $R_C^\#$ since the defining equations of C are

$$\frac{\partial\varphi}{\partial\alpha} = \frac{\partial\varphi}{\partial\beta} = 0.$$

Therefore, hyperbolicity, etc. is an algebraic property of $R_\Lambda^\#$. Q.E.D.

We will now prove:

PROPOSITION 7.5. *For Λ in general position, $S_2(\Lambda)$ is a submanifold of codimension* 3; *the elliptic and hyperbolic points are open subsets of $S_2(\Lambda)$, the parabolic points are a codimension* 1 *submanifold of $S_2(\Lambda)$, the $S_{2,1}$ points a codimension* 2 *submanifold of $S_2(\Lambda)$ and the $S_{2,2}$ points a codimension* 4 *submanifold of $S_2(\Lambda)$.*

Consequently elliptic and hyperbolic points can occur for the first time in dimension 3, *parabolic points for the first time in dimension* 4, *$S_{2,1}$ singularities for the first time in dimension* 5, *and $S_{2,2}$ singularities for the first time in dimension* 7.

PROOF. Let W be the space of all polynomial functions in α and β of degree ≤ 3. Let W_1 be the subspace of polynomials with zero first order term and W_2 the subspace of polynomials with zero, first and second order terms. Let P be the parabolic cubic polynomials and Q the degenerate cubic polynomials.

We will need:

LEMMA. *The set P is a codimension* 1 *submanifold of W_2 and Q a codimension* 2 *submanifold.*

PROOF. Consider in W_2 the open set consisting of polynomials whose α^3 term is non-zero. Restricting such a polynomial to the affine subset of projective one space defined by $\alpha \neq 0$ we get an isomorphism between the set of these polynomials and the set of cubic polynomials on the real line. The parabolic polynomials correspond to polynomials on $\mathbf{R}$ with double roots and the degenerate polynomials to polynomials with triple roots. We can view the polynomials with double roots as a fiber bundle over $\mathbf{R}$ with fiber $\mathbf{R}^2$ (just assign to each polynomial its double root) and the polynomials with triple roots as a fiber bundle over $\mathbf{R}$ with fiber $\mathbf{R}$. Therefore P has dimension 3 and Q has dimension 2. Q.E.D.

Now let φ be a phase function on $X \times \mathbf{R}^2$, and let $\tilde{\varphi}$: $X \times \mathbf{R}^2 \to W$ be the map which assigns to each point (x_0, α_0, β_0) the Taylor series expansion of $\varphi(x_0, \alpha, \beta)$ to order 3 about the point (α_0, β_0). We will say that φ is W generic if $\tilde{\varphi}$ is transversal to W_1, W_2, P and Q. It is clear that if φ is W generic the assertions of Proposition 7.5 are true for the corresponding Lagrangian manifold. However, by the Thom transversality theorem every φ can be perturbed to a W generic φ; so this concludes the proof of Proposition 7.5. Q.E.D.

To each point of Λ we have attached the local ring $R_\Lambda^\#$. We have already seen that from the structure of this local ring alone we can determine whether a $S_{2,0}$ singularity is elliptic, hyperbolic or parabolic. We will now show:

PROPOSITION 7.6. *If (x_0, ζ_0) is elliptic or hyperbolic the dimension of $R_\Lambda^\#$ over the reals is* 4 *and if (x_0, ζ_0) is parabolic the dimension is* ≥ 5.

PROOF. If (x_0, ζ_0) is elliptic we can parametrize Λ in a neighborhood of (x_0, ζ_0) by a phase function of the form

$$\frac{\alpha^3 + \beta^3}{3} + \epsilon(x, \alpha, \beta)$$

where $\epsilon(x, \alpha, \beta)$ vanishes when $x = x_0$. Therefore $R_\Lambda^\#$ is isomorphic to the formal power series ring in α and β divided by the ideal (α^2, β^2). As a vector space over $\mathbf{R}$ this has 1, α, β and $\alpha\beta$ as a basis.

If (x_0, ζ_0) is parabolic we can parametrize Λ in a neighborhood of (x_0, ζ_0) by a phase function of the form $\alpha^2\beta + \epsilon(x, \alpha, \beta)$ where $\epsilon(x_0, \alpha, \beta)$ is of order 4 in α and β. Therefore 1, α, β, β^2, and β^3 are all independent modulo

$$\frac{\partial\varphi}{\partial\alpha} \quad \text{and} \quad \frac{\partial\varphi}{\partial\beta}. \qquad \text{Q.E.D.}$$

DEFINITION 7.3. *We will say a parabolic point $(x_0, \zeta_0) \in S_2(\Lambda)$ is regular if* $\dim R_\Lambda^\# = 5$ *and exceptional if* $\dim R_\Lambda^\# > 5$.

PROPOSITION 7.7. *For Λ in general position the set of exceptional parabolic points is a subset of codimension* 1 *in the set of all parabolic points.*

Therefore in dimension 4 all parabolic points are regular. Proposition 7.7 is an easy consequence of the following proposition which we leave as an exercise for the reader:

PROPOSITION 7.8. *Let (x_0, ξ_0) be a parabolic point on Λ. Let $\varphi = \varphi(x, \alpha, \beta)$ be a phase function parametrizing Λ in a neighborhood of (x_0, ξ_0) and having the form*

$$\varphi(x, \alpha, \beta) = \alpha^2\beta + \epsilon(x, \alpha, \beta)$$

where $\epsilon(x_0, \alpha, \beta)$ *is of order* 4 *in* α *and* β. *Then* (x_0, ξ_0) *is a regular parbolic point if and only if*

$$\frac{\partial^4 \varphi}{\partial \beta^4}(x_0, 0, 0) \neq 0.$$

§8. Proofs of the normal forms.

We must still show that the canonical form theorems we derived in §4 are true C^∞, not just formally. For the proof we will need the Malgrange preparation theorem in its "Grothendieck-Houzel" form. We recall the statement of this theorem as it is given, for example, in Malgrange's book [16], or Golubitsky-Guillemin [9]. Let $\mathcal{E}_n$ be the ring of germs of C^∞ functions at the origin in Euclidean n-space. Let $\mathfrak{M}_n$ be its maximal ideal consisting of the germs of functions vanishing at the origin. Given a mapping ρ of $\mathbf{R}^n$ into $\mathbf{R}^p$ mapping the origin to the origin we get an induced map $\rho^*: \mathcal{E}_p \to \mathcal{E}_n$. Now let $f_1, \ldots, f_k$ be (germs of) functions on $\mathbf{R}^n$ and $(f_1, \ldots, f_k)$ the ideal they generate in $\mathcal{E}_n$. Let $\mathfrak{R}$ be the quotient ring $\mathcal{E}_n/(f_1, \ldots, f_k)$. Because of the map $\rho^*: \mathcal{E}_p \to \mathcal{E}_n$ we can view $\mathfrak{R}$ as an $\mathcal{E}_p$ module.

THEOREM 8.1. *The ring* $\mathfrak{R}$ *is a finitely generated* $\mathcal{E}_p$ *module if and only if* $\mathfrak{R}/\mathfrak{M}_p\mathfrak{R}$ *is a finite dimensional vector space over* $\mathbf{R}$. *Moreover, a collection of elements* $\alpha_1, \ldots, \alpha_k \in \mathfrak{R}$ *is a set of generators for* $\mathfrak{R}$ *as a module over* $\mathcal{E}_p$ *if and only if their images in* $\mathfrak{R}/\mathfrak{M}_p\mathfrak{R}$ *are a spanning set of vectors for* $\mathfrak{R}/\mathfrak{M}_p\mathfrak{R}$ *(as a vector space over* $\mathbf{R}$.)

As an application of this theorem, let us prove the Whitney theorem about even functions quoted in §6. Let $n = p = 1$, let $\mathfrak{R} = \mathcal{E}_1$ and let ρ be the map: $\mathbf{R} \to \mathbf{R}$, $x \to x^2$. Then, since $\mathfrak{R}/(x^2)\mathfrak{R}$ is generated by 1 and x, every smooth function can be written in the form $f(x^2) + xg(x^2)$ for smooth f and g. For the function to be even, g must be 0.

The usual preparation theorem for a function $F = F(x,t)$ on $\mathbf{R}^n \times \mathbf{R}$ can be obtained from the theorem above by letting ρ: $\mathbf{R}^n \times \mathbf{R} \to \mathbf{R}^n$ be the map $(x,t) \to x$ and letting $\mathfrak{R}$ be the ring $\mathcal{E}_{n+1}/(F)$. Conversely it is not hard to prove the theorem above assuming the usual preparation theorem. See Golubitsky-Guillemin [9] for details.

Now let us look at the problem that came up in the preceding section. A phase function $\varphi = \varphi(x,\theta)$ is given to us on $\mathbf{R}^n \times \mathbf{R}^N$ and also a perturbed phase function $\varphi' = \varphi(x,\theta) + \epsilon(x,\theta)$ such that $\epsilon(x,\theta)$ vanishes to infinite order at $x = 0$. We want to find a pair of diffeomorphisms f: $\mathbf{R}^n \to \mathbf{R}^n$ and g: $\mathbf{R}^n \times \mathbf{R}^N \to \mathbf{R}^n \times \mathbf{R}^N$ each defined on a neighborhood of the origin and having the origin as a fixed point such that f and g make the diagram

$$\begin{array}{ccc} \mathbf{R}^n \times \mathbf{R}^N & \xrightarrow{\ g\ } & \mathbf{R}^n \times \mathbf{R}^N \\ \big\downarrow & & \big\downarrow \\ \mathbf{R}^n & \xrightarrow{\ f\ } & \mathbf{R}^n \end{array} \tag{8.1}$$

commute and such that g conjugates φ' into φ in the sense that

$$g^*(\varphi' + \mu) = \varphi \tag{8.2}$$

holds, where $\mu = \mu(x)$ is a function of x alone. Such a result clearly allows us to eliminate the error terms occurring in (7.1), (7.10), (7.11) and (7.12). The presence of μ causes no problems because in each case we may absorb it into the first term on the right hand side of the equations.

The equivalence relation between φ and φ' given by (8.1) and (8.2) was first introduced by Thom in his study of singularities. For Thom, $\varphi(x, \theta)$ is to be regarded as a function of θ depending on the parameter x. If $d_\theta \varphi(0, \theta) = 0$, so that "$\varphi$ has a singularity for $x = 0$", then $\varphi(x, \theta)$ is called an "unfolding of the singularity". Two unfoldings, φ and φ', are regarded as equivalent if there are g and f so that (8.1) and (8.2) hold. An unfolding φ is called (locally) stable if any φ' sufficiently close to φ in the C^∞ topology is actually equivalent to φ in some neighborhood of the origin. We shall show that the right hand sides of (7.1), etc., without the error terms, are stable. We will proceed as in Appendix I of Chapter I and as in Chapter IV, §1, and show that "infinitesimal stability" (to be defined below), implies stability. To check that infinitesimal stability holds we will need the Malgrange preparation theorem as formulated above.

Let $\varphi_t(x, \theta) = \varphi(x, \theta) + t\epsilon(x, \theta)$. We will try to construct f_t, g_t and μ_t with the same properties as f, g and μ such that

$$g_t^*(\varphi_t + \mu_t) = \varphi \qquad \text{for all } t, \tag{8.3}$$

f_t, g_t and μ_t depending smoothly on t and g_t being the identity map when $t = 0$. Near $t = 0$ we can write the coordinates of f_t and g_t in powers of t:

$$f_i(x, t) = x_i + a_i(x)t + O(t^2), \qquad i = 1, \ldots, n,$$

and

$$g_j(x, \theta, t) = \theta_j + b_j(x, \theta)t + O(t^2), \qquad j = 1, \ldots, N.$$

If we plug these expressions into (8.3) and differentiate, setting $t = 0$, we get

$$-\epsilon(x, \theta) = a_0(x) + \sum_i \frac{\partial \varphi}{\partial x^i} a_i + \sum_j \frac{\partial \varphi}{\partial \theta_j} b_j \tag{8.4}$$

with

$$a_0 = -\left.\frac{d\mu_t}{dt}\right|_{t=0}.$$

DEFINITION 8.1. *We will say that* φ *is infinitesimally stable if, for every* $\epsilon(x,\theta)$ *on the left of* (8.4), *there exist* $a_i = a_i(x)$ *and* $b_j = b_j(x,\theta)$ *such that* (8.4) *holds in a neighborhood of the origin.*

If we apply the Malgrange preparation theorem to the map: $\mathbf{R}^n \times \mathbf{R}^N \to \mathbf{R}^n$, $(x,\theta) \to x$, with

$$\mathfrak{R} = \mathcal{E}_{n+N}\Big/\left\{\frac{\partial\varphi}{\partial\theta_1},\ldots,\frac{\partial\varphi}{\partial\theta_N}\right\},$$

we get the following criterion for infinitesimal stability.

PROPOSITION 8.1. *The phase function,* φ, *is infinitesimally stable if and only if for every function* $\lambda = \lambda(\theta)$ *there exist real numbers* $c_0, \ldots, c_n$ *and functions* $\tau_1, \ldots, \tau_N$ *of* θ *such that*

$$\lambda(\theta) = c_0 + \sum_i c_i \frac{\partial\varphi}{\partial x^i}(0,\theta) + \sum_\alpha \tau_\alpha(\theta)\frac{\partial\varphi}{\partial\theta_\alpha}(0,\theta). \tag{8.5}$$

Exercise: Show that this criterion is satisfied for the phase function defining a simple caustic:

$$\varphi(x,\theta) = x\theta - \frac{\theta^3}{3}.$$

In the situation we are considering, our deformed phase function $\varphi_t(x,\theta)$ is of the form $\varphi(x,\theta) + t\epsilon(x,\theta)$ where ϵ vanishes to infinite order when $x = 0$; therefore the criterion (8.5) is the same for φ and for φ_t. This proves:

PROPOSITION 8.2. *If* φ *is infinitesimally stable then* φ_t *is infinitesimally stable for all* t.

The main goal of this section is to prove the following:

THEOREM 8.2. *Suppose* $\varphi = \varphi(x,\theta)$ *is infinitesimally stable. Let* $\varphi'(x,\theta) = \varphi(x,\theta) + \epsilon(x,\theta)$ *where* ϵ *vanishes to infinite order when* $x = 0$. *Then there exist* f, g *and* μ *satisfying* (8.1) *and* (8.2).

We will prove, in fact, that there exist f_t, g_t and μ_t satisfying the condition analogous to (8.1) and satisfying (8.3) on the whole interval $0 \le t \le 1$.

As a first step in the proof we will need:

LEMMA 8.1. *If the* $\epsilon(x,\theta)$ *vanishes in* (8.4) *when* $x = 0$ *then one can choose the a's and b's in* (8.4) *so that they also vanish when* $x = 0$.

PROOF. If $\epsilon(x,\theta) = 0$ when $x = 0$ we can write $\epsilon(x,\theta) = \sum x_i \epsilon_i(x,\theta)$ for smooth ϵ_i. For each i we can solve

$$-\epsilon_i(x,\theta) = a_{0,i} + \frac{\partial\varphi}{\partial x_1} a_{1,i} + \cdots + \frac{\partial\varphi}{\partial\theta_1} b_{1,i}(x,\theta) + \cdots.$$

Setting $a_j = \sum a_{j,i} x_i$ and $b_j = \sum b_{j,i} x_i$ we get a solution of (8.4) that vanishes when $x = 0$. Q. E. D.

We want to determine f_t, g_t and μ_t satisfying (8.3). Differentiating (8.3) with respect to t we get

$$-\epsilon(g_t) = \dot{\mu}_t(g_t) + \sum_i \frac{\partial}{\partial x^i}(\mu_t + \varphi_t)(g_t)\dot{f}_i(x,t)$$
$$+ \sum_\alpha \frac{\partial}{\partial\theta_\alpha}(\mu_t + \varphi_t)(g_t)\dot{g}_\alpha(x,\theta,t).$$

Here the f_i's and g_α's are the coordinates of f_t and g_t, and the dots indicate differentiation with respect to t. If we set

$$\begin{aligned} a_i(x,t) &= \dot{f}_i(f_t^{-1}(x),t), \qquad i = 1,\ldots,n,\\ b_\alpha(x,\theta,t) &= \dot{g}_\alpha(g_t^{-1}(x,\theta),t), \qquad \alpha = 1,\ldots N,\\ a_0(x,t) &= \frac{\partial\mu}{\partial t} + \sum \frac{\partial\mu}{\partial x^i} a_i, \end{aligned} \tag{8.6}$$

the equation above reduces to

$$-\epsilon(x,\theta) = a_0(x,t) + \sum_i \frac{\partial\varphi_t}{\partial x^i} a_i + \sum_\alpha \frac{\partial\varphi_t}{\partial\theta_\alpha} b_\alpha \tag{8.7}$$

which is identical with (8.4) except that all the terms are functions of t. We will try to solve (8.7) for functions a_i and b_α which are smooth over the whole interval $0 \leqslant t \leqslant 1$. We first note that it is enough to do this in a small interval about each point on $[0, 1]$; for, by a partition of unity in t these solutions can be patched together to give a global solution in t. By Proposition 8.2 φ_t is infinitesimally stable so (8.7) can be solved for fixed t such that the solutions are smooth in x and θ. The only question is whether these solutions can be chosen to be smooth in t. To see this we apply the Malgrange preparation theorem to the map: $\mathbf{R}^n \times \mathbf{R}^N \times \mathbf{R} \to \mathbf{R}^n \times \mathbf{R}$, $(x,\theta,t) \to (x,t)$, with

$$\mathcal{R} = \mathcal{E}_{n+N+1}\Big/\left(\frac{\partial\varphi_t}{\partial\theta_1},\ldots,\frac{\partial\varphi_t}{\partial\theta_N}\right).$$

The preparation theorem says that (8.7) can be solved with an arbitrary ϵ if and only if (8.5) holds for φ_t. We have already seen, however, (Proposition 8.2) that this condition is the same for all t, and it holds when $t = 0$ because of the infinitesimal stability of φ.

Finally we note that if $\epsilon(x, \theta)$ vanishes when $x = 0$ as is the case with us, we can choose the a_i's and the b_α's to vanish when $x = 0$ (Lemma 8.1).

To conclude the proof of Theorem 8.2 we must solve the equations (8.6) with initial data:

$$\begin{aligned} f_i(x,0) &= x^i \qquad i = 1, \ldots, n, \\ g_\alpha(x,\theta,0) &= \theta_\alpha \qquad \alpha = 1, \ldots, N, \\ \mu(x,0) &= 0. \end{aligned} \tag{8.8}$$

The first pair of these equations are just ordinary differential equations in f and g. Since the expressions on the left hand side vanish when $x = 0$, they are solvable globally in t for a sufficiently small neighborhood of the origin in (x, θ) space. The last of the equations (8.6) can be solved by linear Hamilton-Jacobi theory, i. e., just by integrating the vector field $(a_1(x,t), \ldots, a_n(x,t))$. This can be done globally in t for the same reasons as before. Q.E.D.

We will now use the results above to derive the canonical form theorems of §7. We will actually prove a theorem which includes these results and is applicable to other singularities besides the elementary ones described in §7. This theorem is a variant of Thom's "universal unfolding theorem."

Let $\psi = \psi(\theta)$ be a smooth function defined on a neighborhood of the origin in $\mathbf{R}^N$. We recall (see Chapter I, Appendix I) that ψ satisfies the *Milnor condition* if the ideal generated by its first partial derivatives, $I_\psi = \{\partial\psi/\partial\theta_1, \ldots, \partial\psi/\partial\theta_N\}$, is of finite codimension in the ring $\mathcal{E}_N$ of germs of smooth functions at the origin of $\mathbf{R}^N$. Examples which we have already seen are:

$$\begin{aligned} &\text{For } N = 1, \\ &\text{(a)} \quad \psi = \theta^{k+1}/(k+1), \\ &\text{For } N = 2, \\ &\text{(b)} \quad \psi = (\theta_1^3 + \theta_2^3)/3, \\ &\text{(c)} \quad \psi = \theta_1^3 - \theta_1\theta_2^2, \\ &\text{(d)} \quad \psi = \theta_1^3\theta_2 + \theta_2^4. \end{aligned} \tag{8.9}$$

Suppose that the codimension of I_ψ in $\mathcal{E}_N$ is $k + 1$. Then we can choose functions $\psi_0, \ldots, \psi_k$ in $\mathcal{E}_N$ so that their images form a basis for the quotient. We may assume that $\psi_0 = 1$. Our main result is:

THEOREM 8.3. *Let* $\varphi = \varphi(x, \theta)$ *be a smooth function defined on a neighborhood of the origin in* $\mathbf{R}^n \times \mathbf{R}^N$ *such that* $\varphi(0, \theta) = \psi(\theta)$. *Then there exist smooth functions* $f_i = f_i(x)$, $i = 0, \ldots, k$ *and* $\bar{\theta}_j(x, \theta)$ *such that* $f_i(0) = 0$ *and* $\bar{\theta}_j(0, \theta) = \theta_j$, $j = 1, \ldots, N$ *and*

$$\varphi(x, \theta) = f_0(x) + f_1(x)\psi_1(\bar{\theta}) + \cdots + f_k(x)\psi_k(\bar{\theta}) + \psi(\bar{\theta}). \tag{8.10}$$

Applying this theorem to example (a) of (8.9), and letting $\psi_i = \theta^i$, $i = 1, \ldots, k - 1$, gives the canonical form (7.1). We leave it as an exercise for the reader to get the canonical forms (7.10), (7.11) and (7.12) by applying this theorem with the examples (b), (c) and (d) of (8.9).

PROOF OF THEOREM 8.3. We first of all prove the assertion formally. The proof is practically identical with the proofs of Propositions 7.3 and 7.4. We will argue by induction that by changing the θ coordinates we can write

$$\varphi(x, \theta) = f_0(x) + \cdots + f_k(x)\psi_k(\theta) + \psi(\theta) + \epsilon(x, \theta) \tag{8.12}$$

where the f_i's are polynomials in x of order $\leq r - 1$ and $\epsilon(x, \theta)$ is $O(|x|^r)$ uniformly in θ. We will try to find a coordinate change

$$\bar{\theta}_i = \theta_i + \sum_{|I|=r} \tau_{i,I}(\theta)x^I, \qquad i = 1, \ldots, N, \tag{8.13}$$

and polynomials

$$f'_i(x) = \sum_{|I|=r} f'_{i,I}x^I \tag{8.14}$$

such that

$$\varphi(x, \theta) = \sum (f_i + f'_i)(x)\psi_i(\bar{\theta}) + \psi(\bar{\theta}) + \bar{\epsilon}(x, \theta) \tag{8.15}$$

where $\bar{\epsilon}(x, \theta)$ is $O(|x|^{r+1})$ uniformly in θ. Let

$$\epsilon(x, \theta) = \sum_{|I|=r} \epsilon_I(\theta)x^I + O(|x|^{r+1}). \tag{8.16}$$

Equating the coefficient of x^I on the right hand side and left hand side of (8.12) we get

$$\sum f_{i,I}\psi_i(\theta) + \sum \frac{\partial \psi}{\partial \theta_i}\tau_{i,I}(\theta) = -\epsilon_I(\theta).$$

By hypothesis these equations are solvable and we can continue the induction. Q.E.D.

We will now prove Theorem 8.3. Because of what we have just proved we can assume that

$$\varphi(x,\theta) = f_0(x) + \sum_{i=1}^{k} f_i(x)\psi_i(\theta) + \psi(\theta) + \epsilon(x,\theta)$$

where $\epsilon(x,\theta)$ vanishes to infinite order when $x = 0$.

Let us also assume for the moment that $n \geq k$ and that $df_1, \ldots, df_k$ are linearly independent at the origin. Then the $\psi_i(\theta)$ are linear combinations (with constant coefficients) of the $(\partial\varphi)/(\partial x_i)(0,\theta)$ and 1, so the hypotheses of Proposition 8.1 are satisfied and $\varphi(x,\theta)$ is infinitesimally stable. Theorem 8.3, therefore, follows from Theorem 8.2.

Suppose on the other hand that the df_i's are not linearly independent. Consider on the space $\mathbf{R}^{n+k} \times \mathbf{R}^N$ the phase function

$$\Phi(x,y,\theta) = f_0(x) + \sum_{i=1}^{k} (f_i(x) + y_i)\psi_i(\theta) + \psi(\theta) + \epsilon.$$

By applying the preceding argument to this phase function we can find a change of coordinates $\tilde{\theta} = \tilde{\theta}(x,y,\theta)$ and $\tilde{f}_i = \tilde{f}_i(x,y)$, $i = 0, \ldots, k$, such that

$$\Phi(x,y,\theta) = f_0(x,y) + \sum_{i=1}^{k} f_i(x,y)\psi_i(\tilde{\theta}) + \psi(\tilde{\theta}).$$

Letting $\bar{\theta} = \tilde{\theta}(x,0,\theta)$, we are done.

§9. Behavior near caustics (continued).

We have seen that seven types of singularities can occur generically in dimension ≤ 4: four types of iterated S_1 singularities and three types of umbilics. For each of these types of singularities we have an associated canonical form for the phase function. We will now use these canonical forms to describe how compound asymptotics behave in the vicinity of a caustic exhibiting one of these singularities. We will begin with the iterated S_1's. This will involve looking at a generalized "Airy function" of the form

$$Y_0(x_1,\ldots,x_{m-2}) = \int \exp\left(i\left(x_1\theta + \cdots + x_{m-2}\theta^{m-2} + \frac{\theta^m}{m}\right)\right)d\theta \quad (9.1)$$

and its first partial derivatives

$$Y_s = \frac{1}{\sqrt{-1}}\frac{\partial Y}{\partial x_s} = \int \theta^s \exp\left(i\left(x_1\theta + \cdots + \frac{\theta^m}{m}\right)\right)d\theta. \quad (9.2)$$

We have to worry a little about whether the expressions on the right are meaningful, i. e., whether the integrals converge (even for the classical Airy function this is not obvious.) We will prove:

LEMMA 9.1. *The integrals above converge in the usual Lebesgue sense (i. e., it is not necessary to define them by regularization).*

PROOF. Restrict $x_1, \ldots, x_{m-2}$ to a compact set. Then we can find $h(\theta)$ so that $h(\theta) \equiv 1$ for large θ and $h(\theta) \equiv 0$ near the zeroes of the polynomial

$$x_1 + 2x_2\theta + \cdots + \theta^{n-1}.$$

Clearly it is enough to show that (9.1) and (9.2) converge after we have multiplied the integrands by $h(\theta)$. The integral we get can be written as:

$$\lim_{R\to\infty} \int_{-R}^{R} \frac{-i\theta^s h(\theta)}{x_1 + 2x_2\theta + \cdots + \theta^{m-1}} \frac{\partial}{\partial\theta} \exp\left(i\left(x_1\theta + \cdots + \frac{\theta^m}{m}\right)\right) d\theta. \quad (9.3)$$

We apply an integration by parts to this and observe that the end terms tend to 0, getting

$$-\int \frac{\partial}{\partial\theta} \frac{-i\theta^s h(\theta)}{x_1 + 2x_2\theta + \cdots + \theta^{m-1}} \exp\left(i\left(x_1\theta + \cdots + \frac{\theta^m}{m}\right)\right) d\theta.$$

The integrand is now of order $O(1/\theta^{m-s})$ for large θ. Since $s \leqslant m-2$, it is integrable. Q. E. D.

Now let Λ be a Lagrangian submanifold of T_X^* which is in general position. Suppose it exhibits an

$$S_{\underbrace{1,\ldots,1}_{m}}$$

type singularity at a point p. Then we can choose coordinates $x_1, \ldots, x_n$ near $\pi(p)$ and a phase function $\varphi = \varphi(x,\theta)$ parametrizing Λ having the form:

$$\varphi(x,\theta) = f_0(x) + x_1\theta + x_2\theta^2 + \cdots + x_{m-2}\theta^{m-2} + \frac{\theta^m}{m}. \quad (9.4)$$

We will prove:

THEOREM 9.1. *A compound asymptotic $\int a(x,\theta)\exp(i\tau\varphi(x,\theta))\,d\theta$ associated with (9.4) has the following asympototic expansion in terms of the generalized Airy function introduced above and its first derivatives*:

$$e^{i\tau f_0} \sum_{i=0}^{m-2} \frac{a_i(x,\tau)}{\tau^{((i+1)/m)}} Y_i(\tau^{(1-1/m)}x_1, \tau^{(1-2/m)}x_2, \ldots, \tau^{(1-(m-2)/m)}x_{m-2}) \quad (9.5)$$

where the $a_j(x,\tau)$ are asymptotic series in τ^{-1} of the form

$$a_i(x) + \tau^{-1}a_{i,1}(x) + \tau^{-2}a_{i,2}(x) + \cdots$$

the a_i's being half densities on X.

PROOF. Using the Malgrange preparation theorem we can write

$$a(x,\theta) = a_0(x) + a_1(x)\theta + \cdots + a_{m-2}(x)\theta^{m-2} + k(x,\theta)(x_1 + 2x_2\theta + \cdots + \theta^{n-1}).$$

Just as in §6, we can get rid of the contribution of the last term by an integration by parts (at the expense of introducing an error of order $O(1/\tau)$). The remaining expression is of the form

$$\sum_{i=0}^{m-2} \exp(i\tau f_0(x)) a_i(x) \int \theta^i \exp(i\tau(x_1\theta + \cdots + \theta^m/m))\, d\theta.$$

Making the coordinate change $\theta \to \tau^{1/m}\theta$ we can rewrite this as

$$e^{i\tau f_0(x)} \sum_{i=0}^{m-2} \frac{a_i(x)/\tau^{i+1}}{\tau^{i+1/m}} Y_i(\tau^{(1-(1/m))}x_1, \ldots, \tau^{(1-(m-2)/m)}x_{m-2}).$$

Now repeat the argument with the remainder term. Repeating it infinitely often we get the asymptotic expansion (9.5). Q.E.D.

Next we will look at a compound asymptotic near an S_2 singularity. For simplicity we will just consider the hyperbolic case. The story for the elliptic and regular parabolic umbilics is roughly the same.

For the hyperbolic umbilic the appropriate "generalized Airy function" is

$$Y(x_1, x_2, x_3) = \int \exp\left(i\left(x_1\alpha + x_2\beta + x_3\alpha\beta + \frac{\alpha^3+\beta^3}{3}\right)\right) d\alpha\, d\beta \tag{9.6}$$

and its first partial derivatives are

$$\frac{1}{i}\frac{\partial Y}{\partial x_1} = \int \alpha \exp\left(i\left(x_1\alpha + x_2\beta + x_3\alpha\beta + \frac{\alpha^3+\beta^3}{3}\right)\right) d\alpha\, d\beta \qquad (9.7)_1$$

$$\frac{1}{i}\frac{\partial Y}{\partial x_2} = \int \beta \exp\left(i\left(x_1\alpha + x_2\beta + x_3\alpha\beta + \frac{\alpha^3+\beta^3}{3}\right)\right) d\alpha\, d\beta \qquad (9.7)_2$$

$$\frac{1}{i}\frac{\partial Y}{\partial x_3} = \int \alpha\beta \exp\left(i\left(x_1\alpha + x_2\beta + x_3\alpha\beta + \frac{\alpha^3+\beta^3}{3}\right)\right) d\alpha\, d\beta. \qquad (9.7)_3$$

The question of convergence for these integrals is a little bit more complicated than for the integrals (9.1) and (9.2). We will prove:

LEMMA 9.2. *Define the integral* (9.6) *as*

$$\lim_{R\to\infty} \int_{\alpha^2+\beta^2 \leqslant R^2} \exp\left(i\left(x_1\alpha + x_2\beta + x_3\alpha\beta + \frac{\alpha^3+\beta^3}{3}\right)\right) d\alpha\, d\beta.$$

Then this limit exists and is a smooth function of x_1, x_2, x_3.

PROOF. The same idea as Lemma 9.1. Denote the phase function in the integral above by $\varphi(x, \alpha, \beta)$. It is easy to see that for x_1, x_2, x_3 restricted to a compact set

$$\left(\frac{\partial\varphi}{\partial\alpha}\right)^2 + \left(\frac{\partial\varphi}{\partial\beta}\right)^2$$

has its zeroes in a compact set of α,β space. Let $h(\alpha, \beta)$ be a function which is zero on this set and 1 for α and β large. Multiplying the integrand above by $h(\alpha, \beta)$ one gets

$$\int_{D_R} h(\alpha, \beta)e^{i\varphi(x,\alpha,\beta)}\,d\alpha\,d\beta$$

$$= \int_{D_R} \frac{-ih(\alpha, \beta)}{\left(\frac{\partial\varphi}{\partial\alpha}\right)^2 + \left(\frac{\partial\varphi}{\partial\beta}\right)^2}\left(\frac{\partial\varphi}{\partial\alpha}\frac{\partial}{\partial\alpha} + \frac{\partial\varphi}{\partial\beta}\frac{\partial}{\partial\beta}\right)e^{i\varphi(x,\alpha,\beta)}\,d\alpha\,d\beta,$$

D_R being the disk of radius R. Integrating by parts one gets an integrand which can be estimated by

$$\frac{h(\alpha, \beta)}{(\alpha^2 + \beta^2)^{3/2}}$$

plus a boundary contribution which goes to zero as $1/R$ when $R \to \infty$. By repeating this argument one can show that the integral is differentiable to arbitrary order. Q.E.D.

Unfortunately the integrals $(9.7)_i$ do not converge in the usual sense; however they can be "regularized" in several ways, for example by using the techniques of Hormander [18, §1.2].

Now let Λ be Lagrangian submanifold of T_X^* which is in general position. Suppose it exhibits a hyperbolic umbilic at a point p. Then we can choose coordinates $x_1, \ldots, x_n$ near $\pi(p)$ and a phase function $\varphi = \varphi(x, \alpha, \beta)$ parametrizing Λ and having the form:

$$\varphi(x, \alpha, \beta) = f_0(x) + x_1\alpha + x_2\beta + x_3\alpha\beta + \frac{\alpha^3 + \beta^3}{3}. \tag{9.8}$$

We will prove:

THEOREM 9.2. *A compound asymptotic* $\int a(x, \alpha, \beta)e^{i\tau\varphi(x,\alpha,\beta)}\,d\alpha\,d\beta$ *associated with* (9.8) *has the following asympototic expansion in terms of the generalized Airy function* (9.6) *and its first partial derivatives*:

$$e^{i\tau f_0(x)}\left\{\frac{a_0(x,\tau)}{\tau^{2/3}}Y(\tau^{2/3}x_1,\tau^{2/3}x_2,\tau^{1/3}x_3)+\frac{a_1(x,\tau)}{i\tau}\frac{\partial Y}{\partial x_1}(\tau^{2/3}x_1,\tau^{2/3}x_2,\tau^{1/3}x_3)\right.$$
$$\left.+\frac{a_2(x,\tau)}{i\tau}\frac{\partial Y}{\partial x_2}(\quad)+\frac{a_3(x,\tau)}{i\tau^{4/3}}\frac{\partial Y}{\partial x_3}(\quad)\right\},\tag{9.9}$$

the a_i' s being asymptotic series in τ^{-1} with half densities as coefficients.

PROOF. Using the Malgrange preparation theorem, we can write

$$a(x,\alpha,\beta)=a_0(x)+a_1(x)\alpha+a_2(x)\beta+a_3(x)\alpha\beta$$
$$+g(x,\alpha,\beta)\frac{\partial\varphi}{\partial\alpha}+h(x,\alpha,\beta)\frac{\partial\varphi}{\partial\beta},$$

φ being the function (9.8). We get rid of the contribution of the last two terms to the integral $\int a(x,\alpha,\beta)e^{i\tau\varphi(\tau,\varphi,\beta)}\,d\alpha\,d\beta$ by an integration by parts, at the expense of introducing an error term of order $O(1/\tau)$. The remaining expression is of the form

$$a_0(x)\int e^{i\tau\varphi(x,\alpha,\beta)}\,d\alpha\,d\beta+a_1(x)\int\alpha e^{i\tau\varphi}\,d\alpha\,d\beta+a_2(x)\int\beta e^{i\tau\varphi}\,d\alpha\,d\beta$$
$$+a_3(x)\int\alpha\beta e^{i\tau\varphi}\,d\alpha\,d\beta.$$

Making the coordinate change $\alpha\to\tau^{1/3}\alpha$, $\beta\to\tau^{1/3}\beta$ this becomes

$$e^{i\tau f_0(x)}\left\{\frac{a_0(x)}{\tau^{2/3}}Y(\tau^{2/3}x_1,\tau^{2/3}x_2,\tau^{1/3}x_3)\right.$$
$$\left.+\frac{a_1(x)}{i\tau}\frac{\partial Y}{\partial x_1}(\quad)+\frac{a_2(x)}{i\tau}\frac{\partial Y}{\partial x_2}(\quad)+\frac{a_3(x)}{i\tau^{4/3}}\frac{\partial Y}{\partial x_3}(\quad)\right\}.$$

Now repeat this argument with the remainder term. Repeating it infinitely often we get the asymptotic expansion (9.9). Q.E.D.

For the elliptic umbilic we get an asymptotic expansion involving the "elliptic Airy function":

$$Y(x_1,x_2,x_3)=\int\exp i(x_1\alpha+x_2\beta+x_3(\alpha^2+\beta^2)+\alpha^3-\alpha\beta^2)d\alpha\,d\beta\tag{9.10}$$

and its first partial derivatives. For the regular parabolic singularity we get an expansion involving the "parabolic Airy function":

$$Y(x_1,x_2,x_3,x_4)=\int\exp i(x_1\alpha+x_2\beta+x_3\alpha^2+x_4\beta^2+\alpha^2\beta+\beta^4)d\alpha\,d\beta\tag{9.11}$$

and its first partial derivatives. These expansions have the same general form as the expansion (9.9).

As far as we know none of the "Airy functions" above have been tabulated or their analytic properties studied in detail. It would be extremely interesting to have some information about these functions in light of the role they seem destined to play in the theory of higher dimensional asymptotics.

From the expansions above we can obtain results like those in §6 for the reduced wave equation in the vicinity of caustics, provided the singularities on the caustics are of the kind studied here. In particular suppose we have an asymptotic solution of $\Delta\mu - \tau^2\mu = 0$ which is normalized so that it is of order $O(1)$ in the region where the approximation of geometric optics is valid. Then we have:

THEOREM 9.3. (1) *In the neighborhood of a regular caustic point* $\mu \cong O(\tau^{1/6})$.

(2) *In the neighborhood of a cusped caustic (i.e., an* $S_{1,1}$ *singularity)* $\mu \cong O(\tau^{1/4})$.

(3) *In the neighborhood of a caustic of the type* $S_{1,1,1}$ *(a "swallow's tail")* $\mu = O(\tau^{3/10})$.

(4) *In the neighborhood of a caustic of type* $S_{1,1,1,1}$ *(a "butterfly")* $\mu \cong O(\tau^{1/3})$.

(5) *In the neighborhood of a hyperbolic umbilic* $\mu \cong O(\tau^{1/3})$.

(6) *In the neighborhood of an elliptic umbilic* $\mu \cong O(\tau^{1/3})$.

(7) *In the neighborhood of a regular parabolic umbilic* $\mu \cong O(\tau^{3/8})$.

PROOF. If the phase function involves one phase variable the asymptotic associated with it must be normalized by multiplying by $\sqrt{\tau}$ so that it will be about of order 1 away from the caustic, and if there are two phase variables it must be normalized by multiplying by τ (since stationary phase applied to an integrand of the form $\int a(x,\theta)\exp(i\tau\varphi(x,\theta))\,d\theta$ introduces a factor $\tau^{-\frac{n}{2}}$ in front of the asymptotic approximation, where n is the number of θ variables). Keeping this fact in mind one gets the approximations 1–7 above as an immediate consequence of the asymptotic formulas. Q.E.D.

It is also possible as in §6 to estimate the width of the boundary layers for "regions of shadow" surrounding the intensely illuminated caustics. (Since the geometry of the caustic is much easier to understand upstairs on Λ rather than downstairs on X, it is easiest to make these estimates in terms of the Λ rather than x coordinates.) One can also without much difficulty write down a set of transport equations satisfied by the coefficients occurring in the asymptotic series (9.5) and (9.9) just as we did for the asymptotic series (6.9).

REFERENCES, CHAPTER VII

1. J. Leray, *Solutions asymptotiques*, Seminar notes, Paris, 1972.
2. J. J. Duistermaat, Comm. Pure Appl. Math. **27** (1974), 207–281.
3. V. I. Arnol'd, Funkcional. Anal. i Priložen. **6** (1972), no. 4, 3–25 = Functional Anal. Appl. **6** (1972), 254–272.
4. S. Sternberg, *Lectures on differential geometry*, Prentice-Hall, Englewood Cliffs, N. J., 1964. MR **33** #1797.

5. J. C. Tougeron, *Idéaux de fonctions différentiables*, Ergebnisse der Mathematik und ihrer Grenzgebiete, Band 71, Springer-Verlag, Berlin and New York, 1972.
6. I. N. Bernšteĭn, Funkcional. Anal. i Priložen. **6** (1972), no. 4, 26–40 = Functional Anal. Appl. **6** (1972), 273–285 (1973). MR **47** #9269.
7. I. M. Gel'and and G. E. Shilov (Šilov), *Generalized functions*. Vol. I: *Operations on them*, Fizmatgiz, Moscow, 1958; English transl., Academic Press, New York, 1964. MR **20** #4182; **29** #3869.
8. O. Zariski and P. Samuel, *Commutative algebra*. Vol. I, University Ser. in Higher Math., Van Nostrand, Princeton, N. J., 1958. MR **19**, 833.
9. M. Golubitski and V. W. Guillemin, *Stable mappings and their singularities*, Graduate Texts in Math., vol. 14, Springer-Verlag, Berlin and New York, 1973. MR **49** #6269.
10. J. M. Boardman, Inst. Hautes Études Sci. Publ. Math. No. 33 (1967), 21–27. MR **37** #6945.
11. R. Thom, *Stabilité structurelle et morphognese*, Benjamin, Reading, Mass., 1972.
12. M. Abramowitz and I. A. Stegun (Editors), *Handbook of mathematical functions with formulas, graphs, and mathematical tables*, Nat. Bur. Standards Appl. Math. Ser., vol. 55, National Bureau of Standards, Washington, D. C., 1964. MR **29** #4914.
13. A. Messiah, *Quantum mechanics*, Interscience, New York, 1961. MR **23** #B2826.
14. D. Ludwig, Comm. Pure Appl. Math. **19** (1966), 215–250. MR **33** #4446.
15. C. Chester, B. Friedman and F. Ursell, Proc. Cambridge Philos. Soc. **53** (1957), 599–611. MR **19**, 853.
16. F. W. J. Olver, *Introduction to asymptotics and special functions*, Academic Press, New York and London, 1974.
17. B. Malgrange, *Ideals of differentiable functions*, Tata Inst. Fund. Res. Studies in Math., no. 3, Tata Institute of Fundamental Research, Bombay; Oxford Univ. Press, London, 1967. MR **35** #3446.
18. L. Hormander, Acta Math. **127** (1971), 79–183.

Appendix II. Various Functorial Constructions

In this appendix we collect a few of the functorial facts that are used throughout the book. We first discuss the notion of morphism of vector bundles, a concept introduced by Atiyah and Bott in their celebrated paper on the fixed point theorem. We then discuss the notion of fiber product, and show how it relates to various constructions that we describe in various chapters.

§1. The category of smooth vector bundles.

In this section we briefly establish notation concerning the calculus in the category of smooth vector bundles. An object in this category consists of a real (resp., complex) C^∞ vector bundle E over the C^∞ manifold X:

$$\begin{array}{c} E \\ \downarrow \\ X \end{array}$$

To describe the morphisms in this category we recall the following: Let Y be a differentiable manifold and $f\colon Y \to X$ a smooth map. Then $f^\# E$ is a vector bundle over Y, whose fibers are given by $(f^\# E)_y = E_{f(y)}$. If F is a vector bundle over Y then $\operatorname{Hom}(f^\# E, F)$ is also a vector bundle over Y whose fiber at y is $\operatorname{Hom}((f^\# E)_y, F_y) = \operatorname{Hom}(E_{f(y)}, F_y)$. It makes sense to talk of a smooth section of this vector bundle. Following Atiyah-Bott [2] of Chapter VI, we define a morphism

$$\begin{array}{ccc} F & & E \\ \downarrow & \xrightarrow[\mathbf{f}=(f,r)]{} & \downarrow \\ Y & & X \end{array}$$

to consist of the pair $\mathbf{f} = (f, r)$ where $f\colon Y \to X$ is a smooth map and r is a smooth section of $\mathrm{Hom}(f^{\#} E, F)$. If $\mathbf{g} = (g, t)$,

$$\mathbf{g}\colon \qquad \begin{matrix} G \\ \downarrow \\ Z \end{matrix} \quad \xrightarrow[\mathbf{g}=(g,t)]{} \quad \begin{matrix} F \\ \downarrow \\ Y \end{matrix}$$

is a morphism from G to F then $\mathbf{f} \circ \mathbf{g}$ is defined by $\mathbf{f} \circ \mathbf{g} = (f \circ g, w)$ where w is the section of

$$\mathrm{Hom}(G, (f \circ g)^{\#} E) = \mathrm{Hom}(G, g^{\#}(f^{\#} E))$$

given by

$$w(z) = r(g(z)) \circ t(z)$$

where

$$t(z) \in \mathrm{Hom}(G_z, F_{g(z)}) \quad \text{and} \quad r(g(z)) \in \mathrm{Hom}(F_{g(z)}, E_{f \circ g(z)})$$

so that

$$w(z) \in \mathrm{Hom}(G_z, E_{(f \circ g)(z)}) = \mathrm{Hom}(G, (f \circ g)^{\#} E)_z.$$

It is easy to see that the various axioms for a category are satisfied. For any vector bundle $E \to X$ we let $\Gamma(E)$ denote the space of C^∞ sections of E and $\Gamma_0(E)$ the space of C^∞ sections with compact support. If $\mathbf{f}\colon F \to E$ is a morphism then it induces a linear map $\mathbf{f}^*\colon \Gamma(E) \to \Gamma(F)$ by

$$(\mathbf{f}^* s)(y) = r(y) s(f(y)) \in F_y$$

for any section s of E. It is clear that

$$\mathbf{f}^*(\varphi s) = (f^* \varphi) \mathbf{f}^* s$$

for any C^∞ function φ on X where $f^* \varphi(y) = \varphi(f(y))$. It is also clear that the map sending f to f^* is a contravariant functor to the category whose objects are $\Gamma(E)$'s.

Similarly, we can consider the subcategory of proper morphisms, consisting of those $\mathbf{f}$ for which $f\colon Y \to X$ is a proper map. Then we get a contravariant functor $E \to \Gamma_0(E)$, $\mathbf{f} \to \mathbf{f}^*$.

We denote by $F \boxtimes E$ the vector bundle over $Y \times X$ whose fiber at (y, x) is $F_y \otimes E_x$. If F and E are fibers over the same space X then $F \otimes E$ denotes the vector bundle over X whose fiber at x is $F_x \otimes E_x$. The map $\mathbf{\Delta} = (\Delta, \mathrm{id})$,

$$\begin{matrix} F \otimes E \\ \downarrow \\ X \end{matrix} \quad \to \quad \begin{matrix} F \boxtimes E \\ \downarrow \\ X \times X, \end{matrix}$$

where $\Delta(x) = (x,x)$ is clearly a morphism, is called a diagonal map. More generally, if G is any vector bundle over X then a morphism of the form

$$\begin{array}{ccc} G & & F \boxtimes E \\ \downarrow & \xrightarrow[(\Delta,r)]{} & \downarrow \\ X & & X \times X \end{array}$$

(so that $r(x)$: $F_x \otimes E_x \to G_x$) will be called a diagonal morphism.

If s_1 and s_2 are sections of F and E then $s_1 \boxtimes s_2$ is the section of $F \boxtimes E$ given by $s_1 \boxtimes s_2(y,x) = s_1(y) \otimes s_2(x)$ and $s_1 \otimes s_2 = \mathbf{\Delta}^*(s_1 \boxtimes s_2)$ is given by $s_1 \otimes s_2(x) = s_1(x) \otimes s_2(x)$.

If $\mathbf{f}_1: F_1 \to E_1$ and $\mathbf{f}_2: F_2 \to E_2$ are morphisms then we get a well-defined morphism $\mathbf{f}_1 \boxtimes \mathbf{f}_2: F_1 \boxtimes F_2 \to E_1 \boxtimes E_2$; and, if F_1, F_2 are vector bundles over the same space Y while E_1, E_2 are vector bundles over the same space X then $\mathbf{f}_1 \otimes \mathbf{f}_2: F_1 \otimes F_2 \to E_1 \otimes E_2$ with

$$\mathbf{\Delta} \circ (\mathbf{f}_1 \otimes \mathbf{f}_2) = (\mathbf{f}_1 \boxtimes \mathbf{f}_2) \circ \mathbf{\Delta}.$$

In particular,

$$\mathbf{f}_1^* s_1 \otimes \mathbf{f}_2^* s_2 = (\mathbf{f}_1 \otimes \mathbf{f}_2)^*(s_1 \otimes s_2).$$

We will not dwell any further on these considerations of a purely functorial nature, leaving the reader to fill in the details.

Any morphism $\mathbf{f} = (f,r)$ has a canonical factorization as

$$\mathbf{f} = \boldsymbol{\pi} \circ \boldsymbol{\iota}.$$

Here

$$\boldsymbol{\iota} = (\iota, \mathrm{id}): \quad \begin{array}{ccc} F & & p^\# F \\ \downarrow & \to & \downarrow \\ Y & & Y \times X \end{array}$$

where p: $Y \times X \to X$ is projection onto the first factor giving $(p^\# F)_{(y,x)} = F_y$ and where $\iota(y) = (y, f(y))$ so that $(\iota^\# p^\# F)_y = F_y$ and so id makes sense. Also $\boldsymbol{\pi} = (\pi, w)$ where π is projection onto the second factor and $w(y,x)$: $E_x \to F_y$ is given by $w(y,x) = r(y)$.

If $\mathbf{f}$: $F \to E$ is a morphism and S is a subset of F then we define the subset $\mathbf{f}_* S$ of E by saying that $e \in E_x$ belongs to $\mathbf{f}_* S$ if and only if $r_y(e) \in S$ for some $y \in f^{-1}(x)$, where $f = (f,r)$. Thus

$$\mathbf{f}_* S = \bigcup_{y \in Y} r_y^{-1}(S \cap F_y).$$

It is easy to check that if $\mathbf{g}\colon G \to F$ is a morphism and R is a subset of G then $f_*(g_* R) = (f \circ g)_* R$.

If $\mathbf{f}\colon F \to E$ is a morphism and W is a subset of E then we define the subset $\mathbf{f}^* W$ of F by setting

$$\mathbf{f}^* W = \bigcup_{y \in Y} r_y(W \cap E_{f(y)})$$

when $\mathbf{f} = (f, r)$. Again, it is easy to check that if $\mathbf{g}\colon G \to F$ is another morphism then $(\mathbf{f} \circ \mathbf{g})^* W = \mathbf{g}^* \mathbf{f}^* W$.

There is a natural functor, T^*, which goes from the category of differentiable manifolds and differentiable maps to the category of vector bundles. It assigns to each differentiable manifold X its cotangent bundle $T^*(X)$ and to each differentiable map $f\colon Y \to X$ the morphism $\mathbf{T}^*(f) = (f, df^t)$ where $df_y\colon T_y(Y) \to T_{f(y)}(X)$ is just the induced map of f on the tangent space (the Jacobian map) and $df_y^t\colon T^*_{f(y)} \to T_y^*(Y)$ is the transpose.

More generally, we can consider the functor which assigns to any vector bundle E over X the vector bundle $E \otimes T^*(X)$ and to any morphism $\mathbf{f}\colon F \to E$ the morphism $\mathbf{f} \otimes \mathbf{T}^*(f)\colon F \otimes T^*(Y) \to E \otimes T^*(X)$. If s is a smooth section of E which vanishes at the point x then $ds(x) \in E_x \otimes T_x^*(X)$ is well defined. (Indeed, compute the differential in terms of some trivialization. The fact that $s(x) = 0$ implies that the result is independent of the trivialization.) If $\mathbf{f}\colon F \to E$ and $y \in f^{-1}(x)$ then $\mathbf{f}^* s(x) = 0$ and $d(\mathbf{f}^* s)(y) = (r_y \otimes df_y^t) ds(x)$.

§2. The fiber product.

Let $f\colon A \to B$ and $g\colon C \to B$ be maps of sets. Then the *fiber product* $D = A \times_B C$ consists of those pairs (a, c) such that $f(a) = g(c)$. We shall usually denote this situation by a diagram such as:

$$\begin{array}{ccc} D & \dashrightarrow & C \\ \downarrow & & \downarrow \\ A & \longrightarrow & B \end{array}$$

If we let $\Delta \subset B \times B$ denote the diagonal, $\Delta = \{(b, b)\}$, and consider $(f, g)\colon A \times C \to B \times B$ where $(f, g)(a, c) = (f(a), g(c))$ then $D \subset A \times C$ is given by

$$D = (f, g)^{-1} \Delta.$$

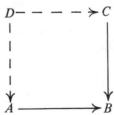

This expression lends itself to a formulation in terms of differentialbe manifolds. If A, C, and B are manifolds with f and g smooth, then f and g are transversal (we write $f \pitchfork g$) if and only if (f, g) is transversal to Δ. In this case $(f, g)^{-1} \Delta$ will be a submanifold so that the fiber product D is again a manifold. Let us give various examples of this construction.

(i) *Pull-back of fiber bundles*. If $f\colon A \to B$ is a submersion then $f \pitchfork g$ for any g. In particular, if A is a fiber bundle over B then D is a fiber bundle over C, the pull-back bundle which we denote by $f^{\#}A$.

(ii) *Push-forward and pull-back under morphism*. Let $E \to X$ and $F \to Y$ be vector bundles and suppose that we are given a morphism from E to F as discussed in §1. This means that we are given a "graph" $\subset E \times F$ consisting of all points of the form $(x, f(x), r(x)u, u)$ where $u \in F_{f(x)}$. Let A denote this graph, which is a submanifold of $B = E \times F$. Let S be a subset of E. Then we may let $C = S \times F$, and form the fiber product D, which consists of all points of the form $(x, f(x), r(x)u, u)$ with $(x, r(x)u) \in S$. We may project D onto F, its image in F will be denoted by $f_* S$. If S is a submanifold and C is transversal to A then $f_* S$ is again a submanifold.

Similarly, suppose that S' is a subset of F and we let $C' = E \times S'$. Then the fiber product D' consists of those points $(x, f(x), r(x)u, u)$ with $(f(x), r(x)u) \in S'$. We can project onto E and obtain a set which we shall denote by $f^* S'$. If S' is a submanifold and C' is transversal to A then $f^* S'$ will be a submanifold of E.

As two examples of this construction, suppose first that $E = X$ and $F = Y$ are the zero bundles. Then $B = \text{graph}\, f \subset X \times Y$ and $f_* S = f(S)$ and $f^* S' = f^{-1} S'$. Here $S \times Y$ is always transversal to graph f but the condition that $X \times S'$ be transversal to B is an honest restriction, indeed it is the condition that S' be transversal to f.

Next let us take $E = T^*X$ and $F = T^*Y$, where $f\colon X \to Y$ induces the standard morphism on the cotangent bundles. In this case $B = \{(x, f(x), df^*\xi, \xi)\}$ is the *anti*-normal bundle to graph f—it becomes the normal bundle if we replace ξ by $-\xi$ and leave $df^*\xi$ alone in the above expression. Notice that B is a Lagrangian submanifold for $T^*X \times T^*Y$ if we use the symplectic structure $\Omega_y - \Omega_x$. Here f_* carries subsets of T^*X to subsets of T^*Y and f^* does the opposite. If the appropriate transversality conditions are satisfied, then f_* and f^* carry submanifolds into submanifolds. If Λ_X is a Lagrangian submanifold of T^*X and $\Lambda_X \times T^*Y$ intersects B transversally then it is proved in Chapter IV that $f_* \Lambda_X$ is Lagrangian. Similarly for f^*. In fact, the arguments given there extend to the case where the fiber product comes from a *clean intersection*: Suppose that

$$\begin{array}{ccc} D & \dashrightarrow & C \\ \downarrow & & \downarrow \\ A & \longrightarrow & B \end{array}$$

is a fiber product diagram in which A, B, and C are manifolds and f and g are smooth maps, and it is *assumed* that the fiber product, D, is a submanifold of $A \times C$ and that the differentials induce a fiber product diagram at all points

$d \in D$. Then we say that we have a clean intersection, or that f and g intersect cleanly. Thus, if f and g intersect transversally then they intersect cleanly; but clean intersections need not be transversal (for instance $A = C$ and $f = g$ is a clean intersection).

Let X and Y be symplectic manifolds; let Γ be a Lagrangian submanifold of $X \times Y$; let Λ be a Lagrangian submanifold of X so that

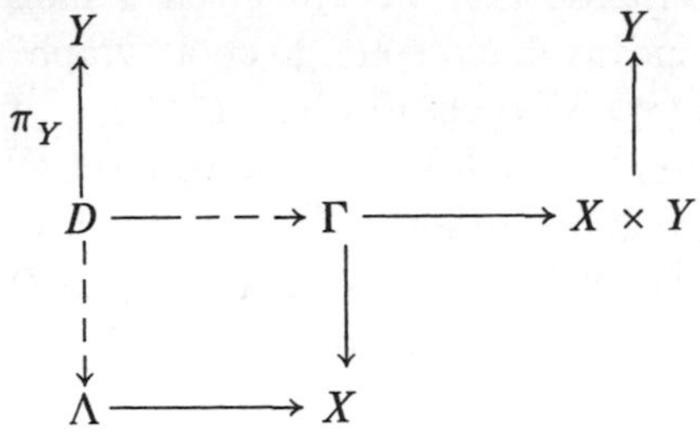

is a clean intersection fiber product. *Then* $\pi_Y\colon D \to Y$ *maps* D *onto a Lagrangian submanifold of* Y.

To prove this assertion for vector spaces: Let V and W be symplectic vector spaces, with C a Lagrangian subspace of $V \times W$ and L a Lagrangian subspace of V. Let F be defined by the fiber product diagram:

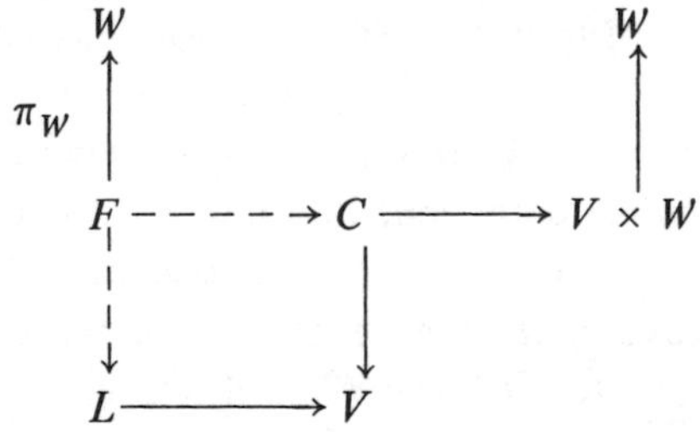

Then $F \to C \xrightarrow{\pi_W} W$ *maps* F *onto a Lagrangian subspace of* W.

PROOF. Let $G = \pi_W F$, so that we have the exact sequence

$$0 \to \ker \rho \to F \xrightarrow{\rho} G \to 0$$

where ρ is the composite map $F \to C \to G$. Here $\ker \rho$ can be identified with the space of all $v \in L$ such that $\{v, 0\} \in C$. Let $(\ ,\ )_V$ and $(\ ,\ )_W$ be the symplectic forms of V and W, and let w_1, w_2 be elements of G. Then $w_i = \pi_W\{v_i, w_i\}$ with $\{v_i, w_i\} \in C$ and $v_i \in L$. Thus

$$(w_1, w_2)_W = (\{v_1, w_1\}, \{v_2, w_2\})_{V \times W} - (v_1, v_2)_V = 0$$

so G is isotropic. We must prove that $\dim G = \frac{1}{2} \dim W$. If the diagram were transversal, we would have $\dim F = \dim C + \dim L - \dim V = \frac{1}{2} \dim V$

$+ \frac{1}{2}(\dim V + \dim W) - \dim V = \frac{1}{2}\dim W$, $\ker \rho = \{0\}$. In general, let $I \subset V$ denote the image of $L \oplus C$ in V, so that I consists of all vectors $u = v_1 + v_2$ where $v_1 \in L$ and $(v_2, w_2) \in C$ for some w_2. If $\{v, 0\} \in \ker \rho$, then $(v, v_1)_V = 0$ since $v \in L$ and $(v, v_2) = 0$ since $\{v, 0\} \in C$. Thus v annihilates I. Conversely, if $(v, v_2) = 0$ for all v_2 with $\{v_2, w_2\} \in C$, then since C is maximally isotropic, $\{v, 0\} \in C$, and, similarly, if $(v, v_1) = 0$ for all $v_1 \in L$ then $v \in L$. Thus $\ker \rho$ consists of those $\{v, 0\}$ with $(v, I)_V = 0$, or $\dim \ker \rho = \dim V - \dim I$, so that

$$\begin{aligned}\dim G &= \dim F - \dim \ker \rho \\ &= \dim C + \dim L - \dim I - (\dim V - \dim I) \\ &= \tfrac{1}{2}\dim W,\end{aligned}$$

as was to be proved.

The corresponding theorem for manifolds now follows by applying the above argument pointwise in each tangent space.

INDEX